D0214674

PROOF, LOGIC, AND CONJECTURE:

THE MATHEMATICIAN'S TOOLBOX

PROOF, LOGIC, AND CONJECTURE:

THE MATHEMATICIAN'S TOOLBOX

Robert S. Wolf

California Polytechnic State University

W. H. Freeman and Company
New York

Acquisitions Editor:	Richard Bonacci
Project Editor:	Penelope Hull
Text Designer:	The Author
Text Illustrations:	Burmar Technical Corp.
Cover Designer:	Marjory Dressler
Cover Illustration:	Michael Minchillo
Production Coordinator:	Julia DeRosa
Composition:	The Author
Manufacturing:	R R Donnelley & Sons Company

Library of Congress Cataloging-in-Publication Data

Wolf, Robert S.

Proof, logic, and conjecture : the mathematician's toolbox / Robert S. Wolf.
p. cm.
Includes bibliographical references (p. -) and index.
ISBN 0-7167-3-5--2 (alk. paper)
1. Mathematics 2. Logic, symbolic and mathematical. 3. Set theory. I. Title.
QA9.W748 1998
511.3--dc21 97-31940
CIP

© 1998 by W. H. Freeman and Company. All rights reserved.

No part of this book may be reproduced by any mechanical, photographic, or electronic process, or in the form of a phonographic recording, nor may it be stored in a retrieval system, transmitted, or otherwise copied for public or private use, without written permission from the publisher.

Printed in the United States of America

First printing 1997

W. H. Freeman and Company
41 Madison Avenue, New York, NewYork 10010
Houndmills, Basingstoke RG21 6XS, England

Contents

Asterisks (*) denote optional sections.

Preface **ix**

Note to the Student xii

Thanks xiv

Unit 1 Logic and Proofs

Chapter 1 Introduction **3**

1.1 Knowledge and Proof 3

1.2 Proofs in Mathematics 7

Chapter 2 Propositional Logic **16**

2.1 The Basics of Propositional Logic 16

2.2 Conditionals and Biconditionals 27

2.3 Propositional Consequence; Introduction to Proofs 35

Chapter 3 Predicate Logic **45**

3.1 The Language and Grammar of Mathematics 45

3.2 Quantifiers 48

3.3 Working with Quantifiers 57

3.4 The Equality Relation; Uniqueness 66

Chapter 4 Mathematical Proofs **72**

4.1 Different Types of Proofs 72

4.2 The Use of Propositional Logic in Proofs 78

4.3 The Use of Quantifiers in Proofs 92
4.4 The Use of Equations in Proofs 107
4.5 Mathematical Induction 114
4.6 Hints for Finding Proofs 126

Unit 2 Sets, Relations and Functions

Chapter 5 Sets 133

5.1 Naive Set Theory and Russell's Paradox 133
5.2 Basic Set Operations 142
5.3 More Advanced Set Operations 153

Chapter 6 Relations 164

6.1 Ordered Pairs, Cartesian Products, and Relations 164
6.2 Equivalence Relations 174
*6.3 Ordering Relations 183

Chapter 7 Functions 193

7.1 Functions and Function Notation 193
7.2 One-to-One and "Onto" Functions; Inverse Functions and Compositions 204
7.3 Proofs Involving Functions 215
7.4 Sequences and Inductive Definitions 223
7.5 Cardinality 230
7.6 Counting and Combinatorics 245
*7.7 The Axiom of Choice and the Continuum Hypothesis 252

Unit 3 Number Systems

Chapter 8 The Integers and the Rationals 261

8.1 The Ring $\mathbb{Z}$ and the Field $\mathbb{Q}$ 261

8.2 Introduction to Number Theory 269
*8.3 More Examples of Rings and Fields 282
*8.4 Isomorphisms 291

Chapter 9 The Real Number System 297

9.1 The Completeness Axiom 297
9.2 Limits of Sequences and Sums of Series 304
9.3 Limits of Functions and Continuity 314
*9.4 Topology of the Real Line 328
*9.5 The Construction of the Real Numbers 339

Chapter 10 The Complex Number System 352

10.1 Complex Numbers 352
*10.2 Additional Algebraic Properties of C 363

Appendix 1 A General-Purpose Axiom System for Mathematics 372

Appendix 2 Elementary Results About Fields and Ordered Fields 377

Appendix 3 Some of the More Useful Tautologies 387

Solutions and Hints to Selected Exercises 389

References 409

List of Symbols and Notation 412

Index 415

Dedicated with much love to my parents and grandparents, who gave me everything, and to my wonderful family:

Laurie, Aaron, Lucian, Fuzzy, Leader, Lego, Machka, and Spanky and the Gang

Preface

Almost all mathematicians will attest to the difficulty of making the transition from the lower division calculus sequence to upper division mathematics courses like abstract algebra and real analysis. One primary reason is that in a typical calculus course, where most of the students are not mathematics majors, the emphasis is on applications rather than theory. As a result, students barely encounter deductive methods and proofs in these courses. Moving from problem solving to the proofs in higher mathematics is so difficult that many students, even some quite talented ones, quit mathematics.

Until the 1970s, very few colleges or universities had a course designed to soften this transition. There seemed to be a sink-or-swim attitude, a belief that the students who really *should* be math majors would be able to handle the transition and learn how to read and write proofs while they learned the material in more advanced courses. This system may work well at some elite universities, but it has obvious drawbacks at colleges and universities that want to make higher mathematics accessible to more than a narrow audience, possibly even including students who are not mathematics majors. The "transition course" or "bridge course," now fairly common, is designed to bridge the gap.

I believe the jump from calculus to higher mathematics is as hard as it is because two things occur simultaneously. First, the material and the concepts being taught become more and more difficult and abstract. Second, since students are expected to read and write proofs in upper-division courses, these courses are *methodologically* much harder than calculus. Therefore, I believe that the most important role of the bridge course is methodological. Simply put, it should be more of a "how" course than a "what" course. This is perhaps what most sets this course apart from other mathematics courses.

About This Book

The Approach

In content, this book is similar to most of the other textbooks designed for this course; it differs in emphasis and method. Chapter 1 familiarizes the reader with the three main processes of mathematical activity: *discovery*, *conjecture*, and *proof*. While the main goal of the course is to learn to read and write proofs, this book views the understanding of the role of discovery and conjecture in mathematics as an important secondary goal and illustrates these processes with examples and exercises throughout. Chapter 1 also includes brief discussions of the way proofs are done in science and in law for the purpose of contrasting these methods with the special meaning the word "proof" has in mathematics.

Chapters 2 and 3 cover the basics of mathematical logic. These chapters emphasize the vital role that logic plays in proofs, and they include numerous *proof previews* that demonstrate the use of particular logical principles in proofs. These chapters also stress the need to pay attention to mathematical *language* and *grammar*. Many of the examples and exercises in these chapters involve analyzing the logical structure of complex English statements (with mathematical or nonmathematical content) and translating them into symbolic language (and vice versa). Unlike many texts that have just one short section on quantifiers, Chapter 3 provides a full explanation of how to understand and work with quantifiers; it includes many examples of alternations of quantifiers and negations of quantified statements. Without studying this material, students can get the impression that constructing truth tables is the main logic-based skill that is important for reading and writing proofs. Clearly, this impression can lead to frustration and failure down the road.

Chapter 4, the last chapter of Unit 1, is a thorough discussion of proofs in mathematics. It carefully explains and illustrates all the standard methods of proof that have a basis in logic, plus mathematical induction. In addition, there are discussions of the meaning of *style* in proofs, including the importance of learning how to find a good balance between formality and informality; the connection between solving equations and doing proofs; and hints for finding proofs, including useful strategies such as examining examples and special cases before tackling the general case of a proof.

The remainder of the book is not directly *about* proofs. Rather, it covers the most basic subject matter of higher mathematics while providing practice at reading and writing proofs. Unit 2 covers the essentials of sets, relations, and functions, including many important special topics such as equivalence relations, sequences and inductive definitions, cardinality, and elementary combinatorics.

Unit 3 discusses the standard number systems of mathematics—the integers, the rationals, the reals and the complex numbers. This unit also includes introductions to *abstract algebra* (primarily in terms of rings and fields rather than groups) and *real analysis*. The material and the treatment in this unit are intentionally more sophisticated than the earlier parts of the book. In fact, nearly half of the sections of this unit are designated "optional." In a one-semester course, it is unlikely that most of this material can be covered; naturally, the intention is to give instructors the opportunity to pick and choose. On the other hand, instructors with the luxury of a one-year course will find that most or all of Unit 3 can be covered, as their students gain more and more confidence with abstract mathematics and proofs.

Unique Features

I would single out *user-friendliness* and *flexibility* as the main features that distinguish this book from the other available bridge course books. User-friendliness could also be called *readability*. One hears continually that reading is a lost art, that students (as well as the general population) don't read any more. I believe people will read books they find enjoyable to read. Every effort has been made to make this book engaging, witty, and thought-provoking. The tone is conversational without being imprecise. New concepts are explained thoroughly from scratch, and complex ideas are often explained in more than one way, with plenty of helpful remarks and pointers. There are abundant examples and exercises, not only mathematical ones but also ones from the real world that show the roles logic and deductive reasoning play in everyday life.

The flexibility of this text is a response to the different approaches to teaching the bridge course. In this course, probably the most important decision the instructor must make is how much emphasis to put on logic and axiomatics. Mathematicians would generally agree that proofs proceed from axioms and that the methods we use in proofs are based on principles of logic. Mathematicians would also generally agree that learning to prove things in mathematics involves much more than learning to follow a set of rules. Constructing proofs is a skill that depends to a great extent on commonsense reasoning, and the formal rules involved must become so ingrained that one is barely aware of them. Different instructors have very different solutions to this dichotomy. Some believe it is necessary to give their students a thorough introduction to logic and to teach the major methods of proof explicitly before this knowledge can be internalized. Others believe the exact opposite—that much coverage of these topics is a waste of time and perhaps even counterproductive to the real purpose of the course. These instructors prefer to start showing their students proofs right away and to discuss logic and rules primarily when questions arise. They believe that reading and writing proofs is a natural skill that, like speech and walking, is best acquired by practice rather than by formal instruction.

I readily admit to being closer to the first point of view. Twenty years of teaching and thinking about this course has convinced me that, while some students are capable of learning how to read and write proofs by osmosis, many other good students are not quite able to do this. Also, if students never see the structure and rules that govern proofs, they might get the impression that writing proofs is a mystical or magical activity or that the correctness of proofs is based solely on the authority of the instructor. Therefore, this text carefully covers the essentials of mathematical logic, the role of logic in proofs, and the axiomatic method. Furthermore, this book is the only one that includes, as an appendix, a mathematically complete axiom system that is meant to be an important reference for students.

At the same time, this text is also an appropriate choice for instructors who prefer not to spend much time discussing logic and its relationship to proofs. Many of the sections in Unit 1 can be skimmed if desired, enabling instructors to spend most of the course teaching (and proving things) about sets, relations, functions, and number systems. The axiom system in Appendix 1 does not need to be covered.

Appendix 2 deserves special mention. It contains many basic results about the real numbers proved from scratch, using the ordered field axioms. If the unit on logic and proofs is covered thoroughly, it is natural to study this appendix in conjunction with the chapter on proofs. It is also possible instead to delay this appendix until the unit on number systems. But an interesting alternative for instructors who prefer to introduce proofs early is to start the course with Appendix 2! The rationale is that all students understand the basic algebraic properties of real numbers, which means that they are familiar with the ordered field axioms even if they do not know them by that name.

Furthermore, many of these proofs, especially those that do not involve inequalities, require very little logic. So Appendix 2 provides an ideal context for introducing students to proofs gently, without needing to explain any abstract concepts or complicated use of logic.

The exercises in this text enhance its flexibility. For one thing, they vary greatly in difficulty. In almost every section, there are some very easy problems and some rather difficult ones (marked with asterisks). There are also many types of exercises. Some problems are straightforward computations. Quite a few problems are intended to encourage the discovery process by asking the student to investigate a situation and then make a conjecture (with or without proof). Since the goal of the bridge course is to teach students to read proofs as well as write them, almost every section (starting with Chapter 4) has exercises that ask the student to critique purported proofs. Of course, in a text of this type, most of the exercises ask the student to prove something or perhaps complete a proof started in the text. The Solutions and Hints to Selected Exercises at the end of the book include a few complete proofs, but they more often provide hints or outlines to help students get started with their proofs. Additional complete proofs and teaching suggestions are provided in the Instructor's Manual.

Every chapter ends with Suggestions for Further Reading that point out several possibilities in the reference list at the end of the book. These suggestions are intended both for students who might be helped by seeing more than one approach to basic material and for students who are interested in pursuing a topic in more depth.

It is my sincere hope that students and instructors will find this text an enjoyable and valuable introduction to higher mathematics and its methodology. I am always interested in any type of honest feedback, including corrections and criticisms. I can be contacted by email at rswolf@calpoly.edu.

Note to the Student

If you are using this book, then I presume that you are a student who has completed most or all of the undergraduate calculus sequence and that your experience in mathematics so far has been satisfying enough that you are now planning to study some "higher" mathematics. This text and the course for which it is written are designed to provide you with a smooth introduction to higher mathematics. The existence of such books and courses should be viewed as a genuine attempt to make abstract mathematics more accessible than ever before. A thorough discussion of this point is found in the preface. (If you have not read the preface, please do so. It outlines the objectives of this course and the approach this textbook takes.)

Based on many years of teaching this course, I have one primary piece of advice for you: approach your study of higher mathematics with a positive and *active* attitude! You have almost certainly heard that post-calculus mathematics is difficult. I would not contradict that opinion. Higher mathematics is not simple. Much of it is abstract and complex and challenging to most students. If you are looking for an easy subject to study, there are better choices. But mathematics is fascinating (in fact, most mathematicians consider it "beautiful"), and learning it can be extremely rewarding. If you have been reasonably successful in mathematics so far, it is likely that you are capable of learning and appreciating much of post-calculus mathematics.

However, your chances of succeeding in higher mathematics are very slim if you wait for it to happen to you. Unfortunately, many students enter a course like this one with an attitude that can only be described as passive, even fearful. They listen passively to lectures and take notes unquestioningly, they wait until assignments are given out before attempting problems from the text, and they wait until just before quizzes and exams to actually read the text. Even if you are somewhat apprehensive about studying abstract mathematics, you will benefit greatly if you can go into it *assertively*.

- If you find something in the text or in a lecture confusing, you may or may not choose to ask your instructor about it right away. But you will probably benefit most if you tackle the point yourself—by thinking about it, reading the text and your notes to try to understand the rationale for it, and by thinking about *examples* that might clarify it.

- Specific, concrete examples are one of the major keys to understanding abstract mathematical concepts. The many examples in this book will help you. But you will benefit even more if you try to construct *your own examples*. When something seems difficult to understand, ask yourself, "Can I come up with an easier version of this or a simple instance or situation that might illustrate this concept?"

- Similarly, you will benefit if you do the homework that is assigned in this course thoughtfully and thoroughly. But you will benefit even more if you view the assigned problems not just as a task to get through quickly but as *investigations* or *stepping-stones to discovery*. What is the purpose of this problem? What points does it illustrate? Why is it worded the way it is? Are the restrictions in it necessary, or could it still be solved with some restrictions loosened? What further questions does it raise? Asking such questions makes a successful mathematics student.

These are a few suggestions that could help *you take control* of your study of higher mathematics rather than the other way around. Am I simply suggesting that you spend lots of time studying? No, not really. In the short term, it is true that approaching mathematics actively takes more time than being passive. But in the long term, an active, inquisitive attitude will actually save you time, because you will develop tools and habits that enable you to study efficiently and get to the core of concepts and problems quickly.

One last piece of advice: in spite of your positive attitude, you should expect some failures. In this course, besides learning some abstract concepts, you will be learning a very special way of gaining knowledge. Unless you worked with proofs in high school or in your calculus courses, you probably have very little experience reading or writing them. Almost no one learns these skills quickly and painlessly. Just as in learning to walk, everyone has to fall down many times and struggle through many halting little steps before mastering proofs. Eventually, a skill that was a major challenge can become so much second nature that it's impossible to remember that it was ever difficult. With work, perseverance and a positive attitude, the ideas of higher mathematics and the language of proofs can become comfortable and familiar to you.

Thanks

Many people have helped me in all sorts of ways with the writing of this book, and I would like to take this opportunity to thank all of them. I am certainly not capable of naming all those who have made helpful suggestions about content, but a partial list would have to include Janet Folina, Barry Langdon-Lassagne, Florence Michelfelder, Mark Stancus, Todor Todorov, John van Eps, and Sally Welsh, with special thanks to Brad Dowden for carefully reading several chapters and providing detailed comments. Several other colleagues and dozens of students at Cal Poly have found errors or suggested improvements. Pat McKaig gave me encouragement at a time when I was thinking of abandoning the project. Other people have helped me with technical tasks such as word processing, proofreading, and printing, including Bill Kalies, Paul Lewis, Patrick Munroe, Joel Shindler, Drew Yearian, and especially Mike Blum for many hours of patient help.

The editors at W. H. Freeman who have worked with me on this project have been astute, conscientious, and pleasant, and I would like to thank them: Richard Bonacci, Penny Hull, and Tim Solomon. I am also grateful to Carol Loomis for her impressively detailed copyediting, and to many other members of the staff at Freeman whose names I do not know. I also wish to thank the reviewers for their thorough work and myriad helpful comments:

David F. Anderson, University of Tennessee

Bill Blubaugh, University of Northern Colorado

William Cherowitzo, University of Colorado at Denver

John W. Emert, Ball State University

Michael Wilson, University of Vermont

Finally, I would like to express my gargantuan gratitude to my Absolutely Incredibly Wonderful Laurie, whose loving devotion and support have nurtured me unflinchingly through every phase of this project.

Unit 1

Logic and Proofs

Chapter 1

Introduction

1.1 Knowledge and Proof

The purpose of many professions and subjects is to gain knowledge about some aspect of reality. Mathematics and science would seem to fit this description. (You might try to think of subjects you have studied that are not in this category. For example, do you think that the main goal of learning to paint or to play tennis is to gain knowledge?) At some point, if you want to become proficient in such a subject, you have to understand how knowledge can be acquired in it. In other words, you have to understand what you mean when you say you "know" something, in a technical subject like mathematics or even in ordinary life.

What do you mean when you say you "know" something? Do you just mean that you believe it or think that it's true? No; clearly, to know something is stronger than just to believe or have an opinion. Somehow, there's more certainty involved when you say you know something, and usually you can also provide some kind of reasons and/or justification for how you know something. How do you acquire enough grounds and/or certainty to say you know something?

Here is a random sample of facts I would say I know:

I like chocolate chip cookies.

Paper burns more easily than steel.

The world's highest mountain is in Nepal.

Mars has two moons.

The Bastille was overrun on July 14, 1789.

If you examine this list, you'll see that there seem to be two obvious sources of this knowledge. One source is firsthand experience; consider the first two statements. The other source is things read in books or heard from other people, such as the last three statements. But how reliable are these sources of knowledge? No one has ever been to Mars. From what I've read, everyone who has ever observed Mars carefully through a good telescope has concluded that it has two moons, and so I confidently believe it. But do I really know it? Would I stake my life on it? Would I be completely devastated and disillusioned if someone announced that a third moon had been discovered or that the storming of the Bastille actually occurred early in the morning of July 15? Regarding the statement about burning, all my experience (and perhaps even some understanding of physics and chemistry) indicates that this statement is true. But do I really know it in any general or universal sense? Do I know that paper burns more easily than steel in subzero temperatures? At altitudes over two miles? Or even on February 29?

A branch of philosophy called **epistemology** studies questions like these. It can be defined as the study of knowledge and how it is acquired. In a sense, this book is about the epistemology of mathematics, but it concentrates on mathematical methods rather than on philosophical issues. The purpose of this chapter is simply to start you thinking about what you mean when you say that you know something, especially in mathematics.

Mathematics is a subject that is supposed to be very exact and certain. Over thousands of years, mathematicians have learned to be extremely careful about what they accept as an established fact. There are several reasons for this. The most obvious is that much of mathematics is very abstract and even the most talented mathematician's intuition can be led astray. As a result, mathematics has evolved into a discipline where nothing is considered to be known unless it has been "proved." In other words, any serious work in mathematics must involve reading and writing mathematical proofs, since they are the only accepted way of definitively establishing new knowledge in the field.

Before we begin our study of proofs in mathematics, let's take a look at what the word "proof" means in some other subjects besides mathematics. There are many other subjects in which people talk about proving things. These include all the natural sciences such as physics, chemistry, biology, and astronomy; disciplines based on the application of science such as medicine and engineering; social sciences like anthropology and sociology; and various other fields such as philosophy and law.

In every subject we can expect to find slightly different criteria for what constitutes a proof. However, it turns out that all of the sciences have a pretty similar standard of what a proof is. So we begin by discussing briefly what proofs are supposed to be in science, since they are quite different from proofs in mathematics. Then we also take a look at what a proof is in law, since it provides a sharp contrast to both mathematical proof and scientific proof.

Proofs in Science

We all have some idea of what scientists do to prove things. When a scientist wants to prove a certain **hypothesis** (an assertion or theory whose truth has not yet been proved), she will usually design some sort of experiment to test the hypothesis. The experiment might consist primarily of observing certain phenomena as they occur naturally, or it might involve a very contrived laboratory setting. In either case, the experiment is used to obtain **data**—factual results observed in the experiment. (In recent years, the word "data" has been borrowed and popularized by the computer industry, which uses the word to refer to any numerical or symbolic information. This is somewhat different from the scientific meaning.) Then comes a process, usually very difficult and sometimes hotly disputed, of trying to determine whether the data support the hypothesis under investigation.

This description of what a scientist does is so oversimplified that it leaves many more questions unanswered than it answers. How do scientists arrive at hypotheses to test in the first place? How do they design an experiment to test a hypothesis? Does it make sense to conduct an experiment without having a particular hypothesis that you're trying to prove? How well do the data from an experiment have to fit a hypothesis in order to prove the hypothesis? Do scientists have to have a logical explanation, as well as supporting experiments, for why their hypotheses are true? And how do scientists handle apparently contradictory experimental results, in which one experiment seems to prove a hypothesis and another seems just as clearly to disprove it?

These are just a few of the difficult questions we could ask about proofs in science. But without straining ourselves to such an extent, we can certainly draw some obvious conclusions. First of all, there is general agreement among scientists that the most important test of a hypothesis is whether it fits real-world events. Therefore, the most common and trusted way to prove something in science is to gather enough supporting data to convince people that this agreement exists. This method of establishing general laws by experimentation and observation is known as the **scientific method** or the **empirical method**. It normally involves **inductive reasoning**, which usually refers to the mental process of "jumping" from the specific to the general, that is, using a number of observations in particular situations to conclude some sort of universal law.

Does pure thought, not connected with observing real-world events, have a role in science? It definitely does. Can you prove something in science by logic or deduction or calculations made on paper without experimental evidence? Well, these methods are definitely important in science, and some of the most important discoveries in science have been brilliantly predicted on paper long before they could be observed. In fields like astronomy, nuclear physics, and microbiology, it's getting so difficult to observe things in a direct, uncomplicated way that the use of theoretical arguments to prove hypotheses is becoming more and more acceptable. An interesting contemporary example in astronomy concerns the existence of black holes in space. These were predicted by very convincing reasoning decades ago, but no one has observed one. Most astronomers are quite sure that black holes exist, but they would probably hesitate to say that their existence has been proven, no matter how ironclad the arguments seem. With few exceptions, scientific theories derived mentally are not considered proved until they are verified empirically. We will see that this type of attitude is very different from what goes on in mathematics.

Proofs in Law

Everyone also has some idea of what it means to prove something in law. First of all, note that a proof in a court of law is a much less objective and permanent thing than a proof in mathematics or science. A proof in mathematics or science must stand the test of time: if it does not stand up under continual scrutiny and criticism by experts in the field, it can be rejected at any time in the future. In contrast, to prove something in a jury trial in a court of law, all you have to do (barring appeals and certain other complications) is convince one particular set of twelve people, just for a little while. The jurors aren't experts in any sense. In fact, they aren't even allowed to know very much in advance about what's going on; and you even have some say in who they are. Furthermore, it doesn't even matter if they change their minds later on!

Now let's consider what kinds of methods are allowable in law proofs. Can a lawyer use the scientific method to convince the jury? In a loose sense, the answer to this is definitely yes. That is, he can certainly present **evidence** to the jury, and evidence usually consists of facts and observations of actual events. A lawyer may also conduct simple experiments, try to convince the jury to make an inductive conclusion, and use various other methods that are similar to what a scientist does. Of course, lawyers are rarely as rigorous as scientists in their argumentation. But at least we can say that most proof methods that are scientifically acceptable would also be allowed in a court of law.

What other methods of proof are available to lawyers? Well, they can certainly use logic and deductive reasoning to sway the jury. As we will see, these are the main tools of the mathematician. Lawyers can also appeal to **precedent** (previous legal decisions) or to the law itself, although such appeals are generally made to the judge, not the jury. This is analogous to the practice in science or mathematics of using a previously established result to prove something new.

Are there any methods of persuasion available to a lawyer that are totally different from scientific and mathematical methods? Again, the answer is yes. A lawyer can use a variety of psychological and emotional tricks that would be completely improper in science or mathematics. The only time that a lawyer can use these psychological tools freely is during opening and closing statements ("Ladies and gentlemen of the jury, look at my client's face. How could this sweet old lady have committed these seventeen grisly..."). However, many psychological ploys can also be used with witnesses, as long as they are used subtly. These include leading questions, attempts to confuse or badger witnesses, clever tricks with words, gestures, facial expressions and tones of voice used to create a mood or impression, and so on. Without going into greater detail, we can see that the guidelines for proofs in law are very broad and freewheeling, for they include almost everything that the scientist and the mathematician can use plus a good deal more.

Exercises 1.1

(1) List six statements that you would say that you know, and explain how you know each one. Pick statements with as much variety as possible.

(2) (a) Briefly discuss the differences (in your own mind) among *believing* that something is true, *thinking* that something is true, and *knowing* that something is true.

(b) Which combinations of these conditions do you think are possible? For example, is it possible to know something is true without believing it is?

(3) Briefly discuss under what circumstances you think it's appropriate to use the inductive method of drawing a general conclusion from a number of specific instances. For example, if someone is chewing gum the first three times you meet him, would you be tempted to say he "always chews gum"?

(4) Mention a few ways in which a lawyer can try to convince a jury to believe something that is not true. Give some specific examples, either made up or from actual cases you have heard about.

1.2 Proofs in Mathematics

The preceding discussions of proofs in science and proofs in law were included primarily to provide a contrast to the main subject of this book. In this section we begin to look at the very special meaning that the word "proof" has in mathematics.

How do we prove something in mathematics? That is, how do we establish the correctness of a mathematical statement? This question was first answered by various Greek scholars well over two thousand years ago. Interestingly, their basic idea of what a mathematical proof should be has been accepted, with relatively minor modifications, right up until this day. This is in sharp contrast to the situation in science, where even in the last three hundred years there have been tremendous changes, advances, and controversy about what constitutes a proof. In part, this is because the range of methods allowed in mathematical proofs is quite a bit more specific and narrow than in other fields.

Basically, almost every mathematician who has ever addressed this issue has agreed that the main mechanism for proving mathematical statements must be logic and deductive reasoning. That is, the reasoning that leads from previously accepted statements to new results in mathematics must be airtight, so that there is no doubt about the conclusion. Inductive reasoning, which is the mainstay of the sciences but by its very nature is not totally certain, is simply never allowed in mathematical proofs.

There are examples that dramatically illustrate this point. In number theory (the branch of mathematics that studies whole numbers) there are some very famous **conjectures**. (Like a hypothesis, a conjecture is a statement that has not been proved, although there is usually evidence for believing it. The word "conjecture" is generally preferred by mathematicians.) One of these is **Goldbach's conjecture**, which claims that every even number greater than 2 can be written as the sum of two prime numbers. In a few minutes, you can easily verify this for numbers up to 100 or so. In fact, it has been verified by computer up into the *trillions*. Yet no finite number of examples can possibly constitute a mathematical proof of this statement, and in fact it is considered unproved! Now imagine such a situation in science, where a proposed law turns out to be true in millions of test cases, without a single failure. It is extremely unlikely that scientists would consider the law unproved, with such overwhelming evidence for it. (By the way, number theory is full of interesting conjectures that have remained unproved for centuries. We encounter more of these in Section 8.2.)

Thus the scientist's most valuable proof method is not considered trustworthy in mathematics. And, as we saw in the previous section, the mathematician's most valuable proof method—deduction—is of only limited use in science. For these reasons, most specialists in the foundations of mathematics do not think that mathematics should be classified as a science. There are some respected scholars who do call it an **exact science**, but then they are careful to distinguish it from the **empirical sciences**.

Discovery and Conjecture in Mathematics

Can we say that the scientific method—observation, experimentation, and the formation of conclusions from data—has no place in mathematics? No, that would be going too far. Even if empirical methods may not be used to prove a mathematical statement, they are used all the time to enable mathematicians to figure out whether a statement is likely to be provable in the first place. This process of **discovery** in mathematics often has a very different flavor from the process of proof. Higher mathematics can be very intimidating, and one of the reasons is that many proofs in mathematics seem extremely sophisticated, abstract, and nonintuitive. Often, this is because most of the real work is hidden from the reader. That five-line, slick proof might well be the result of months or even years of trial and error, guesswork, and dead ends, achieved finally through patience and a little bit of luck. After that it might have been refined many times to get it down from ten pages of grubby steps to five elegant lines. This point is worth remembering when your self-confidence begins to fail. Thomas Edison's famous remark— "Genius is 1 percent inspiration and 99 percent perspiration"—is more true of mathematics than most people realize.

Although the main goal of this book is to help you learn to read and write mathematical proofs, a secondary goal is to acquaint you with how mathematicians investigate problems and formulate conjectures. Examples and exercises relating to discovery and conjecture appear throughout the text. The last seven exercises in this chapter are of this sort.

The process of discovering mathematical truths is sometimes very different from the process of proving them. In many cases, the discovery method is completely useless as a proof method, and vice versa. On the other hand, in many cases these two processes are intimately related. An investigation into *whether* a certain statement is true often leads to an understanding of *why* it is or isn't true. That understanding in turn should normally form the basis for *proving* that the statement is or isn't true.

There is another important use of empirical methods in mathematics. It was stated previously that deduction is the only way to prove new things from old in mathematics. But this raises a big question: Where do you start? How do you prove the "first thing"? Classical Greek scholars such as Eudoxus, Euclid, and Archimedes provided the answer to this question. Since you can't prove things deductively out of thin air, the study of every branch of mathematics must begin by accepting some statements without proof. The idea was to single out a few simple, "obviously true" statements applicable to any given area of mathematics and to state clearly that these statements are assumed without proof. In the great works of Euclid and his contemporaries, some of these assumed statements were called **axioms** and others were called **postulates**. (Axioms were more universal, whereas postulates pertained more to the particular subject.) Today both types are usually called axioms, and this approach is called the **axiomatic method**.

When a new branch of mathematics is developed, it is important to work out the exact list of axioms that will be used for that subject. Once that is done, there should not be any controversy about what constitutes a proof in that system: a proof must be a sequence of irrefutable, logical steps that proceed from axioms and previously proved statements.

Euclid was one of the most important mathematicians of ancient Greece, and yet very little is known of his life. Not even the years of his birth and death or his birthplace are known. As a young man, he probably studied geometry at Plato's academy in Athens. It is known that he spent much of his life in Alexandria and reached his creative prime there around 300 B.C. He is most famous for his ***Elements***, a monumental work consisting of thirteen books, most of which deal with geometry.

The *Elements* are the oldest surviving work in which mathematical subjects were developed from scratch in a thorough, rigorous, and axiomatic way. However, the great majority of the results in Euclid's *Elements* were first proved by someone other than Euclid. Euclid is remembered less for his original contributions to geometry than for the impressive organization and rigor of his work. The *Elements* was viewed as *the* model of mathematical rigor for over two thousand years and is still used as a geometry textbook in some places. Although it became clear in the last century that many of Euclid's definitions and proofs are flawed by modern standards, this does not diminish the importance of his achievement.

How are the axioms for any branch of mathematics determined? Here is where empirical methods come in. Since the axioms are not expected to be proved deductively, the only way to verify that they are true is by intuition and common sense, experience and lots of examples—just the sorts of things a scientist is supposed to use. For example, in the study of the ordinary algebra of the real numbers, two of the usual axioms are the **commutative laws**:

$$x + y = y + x \quad \text{and} \quad xy = yx, \quad \text{for all numbers } x \text{ and } y$$

These are good choices for axioms, for they are extremely simple statements that virtually everyone over the age of eight would agree are clearly true, so clearly true that it would seem pointless even to try to prove them.

The choice of axioms in mathematics is not always such a smooth and uncontroversial affair. There have been cases in which the developers of a subject split into two camps over whether a particular statement should be accepted as an axiom, and in which the disagreement went on for many years. There is usually no single correct answer to such an issue.

The theory of the axiomatic method has been liberalized somewhat in the last two centuries. The classical Greek idea was that the axioms and postulates must be true. Modern mathematics realizes that the idea of truth is often dependent on one's interpretation and that any axiom system that at least fits some consistent interpretation, or **model,** should be an allowable area of study. The most famous example of this liberalization pertains to the **parallel postulate** of Euclid's geometry, which implies the existence of straight lines in a plane that don't meet. This seems to be obviously true; but early in the nineteenth century, it was noted that this postulate is false on the surface of a sphere (with straight lines interpreted as great circles, since arcs of great circles are the shortest paths between points on the surface of a sphere). Any two great circles on a sphere must cross (see Figure 1.1). So if one wants to study the important subject of spherical geometry, this postulate must be rejected and replaced with one that is false in the plane. The subject of **non-Euclidean geometry** may have seemed like a strange curiosity when it was first introduced, but it took on added significance in the twentieth century when Albert Einstein's general theory of relativity showed that our physical universe is actually non-Euclidean.

As another example, consider the equations $1 + 1 = 1$ and $1 + 1 = 0$. At first glance, these are just wrong equations, and it would seem ridiculous to call them axioms. But they are wrong only in our ordinary number systems. They are true (separately, not simultaneously) in some less familiar systems of algebra, in which addition has a different meaning. In fact, the first equation is an axiom of **boolean algebra**, and the second is an axiom in the theory of **fields of characteristic 2**. Both of these subjects are related to the binary arithmetic that is used in designing computer circuits. So it can be very fruitful to have strange-looking statements be axioms in a specialized branch of mathematics. One twentieth-century school of thought, called **formalism**, holds that mathematicians should not worry at all about whether their axioms are "true" or whether the things they study have any relationship at all to the "real world." However, most modern mathematicians would not go quite so far in their loosening of the ancient Greek viewpoint.

Figure 1.1 On a sphere, "straight lines" (great circles) are never parallel

Organization of the Text

The main goal of this book is to teach you about mathematical proofs—how to read, understand, and write them. The rest of Unit 1 includes two chapters on logic, which are intended to provide enough of an understanding of logic to form a foundation for the material on proofs that follows them. The last chapter of this unit is devoted to mathematical proofs. It is perhaps the most important chapter of the book.

Since it has been pointed out that logic and deduction are the only mechanisms for proving new things in mathematics, you might expect this whole book to be about logic. But if you look at the table of contents, you will see that only the first unit is directly devoted to logic and proofs. This is because certain other subject matter is so basic and important in mathematics that you can't understand any branch of mathematics (let alone do proofs in it) unless you understand this core material. This material is covered in the book's two other units.

Unit 2 is about sets, relations, and functions. These are all relatively new concepts in the development of mathematics. The idea of a function is only two or three centuries old, and yet in that time it has become an essential part of just about every branch of mathematics, a concept almost as basic to modern mathematics as the concept of a number. The idea of sets (including relations) and their use in mathematics is only about a hundred years old, and yet this concept has also become indispensable in most parts of contemporary mathematics. Chapter 7, on functions, includes several other important topics such as sequences, cardinality, and counting principles.

Unit 3 is about number systems. The use of numbers and counting is almost certainly the oldest form of mathematics and the one that we all learn first as children. So it should come as no surprise to you that number systems like the integers and the real numbers play an important role in every branch of mathematics, from geometry and calculus to the most advanced and abstract subjects. This unit discusses the most important properties of the natural numbers, the integers, the rational numbers, the real numbers, and the complex numbers. At the same time, it introduces some of the major concepts of abstract algebra, real analysis and topology.

So that's what you will learn about in this book: logic and proofs; sets, relations, and functions; and number systems. I like to think of these three topics as the building blocks or essential tools of mathematical proofs. The viewpoint of this book is that if (and only if!) you learn to understand and use these basic tools will you be well on your way to success in the realm of higher mathematics.

Exercises 1.2

Throughout this text, particularly challenging exercises are marked with asterisks.

For the first three problems, you will probably find it helpful to have a list of all prime numbers up to 200 or so. The most efficient way to get such a list is by a technique called the **sieve of Eratosthenes**: first list all integers (whole numbers) from 2 up to wherever you want to stop, say 200. Now, 2 is the smallest number in the list, so circle it and cross out all larger multiples of 2. Then 3 is the smallest remaining number in the list, so circle it and cross out all larger multiples of 3. Then circle 5 and cross out all larger multiples of 5. Continue in this manner. When you're done, the circled numbers are all the prime numbers up to 200. (If your table goes up to 200, the largest number whose multiples you need to cross out is 13. Can you see why? See Exercise 8.)

(1) (a) Consider the expression $n^2 - n + 41$. Substitute at least a half dozen small nonnegative integers for the variable n in this expression, and in each case test whether the value of the expression turns out to be a prime number. Does it seem plausible that this expression yields a prime number for every nonnegative integer n?

(b) Now find a positive integer value of n for which this expression is not a prime number. ***Hint:*** You probably won't find the right n by trial and error. Instead, try to think the problem through.

(2) Verify Goldbach's conjecture for all the even numbers from 4 to 20 and from 100 to 110.

(3) An interesting variant of Goldbach's conjecture, known as de Polignac's conjecture, is the claim that every positive even number can be written as the *difference* of two prime numbers. As with Goldbach's conjecture, it is not known whether this statement is true or false.

(a) Verify de Polignac's conjecture for each positive even number up to 12.

*(b) In the unlikely event that one or both of these conjectures is actually false, de Polignac's conjecture would probably be much more difficult to disprove than Goldbach's conjecture. Can you explain why?

*(4) Try to prove each of the following statements. Since we have not begun our study of axiomatic mathematics, the word "prove" is being used here in an informal sense. That is, you should try to come up with what you think are convincing arguments or explanations for why these statements are true. Perhaps you can succeed with pictures and/or words. Or, you might need to resort to more sophisticated methods, such as algebra or even calculus. (Don't worry if you feel as if you're groping in the dark in this problem. When we get to Chapter 4, we get much more exact and technical about what constitutes a proof.)

(a) A negative number times a negative number always equals a positive number. (You may assume that the product of two positive numbers is always positive, as well as basic algebraic rules for manipulating minus signs.)

(b) If you add a positive number to its reciprocal, the sum must be at least 2.

(c) The area of a rectangle equals its length times its width. (You may assume that the area of a one-by-one square is one, but this problem is still not easy.)

(d) A straight line and a circle meet in at most two points.

The remaining exercises have to do with the process of discovery in mathematics; as we have discussed, this often precedes proof but is no less important.

(5) (a) Complete the last three equations:

$$\begin{aligned} 1 &= 1 \\ 1+3 &= 4 \\ 1+3+5 &= ? \\ 1+3+5+7 &= ? \\ 1+3+5+7+9 &= ? \end{aligned}$$

(b) On the basis of the equations in part (a), make a conjecture about the sum of the first n odd numbers, where n can be any positive integer.

(c) Test your conjecture for at least four other values of n, including two values that are greater than 10.

(6) Consider the following equations:

$$\begin{aligned} 1^3 &= 1 &&= 1^2 \\ 1^3+2^3 &= 9 &&= (1+2)^2 \\ 1^3+2^3+3^3 &= 36 &&= (1+2+3)^2 \end{aligned}$$

(a) On the basis of these equations, make a conjecture.

(b) Test your conjecture for at least two other cases.

(7) (a) Carefully draw three triangles. Make their shapes quite different from each other.

(b) In each triangle, *carefully* draw all three medians. (A **median** is a line from a vertex of a triangle to the midpoint of the opposite side. Use a ruler to find these midpoints, unless you prefer to use an exact geometric construction!)

(c) On the basis of your figures, make a conjecture about the medians of any triangle.

*(d) After making some careful measurements with a ruler, make a conjecture about how any median of a triangle is cut by the other medians.

(8) (a) If you haven't already done so, construct the sieve of Eratosthenes for numbers up to 200, as described before Exercise 1.

(b) By trial and error, fill in each of the following blanks with the *smallest* number that makes the statement correct:

(i) Every nonprime number less than 100 has a prime factor less than ____ .

(ii) Every nonprime number less than 150 has a prime factor less than ____ .

(iii) Every nonprime number less than 200 has a prime factor less than ____ .

(c) Using your results from part (b), additional investigation if you need it, and some logical analysis of the situation, fill in the following blank with the expression that you think yields the smallest number that makes your conjecture correct:

Every nonprime number n has a prime factor equal to or less than ______ .

(9) The numbers 3, 4, and 5 can be the sides of a right-angled triangle, since they satisfy Pythagoras's theorem (the familiar $a^2 + b^2 = c^2$). Positive integers with this property are called **Pythagorean triples**. The triple 3, 4, 5 also has the property that the largest number of the triple (the hypotenuse) is only one more than the middle number.

(a) Find two more Pythagorean triples with this property.

(b) Could the smallest member of a Pythagorean triple with this property be an even number? Why or why not?

*(c) Try to find a general formula or rule that can be used to list all Pythagorean triples of this type

(d) Can two of the numbers in a Pythagorean triple be equal? Why or why not? (You may use the fact that $\sqrt{2}$ is not equal to any fraction.)

(10) Starting with any positive integer, it is possible to generate a sequence of numbers by these rules: If the current number is even, the next number is half the current number. If the current number is odd, the next number is 1 more than 3 times the current number. For example, one such sequence begins 26, 13, 40, 20, 10, 5, 16, ...

(a) Choose three or four starting numbers, and for each of them generate the sequence just described. Keep going until the sequence stabilizes in a clear-cut way. (A good range for most of your starting numbers would be between 20 and 50.)

(b) On the basis of your results in part (a), make a conjecture about what happens to these sequences, for any starting number. It turns out that a general law does hold here; that is, all such sequences end in exactly the same pattern. However, it is quite difficult to prove this theorem, or even understand intuitively why it should be true.

(11) The ancient game of Nim is very simple to play (in terms of both equipment and rules) but is quite entertaining and challenging. It is also a good setting for learning about the mathematical theory of games. Here are the rules:

Nim is a competitive game between two players. To start the game, the players create two or more piles of match sticks, not necessarily equal in number. One classic starting configuration uses piles of three, four, and five, but the players can agree to any starting configuration (see Figure 1.2).

After the setup, the players take turns. When it is his or her turn, a player must remove at least one match stick from *one* pile. For instance, a player may remove an entire pile at one turn; but a player may not remove parts of more than one pile at one turn. The player who removes the last match stick wins the game.

Once the starting configuration is determined, Nim becomes a "finite two-person win-lose game of perfect information." The most important mathematical result about such a game is that one player (either the one who plays first or the one who plays second) has a strategy that always wins for that player.

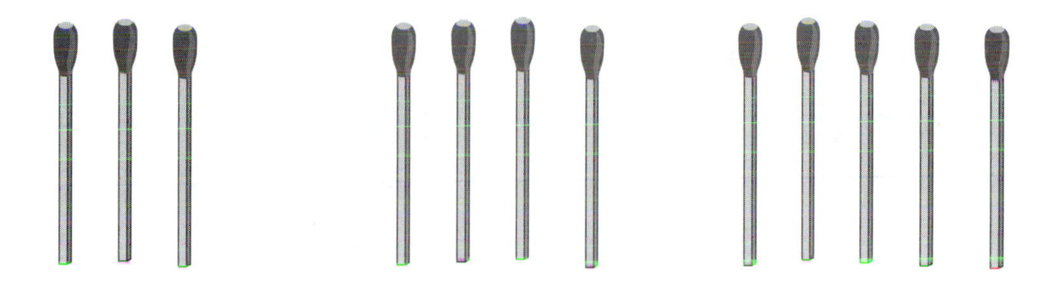

Figure 1.2 One typical starting configuration for Nim

(a) Play several games of Nim (by yourself or with someone else) using only two piles of sticks but of various sizes. On the basis of your experience, devise a rule for determining which player has the winning strategy for which games of this type, and what that strategy is. You will be asked to prove your conjecture in Section 8.2.

(b) An alternate version of Nim states that the one who removes the last match stick *loses*. Repeat part (a) with this alternate rule.

*(c) Repeat part (a), now starting with three piles of sticks but with one of the piles having only one stick.

*(d) Repeat part (c) using the alternate rule of part (b).

Suggestions for Further Reading: Literally thousands of fine books have been written about the subjects touched on in this chapter, including inductive and deductive reasoning, the processes of discovery and proof in science and mathematics, and the history of the axiomatic method. A few of these appear in the References at the end of this text: Davis and Hersh (1980 and 1986), Eves (1995), Kline (1959 and 1980), Lakatos (1976), Polya (1954), and Stabler (1953). For a witty and informative discussion of Goldbach's conjecture and related problems of number theory, see Hofstadter (1989).

Chapter 2

Propositional Logic

2.1 The Basics of Propositional Logic

What is logic? Dictionaries define it to be the study of pure reasoning or the study of valid principles of making inferences and drawing conclusions. As Chapter 1 emphasized, logic plays an extremely important role in mathematics, more so than in the sciences or perhaps in any other subject or profession. The field of **mathematical logic** is divided into the branches of **propositional logic** and **predicate logic**.

This chapter is about propositional logic. This is a very old subject, first developed systematically by the Greek philosopher Aristotle. It has various other names, including the **propositional calculus**, **sentential logic**, and the **sentential calculus**. Basically, propositional logic studies the meaning of various simple words like "and," "or," and "not" and how these words are used in reasoning. Although it is possible to carry out this study without any special terminology or symbols, it's convenient to introduce some.

Definition: A **proposition** is any declarative sentence (including mathematical sentences such as equations) that is true or false.

Example 1: (a) "Snow is white" is a typical example of a proposition. Most people would agree that it's a true one, but in the real world few things are absolute: city dwellers will tell you that snow can be grey, black, or yellow.

(b) "$3 + 2 = 5$" is a simple mathematical proposition. Under the most common interpretation of the symbols in it, it is of course true.

(c) "$3 + 2 = 7$" is also a proposition, even though it is false in the standard number system. Nothing says a proposition can't be false. Also, this equation could be true (and the previous one false) in a nonstandard number system.

(d) "Is anybody home?" is *not* a proposition; questions are not declarative sentences.

(e) "Shut the door!" and "Wow!" are also *not* propositions, because commands and exclamations are not declarative sentences.

(f) "Ludwig van Beethoven sneezed at least 400 times in the year 1800" is a sentence whose truth is presumably hopeless to verify or refute. Nonetheless, such sentences are generally considered to be propositions.

Aristotle (384–322 B.C.), like his teacher Plato, was a philosopher who was very interested in mathematics but did not work in mathematics to any extent. Aristotle was apparently the first person to develop formal logic in a systematic way. His treatment of propositional logic does not differ greatly from the modern approach to the subject, and the study of logic based on truth conditions is still called **Aristotelian logic.**

Besides writing extensively on other humanistic subjects such as ethics and political science, Aristotle also produced the first important works on physics, astronomy, and biology. Some of his claims were rather crude by modern standards and others were simply wrong. For example, Aristotle asserted that heavy objects fall faster than light ones, a belief that was not refuted until the sixteenth century, by Galileo. Still, his scientific work was the starting point of much of modern science. Very few people in the history of humanity have contributed to as many fields as Aristotle.

(g) "$x > 5$" is a mathematical inequality whose truth clearly depends on more information, namely what value is given to the variable x. In a sense, the truth or falsity of this example is much easier to determine than that of example f. Even so, we follow standard practice and call such sentences **predicates** rather than propositions.

(h) "Diane has beautiful eyes" is a sentence whose truth depends not only on getting more information (which Diane is being referred to?) but also on a value judgment about beauty. Most logicians would say that a sentence whose truth involves a value judgment cannot be a proposition.

We use the word **statement** as a more all-encompassing term that includes propositions as well as sentences like the last two examples. Section 3.2 clarifies this terminology further.

(i) "23 is a purple number" has more serious flaws than examples (g) and (h). Neither more information nor a value judgment determines its truth or falsehood. Most people would say this sentence is meaningless and therefore not a statement.

(j) "This sentence is false" is a simple example of a **paradox**. If it's true, then it must be false, and vice versa. So there is no way it could sensibly be called true *or* false, and therefore it is not a statement.

Notation: We use the letters P, Q, R, ... as **propositional variables**. That is, we let these letters stand for or represent statements, in much the same way that a mathematical variable like x represents a number.

Notation: Five symbols, called **connectives**, are used to stand for the following words:

- $\wedge$ for "and"
- $\vee$ for "or"
- $\sim$ for "not"
- $\rightarrow$ for "implies" or "if ... then"
- $\leftrightarrow$ for "if and only if"

The words themselves, as well as the symbols, may be called connectives. Using the connectives, we can build new statements from simpler ones. Specifically, if P and Q are any two statements, then

$$P \wedge Q, \quad P \vee Q, \quad \sim P, \quad P \rightarrow Q, \quad \text{and} \quad P \leftrightarrow Q$$

are also statements.

Definitions: A statement that is *not* built up from simpler ones by connectives and/or quantifiers is called **atomic** or **simple**. (Quantifiers are introduced in Chapter 3.) A statement that is built up from simpler ones is called **compound**.

Example 2: "I am not cold," "Roses are red and violets are blue," and "If a function is continuous, then it's integrable" are compound statements because they contain connectives. On the other hand, the statements in Example 1 are all atomic.

Remarks: That's pretty much all there is to the grammar of propositional logic. However, there are a few other details and subtleties that ought to be mentioned.

(1) Notice that each connective is represented by both a symbol and a word (or phrase). The symbols are handy abbreviations that are useful when studying logic or learning about proofs. ***Otherwise, the usual practice in mathematics is to use the words rather than the symbols. Similarly, propositional variables are seldom used except when studying logic.***

(2) Why do we use these particular five connectives? Is there something special about them or the number five? Not at all. It would be possible to have dozens of connectives. Or we could have fewer than five connectives—even just one—and still keep the full "power" of propositional logic. (This type of reduction is discussed in the exercises for Section 2.3.) But it's pretty standard to use these five, because five seems like a good compromise numerically and because all these connectives correspond to familiar thought processes or words.

(3) When connectives are used to build up symbolic statements, parentheses are often needed to show the order of operations, just as in algebra. For example, it's confusing to write $P \wedge Q \vee R$, since this could mean either $P \wedge (Q \vee R)$ or $(P \wedge Q) \vee R$.

However, just as in algebra, we give the connectives a priority ordering that resolves such ambiguities when parentheses are omitted. The **priority of the connectives**, from highest to lowest, is $\sim, \wedge, \vee, \rightarrow, \leftrightarrow$. (This order is standard, except that some books give $\wedge$ and $\vee$ equal priority.)

How is a statement interpreted when the same connective is repeated and there are no parentheses? In the case of $\wedge$ or $\vee$, this is never a problem. The statement $(P \wedge Q) \wedge R$ has the same meaning as $P \wedge (Q \wedge R)$, so it's perfectly unambiguous and acceptable to write $P \wedge Q \wedge R$; and the same holds for $\vee$. (Note that this is completely analogous to the fact that we don't need to put parentheses in algebraic expressions of the form $a + b + c$ and abc.) On the other hand, repeating $\rightarrow$ or $\leftrightarrow$ can create ambiguity. In practice, when a mathematician writes a statement with the logical form $P \rightarrow Q \rightarrow R$, the intended meaning is probably $(P \rightarrow Q) \wedge (Q \rightarrow R)$, rather than $(P \rightarrow Q) \rightarrow R$ or $P \rightarrow (Q \rightarrow R$.) A similar convention holds for $\leftrightarrow$. This is analogous to the meaning attached to extended equations and inequalities of the forms $x = y = z$, $x < y < z$, and so on. But it's often important to use parentheses or words to clarify the meaning of compound statements.

Example 3

$$P \vee Q \wedge R \text{ means } P \vee (Q \wedge R)$$

$$P \rightarrow Q \leftrightarrow \sim Q \rightarrow \sim P \text{ means } (P \rightarrow Q) \leftrightarrow [(\sim Q) \rightarrow (\sim P)]$$

Terminology: Each of the connectives has a more formal name than the word it stands for, and there are situations in which this formal terminology is useful.

Specifically, the connective $\wedge$ ("and") is also called **conjunction**. A statement of the form $P \wedge Q$ is called the conjunction of P and Q, and the separate statements P and Q are called the **conjuncts** of this compound statement.

Similarly, the connective $\vee$ ("or") is called **disjunction**, and a statement $P \vee Q$ is called the disjunction of the two **disjuncts** P and Q.

The connectives $\sim$, $\rightarrow$, and $\leftrightarrow$ are called **negation, conditional** (or **implication**), and **biconditional** (or **equivalence**), respectively.

Now it's time to talk about what these connectives mean and what can be done with them. In propositional logic, we are primarily interested in determining when statements are true and when they are false. The main tool for doing this is the following.

Definition: The **truth functions** of the connectives are defined as follows:

- $P \wedge Q$ is true provided P and Q are both true.
- $P \vee Q$ is true provided at least one of the statements P and Q is true.

- $\sim P$ is true provided P is false.
- $P \rightarrow Q$ is true provided P is false, or Q is true (or both).
- $P \leftrightarrow Q$ is true provided P and Q are both true or both false.

Note that these truth functions really are functions except that, instead of using numbers for inputs and outputs, they use "truth values," namely "true" and "false." (If you are not very familiar with functions, don't be concerned; we study them from scratch and in depth in Chapter 7.) We usually abbreviate these truth values as T and F.

Since the domain of each truth function is a finite set of combinations of Ts and Fs, we can show the complete definition of each truth function in a **truth table,** similar to the addition and multiplication tables you used in elementary school. The truth tables for the five basic connectives are shown in Table 2.1.

Table 2.1 Truth tables of the connectives

P	Q	$P \wedge Q$
T	T	T
T	F	F
F	T	F
F	F	F

P	Q	$P \vee Q$
T	T	T
T	F	T
F	T	T
F	F	F

P	$\sim P$
T	F
F	T

P	Q	$P \rightarrow Q$
T	T	T
T	F	F
F	T	T
F	F	T

P	Q	$P \leftrightarrow Q$
T	T	T
T	F	F
F	T	F
F	F	T

It is important to understand how the truth functions of the connectives relate to their normal English meanings. In the cases of $\sim$ and $\wedge$, the relationship is very clear, but it is less so with the others. For example, the truth function for $\vee$ might not correspond to the most common English meaning of the word "or." Consider the statement, "Tonight I'll go to the volleyball game or I'll see that movie." Most likely, this means I will do one of these activities but *not* both. This use of the word "or," which excludes the possibility of both disjuncts being true, is called the **exclusive or.** The truth function we have defined for $\vee$ makes it the **inclusive or,** corresponding to "and/or." In English, the word "or" can be used inclusively or exclusively; this can lead to ambiguity. For instance, suppose someone said, "I'm going to take some aspirin or call the doctor." Does this statement leave open the possibility that the person takes aspirin *and* calls the doctor? It may or may not. In mathematics, the word "or" is generally used inclusively. If you want to express an exclusive or in a mathematical statement, you must use extra words, such as “Either P or Q is true, but not both” or “Exactly one of the conditions P and Q is true” (see Exercise 8).

There are enough subtleties involving the connectives $\rightarrow$ and $\leftrightarrow$ that the entire next section is devoted to them.

Using the five basic truth functions repeatedly, it’s simple to work out the truth function or truth table of any symbolic statement. (If you have studied composition of functions, perhaps you can see that the truth function of any statement must be a composition of the five basic truth functions.) Some examples are shown in Table 2.2. Note how systematically these truth tables are constructed. If there are n propositional variables, there must be 2^n lines in the truth table, since this is the number of different ordered n-tuples that can be chosen from a two-element set (Exercise 11). So a truth table with more than four or five variables would get quite cumbersome. Notice that these tables use a simple pattern to achieve all possible combinations of the propositional variables. Also, note that before we can evaluate the output truth values of the entire statement, we have to figure out the truth values of each of its substatements.

We can now define some useful concepts.

Definitions: A **tautology**, or a **law of propositional logic**, is a statement whose truth function has all Ts as outputs.

A **contradiction** is a statement whose truth function has all Fs as outputs (in other words, it’s a statement whose negation is a tautology).

Two statements are called **propositionally equivalent** if a tautology results when the connective $\leftrightarrow$ is put between them. (Exercise 7 provides an alternate definition of this concept.)

Example 4: One simple tautology is the symbolic statement $P \rightarrow P$. This could represent an English sentence like “If I don’t finish, then I don’t finish.” Note that this sentence is obviously true, but it doesn’t convey any information. This is typically the case with such simple tautologies.

One of the simplest and most important contradictions is the statement $P \wedge \sim P$. An English example would be “I love you and I don’t love you.” Although this statement might make sense in a psychological or emotional context, it is still a contradiction. That is, from a logical standpoint it cannot be true.

The statement $\sim P \rightarrow Q$ is propositionally equivalent to $P \vee Q$, as you can easily verify with tables. For instance, if I say, “If I don’t finish this chapter this week, I’m in trouble,” this is equivalent to saying (and so has essentially the same meaning as), “I (must) finish this chapter this week or I’m in trouble.”

For the rest of this chapter, we use “equivalent” for the longer “propositionally equivalent.” Note that statements can be equivalent even if they don’t have the same set of propositional variables. For example, $P \rightarrow (Q \wedge \sim Q)$ is equivalent to $\sim P$, as you can easily verify with truth tables.

Table 2.2 Truth tables of three symbolic statements

Truth Table of $(P \wedge Q) \vee \sim P$

P	Q	$P \wedge Q$	$\sim P$	$(P \wedge Q) \vee \sim P$
T	T	T	F	T
T	F	F	F	F
F	T	F	T	T
F	F	F	T	T

Truth Table of $P \rightarrow [Q \rightarrow (P \wedge Q)]$

P	Q	$P \wedge Q$	$Q \rightarrow (P \wedge Q)$	$P \rightarrow [Q \rightarrow (P \wedge Q)]$
T	T	T	T	T
T	F	F	T	T
F	T	F	F	T
F	F	F	T	T

Truth Table of $(P \rightarrow Q) \leftrightarrow (R \wedge P)$

P	Q	R	$P \rightarrow Q$	$R \wedge P$	$(P \rightarrow Q) \leftrightarrow (R \wedge P)$
T	T	T	T	T	T
T	T	F	T	F	F
T	F	T	F	T	F
T	F	F	F	F	T
F	T	T	T	F	F
F	T	F	T	F	F
F	F	T	T	F	F
F	F	F	T	F	F

The ideas we have been discussing are quite straightforward as long as we restrict ourselves to symbolic statements. They become more challenging when they are applied to English or mathematical statements. Since logic is such a vital part of mathematics, every mathematics student should learn to recognize the logical structure of English and mathematical statements and translate them into symbolic statements. With English statements, there is often more than one reasonable interpretation of their logical structure, but with mathematical statements there rarely is. Here are some examples of how this is done.

Example 5: For each of the following statements, introduce a propositional variable for each of its atomic substatements, and then use these variables and connectives to write the most accurate symbolic translation of the original statement.

(a) I like milk and cheese but not yogurt.

(b) Rain means no soccer practice.
(c) The only number that is neither positive nor negative is zero.
(d) $2 + 2 = 4$.

Solution: (a) Don't be fooled by a phrase like "milk and cheese." Connectives must connect *statements*, and a noun like "milk" is certainly not a statement. To understand its logical structure, the given statement should be viewed as an abbreviation for "I like milk and I like cheese, but I don't like yogurt." So we introduce the following propositional variables:

P for "I like milk."

Q for "I like cheese."

R for "I like yogurt."

The only remaining difficulty is how to deal with the word "but." This word conveys a different emphasis or mood from the word "and," but the basic logical meaning of the two words is the same. In other words, in statements where the word "but" could be replaced by "and" and still make sense grammatically, the right connective for it is $\wedge$. So the best symbolic representation of the original statement is $P \wedge Q \wedge \sim R$.

(b) Once again, connectives must connect entire statements, not single words or noun phrases. So we write:

P for "It is raining."

Q for "There is soccer practice."

How should we interpret the word "means"? Although it would be plausible to think of it as "if and only if," the most sensible interpretation is that if it rains, there's no soccer practice. So we represent the given English statement as $P \to \sim Q$.

(c) Since this statement involves an unspecified number, we can use a mathematical variable like x to represent it. (It is possible to do this problem without using a letter to stand for the unspecified number, but the wording gets a bit awkward.) So we write:

P for "x is positive."

Q for "x is negative."

R for "x is zero."

Now we must interpret various words. A bit of thought should convince you that "neither P nor Q" has the logical meaning $\sim (P \vee Q)$ or its propositional equivalent

$\sim P \wedge \sim Q$. The words "the only" in this statement require a quantifier to interpret precisely, but the gist of the statement seems to be that a number is neither positive nor negative if and only if the number is zero. So the statement can be represented symbolically as $(\sim P \wedge \sim Q) \leftrightarrow R$.

If we allow ourselves mathematical symbols as well as connectives, we would probably prefer to represent the statement in the form

$$[\sim (x > 0) \wedge \sim (x < 0)] \leftrightarrow x = 0$$

or shorter still

$$(x \not> 0 \wedge x \not< 0) \leftrightarrow x = 0$$

(We use the standard convention that a slash through an equal sign, an inequality symbol, and so on, can be used instead of a negation symbol.)

It should be noted that quantifiers are required for a totally accurate translation of this statement.

(d) This is sort of a trick question. The statement contains *no* connectives, so it is atomic. Therefore, the only way to represent it symbolically is simply P, where P represents the whole statement!

It is very tempting just to assume that this simple equation is a tautology. But since its logical form is P, it's not. It's certainly a true statement of arithmetic, and you might even claim that it's a law of arithmetic, but it's *not* a law of propositional logic. Even a statement like $1 = 1$ is technically not a tautology!

Exercises 2.1

(1) Construct the truth tables of the following statements:

(a) $\sim (P \wedge Q)$
(b) $P \leftrightarrow (P \vee Q)$
(c) $P \rightarrow \sim P$
(d) $P \leftrightarrow \sim P$
(e) $P \rightarrow (Q \rightarrow (P \wedge Q))$
(f) $\sim (P \wedge Q) \rightarrow (\sim P \wedge \sim Q)$
(g) $P \wedge (Q \wedge R) \leftrightarrow (P \wedge Q) \wedge R$
(h) $[(P \vee Q) \rightarrow R] \leftrightarrow [(P \rightarrow R) \wedge (Q \rightarrow R)]$
(i) $(P \wedge Q) \vee (\sim P \wedge R)$

(2) For each of the following, state whether it is a proposition, with a brief explanation. If you believe that a particular case is borderline, provide brief pros and cons for whether it should be considered a proposition. For those which are propositions, determine which are true and which are false, if possible.

(a) 10 is a prime number.
(b) Are there any even prime numbers?

(c) Turn off that music or I'll scream.
(d) Life is good.
(e) 3 + 5.
(f) The number π is bigger than 4.
(g) Benjamin Franklin had many friends.
(h) The Chicago Cubs will win the World Series in the year 2106.
(i) I like olives but not very much.
(j) Goldbach's conjecture is true. (This was described in Chapter 1.)

(3) Determine whether each of the following is a tautology, a contradiction, or neither. If you can determine answers by commonsense logic, do so; otherwise, construct truth tables.

(a) $\sim(P \wedge Q) \rightarrow \sim P \wedge \sim Q$
(b) $\sim P \wedge \sim Q \rightarrow \sim(P \wedge Q)$
(c) $(P \leftrightarrow Q) \leftrightarrow (Q \leftrightarrow P)$
(d) $(P \rightarrow Q) \leftrightarrow (Q \rightarrow P)$
(e) $[(P \vee Q) \vee R] \leftrightarrow [P \vee (Q \vee R)]$
(f) $[(P \vee Q) \wedge R] \leftrightarrow [P \vee (Q \wedge R)]$

(4) Determine whether each of the following pairs of statements are propositionally equivalent to each other. If you can determine answers by commonsense logic, do so; otherwise, construct truth tables.

(a) $P \wedge Q$ and $Q \wedge P$
(b) P and $\sim\sim P$
(c) $\sim(P \vee Q)$ and $\sim P \vee \sim Q$
(d) $\sim(P \vee Q)$ and $\sim P \wedge \sim Q$
(e) $P \rightarrow Q$ and $Q \rightarrow P$
(f) $\sim(P \rightarrow Q)$ and $\sim P \rightarrow \sim Q$
(g) $P \leftrightarrow Q$ and $(P \wedge Q) \vee \sim(P \vee Q)$
(h) $P \wedge (Q \vee R)$ and $(P \wedge Q) \vee R$
(i) $P \wedge (Q \wedge R)$ and $(P \wedge Q) \wedge R$
(j) $P \rightarrow (Q \rightarrow R)$ and $(P \rightarrow Q) \rightarrow R$
(k) $P \leftrightarrow (Q \leftrightarrow R)$ and $(P \leftrightarrow Q) \leftrightarrow R$

(5) Match each statement on the left with a propositionally equivalent one on the right. As with the previous problem, see if you can do this without writing out truth tables.

(a) $P \rightarrow \sim Q$
(b) $P \leftrightarrow (P \wedge Q)$
(c) $(P \vee Q) \wedge \sim(P \wedge Q)$
(d) $P \rightarrow \sim P$
(e) $(P \vee Q) \leftrightarrow (P \wedge Q)$

(i) $P \wedge \sim P$
(ii) $P \rightarrow Q$
(iii) $\sim(P \wedge Q)$
(iv) $Q \rightarrow P$
(v) $P \leftrightarrow \sim Q$
(vi) $\sim P$
(vii) $P \leftrightarrow Q$
(viii) $Q \wedge \sim P$

(6) For each of the following, replace the symbol # with a connective so that the resulting symbolic statement is a tautology. If you can, figure these out without using truth tables.

(a) $[\sim (P \# Q)] \leftrightarrow [P \wedge \sim Q]$
(b) $[P \rightarrow (Q \# R)] \leftrightarrow [(P \rightarrow Q) \wedge (P \rightarrow R)]$
(c) $[(P \# Q) \rightarrow R] \leftrightarrow [(P \rightarrow R) \wedge (Q \rightarrow R)]$
(d) $[(P \wedge Q) \leftrightarrow P] \leftrightarrow [P \# Q]$
(e) $[(P \# Q) \rightarrow R] \leftrightarrow [P \rightarrow (Q \rightarrow R)]$

(7) Show, using a commonsense argument, that for two symbolic statements to be propositionally equivalent means precisely that they have the same truth value (both true or both false) for any truth values of the propositional variables in them.

(8) Recall the discussion of the inclusive or and the exclusive or. Let the symbol $\veebar$ represent the latter.

(a) Construct the truth table for $P \veebar Q$.
(b) Write a statement using our five basic connectives that is equivalent to $P \veebar Q$.
(c) Write a statement using only the connectives $\sim$, $\wedge$, and $\veebar$ that is equivalent to $P \vee Q$.
(d) Make up an English sentence in which you feel the word "or" should be interpreted inclusively.
(e) Make up an English sentence in which you feel the word "or" should be interpreted exclusively.
(f) Make up an English sentence in which you feel the word "or" can be interpreted either way.

(9) Let P, Q, and R stand for "Pigs are fish," "2 + 2 = 4," and "Canada is in Asia," respectively. Translate the following symbolic statements into reasonable-sounding English. Also, determine whether each of them is true or false.

(a) $P \vee \sim Q$ (b) $Q \leftrightarrow \sim R$
(c) $\sim Q \rightarrow (R \wedge \sim P)$ (d) $P \rightarrow \sim P$

(10) For each of the following statements, introduce a propositional variable for each of its atomic substatements, and then use these variables and connectives to write the most accurate symbolic translation of the original statement.

(a) I need to go to Oxnard and Lompoc.
(b) If a number is even and bigger than 2, it's not prime.
(c) You're damned if you do and damned if you don't.
(d) If you order from the dinner menu, you get a soup or a salad, an entree, and a beverage or a dessert. (Be careful with the word "or" in this one.)
(e) If it doesn't rain in the next week, we won't have vegetables or flowers, but if it does, we'll at least have flowers.

(f) No shoes, no shirt, no service. (Of course, this is a highly abbreviated sentence. You have to interpret it properly.)

(g) Men or women may apply for this job. (Be careful; this one's a bit tricky.)

(11) (a) If a symbolic statement has just one propositional variable (say P), how many lines are in its truth table?

(b) How many different possible truth functions are there for such a statement? That is, in how many ways can the output column of such a truth table be filled in? Explain.

*(c) Repeat parts (a) and (b) for a symbolic statement with two propositional variables P and Q. Explain.

*(d) On the basis of the previous parts of this problem, make conjectures that generalize them to a symbolic statement with an arbitrary number *n* of propositional variables.

2.2 Conditionals and Biconditionals

The connectives $\rightarrow$ and $\leftrightarrow$ are not only the most subtle of the five connectives; they are also the two most important ones in mathematical work. So it is worthwhile for us to discuss them at some length. We begin this section by considering the meaning of conditional statements.

In the previous section, we linked the connective $\rightarrow$ to the word "implies," but in ordinary language this word is not used very frequently. Probably the most common way of expressing conditionals in English is with the words "If ... then ... ." As we see shortly, there are several other words or combinations of words that also express conditionals.

Conditional and biconditional statements are often called **implications** and **equivalences**, respectively. However, there is a tendency to reserve these latter words for statements that are known to be true. For instance, "2 + 3 = 5 if and only if pigs can fly" is a biconditional statement. But many mathematicians would not call it an equivalence, since it is false.

Regardless of what words are used to represent conditionals, it takes some thought to understand the truth function for this connective. Refer back to Table 2.1 and note that the statement $P \rightarrow Q$ is false in only one of the four cases, specifically when P is true and Q is false.

Example 1: The best way to understand why this makes sense is to think of a conditional as a promise. Not every conditional can be thought of in this way, but many can. So let's pick one at random, like "If you rub my back today, I'll buy you dinner tonight." This is certainly a conditional; it can be represented as $P \rightarrow Q$, where P is "You rub my back today" and Q is "I'll buy you dinner tonight." Under what circumstances is or is not this promise kept?

Two of the four entries in the truth table are clear-cut. If you rub my back and I buy you dinner, I've obviously kept the promise, so the whole conditional is true. On the other hand, if you rub my back and I don't buy you dinner, I've obviously broken my promise and the conditional must be considered false. It requires more thought to understand the two truth table entries for which P is false. Suppose you don't rub my back and I don't take you to dinner. Even though I haven't done anything, no one could say I've broken my promise. Therefore, we define $P \rightarrow Q$ to be true if both P and Q are false.

Finally, we get to the least intuitive case. Suppose you don't rub my back but I go ahead and buy you dinner anyway. Have I broken my promise? If you reflect on this question, you will probably conclude that, although it's unexpected for me to buy you dinner after you didn't rub my back, it's not breaking my promise. To put it another way, although my promise might lead most people to *assume* that if you don't rub my back, I won't buy you dinner, my statement doesn't say anything about what I'll do if you don't rub my back. It is with these considerations in mind that the third entry in the truth table is also a T. A good way to understand these last two cases is to admit that if you don't rub my back, my promise is true by default, because you haven't done anything to obligate me to act one way or the other regarding dinner.

Now here's some useful terminology.

Definitions: In any conditional $P \rightarrow Q$, the statement P is called the **hypothesis** or **antecedent** and Q is called the **conclusion** or **consequent** of the conditional.

Definitions: Given any conditional $P \rightarrow Q$,

- the statement $Q \rightarrow P$ is called its **converse.**
- the statement $\sim P \rightarrow \sim Q$ is called its **inverse.**
- the statement $\sim Q \rightarrow \sim P$ is called its **contrapositive.**

We now come to the first result in this text that is labeled a "theorem." Since our serious study of proofs does not begin until Chapter 4, many of the theorems in this chapter and the next are presented in a very nonrigorous way. In other words, the proofs given for some of these theorems have more of the flavor of intuitive explanations than of mathematical proofs.

Theorem 2.1: (a) Every conditional is equivalent to its own contrapositive.

(b) A conditional is not necessarily equivalent to its converse or its inverse.

(c) However, the converse and the inverse of any conditional are equivalent to each other.

(d) The conjunction of any conditional $P \rightarrow Q$ and its converse is equivalent to the biconditional $P \leftrightarrow Q$.

Proof: This theorem is so elementary that we can prove it rigorously at this point. The proof simply requires constructing several truth tables. For instance, to prove part (a) we only need to show that $(P \rightarrow Q) \leftrightarrow (\sim Q \rightarrow \sim P)$ is a tautology (Exercise 10). ■

Example 2: Consider the conditional "If you live in California, you live in America." This statement is true for all persons. Its converse is "If you live in America, you live in California"; its inverse is "If you don't live in California, you don't live in America." These two statements are not true in general, so they are not equivalent to the original. However, they are equivalent to each other. The contrapositive of the original statement is "If you don't live in America, you don't live in California," which has the same meaning as the original and is always true.

By the way, it's worth bearing in mind that implication is the only connective whose meaning changes when the two substatements being connected are switched. That is, $P \wedge Q$ is equivalent to $Q \wedge P$, and so on.

Let's elaborate a bit on our earlier discussion of conditionals as promises. When someone says "If you rub my back today, I'll buy you dinner tonight," many people would automatically read into it "And if you don't rub my back, I won't buy you dinner." Note that this other promise is just the inverse of the original one. Now, there is no doubt that in ordinary language, when a person states a conditional, the inverse is sometimes also intended. And then again, sometimes it is not. This kind of fuzziness is a normal feature of spoken language, as we have already mentioned regarding the ambiguity of the word "or" (inclusive versus exclusive). But in mathematics and logic, connectives must have precise meanings. The most useful decision is to agree that conditionals in general should *not* include their own inverses, for the simple reason that if they did, there would be no difference between conditionals and biconditionals (by Theorem 2.1 (c) and (d)).

In spoken language, conditionals aren't always promises, but they almost always at least convey some kind of causal connection between the antecedent and the consequent. When we say "P implies Q" or even "If P then Q," we normally mean that the statement P, if true, somehow causes or forces the statement Q to be true. In mathematics, most conditionals convey this kind of causality, but it is *not* a requirement. In logic (and therefore in mathematics), the truth or falsity of a conditional is based strictly on truth values.

Example 3: The following three statements, although they may seem silly or even wrong, must be considered true:

If $2 + 2 = 4$, then ice is cold.

If $2 + 2 = 3$, then ice is cold.

If $2 + 2 = 3$, then ice is hot.

On the other hand, the statement "If $2 + 2 = 4$, then ice is hot" is certainly false.

There are quite a few ways of expressing conditionals in words, especially in mathematics. It is quite important to be familiar with all of them, so let's talk about them for a bit. You will find the most common ones listed in Table 2.3.

Table 2.3 The most common ways to express a conditional $P \to Q$ in words

(1) P implies Q.

(2) If P then Q.

(3) If P, Q.

(4) Q if P.

(5) P only if Q.

(6) P is sufficient for Q.

(7) Q is necessary for P.

(8) Whenever P, Q.

(9) Q whenever P.

Note that statements 1–4 of Table 2.3 contain nothing new—but pay attention to the word order in statement 4. For example, in the sentence "I'll buy you dinner if you rub my back," the hypothesis consists of the last four words and the conclusion is the first four words.

Now consider statement 5. An example of this construction is "You'll see the comet only if you look in the right spot." What is this saying? The answer is open to debate, but the most likely meaning is "If you *don't* look in the right spot, you *won't* (or *can't*) see the comet," which is the contrapositive of "If you (expect to) see the comet, you (have to) look in the right spot." (The words in parentheses have been added to make the sentence read better.) And this is what statement 5 says this sentence should mean. On the other hand, it's possible to believe that the sentence might also be saying, "If you do look in the right spot, you'll see the comet." But we reject this interpretation because it would mean that "only if" would be a synonym for "if and only if." We therefore follow the standard convention that "P only if Q" is the converse of "P if Q," and neither of these means the same as "P if and only if Q."

The pair of words "sufficient" and "necessary," like the words "if" and "only if," express conditionals in the opposite order from each other. Suppose you are told, "Passing the midterm and the final is sufficient to pass this course." This appears to mean that if you pass these exams, you will pass the course. But does it also mean that if you don't pass both these exams, you can't pass the course? Again, that interpretation is possible, but the word "sufficient" seems to allow the possibility that there might be other ways to pass the course. So, as with the words "if" and "only if," we reject this other interpretation so that the word "sufficient" conveys the meaning of a conditional, not a biconditional.

Now, suppose instead that you are told "Passing the midterm and the final is necessary to pass the course." With only one word changed, this sentence has a completely different emphasis from the previous one. This sentence certainly does *not* say that passing the exams is any sort of guarantee of passing the course. Instead, it appears to say that you *must* pass the exams to even have a chance of passing the course, or, more directly, if you *don't* pass the exams, you definitely *won't* pass the course. So, as statements 6 and 7 of Table 2.3 indicate, the word "necessary" is generally considered to express the converse of the word "sufficient."

Statements 8 and 9 indicate that the word "whenever" often expresses a conditional. In the sentence "Whenever a function is continuous, it's integrable," the word "whenever" is essentially a synonym for "if."

English (and all spoken languages) has many ways of expressing the same thought, and even Table 2.3 does not include all the reasonable ways of expressing conditionals. It should also be pointed out that many statements that seem to have no connective in them are really conditionals. For instance, the important theorem, "A differentiable function is continuous," is really saying that if a function is differentiable, it's continuous. "Hidden connectives" are also often conveyed by quantifiers, as Section 3.2 demonstrates.

☞ Without *any* doubt, *the most frequent logical error* made by mathematics students at all levels is confusing a conditional with its converse (or inverse) or assuming that if a conditional is true, its converse must also be true. Learn to avoid this confusion like the plague, and you will spare yourself *much grief!*

Biconditionals

There are various ways to think of biconditionals, one of which was stated in Theorem 2.1(d): $P \leftrightarrow Q$ is equivalent to $(P \rightarrow Q) \wedge (Q \rightarrow P)$. That is, when you assert both a conditional and its converse, you're stating a biconditional. That's why the symbol for a biconditional is a double arrow. That's also why we use the phrase "if and only if" for biconditionals. (By the way, mathematicians often use the abbreviation "iff" for "if and only if.") Table 2.4 shows this and other ways of expressing biconditionals.

We have seen that the words "necessary" and "sufficient" also have converse meanings, and so the phrase "necessary and sufficient" is often used to express biconditionals. For example, if you read that "a necessary and sufficient condition for a number to be rational is that its decimal expansion terminates or repeats," that means that a number is rational if and only if its decimal expansion terminates or repeats. (The noun "condition" is often used in this way with the words "necessary" and/or "sufficient.") Another common way of expressing biconditionals in mathematics is with the word "equivalent." For example, an alternate way of stating the same fact about numbers that was just given would be "Rationality is equivalent to having a decimal expansion that either terminates or repeats." (When mathematicians say that two statements are equivalent, it does not necessarily mean that they are propositionally equivalent. It just means that they can be proved to imply each other, using whatever axioms and previously proved theorems are available in the situation.)

Finally, Table 2.4 indicates that the words "just in case" can also convey a biconditional, as in "A number is rational just in case its decimal expansion either terminates or repeats."

Table 2.4 The most common ways to express a biconditional $P \leftrightarrow Q$ in words

(1) P if and only if Q.

(2) P is necessary and sufficient for Q.

(3) P is equivalent to Q.

(4) P and Q are equivalent.

(5) P (is true) just in case Q (is).

We have already mentioned that, in ordinary speech, statements that on the surface are just one-way conditionals are often understood to be biconditionals. This is partly because there are no fluid-sounding ways of expressing biconditionals in English. All the phrases in Table 2.4 sound fine to a mathematician, but they are somewhat awkward when used in ordinary conversation. If I say "You'll pass this course if and only if you pass the midterm and the final," I'm clearly stating a biconditional, but it sounds strange. Since people are not used to hearing the phrase "if and only if," they might take this statement to mean a biconditional even if the words "if and" are left out. This interpretation could lead to some serious disappointment, since with these two words omitted I would only be stating a conditional.

There are several useful ways of thinking of biconditionals. Most directly, a biconditional represents a two-way conditional. Another way of looking at a biconditional $P \leftrightarrow Q$ is that if either P or Q is true, they both are. That is, either they're both true, or they're both false. So a biconditional between two statements says that they have the same *truth values*. For this reason, the biconditional connective is very similar to an equal sign, except that it is applied to statements rather than to mathematical quantities. To put it even more strongly, when mathematicians assert that two (or more) statements are equivalent, they are more or less saying that these statements are *different ways of saying the same thing*.

We conclude this section with our first proof preview. These are called "previews" because they occur before our in-depth study of proofs. Thus they are not axiomatic or rigorous proofs. But each of them illustrates at least one important proof technique, and we see later that each of them can be fleshed out to a more complete, rigorous proof. Furthermore, the relatively informal style of these proof previews is typical of the way mathematicians write proofs in practice.

In these proof previews, and occasionally elsewhere in proofs in this book, comments in brackets and italics are explanations to the reader that would probably not be included under normal circumstances.

Proof Preview 1

Theorem: (a) An integer n is even if and only if $n + 1$ is odd.

(b) Similarly, n is odd if and only if $n + 1$ is even.

Proof: (a) *[We are asked to prove a biconditional. By Theorem 2.1(d), one way to do this—in fact, the most natural and common way—is to prove two conditional* *statements: a forward direction, and a reverse (or converse) direction. Now, how should we try to prove a conditional statement? Well, a conditional statement has the form "If P, then Q." That is, if P is true, Q is supposed to be true too. Therefore, the logical way to prove such a statement is to assume that P is true, and use this to derive the conclusion that Q is also true.]*

For the forward direction, assume that n is even. By definition of the word "even," that means that n is of the form $2m$, for some integer m. But from the equation $n = 2m$, we can add 1 to both sides and obtain $n + 1 = 2m + 1$. Thus, $n + 1$ is odd *[by the analogous definition of what it means to be odd]*.

Conversely, assume that $n + 1$ is odd. That means $n + 1$ is of the form $2m + 1$, and by subtracting 1 from both sides of the equation $n + 1 = 2m + 1$, we obtain $n = 2m$. So n is even. *[Biconditional (a) is now proved because we have proved both directions of it.]*

(b) For the forward direction, assume that n is odd. So $n = 2m + 1$, for some integer m. From this equation, we get $n + 1 = 2m + 2 = 2(m + 1)$. Therefore, $n + 1$ is even, because it equals 2 times an integer. The reverse direction is left for Exercise 11. ■

The only nonrigorous feature of the previous proof is that it does not properly deal with quantifiers (see Exercise 2 of Section 4.3). The proof is straightforward because of the definition of the word "odd" it uses. If "odd" is defined to mean "not even," this theorem becomes somewhat harder to prove. Exercise 12 covers a slightly different approach to this result.

Exercises 2.2

(1) Consider a conditional statement $P \rightarrow Q$. Write the following symbolic statements. (Whenever you obtain two consecutive negation symbols, delete them).

(a) The converse of the converse of the original statement
(b) The contrapositive of the contrapositive of the original statement
(c) The inverse of the contrapositive of the original statement

(2) Restate each of the following statements in the form of an implication (using the words "If ... then ... "):

(a) Whenever a function is differentiable, it's continuous.
(b) A continuous function must be integrable.
(c) A prime number greater than 2 can't be even.
(d) A nonnegative number necessarily has a square root.
(e) Being nonnegative is a necessary condition for a number to have a square root.
(f) A one-to-one function has an inverse function.

(3) Write the contrapositive of the following statements. (Replace any substatement of the form $\sim\sim P$ with P.)

(a) If John's happy, Mary's happy.
(b) If Mary's not happy, John's happy.

(c) John's not happy only if Mary's not happy.
(d) Mary's lack of happiness is necessary for John's happiness.

(4) Write each of the following conditionals *and* its converse in the indicated forms from Table 2.3. Some answers might be difficult to express in sensible English, but do your best. For instance, statement (a) in form 9 could be "Whenever I read a good book, I'm happy all day," and its converse in that form could be "Whenever I'm happy all day, I must be reading a good book."

(a) Reading a good book is sufficient to keep me happy all day. (Forms 3, 5 and 7)
(b) I will pay you if you apologize. (Forms 1, 3, and 5)
(c) It's necessary to give a baby nourishing food in order for it to grow up healthy. (Forms 2, 6, and 8)

(5) Write each of the following biconditionals in the indicated forms from Table 2.4. Some answers might be difficult to express in sensible English, but do your best.

(a) A triangle is isosceles if and only if it has two equal angles. (Forms 2 and 3)
(b) I'll go for a hike today just in case I finish my paper this morning. (Forms 1 and 4)
(c) The Axiom of Choice is equivalent to Zorn's lemma. (Forms 1 and 5)
(d) Being rich is a necessary and sufficient condition to be allowed in that country club. (Forms 4 and 5)

(6) Restate each of the following statements in the form of a conditional (with the words "If ... then ... "), a biconditional, or the negation of a conditional. If you think there's more than one reasonable interpretation for a statement, you may give more than one answer.

(a) Stop that right now or I'll call the police.
(b) If you clean your room, you can watch TV; otherwise you can't.
(c) You can't have your cake and eat it too.
(d) Thanksgiving must fall on a Thursday.
(e) You can't get what you want unless you ask for it.
*(f) This dog is fat but not lazy.
(g) An integer is odd or even, but not both.
(h) In order to become president, it's necessary to have a good publicity firm.
(i) A person can become a professional tennis player only by hard work.
(j) I won't pay you if you don't apologize.
(k) Math professors aren't boring.

(7) Give an example of each of the following if possible:

(a) A true (that is, *necessarily* true) conditional statement whose converse is false (that is, *not* necessarily true)
(b) A false conditional statement whose contrapositive is true
(c) A false conditional statement whose inverse is true
(d) A false conditional statement whose converse is false

(8) Classify each of the following conditionals as necessarily true, necessarily false, or sometimes true and sometimes false (depending on which number or which person is being referred to). Also, do the same for the converse of each statement. Explain.

(a) If ice is cold, then $2 + 2 = 3$.
(b) If a number is divisible by 2, it's divisible by 6.
(c) If a person lives in Europe, then he or she lives in France.
*(d) If a person lives in Europe, then he or she lives in Brazil.
(e) If $x > 0$, then $x > 0$ or $2 + 2 = 3$.
*(f) If $x > 0$, then $x > 0$ and $2 + 2 = 3$.

(9) Construct a truth table that you think best captures of the meaning of "P unless Q." There may be more than one reasonable way to do this. To help you, you might want to consider a couple of specific examples, like "You can go swimming tomorrow unless you have a temperature." Do you think that the word "unless" usually has the same meaning as the exclusive or?

(10) Prove Theorem 2.1, in the manner indicated in the text.

(11) Prove the converse of part (b) of the theorem in Proof Preview 1.

(12) Proof Preview 1 uses the definition that a number is odd iff it is of the form $2m + 1$. It is just as correct to say that a number is odd iff it is of the form $2m - 1$. Prove the same result, using this alternate definition.

(13) Prove the following, in the manner of Proof Preview 1. ***Hint:*** You will need to use four variables, not just two, in each of these proofs.

(a) The sum of two even numbers must be even.
(b) The sum of two odd numbers must be even.
(c) The product of two odd numbers must be odd.

(14) By experimentation, fill in each blank with a number that you believe yields a correct conjecture. Then prove the conjecture, in the manner of Proof Preview 1.

(a) If n is ____ or ____ more than a multiple of 10, then n^2 is 1 less than a multiple of 10.

(b) If n is ____, ____, or ____ more than a multiple of 6, then there is no number m such that mn is 1 more than a multiple of 6.

2.3 Propositional Consequence; Introduction to Proofs

In Section 2.1 we defined the concepts of tautology and propositional equivalence. Now that we have discussed the various connectives individually, it's time to examine these concepts in more detail.

Why are these notions important? Recall that a tautology is a statement that is always true because of the relationship or pattern of its connectives. Also recall that it's very easy to tell whether a given statement is a tautology; all that's required is a truth table. In other words, tautologies are absolute truths that are easily identifiable. So there is almost universal agreement that all tautologies can be considered axioms in mathematical work.

As far as propositional equivalence is concerned, we have mentioned that if two statements are equivalent, they are essentially two different ways of saying the same thing. If that's so, we should expect equivalent statements to be interchangeable; and in fact one simple but important tool in proofs is to replace one statement with another equivalent one.

Table 2.5 shows some of the more common and useful tautologies. It is certainly not a complete list. In fact there's no such thing: there are an infinite number of tautologies. At the same time, it's important to realize that even Table 2.5 shows an infinite set of tautologies, in a certain sense; remember that our propositional variables can stand for any statement. So a single tautology like the law of the excluded middle actually comprises an infinite number of statements, including purely symbolic ones like $(Q \rightarrow \sim R) \vee \sim (Q \rightarrow \sim R)$, mathematical ones like "$x + y = 3$ or $x + y \neq 3$," and English ones like "Either I'll finish or I won't."

To what extent should you know this list? Well, if there were only thirty tautologies in existence, it might be worthwhile to memorize them. But since there are an infinite number of them, there's not much reason to memorize some finite list. It might be fruitful for you to go through Table 2.5 and try to see (without truth tables, as much as possible) why all the statements in it are tautologies. This would be one way to become familiar with these tautologies for future reference. Some of the statements in Table 2.5, such as the law of the excluded middle and the law of double negation, are very simple to understand. Others, like numbers 26 and 27, are somewhat more complex, and it might take some thought to realize that they are tautologies.

Notice the groupings of the entries in Table 2.5. Most useful tautologies are either implications or equivalences. Remember that an implication is a one-way street that says that if the left side is true, the right side must also be. The usefulness of implications in proofs is based on this fact. For example, tautology number 3 seems to indicate that if we have proved a statement $P \wedge Q$, we should then be allowed to assert the individual statement P. We will see that this type of reasoning is certainly allowed in proofs. (By the way, note that several of the tautologies in Table 2.5 are labeled "Basis for ... ." These tautologies are used to justify specific proof methods discussed in Chapter 4.)

Equivalences are two-way streets asserting that if either side is true, the other must be. So the standard way that equivalences are used in proofs is to replace either side with the other. De Morgan's laws are particularly useful. For example, if you want to prove that a disjunction is false, tautology 18 says that you can do this by proving both the disjuncts false. Also, tautology 19 provides the most useful way of proving that a conditional statement is false. In general, knowing how to rewrite or simplify the negation of a statement is a very important skill (see Exercise 2).

In Section 2.1 it was mentioned that it's not necessary to have five connectives. More precisely, there's quite a bit of redundancy among the standard connectives. For example, tautologies 20 and 22 provide ways of rewriting conditionals and biconditionals in terms of the other three connectives. Also, more equivalences of this sort can be obtained by negating both sides of tautologies 17 through 19. For example,

Table 2.5 Some of the more useful tautologies

(1) $P \vee \sim P$	Law of the excluded middle
(2) $\sim (P \wedge \sim P)$	Law of noncontradiction

Some implications

(3) $(P \wedge Q) \rightarrow P$	Basis for simplification
(4) $(P \wedge Q) \rightarrow Q$	Basis for simplification
(5) $P \rightarrow (P \vee Q)$	Basis for addition
(6) $Q \rightarrow (P \vee Q)$	Basis for addition
(7) $Q \rightarrow (P \rightarrow Q)$	
(8) $\sim P \rightarrow (P \rightarrow Q)$	
(9) $[P \wedge (P \rightarrow Q)] \rightarrow Q$	Basis for modus ponens
(10) $[\sim Q \wedge (P \rightarrow Q)] \rightarrow \sim P$	Basis for modus tollens
(11) $[\sim P \wedge (P \vee Q)] \rightarrow Q$	
(12) $P \rightarrow [Q \rightarrow (P \wedge Q)]$	
(13) $[(P \rightarrow Q) \wedge (Q \rightarrow R)] \rightarrow (P \rightarrow R)$	Transitivity of implication
(14) $(P \rightarrow Q) \rightarrow [(P \vee R) \rightarrow (Q \vee R)]$	
(15) $(P \rightarrow Q) \rightarrow [(P \wedge R) \rightarrow (Q \wedge R)]$	
(16) $[(P \leftrightarrow Q) \wedge (Q \leftrightarrow R)] \rightarrow (P \leftrightarrow R)$	Transitivity of equivalence

Equivalences for rewriting negations

(17) $\sim (P \wedge Q) \leftrightarrow \sim P \vee \sim Q$	De Morgan's law
(18) $\sim (P \vee Q) \leftrightarrow \sim P \wedge \sim Q$	De Morgan's law
(19) $\sim (P \rightarrow Q) \leftrightarrow P \wedge \sim Q$	

Equivalences for replacing connectives

(20) $(P \rightarrow Q) \leftrightarrow (\sim P \vee Q)$

(21) $(P \leftrightarrow Q) \leftrightarrow [(P \rightarrow Q) \wedge (Q \rightarrow P)]$

(22) $(P \leftrightarrow Q) \leftrightarrow [(P \wedge Q) \vee (\sim P \wedge \sim Q)]$

Other equivalences

(23) $\sim \sim P \leftrightarrow P$	Law of double negation
(24) $(P \rightarrow Q) \leftrightarrow (\sim Q \rightarrow \sim P)$	Law of contraposition
(25) $[(P \rightarrow Q) \wedge (P \rightarrow R)] \leftrightarrow [P \rightarrow (Q \wedge R)]$	
(26) $[(P \rightarrow R) \wedge (Q \rightarrow R)] \leftrightarrow [(P \vee Q) \rightarrow R]$	Basis for proof by cases
(27) $[P \rightarrow (Q \rightarrow R)] \leftrightarrow [(P \wedge Q) \rightarrow R]$	
(28) $[P \rightarrow (Q \wedge \sim Q)] \leftrightarrow \sim P$	Basis for indirect proof
(29) $[P \wedge (Q \vee R)] \leftrightarrow [(P \wedge Q) \vee (P \wedge R)]$	Distributive law
(30) $[P \vee (Q \wedge R)] \leftrightarrow [(P \vee Q) \wedge (P \vee R)]$	Distributive law

from the first De Morgan's law we can construct the related equivalence $(P \land Q) \leftrightarrow \sim(\sim P \lor \sim Q)$. In other words, any conjunction can be rewritten in terms of negation and disjunction. In general, knowing when and how to rewrite a connective in terms of specific other ones is a very valuable skill in mathematics. It is also often very useful to rewrite the negation of a given statement; tautologies 17–19 show how this is done.

Exercises 11 through 17 are concerned with rewriting connectives and reducing the number of connectives.

For the remainder of this book, references to "tautology number ... " refer to Table 2.5. For convenient reference, Table 2.5 is repeated as Appendix 3 at the end of the book.

To conclude this chapter, we discuss a method that can be used to analyze everyday, nontechnical arguments for logical correctness. This method is really a simple (but incomplete) framework for doing proofs, so studying it will provide a good preview of Chapter 4.

Definitions: A statement Q is said to be a **propositional consequence** of statements $P_1, P_2, \ldots, P_n$ iff the single statement $(P_1 \land P_2 \land \ldots \land P_n) \rightarrow Q$ is a tautology. (In this section, the word "propositional" may be dropped when discussing this notion.)

The assertion that a statement Q is a consequence of some list of statements is called an **argument.** The statements in the list are called the **premises** or **hypotheses** or **givens** of the argument, and Q is called the **conclusion** of the argument. If Q really is a consequence of the list of statements, the argument is said to be **valid.**

Recall that if a conditional is a tautology, then whenever the hypothesis of that conditional is true, the conclusion must also be true. So the significance of having a valid argument is that whenever the premises are true, the conclusion must be too.

In the definition of propositional consequence, it is possible that $n = 1$. So Q is a propositional consequence of P if $P \rightarrow Q$ is a tautology. With this in mind, note that two statements are equivalent if and only if each is a consequence of the other.

Example 1: Determine whether each of the following arguments is valid:

(a) Premises: $P \rightarrow Q$
$\sim R \rightarrow \sim Q$
$\sim R$

Conclusion: $\sim P$

By the way, this sort of diagram is commonly used for logical arguments, especially ones in which the statements involved are purely symbolic.

(b) Premises: If I'm right, you're wrong. If you're right, I'm wrong.
Conclusion: Therefore, at least one of us is right.

(c) If Al shows up, Betty won't. If Al and Cathy show up, then so will Dave. Betty or Cathy (or both) will show up. But Al and Dave won't both show up. Therefore, Al won't show up.

Solution: (a) To determine whether this argument is valid, we just need to test whether $[(P \rightarrow Q) \wedge (\sim R \rightarrow \sim Q) \wedge \sim R] \rightarrow \sim P$ is a tautology. We leave it to you (Exercise 3) to verify that it is, so the argument is valid.

(b) It's not absolutely required, but such arguments are usually easier to analyze if they are translated into symbolic form. So let P stand for "I'm right" and Q stand for "You're right." Let's also make the reasonable interpretation that "wrong" means "not right." The argument then has the form

Premises: $P \rightarrow \sim Q$
$Q \rightarrow \sim P$

Conclusion: $P \vee Q$

The conditional $[(P \rightarrow \sim Q) \wedge (Q \rightarrow \sim P)] \rightarrow (P \vee Q)$ is *not* a tautology (Exercise 3), so this argument is not valid.

By the way, this is an argument that I actually heard used in a real-life situation. Can you explain why the argument fails? The simplest explanation involves the relationship between the two premises.

(c) As in part (b), let's introduce propositional variables: A for "Al will show up" and similarly B, C, and D, for Betty's, Cathy's and Dave's showing up. It turns out that

$$[(A \rightarrow \sim B) \wedge (A \wedge C \rightarrow D) \wedge (B \vee C) \wedge \sim (A \wedge D)] \rightarrow \sim A$$

is a tautology (Exercise 3), so this argument is valid.

Since this argument involves four propositional variables, the truth table required to validate it contains sixteen lines, which makes it somewhat unwieldy and tedious to construct. So we now introduce a "nicer" method for validating such arguments:

Theorem 2.2: Suppose the statement R is a consequence of premises $P_1, P_2, \ldots,$ P_n, and another statement Q is a consequence of $P_1, P_2, \ldots, P_n$ *and* R. Then Q is a consequence of just $P_1, P_2, \ldots, P_n$.

Proof: Let P be an abbreviation for $(P_1 \wedge P_2 \wedge \ldots \wedge P_n)$. So we are told that $P \rightarrow R$ and $(P \wedge R) \rightarrow Q$ are both tautologies. Now consider what the truth table of $P \rightarrow Q$ must look like. In every row where P is true, R must be too, since $P \rightarrow R$ is always true. But since $(P \wedge R) \rightarrow Q$ is also always true, this guarantees that in every row where P is true, Q must be true too. And remember that when P is false, $P \rightarrow Q$ is true by definition. In other words, $P \rightarrow Q$ must be a tautology; this is what we wanted to show. ■

The practical significance of this theorem is that you can use intermediate steps to show an argument is valid. In other words, if you want to show a statement is a consequence of some premises, you don't have to test whether the entire conditional is a tautology. Instead, if you prefer, you can begin listing statements that are obvious consequences of some or all of the premises. Each time you find such a statement you can use it as a *new* premise to find more consequences. This method can lead easily to the desired conclusion. (Unfortunately, it also can lead you nowhere, even if the argument is valid.)

We now give alternate solutions to Examples 1(a) and 1(c), using this method of intermediate steps. If you have any experience with formal proofs (from high school geometry, for example), you will recognize the similarity. In fact, the derivations that follow are perfectly good mathematical proofs, and except for the need to include principles involving quantifiers, mathematical proofs could be based entirely on propositional consequence.

Alternate Solution: Our solutions consist of a sequence of statements, numbered for easy reference, beginning with the premises and ending with the desired conclusion. Each statement in the derivation, after the premises, is a consequence of the previous lines. Since constructing truth tables is so straightforward, there's no need to explain or justify the steps in these derivations any further. But to help you develop the habit of good proof-writing, we explain each step.

Formal solution to Example 1(a):

(1) $P \to Q$	Premise
(2) $\sim R \to \sim Q$	Premise
(3) $\sim R$	Premise
(4) $\sim Q$	From steps 2 and 3, by tautology 9
(5) $\sim P$	From steps 1 and 4, by tautology 10

Formal solution to Example 1(c):

(1) $A \to \sim B$	Premise
(2) $(A \land C) \to D$	Premise
(3) $B \lor C$	Premise
(4) $\sim (A \land D)$	Premise
(5) $\sim B \to C$	From step 3, by tautology 20, essentially
(6) $A \to C$	From steps 1 and 5, by tautology 13
(7) $A \to (A \land C)$	From step 6
(8) $A \to D$	From steps 7 and 2, by tautology 13
(9) $A \to \sim D$	From step 4, by tautology 19, essentially
(10) $A \to (D \land \sim D)$	From steps 8 and 9, by tautology 25
(11) $\sim A$	From step 10, by tautology 28

Which is the easier solution to this problem: the sixteen-line truth table or the derivation just given? It's hard to say, but there's no doubt that the derivation is more informative and better practice for learning how to do proofs.

On the other hand, neither a sixteen-line truth table nor an eleven-step formal proof is particularly readable. One of the main themes of Chapter 4 is that formal proofs, although having the advantage of encouraging thoroughness and correctness in proofwriting, are cumbersome to write and to read. Mathematicians almost always prefer to write less formal proofs that communicate an outline or synopsis of the full formal proof. With that in mind, here is an informal solution to Example 1(c). Exercise 6 asks you to do the same for Example 1(a).

Informal Solution to Example 1(c): We are given that Al and Dave won't both show up. Therefore, if Al shows up, Dave won't (using tautology 19).

Now, let's assume Al shows up. Then we are told that Betty will not show up. But we also know that Betty or Cathy will show up. Therefore, Cathy must show up. But that means Al and Cathy show up, and we are told that if they both show up, then Dave must show up. So we have shown that if Al shows up, then Dave shows up.

Putting both previous paragraphs together, we have shown that if Al shows up, then Dave will show up and Dave won't show up. That is, if Al shows up, something impossible occurs. Therefore, Al cannot show up (tautology 28).

We close this chapter with two more proof previews. These are also written in an informal style but would not be difficult to turn into formal proofs. Each of them is based on one or two key tautologies from Table 2.5.

Proof Preview 2

Theorem: Given sets A, B, and C, if $A \subseteq B$ and $B \subseteq C$, then $A \subseteq C$. *[The symbol $\subseteq$ is read "is a subset of." This notion is defined and discussed in Section 5.2, but we need to use its definition here to carry out this proof.]*

Proof: *[As in Proof Preview 1 at the end of Section 2.2, we are asked to prove a conditional statement. So, once again, we begin our proof by making an assumption. In the terminology of this section, we could say that $A \subseteq B$ and $B \subseteq C$ are the premises of this proof.]* Assume that $A \subseteq B$ and $B \subseteq C$. By the definition of $\subseteq$, this means that for any object x, $x \in A$ implies $x \in B$, and $x \in B$ implies $x \in C$. Therefore, $x \in A$ implies $x \in C$ *[because, by tautology 13, this latter conditional statement is a consequence of the two in the previous sentence]*. And this is exactly what $A \subseteq C$ means. ■

As with Proof Preview 1, this proof glosses over some points involving quantifiers (see Exercise 1 of Section 4.3).

Proof Preview 3

Theorem: For any real number x, $|x| \geq x$.

Proof: Let the propositional variables Q, R, and P stand for $x \geq 0$, $x < 0$, and $|x| \geq x$, respectively. *[Mathematicians would rarely introduce explicit propositional* *variables in this manner, but it can't hurt to do so.]* We know that x must be positive, zero, or negative; that is, we know Q ∨ R. If $x \geq 0$, we know that $|x| = x$ (by definition of absolute value), which implies $|x| \geq x$. In other words, Q implies P. On the other hand, if $x < 0$, then $|x| > 0 > x$, so we still can conclude $|x| \geq x$. In other words, R implies P. So we have shown that Q implies P, and R implies P. By tautology 26, we can conclude the equivalent statement (Q or R) implies P. But since we also know (Q or R), we obtain (by tautology 9) P; that is, $|x| \geq x$. ■

The argument in Proof Preview 3 is a proof by cases, as we see in Section 4.2.

Exercises 2.3

(1) Replace each of the following statements by an equivalent statement that is as short as possible (in number of symbols). In some cases, the answer may be the given statement.

(a) P ∧ P
(b) ~ (P → ~ Q)
(c) Q ∧ (Q → P)
(d) P → ~ P
(e) (P ∧ Q) ∨ (P ∧ R)
(f) P ∨ Q ∨ R
(g) (P → Q) ↔ (Q → P)
(h) P → (Q → ~ P)

(2) For each of the following statements, express its *negation* in as short and simple a way as possible. You will probably want to use tautologies number 17 through 19 (and possibly others) from Table 2.5.

(a) This function is continuous but not increasing.
(b) Pigs are not blue or dogs are not green.
(c) If x^2 is positive, then x is positive.
(d) Pigs are blue if and only if dogs are not green.
(e) If set A is finite, then set B is finite and not empty.

(3) Construct the truth tables necessary to test the validity of the three arguments in Example 1.

(4) Test each of the following arguments for validity, by directly applying the definition of propositional consequence. In other words, construct just one truth table for each argument.

(a) Premises: P → Q, P → ~ R, Q ↔ R. Conclusion: ~ P.
(b) Premises: P ∨ Q ↔ ~ P ∧ R, R → P. Conclusion: ~ (P ∨ Q ∨ R).
(c) Premises: P ∨ Q, Q ∨ R ↔ ~ P. Conclusion: R ∨ ~ Q.
(d) If Alice is wrong, then Bill is wrong. If Bill is wrong, then Connie is wrong. Connie is wrong. Therefore, Alice is wrong.

(e) If turtles can sing, then artichokes can fly. If artichokes can fly, then turtles can sing and dogs can't play chess. Dogs can play chess if and only if turtles can sing. Therefore, turtles can't sing.

(5) Show that each of the following arguments is valid, using the method employed in the alternate solutions given previously. Do not use any tautologies with more than three propositional variables. Consult your instructor about whether to write formal or informal solutions.

(a) Premises: $Q \to R$, $R \vee S \to P$, $Q \vee S$. Conclusion: P.

*(b) Premises: $P \to (Q \leftrightarrow \sim R)$, $P \vee \sim S$, $R \to S$, $\sim Q \to \sim R$. Conclusion: $\sim R$.

(c) Premises: Babies are illogical. A person who can manage a crocodile is not despised. Illogical persons are not despised. Therefore, babies cannot manage crocodiles. (This example was created by Lewis Carroll.)

*(d) If I oversleep, I will miss the bus. If I miss the bus, I'll be late for work unless Sue gives me a ride. If Sue's car is not working, she won't give me a ride. If I'm late for work, I'll lose my job unless the boss is away. Sue's car is not working. The boss is not away. Therefore, if I oversleep, I'll lose my job.

(6) Turn the formal alternate solution to Example 1(a) into an informal proof, similar to that given for Example 1(c).

(7) Two sets A and B are defined to be equal if they have exactly the same members, that is, if $x \in A$ is equivalent to $x \in B$, for any object x. Prove that $A = B$ if and only if ($A \subseteq B$ and $B \subseteq A$). You may want to refer to Proof Preview 2 in this section, as well as Proof Preview 1 in Section 2.2, to review how biconditionals are normally proved. But don't make this proof harder than it needs to be; there really isn't much to it.

(8) Prove that for any real number x, $|x| \geq -x$.

(9) Prove that if n is an integer, then $n^2 + n$ must be even. ***Hint:*** You may assume that an integer must be even or odd. Then use the technique used in Proof Preview 3.

(10) Prove that if n is an integer which is not a multiple of 3, then n^2 is 1 more than a multiple of 3. ***Hint:*** To do this, you need to find a disjunction that is equivalent to the condition that n is not a multiple of 3. Do *not* try to prove this equivalence; you may assume it.

Exercises 11 through 17 are rather technical and are concerned with material that has not been directly discussed in the text.

*(11) A set of connectives is called **complete** if every truth function can be represented by it; that is, given any truth function, there is a symbolic statement that uses only connectives in the set and has that truth function.

Show that the connectives $\wedge$, $\vee$, and $\sim$ together form a complete set of connectives. ***Hint:*** First consider a truth function with exactly one T in its final output column. Show that any such truth function can be represented by a conjunction of propositional variables and their negations. Then, any truth function at all can be represented by a disjunction of such conjunctions. The resulting statement is called the **disjunctive normal form** of the given truth function. Don't try to make this a very rigorous proof.

(12) Find the disjunctive normal form for each of the following statements:
   (a) $P \leftrightarrow Q$
   (b) $\sim (P \wedge Q)$
   (c) $P \leftrightarrow (Q \rightarrow \sim R)$
   (d) $\sim P \wedge (Q \rightarrow R)$

(13) Show that $\wedge$ and $\sim$ together form a complete set of connectives.

(14) Show that $\vee$ and $\sim$ together form a complete set of connectives.

*(15) Show that $\rightarrow$ and $\sim$ form a complete set of connectives.

*(16) Show that $\wedge$, $\vee$, $\rightarrow$, and $\leftrightarrow$ do *not* form a complete set of connectives.

*(17) Define a connective |, called the **Sheffer stroke**, based on the words "not both." That is, $P|Q$ is true *except* when *both* P and Q are true. Show that the single connective | forms a complete set of connectives.

Suggestions for Further Reading: For a more thorough treatment of mathematical logic at a level that is not much higher than the level of this text, see Copi and Cohen (1997), Hamilton (1988), or Mendelson (1987). For a more advanced treatment, see Enderton (1972) or Shoenfield (1967).

Chapter 3

Predicate Logic

3.1 The Language and Grammar of Mathematics

Propositional logic is important in mathematics, but it is much too limited to capture the full power of mathematical language or reasoning. For one thing, although propositional logic deals with connectives and how they are used to build up statements, it does not concern itself with the structure of *atomic* statements. Remember that we call a statement atomic if it is not built up from any shorter statements. The goal of this section is to examine what atomic statements look like in mathematical language.

Example 1: One important category of atomic mathematical statements are *equations* such as $x + y = 3$. As discussed in Section 2.1, a statement of this sort is called a **predicate**, since its truth depends on the values of variables. It may also be called an **open statement**. You can see that it contains no connectives. Quantifiers are words like "all," "every," and "some" or symbols standing for those words; so our equation contains none of those either. And that makes it atomic.

It's important to see why neither $x + y$ nor $y = 3$ can be considered a substatement of $x + y = 3$. The expression $x + y$ isn't even a sentence; it has no verb. The expression $y = 3$ is a perfectly good sentence, but it makes no sense to say that the equation $x + y = 3$ is built up grammatically from the equation $y = 3$. So this equation, and in fact any equation, is atomic. In many branches of mathematics, equations and inequalities account for virtually all the atomic statements.

In the equation we've been using as an example, the letters x and y are, of course, variables.

Definitions: A **mathematical variable** is a symbol (or combination of symbols like x_1) that stands for an unspecified number or other object.

The collection of objects from which any particular variable can take its values is called the **domain** or the **universe** of that variable. Variables with the same domain are said to be of the same **sort**. (It's generally assumed that the domain of a variable must be nonempty.)

You have undoubtedly been using variables to stand for numbers since junior high school, and you have probably also encountered variables representing functions, sets, points, vectors, and so on. These are all mathematical variables.

Example 2: If you saw the equation $f(x) = 3$, you would probably read this as "f of x equals 3," because you recognize this as an example of function notation. You would probably also think of x as the only variable in this equation. But strictly speaking, this equation contains two variables: x, presumably standing for a number, and f, presumably standing for a function.

There is nothing that says what letters must be used to stand for what in mathematics, but there are certain conventions or traditions that most people stick to avoid unnecessary confusion. In algebra and calculus, for example, the letters x, y, and z almost always stand for real numbers, whereas the letters f and g stand for functions. The fact that almost everyone automatically interprets the equation $f(x) = 3$ in the same way shows how strong a cue is associated with certain letters. On the other hand, if someone wanted to let the letter Q represent an arbitrary triangle, it would be best to inform the reader of this unusual usage.

In Chapter 2 we introduced the idea of a propositional variable—a letter used to stand for a statement. Propositional variables are not normally used in mathematics. They are used primarily in the study of logic.

☞ The difference between propositional variables and mathematical variables is very important, and you should be careful not to confuse them. A propositional variable always stands for a statement—spoken, written, mathematical, English, Swedish, or whatever—that could take on a value of true or false. A mathematical variable can stand for almost any type of quantity or object *except* a statement.

Not every letter that stands for something in mathematics is a variable.

Definition: A symbol (or a combination of symbols) that stands for a *fixed* number or other object is called a **constant symbol** or simply a **constant**.

Example 3: The symbols π and e are constant symbols, not variables, since they stand for specific numbers, not unknown numbers. Constant symbols need not be letters: **numerals** like 2, 73, and 5.3 are also constants.

Starting with variables and constants, mathematicians use a variety of other symbols to build up mathematical expressions and statements. It is possible to describe the structure of mathematical language in great detail. Rather than do that, let's just make one vital point. We've already mentioned that equations and inequalities are two very common types of mathematical statements. Expressions like $x + y$ and $\cos 3z$, on the other hand, are not statements at all because they take on numerical values, not truth values, when we substitute numbers for the mathematical variables in them. We call this kind of mathematical expression, which represents a mathematical value or object, a

term. (Throughout this book, our use of the word "term" is more general than its usual meaning in high school algebra.) The simplest kind of term is a single variable or constant.

The distinction between statements and terms can be made more clear by drawing an analogy to English grammar. One of the first things taught in grammar is that a sentence must have a verb. This is just as true in mathematics as it is in English. The word "equals" is a verb, and the word group "is less than" includes the verb "is" and functions as a verb. So if we say that one quantity equals another or is less than another, we have a complete sentence or statement. Therefore, $=$ and $<$ should be regarded as mathematical verbs that can be used to create symbolic statements. The technical name for such verb symbols is **predicate symbols**. In contrast, the word "plus" is not a verb and so cannot be used to form a statement. Since $x + y$ stands for an object (specifically, a number), it's essentially a mathematical noun. It's no more a complete statement than the phrase "frogs and toads" is a complete English sentence. The technical name for mathematical symbols like $+$, $-$, and $\sqrt{\ }$, which are used to form terms that denote objects, is **function symbols** or **operator symbols**.

Example 4: Let's consider what could be the elements of a symbolic language for high school algebra. There would have to be at least two sorts of variables: real variables, that is, variables whose domain is the set of all real numbers, and function variables, that is, variables whose domain is the set of all real-valued functions. It might also be convenient to have variables whose domain is the set of all integers. In addition, it is normal to have an infinite number of constant symbols (including numerals) representing particular real numbers.

The most basic operator symbols of algebra are the symbols $+$, $-$, $\times$, and $/$. The minus sign can be used syntactically in two different ways: it can be put in front of a single term to make a new term, or it can be put between two terms to make a new term. Technically, there should be two different symbols for these two different operations, but it is standard to use the same one. Some other important operator symbols of algebra are the absolute value and radical symbols.

Exponentiation represents a rather special case in the grammar of algebra. An expression like x^y is certainly a term, built up from two simpler terms. But instead of using a symbol to show exponentiation, we show it by writing the second term to the upper right of the first term. It would perhaps be better to have a specific symbol for exponentiation, but traditionally there isn't one. However, note that most calculators and computer languages do have a specific key or symbol for exponentiation.

For more advanced work, one might want many other operator symbols, for things like logarithms, function inverses and compositions, trigonometric functions, and so on.

It is much easier to list all the predicate symbols of algebra than all the operator symbols. The only atomic predicate symbols are $=$, $<$, and $>$. There are two other standard inequality symbols, $\leq$ and $\geq$, but they are not atomic (their meaning includes an "or"). Also since $x > y$ means the same thing as $y < x$, it is necessary to have only two atomic predicate symbols.

We have just described the sorts of variables and the constant symbols, operator symbols, and predicate symbols required for a symbolic language in which high school algebra can be done. These are the basic ingredients of what is called a **first-order language**.

Example 5: Now let's describe a first-order language for the subject of plane geometry. In the traditional Euclidean approach to this subject, there are three basic, undefined types of objects: points, lines, and "magnitudes" (positive real numbers). So there should be at least these three sorts of variables.

Since some use of arithmetic and algebra is necessary to study geometry, this language should contain numerals and most of the operator and predicate symbols mentioned in the previous example. There should also be a few more operator symbols. Typically, $\overline{AB}$ denotes the line segment between points A and B (and then $|\overline{AB}|$ means the length of that line segment). The symbol $\angle$ represents the angle formed by any three distinct points. Two other notions for which there is no standard operator symbol but for which symbols might be useful are the (two-directional) line formed by two points, and the (one-directional) ray from one point through another point.

In addition, geometry requires one more predicate symbol, used to mean that a certain point is on a certain line. There is no standard symbol for this, and it's not particularly important what symbol is used. We could just as well use the symbol "On." That is, the notation On(A, L) would mean that point A is on line L. This single predicate symbol is all that's needed to talk about parallel lines, triangles, rectangles, and so on (see Exercise 8 of Section 3.4).

Note that operator symbols and even predicate symbols can mix sorts. For example, the angle symbol uses three terms representing points to form a term representing a number. The On symbol uses one term for a point and another term for a line to form an atomic sentence.

By the way, have you ever heard it said that mathematics is a language? If you never thought about this before, now would be a good time to do so. Mathematics definitely includes its own language with its own grammar. When studying mathematical logic or almost any part of higher mathematics, it's essential to understand *and respect* this grammar!

3.2 Quantifiers

Section 3.1 discussed some of the specifics of how symbolic mathematical language is structured. Now it's time to go one more step beyond propositional logic by introducing the concept of quantifiers. The study of quantifiers, together with connectives and the concepts discussed in the previous section, is called **predicate logic, quantifier logic, first-order logic,** or the **predicate calculus.**

Notation: Two symbols, called **quantifiers**, stand for the following words:

- $\forall$ for "for all" or "for every" or "for any"
- $\exists$ for "there exists" or "there is" or "for some"

$\forall$ is called the **universal quantifier;** $\exists$ is called the **existential quantifier.**

The quantifiers are used in symbolic mathematical language as follows: if P is any statement, and x is any mathematical variable (not necessarily a real number variable), then $\forall x$ P and $\exists x$ P are also statements.

Example 1: Quantifiers are used in ordinary life as well as in mathematics. For example, consider the argument: "Susan has to show up at the station *some* day this week at noon to get the key. So if I go there *every* day at noon, I'm bound to meet her." The logical reasoning involved in this conclusion is simple enough, but it has nothing to do with connectives. Rather, it is an example of a deduction based on quantifier logic (see Exercise 3 of Section 4.3).

When using these symbols, it's important to stick to the rule given previously for how they are used. Note that a quantifier *must* be followed immediately by a mathematical variable, which in turn *must* be followed by a statement.

Example 2: Quantifiers often occur in sequence, and this is both legitimate and useful. For instance, consider the statement, "For any numbers x and y, there's a number z that, when added to x, gives a sum equal to y." This would be written symbolically as $\forall x\ \forall y\ \exists z\ (x + z = y)$. This is a perfectly well-formed symbolic statement, because each quantifier is followed by a mathematical variable, which is in turn followed by a statement. Note that the word "and" in the English statement is misleading; there's really no conjunction in it. A symbolic statement may *never* begin "$\forall x \wedge$ ..." or "$\exists x \wedge$ ... ." (By the way, if all the variables have the set of real numbers as their domain, can you tell whether this statement is true or false?)

Notation: When a statement contains a sequence of two or more quantifiers of the same type ($\forall$ or $\exists$), it's permissible to write the quantifier just once and then separate the variables by commas. So the above statement $\forall x\ \forall y\ \exists z$ (...) can also be written $\forall x,y\ \exists z$ (...). This should be viewed as merely an abbreviation for the complete form.

Just as in propositional logic, parentheses are often needed in quantifier logic to make it clear what the **scope** of a quantifier is. For example, $\forall x$ (P $\wedge$ Q) has a different meaning from ($\forall x$ P) $\wedge$ Q. If parentheses are omitted, the usual convention is that a quantifier has higher priority than any connective. So $\forall x$ P $\wedge$ Q would be interpreted as ($\forall x$ P) $\wedge$ Q.

The most common English words for both quantifiers have already been given. When you read a quantified statement in English it is usually necessary to follow each instance of the existential quantifier with the words "such that." For example, $\exists x\ \forall y\ (y + x = y)$ should be read "There is an x such that, for every y, $y + x = y$." It doesn't make sense to read it "There is an x for every y, $y + x = y$." To the nonmathematician, the words "such that" sound awkward. But there's no adequate substitute for them in many cases.

Definitions: A mathematical variable occurring in a symbolic statement is called **free** if it is unquantified and **bound** if it is quantified. If a statement has no free variables it's called **closed.** Otherwise it's called a **predicate**, an **open sentence**, an **open statement,** or a **propositional function**.

Example 3: In the statement $\forall x\ (x^2 \geq 0)$, the variable x is bound, so the statement is closed. In the statement $\forall x\ \exists y\ (x - y = 2z)$, x and y are bound whereas z is free. So this statement is open; it is a propositional function of z.

Example 4: Strictly speaking, it's "legal" for the same variable to occur both bound and free in the same statement. Consider $x = y \vee \exists x\ (2x = z)$. Then x is free in the first disjunct and bound in the second. But most people consider it very awkward and confusing to have the same variable bound and free in a single statement. Furthermore, this awkwardness can always be avoided, because a bound variable can be replaced by any new variable of the same sort, without changing the meaning of the statement. In the above example, the rewritten statement $x = y \vee \exists u\ (2u = z)$ would be more readable and would have the same meaning as the original, as long as u and x have the same domain.

Convention: This text follows the convention that the same variable should not occur both bound and free in the same statement. You should, too.

☞ It is important to develop an understanding of the difference between free and bound variables. A free variable represents a genuine unknown quantity—one whose value you probably need to know to tell whether the statement is true or false. For example, given a simple statement like "$5 + x = 3$," you can't determine whether it's true or false until you know the value of the free variable x. But a bound variable is quantified; this means that the statement is not talking about a single value of that variable. If you are asked whether the statement "$\exists x\ (5 + x = 3)$" is true, it wouldn't make sense to ask what the value of x is; instead, it would make sense to ask what the *domain* of x is. (If the domain were all real numbers, the statement would be true; but if it were just the set of all positive numbers, the statement would be false.) In this way, a bound variable is similar to a dummy variable, like the variable inside a definite integral: it doesn't represent a particular unknown value.

Notation: If P is any propositional variable, it is permissible and often helpful to the reader to show some or all of its free (unquantified) mathematical variables in parentheses. So the notation P(x) (read "P of x") would imply that the variable x is free in P, whereas the notation P(x, y) would imply that both x and y are free in P. Some mathematicians follow the convention that all the free variables of a statement must be shown in parentheses in this manner, but we don't. So, for example, when we write P(x), there could be other free variables besides x in P.

You may notice that this notation strongly resembles function notation $f(x)$. The resemblance is deliberate. An open sentence does define a function of its free variables, namely a truth-valued function. This is why open sentences are also called "propositional functions." (On the other hand, it's important to distinguish between an open sentence and a mathematical function; the latter is a mathematical object, *not* a statement.)

Another way that this new notation is similar to ordinary function notation involves substituting or "plugging in" for free variables. Suppose we introduce the notation $P(x)$ for some statement. If we then write $P(y)$ or $P(2)$ or $P(\sin 3u)$, this means that the term in parentheses is substituted for the free variable x throughout the statement P.

Enough technicalities for now. It's time to talk about the meaning of the quantifiers and then look at some examples of how to use quantifier logic to represent English words and statements symbolically.

Definition: A statement of the form $\forall x\, P(x)$ is defined to be true provided $P(x)$ is true for each particular value of x from its domain. Similarly, $\exists x\, P(x)$ is defined to be true provided $P(x)$ is true for *at least* one value of x from that domain.

Perhaps you object to these definitions on the grounds that they are circular or just don't say anything very useful. In a sense, this objection is valid, but there is no simpler method (such as truth tables) to define or determine the truth of quantified statements.

Note that this definition of the existential quantifier gives it the meaning of "there is at least one." There are also situations in which you want to say things like "There is *exactly* one real number such that ... ." It would be possible to introduce a third quantifier corresponding to these words, but it's not needed. Section 3.4 explains why.

Also note that our interpretation of $\exists$ is analogous to our interpretation of $\vee$ as the inclusive or, since that connective means at least one disjunct is true, rather than exactly one disjunct is true. It is reasonable and often helpful to think of the existential and universal quantifiers as being closely related to disjunction and conjunction, respectively.

Section 2.1 ended with a few examples of how to translate English statements into symbolic statements of propositional logic. When quantifiers are involved, these translations can be somewhat tricky to do correctly, but every mathematician needs to learn this skill. As in the earlier examples, the first step in these translations is to determine the atomic substatements of the given statement and then to assign a propositional variable to each of them. But when quantifiers are involved, it also becomes very important to identify and show the free mathematical variables present.

This process is much easier if you remember some of the grammatical issues we've talked about: propositional variables stand for *whole statements,* each of which *must* contain a *verb.* The free mathematical variables of a given propositional variable should correspond to *nouns* or *pronouns* that appear in that statement. For instance, if you wanted to symbolize a statement that talked about people liking each other, it would be reasonable to use a propositional variable $L(x, y)$ to stand for the sentence "x likes y," where it is understood that x and y represent people. The verb "likes" involves two nouns, so there are two free variables.

☞ The following rule of thumb is also helpful: *The symbolic translation of a statement must have the same free variables as the original statement.*

Example 5: For each of the following, write a completely symbolic statement of predicate logic that captures its meaning.

(a) All gorillas are mammals.
(b) Some lawyers are reasonable.
(c) No artichokes are blue.
(d) Everybody has a father and a mother.
(e) Some teachers are never satisfied.
(f) (The number) x has a cube root.
(g) For any integer greater than 1, there's a prime number strictly between it and its double.

Solution: (a) Certainly, the word "all" indicates a universal quantifier. But if you have never done such problems before, it might not be clear to you how to proceed. The key is to realize that what this proposition says is that *if* something is a gorilla, it must be a mammal. So within the universal quantifier, what we have is an implication. The logical structure of the statement is therefore

$$\forall x\ (x \text{ is a gorilla} \rightarrow x \text{ is a mammal})$$

Of course, this is not a completely symbolic rendition of the original statement. If we want to make it completely symbolic, we have to introduce propositional variables for the atomic substatements. Let $\mathrm{G}(x)$ mean "x is a gorilla" and let $\mathrm{M}(x)$ mean "x is a mammal." Then the original statement can be represented symbolically as $\forall x\ (\mathrm{G}(x) \rightarrow \mathrm{M}(x))$.

We have not specified the domain of the variable x in this solution. This is because we don't want any particular limitations on it. Since the implication inside the quantifier limits things to gorillas anyway, we might as well assume x can stand for any thing whatsoever, or perhaps any animal. It's not uncommon to use a variable whose domain might as well be unlimited.

Note that the given English statement has no free variables, and therefore neither does its symbolic translation. This is true for all the parts of this example except part (f).

Perhaps you see a shorter way of translating the given statement into symbols. Why not specify that the variable x stands for any gorilla, as opposed to a larger set like all animals? Then it appears that the given statement can be represented as

$$\forall x\ (x \text{ is a mammal}) \quad \text{or} \quad \forall x\ \mathrm{M}(x)$$

There is nothing wrong with this approach to the problem, and it does yield a shorter, simpler-looking answer. However, it's not necessarily helpful in mathematics to introduce variables with any old domain that's considered convenient at the time.

Therefore, you should know the long way of doing this problem and especially that this type of wording translates into an implication.

(b) This time, because of the word "some," the solution requires an existential quantifier. Notice that, except for replacing the word "all" by the word "some," the structure of this statement seems the same as the structure of the previous statement. So you might automatically think that an implication is involved here too. But if you give it some thought, you'll realize that this statement says that there is a person who is a lawyer *and* is reasonable. So it's a conjunction, not an implication. With propositional variables $L(x)$ and $R(x)$ standing for "x is a lawyer" and "x is reasonable," the correct symbolic translation is $\exists x\,(L(x) \wedge R(x))$. The same shortcut that was mentioned in part (a)—using a more specific variable—could also be applied to this problem.

☞ Pay close attention to the contrast between parts (a) and (b). Again, the deceptive thing is that the words seem to indicate that the only logical difference between the two is the quantifier. Yet the "hidden connective" turns out to be different too. In general, the words "All ...s are ...s" always represent an implication, whereas "Some ...s are ...s" always translates to a conjunction.

(c) Here we encounter the word "no," which would seem to indicate a negation, perhaps combined with a quantifier. At first thought, it might seem that "No artichokes are blue" is the negation of "All artichokes are blue." But remember that the negation of a statement means that the statement is not true. And "No artichokes are blue" surely does not mean "It's not true that all artichokes are blue." Rather, it means "It's not true that some artichokes are blue." So one way to symbolize this statement is to first symbolize "Some artichokes are blue," as in part (b), and then to stick ~ in front of it. Another correct approach, perhaps less obvious, is to realize that the given statement means the same thing as "All artichokes are nonblue" and to go from there. The details are left for Exercise 2.

This example illustrates some of the subtleties and ambiguities of English. "No artichokes are blue" definitely has a different meaning from "Not all artichokes are blue." How about "All artichokes are not blue"? Do you think the meaning of this is clear, or is it ambiguous?

(d) We can see that "everybody" means "every person." So the symbolic form of this statement should begin with a universal quantifier, and it is convenient to use a variable whose domain is the set of all people. If we then write $M(x)$ and $F(x)$ to represent, respectively, "x has a mother" and "x has a father," we can translate the given statement as

$$\forall x\,(M(x) \wedge F(x))$$

This solution isn't wrong, but it can be criticized as incomplete. A statement like "x has a mother" should not be considered atomic, because it contains a hidden quantifier. That is, it really means "There is somebody who is x's mother." So a better representation of the statement is obtained as follows: Let $M(x, y)$ mean "y is x's mother" and $F(x, y)$ mean "y is x's father." Then the statement can be symbolized as

$$\forall x\, (\exists y\, M(x, y) \land \exists z\, F(x, z))$$

where x, y, and z are people variables. Note that there is a variable for each person under consideration—person x, mother y, and father z. But they are all bound variables.

(e) As before, let x be a variable whose domain is the set of all people. Recall from part (b) that "Some teachers are ..." should be thought of as "There exists someone who is a teacher *and* who is ... ." But how do we say someone is never satisfied? This means that there is no time at which the person is satisfied. So we also need a variable t whose domain is the set of all possible times. Let's define $T(x)$ to mean "x is a teacher" and $S(x, t)$ to mean "x is satisfied at time t." With this notation, the given statement can be represented as

$$\exists x\, (T(x) \land \sim \exists t\, S(x, t))$$

(f) In part (c) we saw that words like "has a cube root" include a hidden quantifier. To say that a number has a cube root is to say that there is a number whose cube is the given number. So what we want is

$$\exists y\, (y \cdot y \cdot y = x) \quad \text{or} \quad \exists y\, (y^3 = x) \quad \text{or} \quad \exists y\, (y = \sqrt[3]{x})$$

Note that, in all of these solutions as well as in the original, x is free whereas y is not.

(g) Let m and n be variables whose domain is the set of all natural numbers (the positive integers 1, 2, 3, and so on). Then if we write $P(n)$ for "n is a prime number," we can translate the given statement as

$$\forall m\, [m > 1 \rightarrow \exists n\, (m < n < 2m \land P(n))]$$

If you wanted to be technical, you could point out that an extended inequality is not really atomic, and so the solution should have $m < n \land n < 2m$ instead of $m < n < 2m$. A more substantial objection would be that the sentence "n is a prime number" is not atomic; it can be written symbolically with quantifiers and connectives (see Exercise 3(d)).

By the way, this statement is true. It is a famous result of number theory, known as **Bertrand's postulate**. See Section 8.2 for additional discussion.

Exercises 3.2

(1) For each of the following, determine whether it is a grammatically correct symbolic statement. (As usual, P, Q, and R are propositional variables, and x, y, and z

are mathematical variables.) For each one that's *not* grammatically correct, explain briefly why not. For each one that is grammatically correct, list its free and bound mathematical variables.

(a) $\forall x\ \mathrm{P}(x, z) \leftrightarrow \exists z\ \mathrm{Q}(y, z)$
(b) $\exists\ (x \wedge y)\ (x > 0 \wedge y < 0)$
(c) $\forall x\ \mathrm{P}(x) \rightarrow \exists x$
(d) $\sim \forall x \sim \forall y \sim \forall z\ (2 + 2 = u)$
(e) $\forall x\ [\mathrm{P}(x) \rightarrow \exists z\ (\mathrm{Q}(z) \rightarrow \forall y\ \mathrm{R}(x, y))]$

(2) Write out both of the symbolic answers described in the solution to Example 5(c).

(3) Translate each of the following into purely symbolic form. For the sake of uniformity, use the variables x, y, and z to stand for real numbers, and m, n, and k for integers. Initially, you may use only equations and inequalities as atomic statements. For instance, to express "n is a multiple of 10" symbolically, you could write "$\exists m\ (n = 10m)$." Then you can introduce new propositional variables as abbreviations for statements that you have written in symbolic form. For example, *after* you do part (d), you can define a propositional variable, perhaps $\mathrm{P}(n)$, to stand for your answer to part (d) when you do parts (e) and (f).

(a) 1 is the smallest positive integer.
(b) There is no largest integer.
(c) m is an odd number.
*(d) n is a prime number.
(e) Every prime number except 2 is odd.
(f) There are an infinite number of prime numbers. ***Hint:*** There's no simple way to express this literally. Instead, say that there's no largest prime number.
*(g) For any nonwhole real number x, there's an integer strictly between x and $x + 1$. ***Hint:*** The difficult part of this problem is that you may not use a variable whose domain is precisely the set of nonwhole real numbers. How can you express symbolically that x is not a whole number?
(h) Between any two (different) real numbers there's another one.

(4) (a) Which of the statements in Exercise 3 are closed?
(b) Name at least three of these closed statements that are true.

(5) As before, in the following statements, x, y, and z denote real numbers, and m, n, and k denote integers. For each statement, first identify its free variable(s); then find one set of values for its free variable(s) that makes the statement true and one set that makes the statement false. (Example: the statement $\exists n\ (m = n^2)$ has only m as a free variable. The statement is true for $m = 9$, and false for $m = 7$.) Justify your answers.

(a) $\exists n\ (n > 5 \wedge m^2 + k^2 = n^2)$
(b) $\forall x, y\ (x < y \leftrightarrow xz > yz)$
(c) $\forall x\ \exists y\ [xz = y \wedge yz = x \wedge (x \neq 0 \rightarrow y \neq x)]$
(d) $\forall x\ (x^2 - x \geq m)$

(6) The following symbolic statements are true in the real number system. Rewrite each of them in reasonable-sounding English.

(a) $\forall x\, [x \geq 0 \rightarrow \exists y\, (y^2 = x)]$
(b) $\forall x\, [x \leq 0 \rightarrow \sim \exists y\, (y = \log x)]$
(c) $\exists x\, \forall y\, (xy = y)$
(d) $\forall a, b\, [a \neq 0 \rightarrow \exists x\, (ax + b = 0)]$

(7) Represent each of the following statements symbolically, starting with only the following atomic statements: P(x, y) for "x is a parent of y," W(x) for "x is female," and $x = y$ (meaning x and y are the same person). All your variables should have the set of all people as their domain. As in Exercise 3, you may introduce new propositional variables for statements that you have already written symbolically. Remember that it is OK to substitute for the free variables of a statement. For example, W(z) would mean that z is female.

(a) x is male.
(b) x is y's father.
(c) x is y's grandmother.
(d) x is y's sibling. (This means that x and y have the same mother and father, but they are not the same person.)
(e) x is an only child. (That is, x has no siblings).
(f) x is y's first cousin.
(g) x has no uncles.
(h) Some people have brothers but no sisters.

(8) For each of the following statements, introduce a propositional variable (with free variables indicated) for each of its atomic substatements, and then write a totally symbolic translation of the given statement. You can define variables with any domain you want. For instance, for part (a), you might let one of your propositional variables be S(x), meaning "x likes spinach" (where x can be any person).

(a) Not everyone likes spinach, and no one likes asparagus.
(b) All crows are black, but not all black things are crows.
(c) If someone kisses the frog, everyone will benefit.
(d) There are people who like all vegetables.
(e) It's possible to fool all of the people some of the time and some of the people all of the time, but not all of the people all of the time.
(f) If everybody bothers me, I can't help anybody.
(g) Anybody who bothers me won't be helped by me.
(h) Every problem in this section is harder than every problem in Chapter 2.
(i) No one is happy all the time.
(j) Everybody loves somebody sometime.
(k) It's not true in all cases that if one person likes another, the second likes the first.
(l) There are days when everyone in my dorm cuts at least one class.

3.3 Working with Quantifiers

In this section we examine some of the methods that mathematicians use to understand and simplify quantified statements. It was mentioned in Section 3.2 that quantifiers often occur in sequence. Usually, quantifiers of the same type (all $\exists$s or all $\forall$s) occurring in sequence are not difficult to understand or to work with, but *alternations* of quantifiers between $\exists$ and $\forall$ (in either order) can make statements confusing. In more advanced studies of the foundations of mathematics, the complexity of statements is measured by how many alternations of quantifiers they contain. (One well-known mathematical logician has expressed the opinion that three or four is the maximum number of alternations of quantifiers that the human brain can deal with.) Let's begin this section by looking at sequences of quantifiers, paying particular attention to statements with a single alternation.

Example 1: Let's assume that x and y are real variables and consider a simple atomic statement like $x + y = 0$. One simple way to quantify this, with no alternations, is $\exists x\, \exists y\, (x + y = 0)$. What does this quantified statement say, and is it true or false? Technically, the statement says that there is a value of x for which $\exists y\, (x + y = 0)$ is true. But there's no need to split up the quantifiers in this way. In Section 3.2 it was mentioned that this statement can be written as $\exists x,y\, (x + y = 0)$, which would be read "There exist x and y such that $x + y = 0$." The point is that the statement simply says that there is some choice of values for the two variables that makes the equation $x + y = 0$ hold. Clearly, this is true; for example, we could take $x = 3$ and $y = -3$. A consequence of this analysis is that there is no difference in meaning between $\exists x,y\, (x + y = 0)$ and $\exists y,x\, (x + y = 0)$.

Example 2: Similarly, consider the statement $\forall x\, \forall y\, (x + y = 0)$. Again, this can be rewritten as $\forall x,y\, (x + y = 0)$, with the practical consequence that the two quantifiers can be considered together. So this statement says that for all choices of values for x and y, $x + y = 0$. This is blatantly false; for example, it fails when $x = y = 29$. As in the previous paragraph, there is no difference in meaning between $\forall x,y\, (x + y = 0)$ and $\forall y,x\, (x + y = 0)$. This is a general fact: the order of the variables in a sequence of *like* quantifiers is unimportant.

Example 3: Now let's look at the more interesting cases with alternations of quantifiers. First, consider $\forall x\, \exists y\, (x + y = 0)$. This says that for every value of x, the statement $\exists y\, (x + y = 0)$ holds. That is, for every choice of a value for x, there must be a value for y that makes the equation hold. You can see that this is always so. When $x = 7$, y would be -7; when $x = -2.68$, y would be 2.68, and so on. Clearly, the correct choice of y can be described in terms of x by the simple formula $y = -x$. This example also illustrates a general situation: for a statement of the form $\forall x\, \exists y\, \mathrm{P}(x,\ y)$ to be true, it must be possible to choose *y in terms of x* (that is, as a *function* of x) so that the inner statement holds for all values of x when y is chosen according to that function.

Example 4: Now let's reverse the quantifiers and consider $\exists y\ \forall x\ (x + y = 0)$. This says that there is a value for y that makes the statement $\forall x\ (x + y = 0)$ hold. That is, there would have to exist a single value of y, *chosen independently of* x, that makes the inner equation work for all values of x. In this situation, it's not enough to define y in terms of x. You can see that there is no such value of y, and so the whole statement is false.

These examples illustrate several points. For one, they show that the order of quantifiers does matter when they are of opposite types. Also, in general, a statement of the form $\exists y\ \forall x\ P(x, y)$ is harder to satisfy (that is, less likely to be true) than the corresponding statement $\forall x\ \exists y\ P(x, y)$. Additionally, the previous paragraph clarifies why the words "such that" are usually needed after an existential quantifier. If the statement $\exists y\ \forall x\ (x + y = 0)$ were read "There is a y for every x ... ," it would seem to have the same meaning as "For every x there is a y ... ," which it doesn't. The wording "There is a y such that, for every x, ..." helps reinforce the difference in meaning.

The next theorem generalizes the previous examples of how to decipher statements with alternating quantifiers. We omit the proof, since it is quite technical (but see Exercise 11).

Theorem 3.1: Suppose a statement begins with a sequence of quantifiers, followed by some inner statement with no quantifiers. Then the statement is true provided each existentially quantified variable is definable as a function of *some or all of* the universally quantified variables to the left of it, in a way that makes the inner statement always true. (A function of no variables means a single, constant value. The words "as a function of" in this theorem could be replaced by "in terms of.")

We just saw how this theorem applies to statements of the form $\forall x\ \exists y\ P(x, y)$ and $\exists y\ \forall x\ P(x, y)$. It can also help decipher statements with more alternations of quantifiers.

Example 5: Suppose we had to work with a monster like $\exists u\ \forall v\ \exists w\ \forall x, y\ \exists z\ (...)$. Our rule says that, to satisfy this, there must be a single value of u, a function defining w in terms of v, and a function defining z in terms of v, x, and y that guarantee that the inner statement is true. Knowing this probably won't make the problem simple, but it ought to help.

Proof Preview 4

Theorem: For any two real numbers, there is a real number greater than both of them.

Proof: In symbols, what we want to prove is $\forall x, y\ \exists z\ (z > x \wedge z > y)$. By Theorem 3.1, to prove this is true, we must appropriately define z in terms of x and y. One concise way to do this is to let $z = |x| + |y| + 1$. We must then show that this makes the conjunction in parentheses true. *[The rest of the proof uses numerous results from high school algebra, including basic properties of the absolute-value function. Most of these are proved in Appendix 2.]* We know that $|x| \geq x$ and $|y| \geq 0$. Therefore $|x| + |y| \geq x + 0 = x$, and so $|x| + |y| + 1 > x$. Similarly, $|x| + |y| + 1 > y$. *[Mathematicians usually* *omit part of a proof that is nearly identical to a previous part and instead make a comment like the previous sentence.]* This completes the proof. ■

Now let's apply these ideas to determine the truth or falsity of various statements in various number systems.

Example 6: For each statement, determine whether it's true in each of these number systems: the set of all natural numbers (positive integers) $\mathbb{N}$, the set of all integers $\mathbb{Z}$, the set of all real numbers $\mathbb{R}$, and the set of all complex numbers $\mathbb{C}$.

(a) $\forall x, y \, \exists z \, (x + z = y)$

(b) $\exists x \, \forall y \, (x < y)$

(c) $\exists x \, \forall y \, \exists z \, (x = y \lor yz = 1)$

Solution: (a) For this statement to be true, it must be possible to define z as a function of x and y so that the equation inside the quantifiers is always true. To accomplish this, let's solve the equation $x + z = y$ for z: it becomes $z = y + (-x)$, or simply $z = y - x$. Now, in the system of natural numbers, subtraction does not necessarily yield an answer in that system, so the statement is false. But in the other three number systems, subtraction is always possible, so the statement is true.

(b) This statement begins with $\exists x$, so we want to know if there's a single value of x that makes the inequality true, whatever y is. The statement says that there is an x that is less than every y, which at first glance might seem to be saying that there is a smallest number in the domain. So we might expect this statement to be true in $\mathbb{N}$, with $x = 1$. But let's be careful! If $x = 1$, then the statement $\forall y \, (x < y)$ is still false in $\mathbb{N}$, because if $y = 1$, the inequality $1 < 1$ is false. What the statement really says is that there is a value of x that is less than every possible value of y, *including* whatever value x has. And this is false in all standard number systems, because a number is never less than itself. The lesson here is that two different variables *are* allowed to have the same value. So if we want a symbolic statement to say that there is a smallest number, we can't have it say that there is an x that is smaller than every y. Rather, it should say that there is an x that is smaller than every *other* y. This could be symbolized as $\exists x \, \forall y \, (y \neq x \rightarrow x < y)$ or, more simply, as $\exists x \, \forall y \, (x \leq y)$. This modified statement is true in $\mathbb{N}$ but false in $\mathbb{R}$. Inequalities between complex numbers are not even defined, so it is best to say that this statement (either version) is meaningless in $\mathbb{C}$.

(c) This statement has two alternations of quantifiers, which makes it more complex than the previous examples. To make it true, we'd have to find a single value of x, plus a function defining z in terms of y, so that the inner statement must be true. It's probably easiest to consider the relationship of z to y before considering the value of x. The inner statement is a disjunction, one of whose disjuncts is $yz = 1$. This equation is equivalent to $z = 1/y$, so it looks like that's how z should be defined from y. But in $\mathbb{N}$ and $\mathbb{Z}$, most numbers don't have reciprocals. Even in $\mathbb{R}$ and $\mathbb{C}$, not all numbers have reciprocals; zero doesn't. Where does this leave us?

Well, let's consider the role of x. The statement says that there's some particular value of x such that every value of y either equals x or has a reciprocal. It should be clear that we want to take $x = 0$. Then, if $y = 0$, the inner statement is automatically true (so we can pick any value for z that we want). On the other hand, if $y \neq 0$, the inner statement is true provided $z = 1/y$. Therefore, the given statement is true in $\mathbb{R}$ and in $\mathbb{C}$, where every nonzero number has a reciprocal. The given statement is false in $\mathbb{N}$ and in $\mathbb{Z}$, however, for whatever value is chosen for x, every other value of y would have to have a reciprocal. This just isn't true in these two number systems.

By the way, this statement has a name. It's the **multiplicative inverse property**, although not in its most common form. It is generally included as an axiom for the real number system.

Definitions: A **law of logic** is a symbolic statement that is true for *all* possible interpretations of the variables, constants, predicate symbols, and operator symbols occurring in it. That is, it must be true no matter what domains are chosen for its bound variables, no matter what values are chosen for its constants and free variables, and so on. (Only the connectives, the quantifiers, and the equal sign are *not* subject to reinterpretation.)

A statement Q is said to be a **logical consequence** of a finite list of statements $P_1, P_2, ..., P_n$ iff the single statement $(P_1 \wedge P_2 \wedge ... \wedge P_n) \rightarrow Q$ is a law of logic.

Two symbolic statements are called **logically equivalent** provided that each of them is a logical consequence of the other.

These definitions are direct parallels to the definitions of the terms "law of propositional logic," "propositional consequence," and "propositionally equivalent" in Chapter 2. It follows directly from the definitions that *every tautology is a law of logic* (but not the other way around). To distinguish them from tautologies, laws of logic are sometimes referred to as laws of *predicate* logic. In the rest of this chapter, "equivalent" always means "logically equivalent."

Although the preceding definitions are analogous to concepts defined in Chapter 2, there is a vast practical difference. Although it is always straightforward (using truth tables) to test for tautologies, contradictions, propositional equivalence and propositional consequence, there is *absolutely no* simple or computational way to decide whether a statement with quantifiers is a law of logic, whether two statements are equivalent, and so on.

Incidentally, all mathematical statements can be represented in predicate logic (but not in propositional logic). So, in effect, what's being said here is that there's no way to write a computer program that will correctly answer all mathematical questions. Of course, if there were such a computer program, the life of a mathematician would be greatly simplified—maybe even downright boring!

Example 7: Determine which of the following statements are laws of logic and explain why.

(a) $2 + 2 = 4$

(b) $\forall x \, \exists y \, (y > x)$
(c) If $x < y$ and $y < z$, then $x < z$.
(d) If $x = y$ and $y = z$, then $x = z$.
(e) $\forall x \, P(x)$ implies $\exists x \, P(x)$.

Solution: (a) This example was used in Chapter 2, where we said it was not a tautology. Neither is it a law of logic. It's a true statement of ordinary arithmetic, only because of the particular interpretation given to the symbols 2, +, and 4, not because of its logical structure.

(b) This says that for every number, it's possible to find a bigger one. This is certainly true in most common number systems, including the real numbers; for example, we could take $y = x + 1$. But it's *not* a law of logic. For one thing, it depends on the interpretation of the symbol >. And even if this symbol is given its usual meaning, the statement is false in a domain with a largest number, like the set of negative integers.

(c) Even this isn't a law of logic; it still depends on how the symbol < is interpreted!

(d) The definition says that the symbol = must be given its standard interpretation. Therefore, this statement is a law of logic: if x and y have the same value, and so do y and z, then clearly x and z must also. This statement is called the **transitive property of equality** and is usually taken as an axiom of mathematics. By the way, this statement is *not* a tautology.

(e) This statement says that if a certain condition is true for all objects in a certain domain, it's true for at least one. Clearly, such an implication must always be true (see Exercise 1). So this is a law of logic (but not a tautology).

In Chapter 4, with an axiom system at our disposal, we are able to solve more complex problems of this type. In the meantime, you are welcome to peek ahead at Table 4.2, which lists some of the more useful laws of logic.

Negations of Statements with Quantifiers

We have just discussed at some length what has to happen in order for a quantified statement to be true. We have not talked about what has to happen for a quantified statement to be false. It may not seem that this should require a separate treatment, but it does. Suppose that P is a statement that begins with a sequence of quantifiers. We've said that P is true provided that certain functions and/or constants (corresponding to the existential quantifiers of P) exist. So we could say that P is false provided that not all these functions and/or constants exist. However, often this view of the situation doesn't help to figure out whether the statement is false.

To say that P is false is of course to say that $\sim$ P is true. The statement $\sim$ P begins with a negation sign, followed by a sequence of quantifiers. It turns out to be useful to be able to move the negation sign from outside the quantifiers (that is, in front of them) to inside the quantifiers. The key to doing this is the following theorem, for which we just provide an informal, commonsense proof.

Theorem 3.2: For any statement P(x):
(a) $\sim \forall x\, P(x)$ is logically equivalent to $\exists x \sim P(x)$.
(b) $\sim \exists x\, P(x)$ is logically equivalent to $\forall x \sim P(x)$.

Proof: (a) The statement $\sim \forall x\, P(x)$ says that it's not true that P(x) holds for every value of x in its domain. But this means that P(x) is false for at least one value of x, which is precisely what $\exists x \sim P(x)$ says.
(b) This argument is similar and is left for Exercise 2. ■

Theorem 3.2 can be thought of as a direct parallel to De Morgan's laws. Recall that those tautologies say that you can distribute a negation into (or factor a negation out of) a conjunction or disjunction, but in doing so you have to change the inner connective from $\wedge$ to $\vee$, or vice versa. Similarly, Theorem 3.2 says you can move a negation across a quantifier, in either direction, provided you reverse the quantifier from $\forall$ to $\exists$, or vice versa. I like to call these quantifier equivalences De Morgan's laws for quantifiers.

Example 8: Simplify each of the following statements by moving negation signs inward as much as possible.
(a) $\sim \exists x, y\, \forall z \sim \exists u\, \forall w\, P$
(b) $\sim \exists x\, \forall e\, [e > 0 \rightarrow \exists d\, (d > 0 \wedge \forall u\, (|x - u| < d \rightarrow |f(x) - f(u)| < e))]$

Solution: (a) By applying Theorem 3.2 three times to the outer negation sign, we get the logically equivalent statement $\forall x, y\, \exists z \sim\sim \exists u\, \forall w\, P$. But we know from Chapter 2 that $\sim\sim Q$ is always equivalent to Q, and therefore the given statement is logically equivalent to the shorter one $\forall x, y\, \exists z, u\, \forall w\, P$. We could also have achieved this answer by moving the inner negation sign outward.

(b) This is a much more complex example than the previous one, and simplifying it requires both Theorem 3.2 and a couple of tautologies. Here are the steps required (but not the result of each step; see Exercise 4):
(1) Use Theorem 3.2 to move the negation sign through the outer pair of quantifiers.
(2) Now the statement inside the outer two quantifiers has the form $\sim (P \rightarrow Q)$. So we can use tautology 19 to change this to the equivalent form $P \wedge \sim Q$.
(3) Now the negation sign can be moved inside the quantifier $\exists d$, using Theorem 3.2 again.
(4) Now the negation sign is in front of a conjunction. Apply the appropriate De Morgan's law to it.

(5) Use Theorem 3.2 for the last time to move the negation sign inside the last quantifier.

(6) Finally, use the same propositional equivalence as in step 2 to move the negation sign inside the innermost implication.

Exercise 4 asks you to write out the results of each step of this transformation. For now, here is the final form after step 6:

$$\forall x\ \exists e\ [e > 0 \wedge \forall d\ (\sim d > 0 \vee \exists u\ (|x - u| < d \wedge \sim |f(x) - f(u)| < e))]$$

This simplified statement is no shorter than the original, but having the negation symbols inside the quantifiers is an important advantage for most purposes. The two remaining negation symbols in the rewritten statement can be eliminated. First, use tautology 20 to rewrite the disjunction $\sim d > 0 \vee \ldots$ as $d > 0 \rightarrow \ldots$. And if we are also permitted to use basic facts about the real number system, the statement $\sim |f(x) - f(u)| < e$ can be shortened to $|f(x) - f(u)| \geq e$. With this last change, we get a statement that is strictly speaking not *logically* equivalent to the original but is equivalent to it for all practical purposes.

Incidentally, this example is not some random concoction. With the beginning symbols $\sim \exists x$ dropped, the given statement is precisely the definition of what it means for the function f to be **continuous** at the number x. So the statement says that f is not continuous at any point. Believe it or not, such functions do exist. A standard example is given in Section 9.3.

Example 9: Consider the statement "Everybody has a friend who is always honest."

(a) Write a symbolic translation of this statement.

(b) Write the negation of this symbolic statement and then simplify it as in the previous examples.

(c) Translate your answer to part (b) back into reasonable-sounding English.

Solution: (a) Since the word "a" in this statement means "at least one," our symbolic translation has to contain three quantifiers based on the words "Everybody," "a" and "always." Two of these involve people, and one involves time; so we need variables for both. Let's use x and y as people variables and t as a time variable. Let's write $\mathrm{F}(x, y)$ as a propositional variable standing for "x is a friend of y." It's tempting to introduce a propositional variable that stands for "x is honest," but note that the given statement indicates that a person's honesty may vary over time. So we write $\mathrm{H}(x, t)$ to stand for "x is honest at time t." With this notation, the given statement can be represented as $\forall x\ \exists y\ (\mathrm{F}(y, x) \wedge \forall \mathrm{t}\ \mathrm{H}(y, t))$.

(b) If we start with the negation of the solution to part (a) and apply Theorem 3.2 repeatedly and then tautology 19, we finally arrive at the statement $\exists x\ \forall y\ (\mathrm{F}(y, x) \rightarrow \exists t \sim \mathrm{H}(y, t))$.

(c) It's not easy to put the solution to part (b) into smooth-sounding English, but the best try might be, "There are some people, all of whose friends are sometimes dishonest." Perhaps you can do better than this. Of course, the original statement can easily be negated in words as "Not everybody has a friend who is always honest." But that's not what the problem asks us to do.

Proof Preview 5

Theorem: There is no smallest positive real number.

Proof: For convenience, let x and y be variables whose domain consists of all *positive* real numbers. *[This is perfectly legitimate!]* In symbols, the statement that there *is* a smallest positive real number would be $\exists x \, \forall y \, (x \leq y)$. So what we want to prove is $\sim \exists x \, \forall y \, (x \leq y)$. Now, by applying Theorem 3.2 to this, we can change it to $\forall x \, \exists y \sim (x \leq y)$, or more simply, $\forall x \, \exists y \, (y < x)$. To verify that this last statement is true, we recall Theorem 3.1: we must define y as a function of x in such a way that the inequality $y < x$ must be true. *[Before reading further, can you see how to do this?]* Let $y = x/2$. Since x is positive, so is $x/2$. And since x is positive and $1/2 < 1$, we can multiply both sides of this inequality by x to obtain $x/2 < x$. This completes the proof. ■

Some Abbreviations for Restricted Quantifiers

We conclude this section with a few useful abbreviations, or shorthand notations, involving quantifiers. Recall from the previous section that a sentence of the form "All ...s are ...s" is an implication, whereas "Some ...s are ...s" generally represents a conjunction. So, for example, "Every nonnegative number has a square root" becomes, in predicate logic, $\forall x \, (x \geq 0 \rightarrow \exists y \, (y^2 = x))$. Statements like this, in which the scope of a quantified real variable is restricted by an inequality, are so common that it's worth having shorter ways of writing them.

Notation: Let P be any statement, x any variable whose domain has an ordering (for example, real numbers, integers, and so on), and t any term denoting a member of that domain. (So x has to be a single letter, but t could be a single letter, a constant like -4, or a more complicated expression like $3y + 7$.) Then

- $\forall x < \mathrm{t} \; \mathrm{P}$ is an abbreviation for $\forall x \, (x < \mathrm{t} \rightarrow \mathrm{P})$.
- $\exists x < \mathrm{t} \; \mathrm{P}$ is an abbreviation for $\exists x \, (x < \mathrm{t} \wedge \mathrm{P})$.

Similar abbreviations are used with the symbol $<$ replaced by any of the three symbols $>$, $\leq$, or $\geq$.

Even though sets are not discussed in detail until Chapter 5, let's introduce some abbreviations for a variable that is restricted to a set, since this notation is very similar to the notation just introduced.

Notation: Let P be any statement, x any mathematical variable, and t any term that denotes a *set.* (So t could be a single letter standing for a set, or a more complicated expression like $A \cup B$.) Then

- $\forall x \in t$ P is an abbreviation for $\forall x\ (x \in t \rightarrow$ P).
- $\exists x \in t$ P is an abbreviation for $\exists x\ (x \in t \wedge$ P).

Example 10: Write the following statements in symbolic form, using the abbreviations that have just been defined:

(a) Every positive number has a positive cube root, and every negative number has a negative cube root.

(b) For every nonnegative number x, there's an element of set B strictly between x and $x + 1$.

Solution: (a) $\forall x > 0\ \exists y > 0\ (y^3 = x) \wedge \forall x < 0\ \exists y < 0\ (y^3 = x)$

(b) $\forall x \geq 0\ \exists y \in B\ (x < y < x + 1)$

Here are some useful equivalences that are similar to Theorem 3.2 but adapted to restricted-quantifier notation. Their proofs are left for Exercise 3.

Theorem 3.3: (a) $\sim \exists x < t$ P is logically equivalent to $\forall x < t \sim$ P.

(b) $\sim \forall x < t$ P is logically equivalent to $\exists x < t \sim$ P.

Furthermore, both of these equivalences still hold with the symbol $<$ replaced by $>$, $\leq$, $\geq$, or $\in$.

Exercises 3.3

(1) What assumption must be made about the domain of the variable x for Example 7(e) to be correct? Has this assumption about domains been made in this chapter?

(2) Prove Theorem 3.2(b).

(3) Prove Theorem 3.3. Instead of doing this by mimicking the *proof* of Theorem 3.2, use the *result* of that theorem, the definitions of restricted quantifiers, and some tautologies to provide a more rigorous proof.

(4) Write out each step of the transformation described in Example 8(b).

(5) Determine whether each of the following statements is true or false if all variables have the set of real numbers as their domain. Explain briefly.

(a) $\forall x\ \exists y\ (x^2 = y)$
(b) $\forall y\ \exists x\ (x^2 = y)$
(c) $\exists x\ \forall y\ (x + 5 = y)$
(d) $\forall x\ \forall y\ \exists z\ \forall u\ (x + z = y + u)$
(e) $\forall x\ \forall y\ \exists z\ (x^2 + y^2 = z^2)$
(f) $\exists x\ [\forall y\ (yx^2 = y) \wedge \sim\forall y\ (yx = y)]$

(6) Repeat Exercise 5 with all variables having the set of nonnegative integers as their domain.

*(7) Determine whether the following statements are laws of logic. Explain.

(a) $\exists x\, P(x) \rightarrow \forall x\, P(x)$

(b) $[\exists x\, \forall y\, P(x, y)] \rightarrow [\forall y\, \exists x\, P(x, y)]$

(c) $[\forall y\, \exists x\, P(x, y)] \rightarrow [\exists x\, \forall y\, P(x, y)]$

(d) $[\exists x\, (P(x) \vee Q(x))] \leftrightarrow [\exists x\, P(x) \vee \exists x\, Q(x)]$

(e) $[\exists x\, (P(x) \wedge Q(x))] \leftrightarrow [\exists x\, P(x) \wedge \exists x\, Q(x)]$

(f) $\forall x, y, z, u\, [x = y \wedge z = u \rightarrow (P(x, z) \leftrightarrow P(y, u))]$

(8) Simplify each of the following statements by moving negation signs inward as much as possible.

(a) $\sim \forall x, y\, \exists z\, (P \vee \sim \forall u\, Q)$

(b) $\sim (\sim \exists x\, P \rightarrow \forall y \sim Q)$

(c) $\sim \forall x \sim \exists y \sim \forall z\, (P \wedge \sim Q)$

(9) Write each of the following statements in symbolic form, using the restricted quantifier notation introduced in this section.

(a) Every number in set A has a positive square root.

(b) Given any real number, there are integers bigger than it and integers smaller than it.

(c) Every member of a member of A is a member of A.

(d) No positive number equals any negative number.

(10) Prove the following. Your proofs can be based on the proof previews in this section.

(a) For any real number x, there's a number that is larger than both x and x^2.

(b) Given any two unequal real numbers, there's a number between them.

(c) There is no largest real number.

(d) There is no largest negative real number.

*(11) Prove Theorem 3.1 for the special case of statements with only one existential quantifier. Since we haven't studied functions yet, don't expect to do this very rigorously. Just try to give a commonsense argument.

3.4 The Equality Relation; Uniqueness

In Section 3.1, the idea of predicate symbols was introduced. Recall that these are symbols that act as mathematical verbs and are used to form atomic statements in predicate logic. Of course, different branches of mathematics require different predicate symbols. However, whatever differences may exist in the languages of different branches of mathematics, there is one predicate symbol common to all of them, and that is the equal sign. In other words, every branch of mathematics (as well as all of science and many other subjects) makes use of equations. Furthermore, the rules for working with equations do not change between different areas of mathematics and science. Because of this universality of the use of equations, the principles involved are usually included in the study of predicate logic.

You are familiar with equations and how to use them, and there will be no new tricks unleashed on you regarding them. In the next chapter, we begin using the axiom system contained in Appendix 1. But it won't hurt to take a look now at the standard axioms pertaining to equations, which are group III of the axioms. You can see that there are only four of them, and they are all very straightforward. The first one, reflexivity, says that anything equals itself. The second, symmetry, says that equations are reversible. (Symmetry is normally stated as a conditional, but it can be thought of as a biconditional. You might want to think about why this must be so.) The third, transitivity, says that "two things equal to a third thing are equal to each other." It is this axiom that allows you to write a long sequence of equations and then conclude that the first expression in the whole sequence equals the last one. The last axiom, substitution of equals, is a bit more involved. What it says is that if two things are equal, then they are *completely interchangeable.* It is this axiom that also implies that it's OK to change both sides of an equation, as long as the same thing is done to both sides. A more thorough discussion of these axioms and how to use them appears in Section 4.4.

Uniqueness

Recall that the existential quantifier has the meaning "there is at least one," which makes it analogous to the "inclusive-or" meaning of the disjunction connective. In mathematics we often want to say that there is *exactly one* number (or other object) satisfying a certain condition. In mathematics, the word "unique" is used to mean "exactly one." Should we introduce a third quantifier with this meaning? There is nothing wrong with doing so, but it's important to realize that this meaning can be captured with the symbols already defined, just as the exclusive or can be defined or written in terms of the other connectives.

There are several different-looking but equivalent ways to say that there's a unique object satisfying a certain condition. All these versions use the equality symbol; in fact, the desired meaning cannot be captured without it. For example, one way to express uniqueness is to say "There's an object that satisfies the condition *and* that equals *every* object that satisfies the same condition." Another way is "There's an object that satisfies the condition, *and* there are not two different ones satisfying it." A third way, closely related to the previous one, is "There's an object that satisfies the condition, *and* if any two objects satisfy it, they must be equal." Finally, a very concise way is "There's an object such that satisfying the condition is equivalent to being that object." Let's state the content of this paragraph more formally.

Theorem 3.4: The following four statements are equivalent, for any statement $\mathrm{P}(x)$ and any mathematical variables x and y.

(a) $\exists x\,(\,\mathrm{P}(x) \land \forall y\,(\,\mathrm{P}(y) \rightarrow x = y\,))$

(b) $\exists x\,\mathrm{P}(x) \land \sim \exists x, y\,(\,\mathrm{P}(x) \land \mathrm{P}(y) \land x \neq y)$

(c) $\exists x\, P(x) \wedge \forall x,y\, (P(x) \wedge P(y) \rightarrow x = y)$
(d) $\exists x\, \forall y\, (P(y) \leftrightarrow y = x)$

Proof: We give a relatively informal proof of this theorem that is still rigorous enough to illustrate several of the proof methods that are introduced in the next chapter. To prove that three or more statements are equivalent, the most common procedure is to prove a cycle of implications. So we show that statement (a) implies statement (b), then that (b) implies (c), that (c) implies (d), and finally that (d) implies (a).

(a) implies (b): Assume that statement (a) is true. Since there exists an x satisfying the statement after the first quantifier, let's say (for "definiteness") that k is an object satisfying it. Then $P(k)$ is true; this implies that the first conjunct of statement (b) holds. Also, for any x and y, if $P(x)$ and $P(y)$ both hold, then we know that $x = k$ and $y = k$. By transitivity, this implies $x = y$. So there cannot be two different objects satisfying P, and that is what the second conjunct of (b) says.

(b) implies (c): See Exercise 1.

(c) implies (d): Assume statement (c) is true. Since $\exists x\, P(x)$ holds, let's say that k is an object satisfying $P(k)$. We are done if we can show that, for this k, $\forall y\, (P(y) \leftrightarrow y = k)$. Consider any y. If $y = k$, then $P(y)$ holds, since we know $P(k)$. Conversely, if $P(y)$ holds, then we have both $P(k)$ and $P(y)$, and so by the second part of statement (c), $y = k$. So we have established that $P(y) \leftrightarrow y = k$, and since this is for any y, we are done.

(d) implies (a): See Exercise 1. ■

Notation: For any statement P and any mathematical variable x, we write $\exists! x\, P$, read "There is a unique x such that P," to stand for any one of the equivalent statements of Theorem 3.4.

It is important to keep in mind that this notation, like the restricted-quantifier notation defined in Section 3.3, is just an abbreviation for a longer form. In particular, when you want to prove that there is a unique object satisfying some condition, you must prove one of the forms listed in Theorem 3.4. Form (c) tends to be the easiest to work with.

Proof Preview 6

Theorem: If a and b are real numbers with $a \neq 0$, then the equation $ax + b = 0$ has a unique solution.

Proof: Assume that a and b are real numbers and $a \neq 0$. We must show that the equation $ax + b = 0$ has a unique solution. *[We work with form (c) as given in Theorem 3.4. Think of $P(x)$ as the equation $ax + b = 0$.]* First we must prove *existence*—that there is at least one solution. Let $x = -b/a$. *[Note that we do need the condition that $a \neq 0$.]* A little elementary algebra makes it clear that this value satisfies the equation. Now we must prove *uniqueness*—that if there are two solutions, they must be equal. So assume that x and y are both solutions. But if $ax + b = 0$ and $ay + b = 0$, then $ax + b = ay + b$, by the transitive property of equality. Subtracting b from both sides yields $ax = ay$, and then dividing both sides by a gives $x = y$. This completes the proof. ■

Uniqueness plays an important role with respect to definitions in mathematics. Normally, it makes no sense to define something in a permanent way unless we know the object being defined is unique.

Example 1: To illustrate this, suppose we know that for every real number x, there is a larger number. It would be silly to write a definition that says, "Given x, let y be the number that is larger than x," because there are many such numbers. It would make more sense to say "Given x, let y be *some* number that is larger than x." This is fine as a *temporary* definition within a proof; in Chapter 4 we call this type of naming **existential specification**. But it's not appropriate as a permanent definition.

On the other hand, suppose we know that for every real number x, there is a *unique* number y such that $x + y = 0$. (Theorem A-3 in Appendix 2 proves this.) Then it makes sense to have a permanent definition saying "Given x, let $-x$ be the number such that $x + (-x) = 0$." Note that having the variable x appear in the notation $-x$ conveys the fact that this number depends on x.

Exercises 3.4

(1) (a) Prove the (b) implies (c) part of Theorem 3.4.
(b) Prove the (d) implies (a) part of Theorem 3.4.

(2) Write symbolic statements that say:
(a) There are at least two objects such that P(x).
(b) There are at least three objects such that P(x).
*(c) There are at least n objects such that P(x). Here, n is any unspecified positive integer. Since you don't know its value, you need to use at least one "..." in your answer.

(3) Write symbolic statements that say:
(a) There are exactly two objects such that P(x).
(b) There are exactly three objects such that P(x).
*(c) There are exactly n objects such that P(x) (see the comments for Exercise 2(c)).

*(4) Using the method of Exercise 2, do you think it's possible to write a single symbolic statement that says that there are an *infinite* number of objects such that P(x)?

(5) Redo Exercise 5 of Section 3.3 replacing every $\exists$ with $\exists!$

(6) Redo Exercise 6 of Section 3.3 replacing every $\exists$ with $\exists!$

(7) Translate each of the following into symbolic form, using the instructions for Exercise 7 of Section 3.2. You can use the abbreviation $\exists!$; in fact you should use this quantifier (as opposed to $\exists$) whenever you think it's the intended meaning of the statement.

(a) Everybody has a father and a mother.
(b) Not everybody has a sister.
(c) Nobody has more than two grandmothers.
(d) Some people are only children.
(e) Some people have only one uncle.
*(f) Two people can have a common cousin without being cousins.

(8) This exercise relates to Example 5 of Section 3.1. Translate each of the following into symbolic form, following the instructions of Exercise 7. You need to make frequent use the predicate symbol On. For uniformity, use the variables A, B, and C to represent points and L, M, and N to represent lines.

(a) Lines L and M are parallel (that is, they have no point in common).
(b) Any two *distinct* lines meet in at most one point.
(c) Given any two distinct points, there's a unique line that they're both on.
(d) If lines L and M are parallel, then any line that is parallel to L (except for M) is also parallel to M.
(e) Pythagoras's theorem (use the symbols for angle and distance referred to in Section 3.1).
(f) Given any line and any point not on that line, there's a unique line through that point that is parallel to the given line. (This is one version of the famous Parallel Postulate of plane geometry.)
(g) Points A, B, and C are collinear, and B is between A and C.
(h) C is the midpoint of the line segment $\overline{AB}$.

(9) (a) Which of the equality axioms remain true if the symbol = is replaced throughout with the symbol ≤ (and all variables are assumed to represent real numbers)?
(b) Repeat part (a) using the symbol <.

(10) Which of the equality axioms remain true if all the variables are assumed to represent triangles, and the symbol = is replaced by the words "is similar to." Recall that two triangles are called similar if they have the same angles.

(11) Prove:

(a) If x and y are real numbers, there is a unique number z such that $z - x = y - z$.
*(b) If x and y are unequal real numbers, there is a unique number z such that $|z - x| = |z - y|$.
*(c) If m and n are unequal odd integers, there is a unique integer k such that $|k - m| = |k - n|$.

In parts (b) and (c), proving uniqueness requires extra care because of the absolute value signs. A picture might help you to see what's going on.

(12) Prove the following. You need to use some standard results from first-year calculus. You also need to analyze the quantifier structure of the statement you are proving.

(a) Every graph of the form $y = ax^2 + bx + c$, with $a > 0$, has a unique minimum point.

(b) Every graph of the form $y = ax^3 + bx^2 + cx + d$, with $a \neq 0$, has a unique point of inflection.

Suggestions for Further Reading: The same references that were suggested at the end of Chapter 2 apply to this chapter as well.

Chapter 4

Mathematical Proofs

4.1 Different Types of Proofs

Now that we have looked at the basics of propositional and quantifier logic, we are ready to study mathematical proofs in depth. Recall, from Section 1.2, that the primary means for establishing new results in mathematics is called the **axiomatic method**. Certain statements are taken as axioms; this means that they are accepted without proof. A statement may be included as an axiom because it is obviously true and quite simple or (in more specialized studies) because one or more mathematicians think it may have interesting consequences.

In this book we do not use any specialized axioms. Rather, we confine ourselves to standard, "obviously true" ones, that is, axioms almost all mathematicians and you would agree are correct. Axioms fall into two categories. Those based on logic are called **logical axioms**. Axioms based on properties of a particular type of mathematical object (integers, real numbers, sets, and so on) rather than on logic are called **proper axioms**; in Euclid's day they were called **postulates**.

Recall also that logical deduction is the only acceptable way to prove new things in mathematics. Traditionally, mathematicians have not spelled out exactly what types of reasoning should be allowed in proofs. It was generally assumed that everyone doing mathematics would have a feel for this that would be consistent with everyone else's. But in the past only a very small percentage of the population came into contact with advanced mathematics. In today's technological society, it's considered valuable to make mathematics comprehensible to a wider audience than ever before. Furthermore, for various kinds of theoretical studies, including those involving the use of computers to do mathematics, it's important to have an exact definition of what constitutes a proof.

Accordingly, this book not only presents a set of axioms but also explains numerous rules of inference that are commonly used in mathematics. A **rule of inference** is a precise rule that describes how a new statement may be asserted in a proof on the basis of its relationship to previous statements in the proof. Usually, when mathematicians refer to a proof method, they mean a rule of inference.

Example 1: Here is a rule of inference that we call the **conjunction rule:** if two statements P and Q appear as separate steps in a proof, then it's allowable to conclude the single statement $P \land Q$ in the proof. This is a particularly simple and obvious rule of inference, but many of the ones we introduce are not much more complex than this.

Formal Proofs

Definition: An **axiom system** consists of two parts: a list of statements that are to be considered axioms, and a list of rules of inference.

The lists mentioned in this definition may be finite or infinite. But in either case, the axioms and rules of inference must be clearly and unambiguously defined, so that it's always possible to determine whether any given statement is an axiom or follows from certain other statements by a rule of inference.

In the next three definitions, we assume that we have a particular axiom system in mind.

Definition: A **formal proof** is a finite sequence of statements in which every statement (or **step**) is either (1) an axiom, (2) a previously proven statement, (3) a definition, or (4) the result of applying a rule of inference to previous steps in the proof.

Definition: A **theorem** is a statement that can be formally proved. That is, it's a statement for which there's a formal proof whose last step is that statement.

Remarks: (1) There are several other words with more or less the same meaning. A relatively simple theorem may be called a **proposition**. (This usage of this word is clearly quite different from the meaning we defined for it in Chapter 2.) A theorem that is not considered very important on its own but is useful for proving a more important result is usually called a **lemma**. And a theorem that is easily proved from another theorem is usually called a **corollary** to the other theorem. There are no hard-and-fast rules for which of these words to apply to a given result. Some important results in mathematics have been labeled propositions or lemmas, perhaps because their authors were on the modest side.

(2) Sometimes it is appropriate to begin a proof with one or more **assumptions** (also called **hypotheses** or **premises** or **givens**. This usage of the word "hypotheses" is very different from its usage in the sciences, as described in Chapter 1). It is important to understand the distinction between an axiom and an assumption. An axiom is a statement that is agreed on and available for use in proofs permanently, at least within a particular subject. An assumption is a statement that is available for use only in the proof being attempted. Assumptions were made in several of the proof previews in Chapters 2 and 3, in which the goal was to prove an implication. In this chapter, we see that there are only a handful of situations in which it's legitimate to make assumptions in proofs. In fact, proving implications is essentially the *only* situation that justifies assumptions in mathematical proofs. It is *extremely important* to learn when it is appropriate to use the word "assume" and when it is not.

Example 2: In Section 2.3, whenever we verified that an argument was valid, we were essentially doing proofs from premises in an axiom system. This axiom system is very simple, having just one rule of inference: you may assert any statement that is a propositional consequence of the previous statements in the proof. Technically, this axiom system has no axioms, but practically speaking all tautologies are axioms, since every tautology is a propositional consequence of *any* set of statements.

Remarks: (1) Our definition of a formal proof allows four types of steps. Two of those—previously proven statements and definitions—are never needed to prove anything. Quoting a previously proven theorem in a proof just saves the trouble of reproving it, so while it can be a substantial time-saver, it never allows you to prove anything that you couldn't prove without quoting it. Unless specifically disallowed (which could occur on test or homework problems), it's acceptable to save time in this way when doing proofs in mathematics.

The role of definitions is more subtle, but a definition just introduces a shorthand or abbreviated way of saying something. So definitions also save time and can be very enlightening, but they don't allow you to prove anything you couldn't prove without them. You can use two sorts of definitions in proofs. You can make your own definitions, which create abbreviations (temporary or permanent) for your own convenience. But you can also quote any definition that has been given (for example, in whatever text you are using), as if it were an axiom. So if we wanted to give a bare-bones definition of what a formal proof is, we could have limited the possibilities to parts 1 and 4 of the definition.

(2) It is important to understand the difference between an axiom and a rule of inference. An axiom is a *single* statement that we agree to accept without proof and therefore may be asserted at any step in a proof. A rule of inference is never a single statement; rather, it describes some procedure for going from old statements to new ones. However, in the less formal proofs that mathematicians normally write, this distinction often gets blurred, as we soon see.

Except for very specialized and unusual systems, rules of inference are always based on logic. The ones in this book are all based on logic and, like our axioms, are quite standard.

A General-Purpose Axiom System for Mathematics

Appendix 1 consists of a detailed axiom system that we refer to throughout this book. The first part of it (all the rules of inference and groups I, II, and III of axioms) is based on logic and is **logically complete** and **sound**, meaning that its power to prove statements corresponds exactly to logical consequence. The rest of the axiom system consists of the generally accepted axioms about sets, the real numbers, and the natural numbers. Taken as a whole, our axiom system is powerful enough to derive all currently accepted theorems of mathematics, even in the most advanced subjects.

You might be surprised that it's possible to write an axiom system that encompasses all of mathematics but that is only three or four pages long. In fact, the axioms for logic and set theory alone are sufficient for the development of all of mathematics. So if I wanted to be very economical, I could have omitted the entire sections of real number axioms and natural number axioms, and Appendix 1 would still encompass all of mathematics. These extra axioms have been included to make the system easier to use, since it is rather difficult to develop the theory of these number systems from logic and set theory alone.

You might also wonder whether this system is at all standard or if it's just a personal creation of one author. As an experiment, you might ask your favorite mathematics professor what axioms she uses in her work. Her response will probably be a smile or a puzzled expression, with words to the effect, "I don't use axioms when I do mathematics. I just use intuition and deductive reasoning, plus a few well-known principles." This type of response is probably honest, but don't let it mislead you. By the time someone becomes a professional mathematician, she has had so much experience with the usual axioms and rules of inference of mathematics that they have become second nature to her. Even if she has never had a course like this one, she is just about as comfortable using the principles discussed in this book as most people are driving a car or reciting the alphabet. Furthermore, just as these basic skills become unconscious for all of us, an experienced mathematician may not even be aware that she uses a particular set of principles that has been subtly taught to her and is quite universal. But if you were to go beyond that first response and ask her to think in detail about how she does proofs, it would almost certainly turn out that she uses a combination of principles that are exactly equivalent to the list presented here.

Informal Proofs

One major reason why this book discusses formal proofs is to help you understand that there is nothing mysterious or magical about what constitutes a proof in mathematics. It may take a great deal of ingenuity to *find* a formal proof of a given statement, but once one is produced and written down, there would normally not be a controversy about whether it's correct. Any reasonable person who is willing to be very careful and take enough time ought to be able to check a formal proof for correctness. Better yet, computer programs can be written to check them. When mathematics is done formally, it becomes a sort of a game, with exact rules like chess or tic-tac-toe. However, although the rules of a game like chess are arbitrary and so must be learned specifically, the rules of mathematics are directly based on logic and common sense, so that it should be a natural process to become fluent with them.

Now here comes the catch. In spite of the order and precision that could be brought to mathematics by sticking to formal proofs, this type of proof is almost *never* used by mathematicians. If you randomly went through a dozen mathematics books, you would probably not find a single formal proof. We'll soon see some formal proofs, and you'll easily see why they are avoided. They are often extremely long and tedious to write and even worse to read. A complete formal proof usually consists of pages of symbols, even if it is based on just one or two simple ideas.

So if mathematicians don't write formal proofs, what do they write? Naturally, we may say that they write **informal proofs**. Unfortunately, it's not possible to say exactly what is meant by an informal proof. Furthermore, it's inaccurate to think of formal proofs and informal proofs as two, clearly separate categories. The true situation is more like a whole spectrum. On one extreme are strictly formal proofs. On the other extreme are completely informal proofs. Informal proofs are not based on any specific axiom system, generally use English more than mathematical notation, skip and/or lump together many steps, and base most of their logical assertions on commonsense reasoning. They often are laced with words like "obviously," "clearly," "it is easily seen that ... ," and in the case of one well-known mathematician, "it is intuitively obvious to the most casual observer that ... ."

Informal proofs are much easier to write and read than formal ones, and a well-written informal proof conveys information better than any formal one can. The problem with completely informal proofs, especially when used by less experienced proof writers, is that they open the door to sloppy thinking and errors.

To make all this clearer, here is a list of some of the ways in which most of the proofs written by mathematicians do not fit the definition of a formal proof:

(1) *Use of English:* Normally, when an axiom system is precisely defined, the axioms and rules of inference are stated in symbols (that is, mathematical and logical notation). It would then follow that a formal proof in that system would consist of symbolic statements, not English ones. But most mathematics proofs flow better if there are words as well as symbols, and so most mathematicians use a liberal mixture of words and symbols in their proofs. As long as these words strictly correspond to the axioms and valid logical principles, the use of English in a proof does not make the proof informal. But often, words are used to gloss over gaps in a proof, and in that case the proof must be considered informal.

(2) *Lack of an Axiom System:* We have already mentioned that most mathematicians do not consciously have an axiom system in mind when they write proofs; but unconsciously, almost all of them do follow a system that is equivalent to the one given in Appendix 1. However, a mathematician may occasionally write a proof that is not based, even unconsciously, on a clear-cut list of axioms. Such a proof would have to be called informal. On the other hand, many mathematicians would say that a nonaxiomatic proof cannot be a correct mathematical proof.

(3) *Skipping Steps:* Almost all mathematicians simply skip whatever steps they deem to be obvious. This is acceptable if the skipped steps really *are* obvious to whomever reads the proof. But this gets tricky: when you write a proof, how do you know who will be reading it and what will be obvious to whom? Something that would be obvious to most professional mathematicians would not necessarily be obvious to others. In practice, proofs are written differently for different audiences.

How should you handle this subtle point? Under what circumstances and to what extent should you leave out obvious steps? There is no pat answer to this question. Your instructor and the remainder of this text will constantly try to give you a feel for this. In the meantime, here is a good rule of thumb:

☞ *Do not omit any steps in a proof unless you can see clearly how to fill in all the gaps completely.* Nothing gets you into trouble more surely than skipping steps and calling them obvious, without knowing precisely how to carry out all the omitted steps.

(4) *Combining Steps:* This is a variant of skipping steps. Mathematicians often lump several easy steps into a single sentence, glossing over them rather than leaving them out entirely. The same guidelines that were described for skipping steps also apply to this practice.

(5) *Reverse Proofs:* This is a fairly common practice among mathematicians and one that can confuse the inexperienced reader. Technically speaking, when you prove a statement in mathematics, the statement you are proving should be the last step in the proof. That is, you start with things you are given (axioms and/or assumptions) and try to get to what you want. But sometimes the easiest way to figure out how to prove something is to do it in reverse, starting with what you want to prove, then looking for some statement that implies what you want, then looking for some statement that implies that statement, and finally reaching a known statement. A correct reverse proof can always be turned into a formal proof by writing the steps in the standard, forward order. But often a mathematician decides that a proof reads better in reverse and so keeps the final version that way. There is no harm in this informality if done properly. But whenever you do a proof in reverse, make sure it works forward; otherwise, you're probably proving the converse of what you should be proving.

This technique is discussed further in Section 4.6 under the heading "Reverse Reasoning," and we will see many examples of this important idea.

Good Proofs

How formal or informal should your proofs be? There is no pat answer to this. The dangers of both extremes have already been pointed out. While you are learning to write proofs, it is probably better to play it safe by keeping your proofs relatively complete. As you start to gain confidence, you can start to write more informally and skip a bit more. You should feel free to ask for guidance from your instructor and other experienced mathematicians regarding these issues, since it's quite hard to learn good proof writing without frequent feedback.

Here's another rule of thumb: *A good proof should be a clearly written outline or summary of a formal proof.* This means that when you write a proof, each statement you write (especially if it's an English sentence) should describe or indicate one or more steps that you would include in a formal proof if you were writing a formal proof. You can't do this unless you see how the formal proof should go. Once you've done that, you need to outline the formal proof in such a way that any reasonably intelligent reader, *including yourself,* should be able to understand the outline well enough to reconstruct the formal proof from it. This outlining process can require considerable thought and is what is meant by *style* in mathematics.

We have already encountered two very different styles of proofs. In the "Alternate Solutions" to Example 1 of Section 2.3, formal proofs were given for parts (a) and (c). For contrast, a less formal proof of part (c) was also included. All the proof previews in Chapters 2 and 3 are also written in good, nonformal style, as are the great majority of proofs from this point on. Occasionally we give a formal proof in the text, and there are several of them in Appendix 2. When you encounter these formal proofs, you should find it easy to understand their main advantage (that they encourage clear, correct step-by-step thinking) and their main disadvantage (that they are unwieldy, both to write and to read).

We occasionally use the term **semiformal** to describe a proof that directly parallels and summarizes a particular formal proof.

4.2 The Use of Propositional Logic in Proofs

Example 2 of Section 4.1 explained that the notion of propositional consequence introduced in Section 2.3 provides the basis for an important rule of inference. In fact, this one rule of inference is completely general with respect to propositional logic, in the sense that it includes every valid proof method based on propositional logic. Therefore, we discuss this rule of inference first and view it as the basis for all the material in this section.

Rule of Inference: Propositional Consequence

In a proof, you may assert any statement that is a propositional consequence of previous steps in the proof.

We sometimes shorten "propositional consequence" to "**prop. cons.**" or simply "**PC.**" Equivalently, we often just say that a step in a proof follows from previous steps "by propositional logic."

Example 1: Suppose we are talking about a real number x. We know (from axiom V-15) that either $x > 0$, $x = 0$, or $x < 0$. Suppose we also know, somehow, that x is nonzero. Then PC allows us to conclude that $x > 0$ or $x < 0$. This use of PC is based on the tautology $[(P \lor Q \lor R) \land \sim Q] \rightarrow (P \lor R)$. This exact tautology does not appear in Appendix 3, but it is quite close to tautology 11. At any rate, you should try to reach a point where you don't need to refer to Appendix 3 very often to check conclusions of this sort, because your own feel for logic makes it unnecessary.

Example 2: It is a theorem of calculus that if a function is differentiable, it is continuous. Suppose that we know this result, and we want to assert its contrapositive in a proof; that is, if a function is not continuous, then it is not differentiable. The rule PC allows us to do this, using the simple tautology $(P \rightarrow Q) \leftrightarrow ({\sim} Q \rightarrow {\sim} P)$.

For some more substantial examples of proofs based on propositional consequence, refer back to the examples and proof previews in Section 2.3.

As we mentioned in the previous section, having the rule of inference PC essentially makes all tautologies axioms. We now make this explicit.

All tautologies are axioms.

Note that a tautology is, by definition, always true and it's also straightforward to determine whether a given statement is a tautology. These are ideal characteristics of axioms.

Example 3: Suppose that we are trying to prove something about a real number x. If we want to, we can assert the statement that either $x = 0$ or $x \neq 0$, since this is of the form $P \vee {\sim} P$ (tautology 1, the law of the excluded middle). It might seem pointless to assert this disjunction, but this step might be used to set up a proof by cases, which could substantially simplify the proof.

On the basis of what we said at the beginning of this section—that propositional consequence includes all valid proof methods based on propositional logic—we could technically end this section at this point. However, PC is too general to be very convenient in most situations. Instead, mathematicians commonly use at least a half dozen more specific rules of inference. So let's now examine some of the most important of these so-called **derived rules of inference.**

Rule of Inference: Modus Ponens

If you have a step P and another step of the form $P \rightarrow Q$, you may then conclude the statement Q.

This rule of inference can be diagrammed (in the style of Section 2.3) as follows:

$$\begin{array}{l} P \\ P \rightarrow Q \\ \hline \therefore\ Q \end{array}$$

Despite the Latin name (which means "method of affirming"), this is a very simple rule of inference. Hopefully, you can see that modus ponens is logically correct: if you know P and also that P implies Q, then Q must follow. This is more or less the definition of implication.

A bit more formally, it is tautology 9 that justifies modus ponens. Exercise 8 asks you to derive modus ponens from propositional consequence.

Example 4: One important theorem of calculus is that if a function is differentiable, it must be continuous. Another basic result is the derivative formula for polynomials, which guarantees that polynomial functions are differentiable. Applying modus ponens to these steps yields that any given polynomial function, such as $3x^2 - 6x + 2$, must be continuous.

Starting with the next example, we occasionally illustrate a method of proof by referring to a proof of a theorem in Appendix 2 at the end of this book. Even though we have not discussed Appendix 2, you need not be intimidated by these references, because the proofs in Appendix 2 are based on the standard properties of the real number system, which are quite familiar to you from high school algebra.

Example 5: Every proof in Appendix 2 contains uses of modus ponens. A typical instance, although it's not specifically mentioned, occurs in the proof of Theorem A-7. In that proof, we have the step $z \neq 0$, since it's an assumption in the proof. We also have, from axiom V-11, the implication $z \neq 0 \rightarrow zz^{-1} = 1$. Modus ponens applied to these two statements yields the step $zz^{-1} = 1$.

Example 6: Let's redo Example 1(a) of Section 2.3, using a formal proof from hypotheses, with tautologies as our only axioms and modus ponens as our only rule of inference:

(1)	$P \rightarrow Q$	Premise
(2)	$\sim R \rightarrow \sim Q$	Premise
(3)	$\sim R$	Premise
(4)	$\sim Q$	Modus ponens applied to steps 3 and 2
(5)	$(P \rightarrow Q) \rightarrow (\sim Q \rightarrow \sim P)$	Tautology
(6)	$\sim Q \rightarrow \sim P$	Modus ponens applied to steps 1 and 5
(7)	$\sim P$	Modus ponens applied to steps 4 and 6

Rule of Inference: Conditional Proof

If you can produce a proof of Q from the assumption P, you may conclude the single statement $P \rightarrow Q$ (without considering P an assumption of the proof!).

We have been using conditional proof, without calling it that, ever since our first proof preview in Chapter 2. As we mentioned then, this is by far the most common and natural way to prove implications: to prove any statement of the form $P \rightarrow Q$, start by assuming P and try to prove Q. If you succeed, you've also succeeded in proving the conditional.

Even though conditional proof is derivable from propositional consequence, it is so important that we have included it separately in our axiom system. Conditional proofs are often referred to as **direct proofs** of implications. The diagram for conditional proof looks like this:

Assume P
[Some correct intermediate steps]
Q

$\therefore$ $P \rightarrow Q$

Remarks: (1) Technically speaking, conditional proof is not a rule of inference as we have defined it. A rule of inference is supposed to say that if you have certain *steps* in a proof, you can conclude some statement. Conditional proof says that if you can produce a certain *proof*, you can conclude some statement. This may sound like a minor distinction, but you should be aware that conditional proof has a very different flavor from normal rules of inference.

(2) Many students initially confuse modus ponens with conditional proof. The best way to keep them apart in your mind is to realize that they are essentially *opposites*. Modus ponens gives you a way of *using* an implication: it says that if you *know* $P \rightarrow Q$, then you can go from P to Q. Conditional proof gives you a way of *proving* an implication: it says if you can show how to go from P to Q, you can *conclude* $P \rightarrow Q$.

(3) When writing a conditional proof, you need to be clear about what assumption(s) are in effect at each point in your proof. If you decide, in the middle of a proof, to prove an implication by conditional proof, you have to do a certain proof from an assumption, which may be regarded as a subproof of the main proof. As long as you're within the subproof, there's an assumption being made, but when the subproof

is finished, you assert the implication and go back to the main proof without the assumption. This is not particularly confusing as long as there's only one use of conditional proof at a time in the proof, but it gets more involved if there are nested uses of conditional proof, which can occur.

Example 7: Here is an example of nested uses of conditional proof. Suppose we want to prove that if a certain statement P holds, then some set A is a subset of some set B. So we would begin by assuming P, since P is the hypothesis of the implication we are trying to prove. From this assumption, we then need to prove that $A \subseteq B$. But to prove this, we need to show that any member of A must also be a member of B. Therefore, the second line of our proof would probably be "Assume x is any member of A." This begins another conditional proof within the outer one. The goal of the whole proof is then to prove that x is a member of B, using both assumptions.

Rule of Inference: Indirect Proof

If you can produce a proof of *any* contradiction from the assumption ~ P, you may conclude P.

The diagram for this rule of inference is

Assume ~P
[Some correct intermediate steps]
Any contradiction

∴ P

Indirect proof is sometimes called **proof by contradiction** or **reductio ad absurdum** ("reduction to the impossible").

Remarks: (1) Section 4.1 cautioned you against overusing the word "assume" in proofs. Conditional proof and indirect proof are among the few proof methods in mathematics in which it's appropriate to assume something. We see below that indirect proof is really an offshoot of conditional proof, and it can be argued that the *only* situations in mathematics in which assumptions are permitted are those using conditional proof and its offshoots.

(2) You can try to prove any statement by indirect proof, no matter what its logical form is. But indirect proof is a particularly good proof method to try when the statement you're attempting to prove is a *negation*. If you want to prove $\sim$ P by indirect proof, you assume P and try to derive a contradiction.

(3) The most common form of contradiction is a statement of the form Q $\wedge \sim$ Q, but any contradiction will do. The contradiction derived need not involve the original statement P. On the other hand, you may find that from the assumption of $\sim$ P, you are able to prove P. This constitutes a successful indirect proof of P, because the next step can be the contradiction P $\wedge \sim$ P.

Example 8: Theorem A-8 in Appendix 2 is a typical example of an indirect proof. We want to prove the statement $\sim (x < y \wedge y < x)$, which is in the form of a negation. This is the ideal type of statement to prove indirectly. So we assume $x < y \wedge y < x$. Using the transitivity of < (axiom V-14) plus modus ponens, this yields $x < x$, which produces a contradiction in conjunction with the axiom $x \not< x$ (V-13).

Example 9: Indirect proof is the most efficient way to prove that the sum of a rational number and an irrational number must be irrational. (Recall that a rational number is one that can be written as a fraction of integers.) Here is a proof:

Assume the claim is false. Then we have $a + b = c$, for some numbers a, b, and c, with a and c rational and b irrational. Simply subtract a from both sides to obtain $b = c - a$. Since the difference of two fractions can always be written as a single fraction, this makes b rational, a contradiction.

We have been emphasizing that almost all the rules of inference in this section are derivable from propositional consequence, without going into much detail. Here is a proof of a similar fact.

Theorem 4.1: The inference rule of indirect proof is derivable from the rules in our axiom system, namely propositional consequence and conditional proof. That is, anything that can be proved using indirect proof (and perhaps the other two rules as well) can be proved with just those other two rules.

Proof: Suppose we have proved statement P by indirect proof. That means we have a proof that begins with the step "Assume $\sim$ P" and, after a certain number of correct steps, reaches a contradiction. Then simply take this proof and add two more steps:

Therefore, $\sim$ P $\rightarrow$ some contradiction (conditional proof).

Therefore, P (from the previous step, by propositional logic). ∎

Again, it doesn't matter what form of contradiction is obtained in Theorem 4.1; the statements $\sim$ P $\rightarrow$ some contradiction and P must be propositionally equivalent. One way of interpreting this theorem is that indirect proof is a special type or offshoot of conditional proof.

Rule of Inference: Proof by Cases

If you have a step of the form $Q \vee R$ and the two implications $Q \rightarrow P$ and $R \rightarrow P$, you may conclude the statement P.

In practice, the usual format of a proof by cases is as follows: first you establish a disjunction that you think divides the problem up into cases that are easier to handle than the whole problem at once. Often this disjunction is very simple, perhaps a tautology or other axiom, and is not explicitly stated. Then you must show that each disjunct, or case, implies the statement that is to be proved; this is normally done by conditional proof. To keep the reader clear about what's going on, you might say "Case 1: Assume Q" and derive P, and then say "Case 2: Assume R" and derive P from that assumption. This rule of inference can be diagrammed as follows:

$Q \vee R$	(Proved somehow)
Case 1: Assume Q	
[Some correct intermediate steps]	
P	(End of Case 1)
Case 2: Assume R	
[Some correct intermediate steps]	
P	(End of Case 2)

$\therefore$ P

Specifically, this is the diagram for proof by cases with *two* cases. It's fine to have more than two cases. For example, if you want to use a disjunction $Q \vee R \vee S$ to prove P by cases, you have to show that each of the three cases (Q, R, and S) implies P.

Derivation of proof by cases: Proof by cases is a special case of the rule PC, because P is a propositional consequence of the three statements $Q \vee R$, $Q \rightarrow P$, and $R \rightarrow P$. To see this without constructing a truth table, notice that the two implications $Q \rightarrow P$ and $R \rightarrow P$ are together equivalent to the single implication $(Q \vee R) \rightarrow P$, by tautology 26. Using the last implication, we can assert P by modus ponens. A similar derivation can be used to justify proofs by three or more cases (see Exercise 6).

A proof by cases using tautology 26 and propositional consequence was done in Proof Preview 3 in Section 2.3 (see Exercise 7). Proofs by cases in which the cases involve the sign of a real number are extremely common. Here is another example.

Example 10: Theorem A-11 of Appendix 2 shows a very common type of proof by cases. The goal is to prove the inequality $x^2 \geq 0$, for an arbitrary real number x. It is not easy to do this all at once. But if we look separately at the three cases $x = 0$, $x > 0$, and $x < 0$, it is pretty easy to show the desired conclusion holds in each case. (The disjunction of the three cases, which is a necessary part of the proof, is based on axiom V-15.)

Example 11: Let's consider a real-life example. Imagine that your girlfriend has told you that she doesn't want to see you tonight because she needs to stay home and study all evening. You want to believe but you're suspicious, so you're tempted to phone. Suppose you reason: "Well, either she's home or she's not. If she is, I'll be better off if I call because I'll be reassured. If she's not, I'll also be better off if I call because at least I'll know the score. So I should call." You are using proof by cases. The disjunction being used to define the two cases is an instance of the law of the excluded middle, $Q \vee \sim Q$, with Q representing "She's home."

Rule of Inference: Biconditional Rule

If you have implications $P \rightarrow Q$ and $Q \rightarrow P$, you may conclude the biconditional $P \leftrightarrow Q$.

The biconditional rule can be diagrammed

$$\begin{array}{l} P \rightarrow Q \\ Q \rightarrow P \\ \hline \therefore\ P \leftrightarrow Q \end{array}$$

In practice, this is by far the most common way to prove a biconditional, as we did in Proof Preview 1 in Section 2.2. A proof by this rule has two separate parts, or directions, which should *not* be called "cases" since that term is best reserved for proofs by cases. Each direction is an implication, which of course can be proved by conditional proof.

Some mathematicians use arrows to indicate the directions in this type of proof. So a proof of some statement $P \leftrightarrow Q$ could begin "$\rightarrow$: Assume P" and derive Q. It would then say "$\leftarrow$: Assume Q" and derive P.

Example 12: Theorem A-13 of Appendix 2 illustrates the typical use of this rule. We want to prove a biconditional: a number is positive iff its reciprocal is positive. So for the forward direction, we must prove $x^{-1} > 0$ from the assumption $x > 0$; for the reverse direction, we must prove $x > 0$ from the assumption $x^{-1} > 0$.

In simple proofs of this type, it may be obvious that the two directions of the proof are exactly the reverse of each other, step by step. In this situation, it's considered fine to show just one direction and to mention that the reverse direction can be proved by exactly the reverse sequence of steps. What this means is that you've established a sequence of iffs between the two statements whose biconditional you are trying to prove. We use this shortcut frequently, notably in Section 5.2.

Before we state our next rule of inference, we need some explanation and a bit of notation. Suppose that in a proof we have a step of the form P ↔ Q. This says that P and Q are equivalent, which ought to mean that P and Q are *interchangeable.* So suppose we also have some long statement that contains P as a substatement, such as one of the form (R → (~ S ∧ P)) ∨ (P → T). Should we be allowed to conclude the same statement with one or both of the occurrences of P replaced by Q? Intuitively, we should be. Can this conclusion be justified by propositional consequence? Yes, it certainly can. But note that to do so would require a 32-line truth table! The next rule of inference lets you draw such conclusions without having to construct a truth table.

Notation: The notation S[P] denotes a statement S that contains some statement P as a substatement (which could be the whole statement S).

The notation S[P/Q] denotes a statement that results from S[P] by replacing *one or more* of the occurrences of the statement P within S[P] by the statement Q.

Rule of Inference: Substitution

From statements P ↔ Q and S[P], you may conclude S[P/Q] as long as no free variable of P or Q becomes quantified in S[P] or S[P/Q].

The diagram for substitution is

P ↔ Q
S[P]

∴ S[P/Q]

Remarks: (1) The restriction on quantifiers in the use of substitution is meant to guarantee that S[P] is built up from P using connectives only, not quantifiers (or at least not quantifiers that matter). In practice, it is not necessary to worry about this restriction very often. For a specific example, see Exercise 18 of Section 4.3.

(2) There is some deliberate ambiguity in the notation introduced here. Even when S, P, and Q are known, there may be more than one possibility for what statement S[P/Q] represents, for if P occurs within S more than once, then there's a choice as to which occurrence(s) of P within S are to be replaced by Q. The example in the paragraph before the definition of substitution illustrates this.

(3) Why is the notation S[P], rather that S(P), being used here? The reason is that P is *not* a mathematical variable appearing in the statement S. It's a substatement of S, which is totally different. To emphasize this, it makes sense to introduce a different-looking notation.

(4) Do not confuse substitution with the familiar principle of substitution of equals, which says that if two numbers or other mathematical quantities are equal, then they are interchangeable. This principle appears in our axiom system as axiom III-4. In flavor, it is certainly very similar to substitution. The main difference between them is what they talk about: Substitution is about the interchangeability of *statements*, whereas Axiom III-4 is about the interchangeability of *numbers* or other mathematical *objects*.

As we show in Section 4.4, the principle of substitution of equals is what allows us to do the same thing to both sides of an equation. Similarly, with the rule of inference substitution, it becomes permissible in most situations to do the same thing to both sides of an equivalence.

Example 13: Here is a simple example of substitution from real life. Suppose you say, "If Harry shows up at my party, I'll call the police." Then your friend says, "But Harry and your boss do everything together; if one shows up, so will the other." Then you say, "Well, I guess that means that if my boss shows up, I'll call the police." You have just used substitution, because the second part of your friend's statement means, "Harry will show up if and only if your boss does."

Example 14: Suppose we know some biconditional $P \leftrightarrow Q$. Then, if we also know P, we can use substitution to conclude Q. On the other hand, if we know $\sim P$, we can conclude $\sim Q$. If we know a conjunction $P \wedge R$, we can conclude $Q \wedge R$. If we know $P \vee R$, $P \rightarrow R$, $R \rightarrow P$, or $P \leftrightarrow R$, we can conclude, respectively, $Q \vee R$, $Q \rightarrow R$, $R \rightarrow Q$, or $Q \leftrightarrow R$. Of course, these conclusions also follow by propositional logic (see Exercise 1).

We have now discussed all the important propositional rules of inference. Here are a few more rules, all of which are pretty obvious and follow easily from the ones we have listed so far. We won't give derivations or examples of these because of their simplicity.

Rule of Inference: Conjunction

If you have, as separate steps, any two statements P and Q, you may conclude the single statement $P \land Q$.

This rule of inference follows trivially from propositional consequence. See Exercise 9(a) for a related and not-so-easy problem.

Rule of Inference: Modus Tollens

If you have a step of the form $P \rightarrow Q$ and also have $\sim Q$, you may assert $\sim P$.

"Modus tollens" is Latin for "method of denying." It is based on tautology 10, but it can also be thought of as an offshoot of modus ponens, based on the equivalence of $P \rightarrow Q$ and its contrapositive $\sim Q \rightarrow \sim P$. Similarly, the next rule is a straightforward contrapositive offshoot of conditional proof.

Rule of Inference: Contrapositive Conditional Proof

If you can produce a proof of $\sim P$ from the assumption $\sim Q$, you may conclude the single statement $P \rightarrow Q$.

☞ We have now discussed two ways to prove an implication $P \rightarrow Q$, but there are at least three common ways:

(1) Direct conditional proof: Assume P, and derive Q.

(2) Contrapositive conditional proof: Assume $\sim Q$, and derive $\sim P$

(3) Indirect proof: Assume $\sim (P \rightarrow Q)$, or equivalently assume $P \land \sim Q$, and derive a contradiction.

Table 4.1 Summary of propositional proof methods

Statement	Way(s) to Prove	Way(s) to Use
A negation $\sim P$	Indirectly: Assume P, and derive a contradiction.	Move negation sign inward.
A conjunction $P \wedge Q$	Conjunction rule: Prove P and also prove Q.	Assert P and/or Q, separately.
A disjunction $P \vee Q$	(1) Prove P or prove Q. (2) Indirectly, by De Morgan's laws: Assume $\sim P \wedge \sim Q$, and derive a contradiction.	Proof by cases.
A conditional $P \rightarrow Q$	Three methods, listed on Page 88 .	(1) Modus ponens. (2) Modus tollens.
A biconditional $P \leftrightarrow Q$	Biconditional rule: Prove $P \rightarrow Q$ and prove $Q \rightarrow P$.	Substitution.

In a sense, the last method is the most powerful, because it lets you begin with two assumptions instead of one.

We conclude this section with a chart summarizing the most important propositional proof methods (see Table 4.1). This chart shows the most natural ways to *prove* and to *use* each type of statement (with type based on the connective). Table 4.1 is meant to help you when doing proofs, and you are urged to study it carefully. At the same time, don't fall into the trap of thinking that any single chart can teach you everything there is to know about proofs in mathematics or even everything about the proof methods based on propositional logic. There may be only a finite number of rules of inference in common use, but there are literally an infinite number of different ways to apply them.

Exercises 4.2

(1) Show that each of the conclusions in the first paragraph of Example 14 could have been made using propositional consequence instead of substitution.

(2) Suppose that we have proved steps of the form $P \leftrightarrow Q$ and $(P \wedge R) \vee (\sim P \wedge S)$ in a proof. State the three different conclusions that may be made from these two steps using substitution.

(3) Redo Exercise 5 of Section 2.3 as formal proofs from hypotheses. You may use all tautologies as axioms, and you may use all the rules of inference discussed in this section, *except* propositional consequence.

(4) For each of the following, state what seems to be the logical conclusion and also state which rule of inference (other than propositional consequence) could be used to reach that conclusion:

(a) To make you happy today, I'd have to be in two places at once, which is impossible. Therefore, ...

(b) If it's Saturday, I don't go to school. If I don't go to school, I'm very sad. It's Saturday. Therefore, ...

(c) If the Mets win, I'll come out ahead in my bets. If the Mets don't win, I'll also come out ahead in my bets. Therefore, ...

(d) If a function is continuous, it's integrable. This function is not integrable. Therefore, ...

(5) Turn each of the arguments in Exercise 4 into formal proofs. To do this, you should introduce propositional variables for the atomic substatements and rewrite each given statement symbolically, as in Section 2.1. Then write formal proofs to prove each conclusion from the givens.

(6) Precisely state the inference rule of proof by cases with *three* cases, and derive this rule from propositional consequence.

(7) Redo Proof Preview 3 (Section 2.3), explicitly using the method of proof by cases.

(8) Derive the rule modus ponens from propositional consequence.

*(9) In the text, we have shown that all tautologies and the modus ponens rule of inference are derivable from the propositional consequence rule. This exercise establishes the converse result that propositional consequence is derivable from all tautologies and modus ponens. Consider an axiom system that has modus ponens as a rule of inference instead of propositional consequence and has all tautologies as axioms.

(a) Show that the conjunction rule of inference is derivable from this axiom system. (You must *formally* prove $P \land Q$ from the two assumptions P and Q. This can be done in very few steps, but it's tricky to find the proof.)

(b) Using the result of part (a), whether or not you were able to show it, show that propositional consequence is derivable from this axiom system.

Exercises 10 through 12 ask you to prove various results. Since we have not yet discussed methods of proof involving quantifiers, do not attempt to make these proofs rigorous or formal. For the most part, you may assume familiar results from high school algebra about integers, equations and inequalities. However, where *real* numbers are involved, you should not assume anything beyond the axioms in group V of our axiom system, unless instructed to do so.

(10) Prove: If n is any integer, then $n^2 - 3n$ must be even.

(11) Prove: For any real numbers x, y, and z, if $x < y$ and $y \leq z$, then $x < z$.

(12) Prove: For any *positive* real number x, $x + 1/x \geq 2$. It might be tempting to use calculus to prove this, but you may not do so; use information from high school algebra only. ***Hint:*** You might try a *reverse proof*, as explained in the previous section. That is, start with the inequality you're trying to prove and try to transform it into something that you know to be true. But then you have to make sure you can turn this into a *forward* proof. You may assume, without proof, that the square of any real number is at least 0. By the way, is there any point in your proof where you need to know that x is positive?

The rest of the exercises in this section are of a type that occurs frequently throughout the rest of this book. These problems ask you to "critique the following proof," and then give a supposed theorem and proof. To do such an exercise, you should carefully read and consider the given proof and come to one of four conclusions:

(i) The theorem and its proof are correct (and the proof has no major omissions).

(ii) The theorem is correct; the proof has no mistakes but does omit one or more substantial step(s). In this situation, you should supply the missing step(s) in the proof.

(iii) The theorem is true, but the proof is substantially flawed. In this situation, you should point out the error(s) in the given proof *and* provide a correct proof of the theorem.

(iv) The stated theorem is false. In this situation, you should point out the error(s) in the given proof *and*, if appropriate, provide a concrete example to show that the claimed theorem is in fact false. (Such examples are called **counterexamples**; see Section 4.3.)

There may be problems of this type that you believe to be borderline between two of the four choices, for example, between (i) and (ii). If that occurs, feel free to say so.

Critique the proofs in Exercises 13 through 17.

(13) **Theorem:** If x is any real number, then $x^2 \geq x$.

Proof: Assume x is any real number. We proceed by cases:

Case 1: Assume $x \leq 0$. Then $x^2 \geq 0$, while $x \leq 0$. Therefore, $x^2 \geq x$.

Case 2: Assume $x \geq 1$. Then multiply both sides of this inequality by x to obtain $x^2 \geq x$. Since both possible cases lead to the desired conclusion, we have proved it.

(14) **Theorem:** For any real numbers x and y, $x < y$ implies $x^2 < y^2$.

Proof: We use conditional proof. So assume that $x < y$. Then simply square both sides of the inequality. This gives us the desired conclusion $x^2 < y^2$.

(15) **Theorem:** For any real number x, if $x^2 = 0$ then $x = 0$.

Proof: We use indirect proof. Assume the entire implication is false. By tautology 19, that means $x^2 = 0$ and $x \neq 0$. Then we can multiply both sides of the equation $x^2 = 0$ by $1/x$. The left side becomes $x^2(1/x)$, which equals x; and the right side becomes $0(1/x)$, which equals 0. Thus $x = 0$. This contradicts our assumption that $x \neq 0$, and so we are done.

(16) **Theorem:** For any integer n, if n is even, so is n^2.

Proof: We use indirect proof. Assume n is *not* even; from this we must prove that n^2 is not even either. To say that n is not even means that it is odd. So we have $n = 2m + 1$, for some integer m. Therefore, $n^2 = (2m + 1)^2 = 4m^2 + 4m + 1 = 2(2m^2 + 2m) + 1$, which by definition is an odd number.

*(17) **Theorem:** For any positive integer n, if $2^n - 1$ is prime, then so is n.

Proof: We use contrapositive conditional proof. That is, we begin by assuming that n is *not* prime.

Case 1: Assume $n = 1$. Then $2^n - 1 = 2^1 - 1 = 1$, which is not prime.

Case 2: Assume $n > 1$. Then n must be a composite number, so $n = ab$, where a and b are both integers greater than 1. Let $m = 2^a$. So $2^n - 1 = 2^{ab} - 1 = (2^a)^b - 1 = m^b - 1$. From algebra, $m^b - 1$ is divisible by $m - 1$. (Specifically, a simple computation shows that $m^b - 1 = (m - 1)(1 + m + m^2 + m^3 + ... + m^{b-1})$.) Now, since $a > 1$, $2^a > 2$, and thus $m - 1 > 1$. Also, since $m > 1$ and $b > 1$, $m - 1$ is certainly less than $m^b - 1$. Therefore, $2^n - 1$ (which equals $m^b - 1$) has a factor strictly between itself and 1, and so it is not prime. This completes the proof.

4.3 The Use of Quantifiers in Proofs

Recall that we have emphasized propositional consequence as the most important rule of inference for propositional logic; in fact, we said it could be viewed as the *only* propositional rule of inference. Why don't we do the same thing for quantifier logic? That is, why not have a rule of inference that says you may assert any statement that is a *logical* consequence of previous steps in the proof? This question was answered in Section 3.3. There is no simple, mechanical procedure (like truth tables) for testing whether a given statement is a logical consequence of certain other statements. Therefore, the abstract concept of logical consequence cannot be used to define axioms or rules of inference.

Before proceeding to specific axioms and rules of inference, we need to state an important convention that is standard throughout all of mathematics.

☞ **Convention:** Whenever an *axiom* or *theorem* contains free mathematical variables, the statement is understood to begin with universal quantifiers for those variables. (This convention applies to axioms and theorems *only*. It does not apply to definitions or to steps within a proof.)

Axioms: De Morgan's Laws for Quantifiers

$$\sim \forall x\, P(x) \leftrightarrow \exists x \sim P(x)$$
$$\sim \exists x\, P(x) \leftrightarrow \forall x \sim P(x)$$

These important equivalences appeared earlier as Theorem 3.2, but that was in an informal setting. Now, we more rigorously classify them as axioms. As we will soon see (Theorem 4.2 and Exercise 15), either of them can be proved from the other, so only one of them *needs* to be an axiom.

Examples showing why these quantifier laws are useful and how to use them were also given in Chapter 3, so we won't give more examples here. Remember that it's useful to be able to move negation signs through quantifiers, especially from outside to inside, but such a move must be accompanied by changing every quantifier through which the negation sign is moved.

De Morgan's laws for quantifiers can be useful for indirect proofs of quantified statements. For instance, if you want to prove a statement of the form $\exists x\, P(x)$, you can assume $\forall x \sim P(x)$ and derive a contradiction. However, indirect methods are considered a bit less attractive than the more direct methods (UG and EG) to be discussed shortly, especially in the case of statements of the form $\exists x\, P(x)$.

Our next axiom is an obvious enough principle. If we know a statement is true for all members of a certain domain (real numbers, integers, functions, sets, and so on), then it's true for any particular one. This is a direct consequence of what the universal quantifier means.

Axiom: Universal Specification or US

$$\forall x\, P(x) \rightarrow P(t)$$

where the letter t here denotes any term or expression (*not* necessarily a single letter like a variable or constant) that represents an object in the domain of the variable x.

☞ Mathematicians rarely use the terminology "universal specification" and the similar names for the next three proof methods (as well as some of the terminology in Section 4.2). In fact, many fine mathematicians probably don't even know some of these terms. However, *all* mathematicians understand these proof methods and how to use them extremely well. The reason we introduce this terminology is to help you keep these proof methods straight while you are first learning about them. But it is vital that you concentrate on the *content* of these methods, not the words we use to classify them!

Remarks: (1) US allows plenty of freedom in the choice of the term t. It could be a variable, either the same one as in the quantifier or a different one. It could be a constant. Or it could be a more complicated expression.

(2) Technically, there is another restriction in the use of US: no free variable of t may become bound when P(t) is formed. This situation occurs so rarely that we didn't mention it in the definition of US above (see Example 4).

(3) You may wonder why some principles are stated as axioms and others as rules of inference. In some cases the distinction is less important than in others. In particular, any axiom that is an implication can also be viewed (in combination with modus ponens) as a rule of inference. For example, US is primarily used as a rule of inference: if you have a step of the form $\forall x$ P(x), you may then assert P(t), where t is any ... . Usually, if a principle can be considered either an axiom or a rule of inference, we set it up as an axiom.

Example 1: You have been using US ever since you first studied algebra, even if you didn't have a name for it. For instance, consider a typical algebra formula such as $(x + y)^2 = x^2 + 2xy + y^2$. By the convention stated at the beginning of this section, the variables x and y in this formula are assumed to be universally quantified. So when you learned this formula in high school, you learned that it was true for all numbers, and that therefore you could substitute *any* expression for x and/or y. So you knew that you could write

$$(3a + 2)^2 = 9a^2 + 12a + 4$$

$$(x^2 + y^3)^2 = x^4 + 2x^2y^3 + y^6$$

$$(\sin x + \cos x)^2 = \sin^2 x + 2 \sin x \cos x + \cos^2 x, \text{ and so on.}$$

Every time you make this type of substitution for a variable, you are using US (plus modus ponens).

Example 2: Just about every proof in Appendix 2 uses US. A typical but clever use occurs in the proof of Theorem A-1. In the uniqueness part of that proof, we get to assume $\forall x\ (x + y = x)$ and $\forall x\ (x + z = x)$. To use these assumptions to prove that $y = z$, the trick is to specify the first quantified x to be z, and the second one to be y.

Example 3: Here is a simple instance of the necessity to have t be of the same sort as x in the use of US. Suppose we know the formula $\forall x\ (x^2 \geq 0)$, where x is a real variable. Then we cannot conclude that $i^2 \geq 0$, where i is the imaginary unit, since i is not a real number. In fact, $i^2 = -1$, so the conclusion would be false.

Example 4: Here is an example of why the other restriction on t (mentioned in Remark 2) is necessary. Suppose we have proved the true statement (about real numbers) that $\forall x\ \exists y\ (y > x)$. Without the restriction we could substitute the term "$y + 1$" for x and derive the false conclusion $\exists y\ (y > y + 1)$.

Our next rule of inference is by far the most common way to prove a "For all ... " statement. You may well be familiar with it and know how to use it, even if you don't know its name. Here's what it says: Suppose you want to prove a statement of the form $\forall x\ P(x)$. Then it is sufficient to prove the simpler statement $P(x)$, where x is just a variable representing an arbitrary member of the domain in question. It is important that *no assumptions* are made about x except that it is a member of that domain. If you can produce such a proof of $P(x)$, you may conclude $\forall x\ P(x)$.

Rule of Inference: Universal Generalization or UG

If you can produce a proof of $P(x)$, where x is a free variable representing an arbitrary member of a certain domain, you may then conclude $\forall x\ P(x)$.

Universal generalization can be diagrammed as follows:

[Proof with no assumptions about x]
$P(x)$

$\therefore\ \forall x\ P(x)$

Remarks: (1) Although it is not required, it is helpful to let the reader know that you are intending to use UG by saying something like "Let x be any ..." or "Let x be an arbitrary ..." or "Let a ... x be given" at the beginning of the proof.

(2) People sometimes think that the logical way to prove that something is true for all members of some domain is to prove it for each member individually. That works fine if the domain happens to be finite and small but it's impossible if the domain is infinite. Unfortunately, most mathematical variables have infinite domains, such as the set of all real numbers.

(3) UG is used in the vast majority of theorems in mathematics, although it is almost never mentioned. By the convention stated at the beginning of this section, free variables in theorems are always understood to be universally quantified. In the proofs of such theorems, neither those quantifiers nor the use of UG are likely to be explicitly mentioned. The practical meaning of UG is that if you want to prove a statement that begins with one or more universal quantifiers, you can essentially ignore those quantifiers. This is a handy thing to know!

Example 5: Recall Proof Preview 1 in Section 2.2. There is nothing in the proof about quantifiers or UG. But the variable n is free, so the theorem must be about *all* integers. The proof is carried out for an arbitrary integer n, and by UG, this establishes the result for all integers. Similarly, in Proof Preview 2 in Section 2.3, the set variables A, B, and C are free. So this theorem is technically about *all* sets A, B and C, and the proof may be viewed as including an implicit use of UG to quantify these variables. To be fully rigorous, Proof Preview 2 should also have dealt with the quantification of the variable x (see Exercise 1).

Example 6: Suppose we want to prove the important theorem that every nonnegative real number has a square root. We would probably begin the proof with the words "Let x be an arbitrary nonnegative real number." This tells the reader that we intend to prove that this one unspecified x has a square root, which by UG establishes that every such number has one.

It may seem to you that the use of UG in Example 6 is pretty similar to conditional proof and that what's really being proved is the implication "If x is a nonnegative real number, then x has a square root." Conditional proof and UG are in fact related, and many mathematicians mix them and blur the distinction between them. They might start the above proof with the words "Assume $x \geq 0$," as if it were a conditional proof.

Here is the source of the relationship between conditional proof and UG: If a mathematical variable x has some set D as its domain, then the statement $\forall x\ \mathrm{P}(x)$ really means $\forall x \in D\ \mathrm{P}(x)$. But in Section 3.3 we noted that $\forall x \in D\ \mathrm{P}(x)$ is an abbreviation for $\forall x\ (x \in D \rightarrow \mathrm{P}(x))$. In other words, just about every statement that begins with a universal quantifier also contains an implication; so when UG is used, conditional proof is usually used with it. The only exceptions would be when the variable x can denote any object whatsoever, so that it is not restricted to any domain.

In spite of this closeness between conditional proof and UG, they are different and should not be confused. Conditional proof is a rule of inference that involves implications, not quantifiers; UG is a rule of inference that is about universal quantifiers, even though it indirectly involves implications too. Another difference is that when you set up a conditional proof, the assumption made can be any type of statement, even a false one. But when you set up a UG proof, the only thing you get to assume is that some variable is a member of some domain. I usually say "Let x be a ... ," rather than "Assume x is a ... ," when I start a UG proof, because I'm not really assuming anything; I'm just specifying how I'm going to use a certain letter. But as I said above, most mathematicians ignore this distinction, and doing so normally creates no problems.

This discussion is related to the comments made in Section 2.2 about the word "whenever." This word is very close to the word "if" in mathematics, but there is a difference, namely that the word "whenever" actually combines an implication with a universal quantifier. For instance, the words "Whenever a function is continuous, it's integrable" should be interpreted as meaning "*every* continuous function is integrable," which can be written "$\forall f$ (f is a continuous function $\rightarrow f$ is integrable)." To be sure, a *theorem* that says "If a function is continuous, it's integrable" should be interpreted the same way, because of the convention stated earlier in this section. But the word "if" conveys a universal quantifier only some of the time, whereas the word "whenever" *always* does.

Example 7: Here is a simple example of incorrect and correct uses of UG. Suppose that we start a proof by assuming that $x > 7$, and from this we prove that $x > 0$. Can we now apply UG to get the step $\forall x\ (x > 0)$, followed by conditional proof to reach the conclusion $x > 7 \rightarrow \forall x\ (x > 0)$? Definitely not! The rule UG may not be applied to the step $x > 0$ because there is an assumption about x in effect at that point in the proof. Furthermore, the conclusion obtained can't be right, since it would be nonsense to claim that information about one number implies an incorrect general statement about all numbers. Instead, after the step $x > 0$, we can assert $x > 7 \rightarrow x > 0$ by conditional proof and then apply UG to reach the correct conclusion $\forall x\ (x > 7 \rightarrow x > 0)$.

Let us now turn to proof methods involving the existential quantifier. One very important one is based on the following idea. Suppose we know that some object exists. In other words suppose that, in a proof, we have a step of the form $\exists x\ P(x)$. If we don't know a specific value of x that makes $P(x)$ true, it's convenient (and harmless) to introduce a temporary name for some unknown object satisfying $P(x)$.

Rule of Inference: Existential Specification or ES

If you have a step of the form $\exists x\ P(x)$, you may assert $P(c)$, where c is a new, temporary constant symbol.

Remarks: (1) It is important to understand the conditions required for the correct use of this rule. The name c introduced for the unknown object represents *one particular* object. Therefore, it must be a constant, not a variable; it cannot be quantified. Since we don't know anything else about this object except that it satisfies $P(x)$, it must be a *new* symbol; that is, it may not appear earlier in the proof. Finally, this rule is meant to be a temporary convenience; the new symbol should not appear in the final conclusion of the proof.

To view it another way, a constant introduced by ES may be viewed as a temporary *definition*. Section 3.4 pointed out that mathematicians normally make a *permanent* definition only when they know that some object exists *uniquely*. The rule ES provides us with a more limited course of action we can take when the uniqueness condition is lacking.

(2) Mathematicians often seem to violate the restrictions on ES. They go from $\exists x\ P(x)$ to $P(x)$, using the same letter (apparently a variable and not a new symbol) that appeared in the quantifier when they apply ES. However, if you carefully examine these proofs, you see that after they apply ES in this way, they treat x as though it was a constant, not a variable. This avoids any danger of faulty logic, but it takes some

experience to do it this way and keep things straight. Until you are very familiar with doing proofs, I suggest that you always use a new letter whenever you apply ES.

(3) By now you have perhaps figured out the terminology being used for these quantifier proof methods. Specification (also called **instantiation**) means *using* a known quantified statement to assert a statement with the quantifier *removed*. Generalization means *proving* a quantified statement from a known *unquantified* statement.

Example 8: Refer to Proof Preview 1 in Section 2.2. The rigorous definition of "n is even" is $\exists m\ (n = 2m)$, where m also stands for an integer. So a more rigorous version of that proof would have used ES to go from this quantified statement to the unquantified statement $n = 2m$ or, more correctly, $n = 2c$ (see Exercise 2).

Many important theorems are so-called **existence theorems**, which means that they say that something exists without telling you how to find it. The rule ES is the main tool for making use of an existence theorem.

Example 9: An existence theorem that is extremely important in calculus is the **mean value theorem for derivatives.** It says that if a function f is continuous on the closed interval $[a, b]$ and differentiable on the open interval (a, b), then there is a number x strictly between a and b such that $f'(x) = [f(b) - f(a)]/(b - a)$. Let's see how this might be applied in practice. First of all, the unquantified variables a, b, and f are understood to be universally quantified. Therefore, by applying US, we can give them any particular values we want. For instance, we could let $a = 0$, $b = \pi$, and $f(x) = x^2 + \cos x$. (Note how a function is specified by giving a defining equation for it, rather than a numerical value.) This function is differentiable (and therefore continuous) on the whole real line. So by modus ponens, we get $\exists x\ (0 < x < \pi \text{ and } f'(x) = \pi - 2/\pi)$. By taking the derivative of $f(x)$, we obtain $\exists x\ (0 < x < \pi \text{ and } 2x - \sin x = \pi - 2/\pi)$.

If you try to solve this equation for x, you quickly run out of things to do; it's essentially impossible to solve. (One could use Newton's method or some other approximation technique to compute a solution to as many decimal places as desired, but that is not our purpose here.) So the only way to eliminate the existential quantifier is to use ES; we can say "Let c be a number such that $0 < c < \pi$ and $2c - \sin c = \pi - 2/\pi$." (The use of ES is usually accompanied by the word "let" in this manner.) Then, even though we don't know the exact value of c, we have a convenient temporary symbol that denotes a solution to this equation between 0 and π.

Example 10: Here is an incorrect use of ES. Suppose we have a step of the form $\exists x\ P(x)$. Consider this proof:

(1) $\exists x\ P(x)$	[Proved somehow]
(2) $P(x)$	ES on step 1
(3) $\forall x\ P(x)$	UG on step 2

Of course, the mistake here is that the letter x in step 2 gets treated as a variable in step 3, but after the use of ES it has to be treated as a constant. Note that if the above proof were correct, it would (by conditional proof) yield an absurd conclusion: that if there's one object with a certain property, then all objects have that property.

Example 11: Here is another, more subtle, error to watch out for when using ES. Suppose that we have a step of the form $\exists x\,(P(x) \wedge Q(x))$. Then we can certainly apply ES to get $P(c) \wedge Q(c)$. But suppose instead that we have the step $\exists x\, P(x) \wedge \exists x\, Q(x)$. Can we still apply ES to obtain $P(c) \wedge Q(c)$? No, this is not allowed! The rule ES can only be applied to one quantifier at a time. To eliminate the quantifiers from this latter statement, we must break it up as follows:

(1)	$\exists x\, P(x) \wedge \exists x\, Q(x)$	[Proved somehow]
(2)	$\exists x\, P(x)$	From step 1
(3)	$P(c)$	ES on step 2
(4)	$\exists x\, Q(x)$	From step 1
(5)	$Q(b)$	ES on step 4
(6)	$P(c) \wedge Q(b)$	Conjunction on steps 3 and 5

Note that we can't write $Q(c)$ at step 5 because c would not be a new constant symbol. In fact, there is no way to prove $P(c) \wedge Q(c)$ from step 1, because step 1 does not say that there is a single object that satisfies *both* P and Q simultaneously (even though the same variable x is used in both conjuncts of step 1). To make this more concrete, let $P(x)$ be "$x > 0$" and let $Q(x)$ be "$x < 0$." Then step 1 of the above argument is true in the real number system, but $c > 0 \wedge c < 0$ cannot be true.

We have one more important quantifier proof method to discuss.

Axiom: Existential Generalization or EG

$$P(t) \rightarrow \exists x\, P(x)$$

where t is a term with the same restrictions as in the rule US.

It is no accident that US and EG both mention a term t with exactly the same restrictions. In fact, either of them can be derived from the other (see Theorem 4.3 and Exercise 16).

Since EG is an implication, it can be combined with modus ponens to form a new rule of inference. In this form, it is *by far* the most natural and common way to prove an existential statement. A more down-to-earth name for this proof method would be

proof by example, since what it says is that if you want to prove that something exists, it suffices to find one actual example of whatever it is. For instance, if you want to show there's a real number with a certain property, the cleanest way by far is to find a specific number with that property.

However, recall Theorem 3.1, which said that existential quantifiers normally describe functions of the universally quantified variables to the outside of them. For this reason, the example you find in a proof by example won't necessarily be a constant; it might have to depend on some variables. That is why EG refers to a term or expression t, rather than a constant. In a sense, EG is just a restatement of Theorem 3.1.

Example 12: Proof Preview 4 (in Section 3.3) provides a typical example of the use of EG. We want to prove $\forall x, y\ \exists z\ (z > x \wedge z > y)$, with all three variables being real variables. We could begin the proof with the words "Let x and y be given," reflecting the fact that we can ignore the universal quantifiers, by UG. Then we need to prove the statement $\exists z\ (z > x \wedge z > y)$. To prove this by EG, we must find a term or expression that works when substituted for the variable z, and this term may involve x and y. As we showed in Proof Preview 4, the term $z = |x| + |y| + 1$ works. In standard mathematical writing, the second sentence of this proof would probably be simply "Let $z = |x| + |y| + 1$." (In practice mathematicians don't use the letter t to represent a term in such proofs, any more than they use the letter P to represent a statement.)

Another way to prove this statement is by cases (see Exercise 4).

Example 13: Let's prove, assuming basic results from calculus, that given any number, there's a real-valued function (other than the zero function) whose derivative is that (constant) number times itself. The statement to be proved has the form $\forall k\ \exists f\ (f \neq 0 \wedge f' = kf)$. Note that, in terms of the logical structure of this statement, both k and f are mathematical variables, even though k has been called a constant (because it's not a variable of the function f), and f represents a function, not a number. So let k be any real number. By EG, we just need to find a nonzero function $f(x)$, which can depend on k, with the desired property. Perhaps you've already figured out that we should let $f(x) = e^{kx}$. It can then be easily be verified by differentiation that $f'(x) = kf(x)$.

Most of the time, when one uses EG or Theorem 3.1, the term found for the existentially quantified variable must involve all the universally quantified variables to the left of it. So, in Example 12, it's not possible to find an expression for z that does not involve both x and y. Similarly, in Example 13, the expression we find for f must involve k. But it's permissible to omit a variable, as the next example shows.

Example 14: Suppose we want to prove $\forall x\ \exists y\ (x + y = x)$, where x and y are real variables. The simplest expression that works for y in the equation $x + y = x$ is 0. This certainly satisfies the requirement of Theorem 3.1. So, assuming that we know that $x + 0 = x$, we can consider this result proved, by EG and UG. The fact that we could find an expression for y that does not involve x tells us that we could just as easily prove the stronger (that is, better) result $\exists y\ \forall x\ (x + y = x)$.

Counterexamples

The *words* (as opposed to the methods) "existential generalization" and "proof by example" are rarely used by mathematicians. In contrast, mathematicians frequently talk about **counterexamples,** primarily as a method of *disproving* statements. This method is just a special case of EG, but it is used so often that it deserves separate discussion.

One standard type of mathematics problem asks the reader to prove or disprove some statement. This often involves more work than a problem that just asks the reader to prove a statement: first you have to determine (or at least guess) whether the statement is true or false; then you must prove the statement or its negation.

Now, imagine that you are asked to prove or disprove a statement of the form $\forall x\, P(x)$. If you think the statement is true, you probably try to prove it by UG. But if you think it's false, how do you disprove it? Well, *disproving* $\forall x\, P(x)$ means *proving* $\sim \forall x\, P(x)$, which is equivalent to $\exists x \sim P(x)$. And by EG, we know we can prove this if we can find a term t such that $\sim P(t)$ holds. That is, we want to find a specific example of an object for which P is false. Such an example is called a **counterexample** to the statement $\forall x\, P(x)$.

When a statement of the form $\forall x\, P(x)$ is involved, the words "prove or find a counterexample" are more commonly used than "prove or disprove."

Example 15: Suppose we are asked to prove or disprove that $n^2 - n + 41$ is prime for every nonnegative integer n (recall Exercise 5 of Section 1.3). If this were true, it might be very difficult to prove. But it's not true, and all it takes to show this is a single counterexample. Interestingly, $n^2 - n + 41$ is prime for all integers from 0 to 40, but 41 is obviously (in retrospect, anyway) a counterexample.

Example 16: Chapter 1 discussed Goldbach's conjecture and de Polignac's conjecture. Both of these are almost certainly true, but neither has been proved. What would it take to disprove these conjectures? A counterexample to Goldbach's conjecture would be a positive even number (greater than 2) that is not the sum of any two prime numbers. Since there are only a *finite* number of ways to express a given integer as a sum of two positive integers, a counterexample to Goldbach's conjecture, if it exists, could be identified by simple arithmetic (but lots of it! A powerful computer would probably be required).

Similarly, a counterexample to de Polignac's conjecture would be an even number that is not the *difference* of any two prime numbers. But there are an infinite number of ways to express a given integer as a difference of two integers, and there are also an infinite number of primes. So there would be no way to verify that a given number was a counterexample merely by arithmetic computation. For instance, suppose that you believed that 6 was not the difference between any two prime numbers. How would you attempt to verify this, even with a powerful computer?

We conclude the main part of this section with two tables that you should look through now and refer to as needed in the future. Table 4.2 contains a list of some laws of logic, analogous to the list of tautologies in Appendix 3. All these are provable from

Table 4.2 Some useful laws of logic

In the following statements, it is always assumed that the proposition R does not contain the variable x. The propositions P and Q may be assumed to contain, as free variables, the variables in the quantifiers that precede them. The restrictions on the term t are as described previously in the definition of universal specification.

(1) $\forall x\, P \rightarrow \exists x\, P$
(2) $\forall x\, \forall y\, P \leftrightarrow \forall y\, \forall x\, P$
(3) $\exists x\, \exists y\, P \leftrightarrow \exists y\, \exists x\, P$
(4) $\exists x\, \forall y\, P \rightarrow \forall y\, \exists x\, P$
(5) $\forall x\, R \leftrightarrow R$
(6) $\exists x\, R \leftrightarrow R$
(7) $\forall x\, P(x) \rightarrow P(t)$ (Universal specification)
(8) $P(t) \rightarrow \exists x\, P(x)$ (Existential generalization)

De Morgan's laws for quantifiers

(9) $\sim \forall x\, P \leftrightarrow \exists x \sim P$
(10) $\sim \exists x\, P \leftrightarrow \forall x \sim P$

Replacement of one quantifier by the other

(11) $\forall x\, P \leftrightarrow \sim \exists x \sim P$
(12) $\exists x\, P \leftrightarrow \sim \forall x \sim P$

Distributing quantifiers over connectives

(13) $\forall x\, (P \wedge Q) \leftrightarrow (\forall x\, P \wedge \forall x\, Q)$
(14) $\exists x\, (P \vee Q) \leftrightarrow (\exists x\, P \vee \exists x\, Q)$
(15) $(\forall x\, P \vee \forall x\, Q) \rightarrow \forall x\, (P \vee Q)$
(16) $\exists x\, (P \wedge Q) \rightarrow (\exists x\, P \wedge \exists x\, Q)$
(17) $\forall x\, (P \vee R) \leftrightarrow (\forall x\, P \vee R)$
(18) $\exists x\, (P \wedge R) \leftrightarrow (\exists x\, P \wedge R)$
(19) $\exists x\, (P \rightarrow Q) \leftrightarrow (\forall x\, P \rightarrow \exists x\, Q)$
(20) $\forall x\, (R \rightarrow Q) \leftrightarrow (R \rightarrow \forall x\, Q)$
(21) $\forall x\, (P \rightarrow R) \leftrightarrow (\exists x\, P \rightarrow R)$

Table 4.3 Summary of quantifier proof methods

Statement	Ways to Prove	Ways to Use
A universally quantified statement $\forall x\, P(x)$	UG: Prove $P(x)$, for an arbitrary x. Indirectly: Assume $\exists x \sim P(x)$ and derive a contradiction.	US: Assert P(t), for any appropriate term t.
An existentially quantified statement $\exists x\, P(x)$	EG: Prove P(t), for some appropriate term t. Indirectly: Assume $\forall x \sim P(x)$, and derive a contradiction.	ES: Assert $P(c)$ for a new constant c.

our axiom system, and the exercises ask you to prove a few of them. Better yet, look through the list yourself, pick a couple of statements in the list that look the *least* obvious to you, and see if you can prove them.

Even though most of the laws of logic in Table 4.2 are not included as axioms in our system, you should feel free to consider them as known results or theorems, unless, of course, you're asked to do a proof using only what's in the axiom system. Pay particular attention to the laws in Table 4.2 that involve the propositional variable R. It is specified that, in these laws, R does not contain x as a free variable, and this restriction allows steps that otherwise would not be correct.

Example 17: An important feature of the real number system is that between any two distinct real numbers there's another one. Suppose we want to prove this. The statement can be symbolized, a bit loosely, as $\forall x,y\ (x < y \rightarrow \exists z\ (x < z < y))$.

It is not difficult to prove this statement as it stands, using methods of proof from this section and the previous one. But a slightly different approach is to note that the variable z does not occur in the inequality $x < y$. This means we can use laws 5 and 19 of Table 4.2 to rewrite $(x < y \rightarrow \exists z\ (x < z < y))$ in the equivalent form $\exists z\ (x < y \rightarrow x < z < y)$. So the statement we want to prove becomes $\forall x,y\ \exists z\ (x < y \rightarrow x < z < y)$. The point is that this revised statement fits the conditions needed to apply Theorem 3.1, which provides a straightforward way of completing the proof (see Exercise 10).

Table 4.3 is a quantifier version of Table 4.1 in the previous section. Since there are only two quantifiers as compared to five connectives, Table 4.3 is shorter and simpler than Table 4.1. Probably the most important thing to get from Table 4.3 is what the roles of the four principles US, UG, ES, and EG are.

Some Theorems Involving Quantifiers (Optional Material)

We now prove some simple theorems or laws of logic, all of which appear in Table 4.2. The first two provide the derivations of the two asterisked quantifier axioms, and the next two prove extremely simple laws of logic. Theorem 4.6 is probably the most useful of these results, since it clarifies some of the subtleties involved when quantifiers are combined with connectives. The fact that the content of these theorems is pure logic gives their proofs a rather artificial, nonmathematical flavor.

Theorem 4.2: Quantifier axiom II-3, $\sim \exists x\, P(x) \leftrightarrow \forall x \sim P(x)$, is derivable from the nonasterisked portion of our axiom system.

Proof: For the forward direction, assume $\sim \exists x\, P(x)$. Using tautology 23 (and substitution), we can replace $P(x)$ by $\sim\sim P(x)$. This gives us the step $\sim \exists x \sim\sim P(x)$. But this step contains a substatement of the form $\exists x \sim \ldots$. Therefore, we can apply axiom II-2, plus substitution, to yield the step $\sim\sim \forall x \sim P(x)$. Applying tautology 23 again lets us delete the double not in front of this statement and gives us $\forall x \sim P(x)$. This completes the conditional proof of the implication $\sim \exists x\, P(x) \rightarrow \forall x \sim P(x)$.

For the reverse direction, assume $\forall x \sim P(x)$. Again applying tautology 23 plus substitution, we get $\sim\sim \forall x \sim P(x)$. Then by applying axiom II-2 plus substitution to the substatement $\sim \forall x \ldots$, we get the step $\sim \exists x \sim\sim P(x)$. With tautology 23 plus substitution used one final time, we obtain $\sim \exists x\, P(x)$. Therefore, by conditional proof, the reverse implication is also established. By the biconditional rule, we conclude $\sim \exists x\, P(x) \leftrightarrow \forall x \sim P(x)$. (See Exercise 15 for the converse of this theorem.) ■

Theorem 4.3: Existential Generalization, $P(t) \rightarrow \exists x\, P(x)$, is derivable from the nonasterisked portion of our axiom system.

Proof: We prove the contrapositive of the desired statement. So assume $\sim \exists x\, P(x)$. By Theorem 4.2, this gives us $\forall x \sim P(x)$. By US, this implies $\sim P(t)$, which is what we want. (See Exercise 16 for the converse result.) ■

Theorem 4.4: For any statement $P(x)$,

(a) $\forall x\, P(x) \leftrightarrow \sim \exists x \sim P(x)$

(b) $\exists x\, P(x) \leftrightarrow \sim \forall x \sim P(x)$

Proof: These statements are just the result of negating both sides of De Morgan's Laws for quantifiers (see Exercise 11). ■

Theorem 4.5: (a) For any statement $P(x)$, $\forall x\, P(x) \rightarrow \exists x\, P(x)$.

(b) For any statement $P(x,y)$, $\exists x\, \forall y\, P(x,y) \rightarrow \forall y\, \exists x\, P(x,y)$.

Proof: (a) Assume $\forall x\, P(x)$. By US, $P(x)$. Then by EG, $\exists x\, P(x)$. By conditional proof, we are done.

(b) Assume $\exists x\, \forall y\, P(x,y)$. Then by ES, let c be an object satisfying $\forall y\, P(c,y)$. To prove the right side of the implication, let y be given. By US applied to $\forall y\, P(c,y)$, we have $P(c,y)$. Then we can use EG to get $\exists x\, P(x,y)$. Finally, since y was arbitrary, we can apply UG to get $\forall y\, \exists x\, P(x,y)$. ■

You might get more out of the proof of Theorem 4.5(b) if you try to use similar logic to prove its converse. Recall from Section 3.3 that the converse of this statement is not true in general, and so of course it can't be proved.

Example 18: In the real number system, $\exists x\, \forall y\, (x + y = y)$ is true, since the constant value 0 works for x. So, by Theorem 4.5(b), $\forall y\, \exists x\, (x + y = y)$ must also be true.

In the real number system, the statement $\forall x\, \exists y\, (x + y = 0)$ is also true, since for any x we can let $y = -x$. But since this value of y depends on x and there is no constant value of y that works for all x, the statement $\exists y\, \forall x\, (x + y = 0)$ is false.

Theorem 4.6: For any statements $P(x)$ and $Q(x)$:

(a) $\forall x\, P(x) \land \forall x\, Q(x) \leftrightarrow \forall x\, [P(x) \land Q(x)]$

(b) $\forall x\, P(x) \lor \forall x\, Q(x) \rightarrow \forall x\, [P(x) \lor Q(x)]$

(c) $\exists x\, P(x) \lor \exists x\, Q(x) \leftrightarrow \exists x\, [P(x) \lor Q(x)]$

(d) $\exists x\, [P(x) \land Q(x)] \rightarrow \exists x\, P(x) \land \exists x\, Q(x)$

Proof: (a) See Exercise 12.

(b) Assume the left side. Since this is a disjunction, we can do a proof by cases. Case 1: Assume $\forall x\, P(x)$. Let x be given. By US, we have $P(x)$. By propositional logic, we then get $P(x) \land Q(x)$. Finally, UG yields $\forall x\, [P(x) \land Q(x)]$. Case 2: Assume $\forall x\, Q(x)$. The argument for this case is almost identical to the other case and so we omit it. (It is common practice to say something like this instead of practically repeating a proof.)

(c) See Exercise 13.

(d) Assume the left side. Using ES, we can write $P(c) \land Q(c)$. By propositional logic, we then get $P(c)$. Then EG yields $\exists x\, P(x)$. An almost identical argument proves $\exists x\, Q(x)$. Finally, the conjunction rule yields the desired statement. ■

Note that Theorem 4.6(b) and (d) are just conditionals, not biconditionals. Exercise 14 asks you to show that the converses of these conditionals are not valid.

Exercises 4.3

(1) The rigorous definition of $A \subseteq B$ is $\forall x\, (x \in A \rightarrow x \in B)$. Using this definition, write a more correct version of the proof in Proof Preview 2 (Section 2.3), dealing properly with the quantification of x as well as that of A, B, and C.

(2) Write a more rigorous version of the proof in Proof Preview 1 (Section 2.2). Use the definition of "n is even" given in Example 8 of this section, and write a similar definition of "n is odd." Because of the existential quantifiers involved, you need to use ES and EG in your proof.

(3) Carefully prove the argument given in Example 1 of Section 3.2.

(4) Prove the result discussed in Example 12, using cases that are based on the relationship between x and y.

(5) In this exercise, n and k denote integers.

(a) Write a symbolic statement that captures the meaning of "n is divisible by k," or, equivalently, "n is a multiple of k." Your solution should include a quantifier and should mention multiplication rather than division or fractions.

(b) Prove that if n is a multiple of k, then so is n^2. Make sure to treat quantifiers rigorously.

(c) Prove that if n is one greater than a multiple of k, then so is n^2. (Same warning as in part (b).)

(d) Prove or find a counterexample to the converse of part (b).

(6) Redo the argument of Example 4 of the previous section, carefully including all the implicit quantifiers and the reasoning needed to deal with them.

(7) Give a more rigorous, axiomatic proof of Theorem 3.4 than was given on page 68.

*(8) Make up at least three symbolic statements, *not* already in Table 4.2 and *not* tautologies, that you believe are laws of logic. Then prove them from our axiom system.

(9) Prove the following laws from Table 4.2: 2, 19, and 20. Do not make these informal proofs; rather, make them axiomatic and quite formal. You may use any theorems in your proofs, but you may not use any of the laws in Table 4.2.

(10) (a) Prove the result discussed in Example 17 by the first method outlined in that example (without directly referring to any of the laws in Table 4.2).

(b) Prove this result by the second method outlined in Example 17 (using laws 5 and 19 of Table 4.2, and Theorem 3.1). Don't try to make this proof very rigorous.

Exercises 11 through 16 pertain to the optional material at the end of this section.

(11) Prove Theorem 4.4.

(12) Prove Theorem 4.6(a).

(13) Prove Theorem 4.6(c).

(14) Give examples to show that the converses of Theorem 4.6(b) and (d) are *not* laws of logic. An example in this situation should consist of a specific domain for the variable x and specific statements for P and Q.

(15) Prove the converse of Theorem 4.2. In other words, derive axiom II-2 from the rest of the axiom system, including axiom II-3.

(16) Prove the converse of Theorem 4.3. In other words, derive axiom II-1 from the rest of the axiom system, including axiom II-4.

Critique the following proofs. (If necessary, review the instructions for such problems in Exercises 4.2.)

(17) **Theorem:** In the real number system, $\exists x\, \forall y\, (y + x = 3)$.

Proof: Let y be any real number. We know from basic algebra that $y + (3 - y) = 3$, and by UG we can conclude that $\forall y\, (y + (3 - y) = 3)$. Now, the expression $3 - y$ certainly denotes a real number, so we can apply EG to obtain $\exists x\, \forall y\, (y + x = 3)$. (Technically, we are applying EG with P(x) being $\forall y\, (y + x = 3)$ and the term t being $3 - y$.)

(18) The purpose of this problem is to illustrate the quantifier-related restriction in the definition of the substitution rule of inference.

Theorem: If x is a positive real number, then all real numbers are positive.

Proof: Assume $x > 0$. We also know that $1 > 0$, and therefore $x > 0$ iff $1 > 0$. By substitution, we can do the same thing to both sides of this equivalence; in particular, we can say $\forall x\, (x > 0)$ iff $\forall x\, (1 > 0)$. But the inequality $1 > 0$ is true for every value of x, so $\forall x\, (1 > 0)$ is true. Therefore, we can conclude $\forall x\, (x > 0)$.

4.4 The Use of Equations in Proofs

The last logic-based part of our axiom system consists of the equality axioms. These were discussed briefly in Section 3.4. Now we examine them, and their use in proofs, in detail. Axioms III-1 and III-2 (the reflexive and symmetric properties of equality) are so simple that they are not often used overtly in proofs, and there isn't a whole lot to say about them. Here is a small point involving symmetry. Remember that every equation in mathematics is a two-way street (the same phrase that we applied to biconditionals) and that even if a particular equation is usually applied in one direction, it must be allowable to apply it in the other direction as well.

Example 1: An equation that is frequently used in both directions is the distributive law for real numbers: $x(y + z) = xy + xz$. When you see an algebraic law like this, it's natural to think that the equation is usually used to change expressions that look like the left side into expressions that look like the right side. In this direction, the distributive law provides the main rule for multiplying out algebraic expressions. But the distributive law is equally important when used from right to left, for in that direction it provides the main rule for *factoring* algebraic expressions.

The transitive property of equality, axiom III-3, is more substantial than the previous two axioms. It is this axiom that allows you to write an extended equation and then conclude that the first expression equals the last expression. This technique is used in proofs as well as in problem solving, as the following example illustrates.

Example 2: Suppose you want to factor the expression $x^3 - 6a^2 - 2ax^2 + 3ax$, and you decide to try factoring by grouping. Then you might write

$$\begin{aligned} x^3 - 6a^2 - 2ax^2 + 3ax &= x^3 - 2ax^2 + 3ax - 6a^2 \\ &= x^2(x - 2a) + 3a(x - 2a) \\ &= (x^2 + 3a)(x - 2a) \end{aligned}$$

Of course, when you write this, you mean to say that the final, factored expression is equal to the original expression, and almost anyone reading it would interpret it in that way. But how can this conclusion be proved rigorously? Let's take a look.

The extended equation above has the structure $A = B = C = D$, where A, B, C, and D are particular expressions. To turn this into a formal proof, we would first need to prove the three separate equations $A = B$, $B = C$, and $C = D$. For now, we won't concern ourselves with how to prove these separate equations (see Exercise 3). From the two separate equations $A = B$ and $B = C$, we can use transitivity (plus propositional logic) to get $A = C$. From $A = C$ and $C = D$, we can then prove $A = D$ in a similar manner.

By the way, this example illustrates another point. Many people think of doing proofs in mathematics as a very different (and much harder) activity from ordinary problem solving. But it's really a false distinction. Every time you solve a problem in mathematics, you must have some justification or rationale for your steps; this means that you must have some sort of proof for your solution. Realizing this may help you to see that writing proofs is not such a strange activity.

To illustrate this, suppose you were given the task to find the value of such-and-such. You would probably view this as a problem, as opposed to a proof, and would use whatever means you could think of to simplify or evaluate the given expression. But suppose instead that you were asked to prove that such-and-such equals 2. You would probably think of this as a proof, and thus harder than an ordinary problem. The irony of this is that the second version should be easier than the first, because it tells you what the answer is.

Example 3: The transitive property of equality is used in just about every proof in Appendix 2. A typical example occurs in Theorem A-5. Step 2 of the formal proof is the equation $x(x + 0) = x{\cdot}x$, and step 3 says that $x(x + 0) = x{\cdot}x + x{\cdot}0$. By transitivity of equality (plus axiom III-2 and a bit of propositional logic), this enables us to assert that $x{\cdot}x + x{\cdot}0 = x{\cdot}x$.

The last equality axiom, substitution of equals, has already been discussed a bit. Very simply, if two things are equal, then they are *interchangeable*. Remember that this axiom and the propositional rule of inference substitution (described in Section 4.2) are based on a similar idea, but they are grammatically very different.

Example 4: Suppose we start with the known fact that $\sin x \leq 1$, for any number x. Then, by using US, we can derive $\sin (x + y) \leq 1$. Now, we also have the trigonometric formula

$$\sin (x + y) = \sin x \cos y + \cos x \sin y$$

Then we can use axiom III-4 to replace the expression $\sin (x + y)$ with the right side of this equation, to obtain

$$\sin x \cos y + \cos x \sin y \leq 1$$

Since x and y were arbitrary in this derivation, this inequality can be concluded for all real numbers x and y.

The following theorem often gives us a more usable form of axiom III-4.

Theorem 4.7: Let $\mathrm{t}(x)$ be an expression or term containing the free variable x, and let $\mathrm{t}(y)$ denote the same term with *some or all* of the occurrences of x replaced with the variable y. Then $x = y$ implies $\mathrm{t}(x) = \mathrm{t}(y)$.

Proof: See Exercise 4. ■

A term or expression in mathematics may be undefined, in the sense that it can't denote an actual value or object. The most common examples of this are fractions with denominator 0. The implication in Theorem 4.7 makes no sense if the terms $\mathrm{t}(x)$ and $\mathrm{t}(y)$ are undefined (unless we want to say that two undefined things equal each other, which is a consistent viewpoint but best avoided since it can create confusion). Therefore, the usual convention is that *Theorem 4.7 applies only when the expressions on the right side of the implication are defined*. The significance of this restriction becomes apparent in Example 11 at the end of this section.

Doing the Same Thing to Both Sides of an Equation

As it's stated, it's hard to see just how powerful Theorem 4.7 is. But in fact it is the basis of the fundamental rule that you can do just about anything to both sides of an equation, provided that you do precisely the same thing to both sides. To see how Theorem 4.7 says this, remember that the variables x and y in it are universally quantified. Therefore, using US, they can be replaced by *any expressions whatsoever*. So the practical significance of Theorem 4.7 is that whenever we know any equation, we can conclude any other equation in which the two sides of the original equation appear in the same way within the two sides of the new equation. The next few examples illustrate how this works.

Example 5: Let $\mathrm{t}(x)$ be the expression "$x + z$," so that $\mathrm{t}(y)$ is the expression "$y + z$." Then the theorem for this t reads: $x = y \rightarrow x + z = y + z$. Since x, y, and z are all understood to be universally quantified, this implication holds with any three expressions in their place. In other words, an immediate corollary of Theorem 4.7 is that *you can add any expression you want to both sides of an equation*. If the original equation is true, the new one must be also. Similar reasoning shows that you can subtract, multiply, or divide both sides of an equation by the same thing.

Example 6: Let $\mathrm{t}(x)$ be the expression x^2. Then Theorem 4.7, with this particular t, becomes $x = y \rightarrow x^2 = y^2$. In other words, you can square both sides of an equation.

Example 7: Now let's use the variables A, B, and C to stand for sets. We can let $\mathrm{t}(A)$ be the expression $A \cup C$, and $\mathrm{t}(B)$ be the expression $B \cup C$. Theorem 4.7 then yields the implication $A = B \to A \cup C = B \cup C$. In words, you can form the union of both sides of an equation between sets, with the same set.

Example 8: There are places in Appendix 2 where axiom III-4 is stated as a reason for a step, but it might be more to the point to use Theorem 4.7 as the justification. For instance, step 2 of the formal proof of Theorem A-5 is obtained by multiplying both sides of step 1 by x.

Reversibility

Our discussion so far has indicated that you can do anything you want to both sides of an equation, but there are some subtleties involved with this rule. You may remember these subtleties from precalculus algebra. They have to do with reversibility of steps used in solving or simplifying an equation.

Not all steps allowed by Theorem 4.7 are reversible; sometimes that matters, and sometimes it doesn't. Let's clarify this with some examples.

Example 9: Suppose we want to solve the equation $x + 2 = 8 - 2x$. Our solution might look like this:

$x + 2 = 8 - 2x$	
$2x + x + 2 = 8$	Adding $2x$ to both sides
$3x + 2 = 8$	Combining terms
$3x = 8 - 2 = 6$	Subtracting 2 from both sides
$x = 6/3 = 2$	Dividing both sides by 3

So we would say that $x = 2$ is the solution of the equation. But if we view the above solution as a sort of proof, then what exactly have we proved? Have we proved that if a number x satisfies the given equation, then $x = 2$? Or have we proved the converse, that if $x = 2$, then x satisfies the given equation? Or have we proved both directions, that is, a biconditional: x satisfies the equation if and only if $x = 2$?

Whenever you solve an equation, your solution should establish a biconditional if possible. In other words, to solve an equation (in one variable) means to find a set of numbers (the **solution set**) such that any number in that set satisfies the equation, *and* no other numbers do. In the above example, we're not just saying that 2 works in the equation; we're also saying that no others work.

Thus, when solving or simplifying an equation, in a sense more care is required than when doing an ordinary proof: you have to make sure your steps are *reversible*. In the above example, the steps are definitely reversible: adding something to both sides can be reversed by subtracting that thing from both sides, dividing both sides by 3 can be reversed by tripling both sides, and so on. So the solution shown establishes a biconditional, as it should.

What sorts of steps used to solve equations would not be reversible? Two standard ones are squaring both sides and multiplying both sides by an expression that could equal zero, as the next two examples illustrate.

Example 10: Suppose we want to solve the equation $\sqrt{x+3} + x = 3$. The obvious steps to solve this equation are as follows:

$\sqrt{x+3} + x = 3$	
$\sqrt{x+3} = 3 - x$	Subtracting x from both sides
$x + 3 = 9 - 6x + x^2$	Squaring both sides
$0 = x^2 - 7x + 6$	Subtracting $x + 3$ from both sides
$0 = (x - 1)(x - 6)$	Factoring
$x = 1$ or $x = 6$	See Exercise 12

But on checking these numbers in the original equation, we find that $x = 1$ works, but $x = 6$ doesn't. (The value $x = 6$ works if we set $\sqrt{9} = -3$, but that's not how the symbol $\sqrt{\ }$ is normally used.) You might remember that 6 is called an **extraneous solution** to this equation; but why does a "wrong solution" crop up here?

The reason that our solution method to this equation can lead to extraneous solutions is that squaring both sides of an equation is not reversible. It might seem that the reverse of squaring both sides would be taking the square root of both sides; but it's not, because if you start with a negative number, square it, and then take the square root of that number, you don't get the original number back. To put it another way, if the equation $A^2 = B^2$ is true, we can't conclude that $A = B$. We can only conclude that $|A| = |B|$, or $A = \pm B$.

Because one step is not reversible, the above solution to this equation only shows a forward implication, not a biconditional. So if a number satisfies the equation, it must be either 1 or 6. But we can't conclude from the steps shown that 1 and 6 do satisfy the equation. Note that this is not such a terrible situation: it just means we need to check for extraneous solutions. On the other hand, a solution to an equation that represented a reverse implication but not a forward one would be useless, since it would mean that you might not have found all the correct solutions.

For example, if you try to solve the equation $x^2 = 4$ by taking the square root of both sides to yield $x = 2$, you've made a mistake. The correct solution is $x = \pm 2$. In general, while squaring both sides of an equation is OK (provided you remember to check for extraneous solutions), taking the square root of both sides of an equation is a step that can lead to trouble. By the way, none of this complication would occur with cubing or taking the cube root of both sides of an equation, since these operations are truly the reverse (or *inverse*) of each other.

Example 11: Suppose you were asked to solve the equation $(\sin x)/x = 1$. You would probably multiply both sides by x and obtain $\sin x = x$. It can be shown that the only number satisfying this latter equation is 0. But you can't have a denominator of 0, so 0 is an extraneous solution, and the given equation has no solution.

Why does an extraneous solution occur here? The reason is that, in our solution, we multiplied both sides by an expression that turns out to be zero. This step is not reversible, since division by zero is impossible. (Remember the remark after Theorem 4.7, which said that what you do to both sides of an equation must keep both sides defined.) Therefore, a solution that includes multiplying both sides of an equation by an expression that could be zero is only a forward implication, and extraneous solutions can appear.

Like squaring both sides of an equation, multiplying both sides by an expression that could be zero is fine, provided you check for extraneous solutions at the end. But dividing both sides by such an expression is dangerous and should be avoided if possible. For example, it's wrong to try to solve the equation $x^2 = 5x$ by dividing both sides by x. The correct method is to put everything on one side and then factor.

We conclude this section by justifying the fact that axioms III-2 and III-3 are marked with asterisks.

Theorem 4.8: Axioms III-2 and III-3 are superfluous; that is, they can be proved from the rest of the axiom system.

Proof: First let's prove symmetry: assume $x = y$. Applying axiom III-4 with $S(x)$ being the statement $x = x$, we get $x = x \leftrightarrow y = x$. But we know $x = x$ (from axiom III-1 and US), and so $y = x$ follows by propositional logic. Thus we have proved that $x = y$ implies $y = x$. ■

Exercises 4.4

(1) Complete the proof of Theorem 4.8, by proving axiom III-3, transitivity, from the rest of our axiom system.

(2) Prove: if $x = y$ and $u = v$, then $x + u = y + v$.

(3) Prove the three separate equations involved in Example 2. You may use any of the results of Appendix 2, as well as what's in the axiom system.

(4) Prove Theorem 4.7.

The remaining exercises of this section do not pertain specifically to the equality axioms. Rather, they involve the material in Appendix 2 and are intended to give you practice with all the proof methods that have been discussed in this chapter.

(5) Prove Theorem A-2.

(6) Prove Theorem A-4.

(7) Redo the formal proof given for Theorem A-8 in good, nonformal style.

(8) Redo the formal proof given for the forward direction of Theorem A-13 in good, nonformal style.

(9) Identify the omissions in the proof of Theorem A-7, and complete the proof by filling in the gaps.

In Exercises 10 through 17, prove the statement from the field axioms. (When such a statement is made, it is understood that you may also use all rules of inference and all axioms that are based on logic and equality.) You may also use any results in Appendix 2, unless stated otherwise. Also remember that all free variables in these statements are understood to be universally quantified. Consult your instructor if you have questions about style, format, or how much detail to show.

(10) The number 0 has no multiplicative inverse. (First write the statement in symbols.)

(11) $(-1)x = -x$

(12) $xy = 0$ iff ($x = 0$ or $y = 0$)

(13) $x \neq 0$ iff $x^{-1} \neq 0$

(14) If $x \neq 0$ and $y \neq 0$, then $(xy)^{-1} = y^{-1}x^{-1}$

(15) If $x \neq 0$ and $y \neq 0$, then $u/x + v/y = (uy + vx)/xy$

(16) $\forall x, y\, \exists z\, (x + z = y)$

(17) $(x + y)^2 = x^2 + 2xy + y^2$

(18) Replace field axiom V-12 with the statement that $0 = 1$, and prove from this alternate set of axioms that there is only one number.

(19) Prove Theorem A-10(b).

(20) Prove Theorem A-10(c).

(21) Prove the last case ($x < 0$) of Theorem A-11.

(22) Prove Corollary A-12.

(23) Prove the right-to-left implication of Theorem A-13. Your proof may be formal, but it doesn't have to be.

(24) (a) Fill in the details of the proof of Theorem A-15(a). In particular, show that the cases used in that proof really do cover all possible cases.

(b) Prove Theorem A-15(b).

(c) Prove Theorem A-15(c).

*(25) Fill in the missing details in the proof of Theorem A-16.

The instructions for Exercises 26 through 31 are the same as those for Exercises 10 through 17, except that now you may use all the ordered field axioms.

(26) If $x \leq y$ and $y \leq x$, then $x = y$. This is called the **antisymmetry property** of $\leq$.

(27) (a) $x < y$ iff $-x > -y$
(b) $x > 0$ iff $-x < 0$

(28) If $x < y$ and $z < 0$, then $xz > yz$. Note that this is the usual rule about multiplying both sides of an inequality by a negative number.

(29) $1 + 1 \neq 0$.

(30) If $0 < x < y$, then $x^2 < y^2$. ($0 < x < y$ is an abbreviation for $0 < x$ and $x < y$.)

*(31) If $x < y$, then $x^3 < y^3$.

(32) Rewrite axioms V-13 through V-17 and the definitions that follow them, so that the only inequality symbol mentioned in the axioms is $\geq$, and the other three symbols are defined in terms of this one. Of course, make sure all your axioms and definitions are correct.

(33) Critique the following proof. If necessary, review the instructions for such problems in Exercises 4.2.

Theorem: $1 + 1 \neq 0$.

Proof: Assume, on the contrary that $1 + 1 = 0$, that is, $2 = 0$. By Theorem 4.7, we can multiply both sides of this equation by 1/2 to obtain $2(1/2) = 0(1/2)$. The left side equals 1, by axiom V-11; and the right side equals 0, by Theorem A-5. Therefore, $1 = 0$. But this is a contradiction (in conjunction with axiom V-12), so we are done, by indirect proof.

4.5 Mathematical Induction

This section is devoted to a single method of proof, known as the principle of mathematical induction (PMI), or simply induction. Induction is undoubtedly one of the most important proof techniques in mathematics. Mathematical induction is quite different from the axioms and rules of inference described in the previous three sections. It is not based on logic. Technically, it is an axiom for just one particular number system, the natural numbers. This would presumably make it less useful than very general methods like conditional proof, indirect proof, and so on, but natural numbers occur so universally in mathematics that proofs by induction are a vital part of every branch of the subject.

The term **natural numbers** refers to the numbers we all study first in grade school: 1, 2, 3, and so on. These, the positive integers, are the simplest type of number. (Some books, particularly older ones, include 0 as a natural number and use the term **counting numbers** to denote the positive integers.)

Notation: The letter ℕ denotes the set of all natural numbers. For now, we use the letters m, n, k, and j as natural number variables, that is, variables whose domain is ℕ. (Later, we may use one or more of these letters to stand for *any* integer, not necessarily a positive one.)

Neither the notation just introduced nor the paragraph preceding it constitutes a mathematical definition of the set of natural numbers. Therefore, this notation and discussion may not be used to *prove* things about ℕ. The only basis for proving things about the natural numbers is the axioms pertaining to them. For example, to prove even the "obvious" fact that every natural number is positive, induction is required. You can't prove it "by definition" because no definition has been given.

Axioms for the Natural Numbers

Our axioms for ℕ comprise group VI of Appendix 1. There are only three axioms in this group, and the first two are extremely simple. Axiom VI-1 just says that the number 1 is a natural number. Axiom VI-2 says that if you add 1 to any natural number, the result is still a natural number. These two axioms, taken together, can be thought of as describing how ℕ is generated.

The principle of mathematical induction can be stated in two equivalent forms, the set form and the statement form. In the axiom system, we have listed both. Since we have not studied sets yet, this section concentrates on the statement form.

Definition: The **principle of mathematical induction** (statement form) consists of all statements of the form

$$[\mathrm{P}(1) \land \forall n\, (\mathrm{P}(n) \to \mathrm{P}(n+1))] \to \forall n\, \mathrm{P}(n)$$

where $\mathrm{P}(n)$ is any statement containing a free natural number variable n.

The Meaning of Mathematical Induction

Some people learn only the *method* of induction proofs without ever learning the reasoning behind induction. This approach can be successful with straightforward induction proofs, but it falls apart when the problems get more involved. So let's spend some effort now to analyze the content of this principle.

To visualize what induction says, imagine that the natural numbers 1, 2, 3, ... are arranged vertically, in a sort of infinite ladder (see Figure 4.1). Let $\mathrm{P}(n)$ be the statement that it's possible to reach the nth step of the ladder. Then what does induction claim with this $\mathrm{P}(n)$? Well, $\mathrm{P}(1)$ says it's possible to reach the first step, and $\forall n\, (\mathrm{P}(n) \to \mathrm{P}(n+1))$ says that, for any n, if you can reach the nth step, you can also

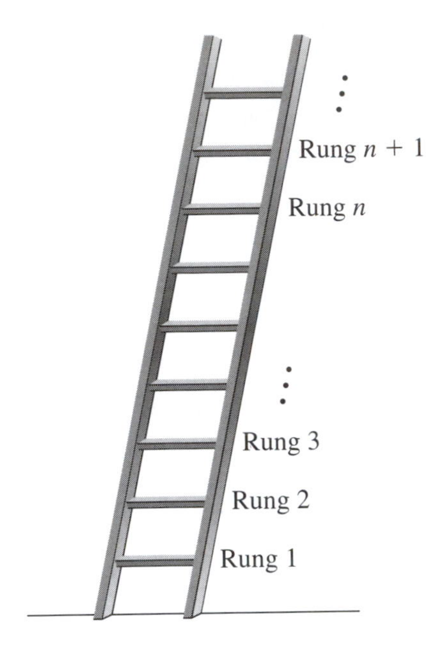

Figure 4.1 Ladder image for mathematical induction

reach the $(n + 1)$-th step. In other words, it says that it is always possible to go one step higher than you already are. So the entire implication can be paraphrased as follows: if you can get to the first step of the ladder and you can always take one more step, then you can go as high as you want.

PMI can also be illustrated nicely with a horizontal image instead of a vertical one. Imagine an infinite row of standing dominos, as in Figure 4.2. Let $P(n)$ be the statement that the nth domino falls. Suppose we know that the first domino will be knocked over and that each domino is close enough to the next one so that, when it falls, the next one will also fall. Then induction, applied to this situation, states that every domino will fall—a fact that even most young children are aware of.

Now let's see what PMI says for an arbitrary statement $P(n)$. The hypothesis of induction says two things: first, it says that P is true for the particular number 1. Second, it says that whenever P is true for a number n, it must also be true for $n + 1$. Induction says that if these two things both hold, then P must be true for every natural number n. To see that this is valid, assume the hypothesis is true, and let's start listing numbers for which P must be true. It must be true for 1, because that's specifically stated. But then, since it's true for 1, it must be true for 2. Then, since it's true for 2, it must be true for 3, and so on. Continuing in this way, we see that P must be true for every value of n.

Remember that mathematical induction is an axiom based on the particular way that the natural numbers are structured. It holds only because $\mathbb{N}$ consists of a single infinite sequence of numbers. For example, there's no such thing as a direct induction proof on the set of all real numbers.

It is instructive to think about the relationship between induction and the other natural number axioms. It is similar to the relationship between modus ponens and conditional proof: induction is essentially the *converse* of axioms VI-1 and VI-2. Axioms VI-1 and VI-2 say that if you start at 1 and count by ones, you stay within $\mathbb{N}$. Induction says that if you start at 1 and count by ones, you eventually reach *every* member of $\mathbb{N}$.

This distinction can be made even clearer by referring to sets. Let A denote the set of all numbers that can be reached by starting at 1 and counting by ones. Then axioms VI-1 and VI-2 together say that A is a subset of $\mathbb{N}$, whereas PMI says that $\mathbb{N}$ is a subset of A. The combined meaning of all the natural number axioms is that $\mathbb{N} = A$. This makes sense, because the way we defined the set A is the most sensible way to describe the natural numbers rigorously.

While we're at it, let's clarify another potential source of confusion. In Chapter 1, we talked about inductive reasoning (or the inductive method), which is the main tool for acquiring knowledge in science. What is the relationship between inductive reasoning and mathematical induction? The answer is simple: *There is no connection between inductive reasoning and mathematical induction*. It's best to view the similarity in wording as an historical accident and leave it at that. When mathematicians use the term "induction," they almost always mean mathematical induction, not inductive reasoning.

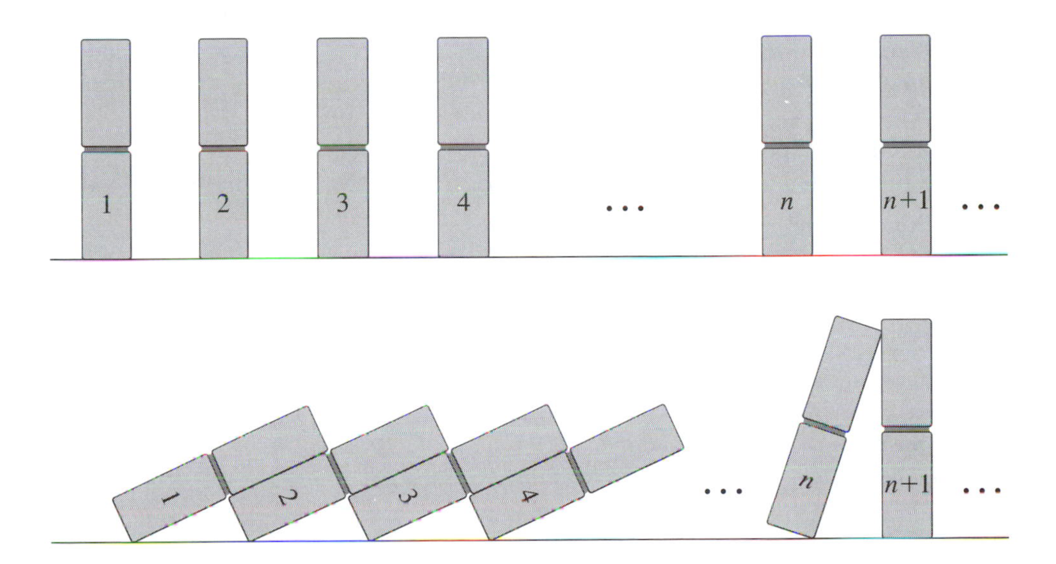

Figure 4.2 Domino image for mathematical induction

The Structure of Proofs by Mathematical Induction

In a proof by induction, the goal is to prove a statement of the form $\forall n\ P(n)$. By PMI and modus ponens, it suffices to prove P(1) and $\forall n\ (P(n) \rightarrow P(n+1))$. So the first part of an induction proof is usually to prove P(1). This is often a very short, obvious step. The second and major part is to prove $\forall n\ (P(n) \rightarrow P(n+1))$; this is called the **induction step** of the proof. As usual, to prove this quantified statement, it suffices to prove the unquantified implication $P(n) \rightarrow P(n+1)$. And this will usually be proved by conditional proof, which means the induction step starts with assuming $P(n)$. (This assumption of $P(n)$ is sometimes called the **induction hypothesis** of the proof). It is then required to prove $P(n+1)$ from this assumption. If these two parts can be done, then the desired statement can be asserted.

People sometimes are surprised by the fact that, in an induction proof, you get to assume $P(n)$, which is very similar to what you are trying to prove. But note that the statement to be proved is $\forall n\ P(n)$, whereas the induction hypothesis is just $P(n)$. It's vital that you *don't* put the quantifier $\forall n$ in your assumption for the induction step. Still, what you get to assume is quite close to what you are trying to prove; there is no other type of proof in mathematics where you can assume something so close to what you are trying to prove. But, as has already been pointed out, mathematical induction is a very special axiom, based on the particular arrangement of $\mathbb{N}$. If you understand the logic behind PMI, then there should be nothing surprising about how induction proofs are structured.

Now let's look at some examples of theorems proved by mathematical induction.

Theorem 4.9: Every natural number is a real number; that is, $\mathbb{N}$ is a subset of $\mathbb{R}$.

Proof: We want to prove the statement $\forall n\ (n \in \mathbb{R})$. *[Remember that n automatically denotes a natural number.]* We prove this by induction, with $P(n)$ being "$n \in \mathbb{R}$."

We first must prove $1 \in \mathbb{R}$. This is implied by axiom V-9.

To prove the induction step, assume $n \in \mathbb{R}$. We already know that $1 \in \mathbb{R}$, and so by axiom V-1 (plus US), $n + 1 \in \mathbb{R}$. ■

Theorem 4.10: The set $\mathbb{N}$ is closed under addition; that is, $\forall m, n\ (m + n \in \mathbb{N})$.

[We've said that induction is used to prove statements of the form $\forall n\ P(n)$. But now we are asked to prove a statement that begins with two universally quantified natural number variables. How should we approach that? Do we have to do a separate induction proof for each of the variables m and n? There are proofs in which it is necessary to do a double induction; but this is a complicated technique that is not required very often. Even with more than one variable present, it is permissible to use induction on just one of them. This simple approach works quite often, and we use it here.]

Proof: Let m be any natural number. We do induction on n only, with $P(n)$ being the statement "$m + n \in \mathbb{N}$." (So m is viewed as *fixed*.)

P(1) is just a special case of axiom VI-2, so it holds.

To prove the induction step, assume $m + n \in \mathbb{N}$. We want to show that $m + (n + 1)$ is also a natural number. But note that m, n, and 1 are all real numbers, by Theorem 4.9. Therefore $m + (n + 1) = (m + n) + 1$ by axiom V-3 (associativity). It follows, by axiom VI-2 (and axiom III-4) that $m + (n + 1)$ is a natural number.

So we have $\forall n\,(m + n \in \mathbb{N})$, by induction. Since m was arbitrary, we can use UG to conclude $\forall m, n\,(m + n \in \mathbb{N})$, as desired. ■

Several of the exercises in this section are related to Theorem 4.10. Exercise 3 asks you to prove that $\mathbb{N}$ is closed under multiplication, a proof very similar to Theorem 4.10's. And Exercise 22 asks you to critique an alternative proof of Theorem 4.10.

In our proof of Theorem 4.10, we used induction on n for any one given value of m; so m was unquantified in the statement $P(n)$. It is possible to take a different approach, in which $P(n)$ is $\forall m\,(m + n \in \mathbb{N})$. Exercise 2 asks you to do this. Usually, it doesn't matter whether $P(n)$ contains universal quantifiers of variables besides n, but occasionally an induction proof becomes much easier if an extra universal quantifier is included in $P(n)$.

Mathematical induction is the main tool used to prove formulas for sums and products of sequences of numbers. We turn now to several examples of this. A rigorous treatment of this material requires the use of functions. For now we instead use the familiar ellipsis notation (three dots), which most mathematicians consider rigorous enough.

Theorem 4.11: $1 + 2 + 3 + \ldots + n = n(n + 1)/2$ (that is, $\sum_{i=1}^{n} i = n(n + 1)/2$)

Proof: Since the variable n is unquantified in the statement of this theorem, we're supposed to prove it for all n, and we do so by induction.

For $n = 1$, the left side equals 1. *[You have to be sensible about how you read something like 1 + 2 + 3 + ... + 1. The numbers to be added only go up to n, so the 2 and the 3 would not be included here.]* And the right side equals $1(1 + 1)/2$, which also equals 1, so the equation holds.

For the induction step, assume

$$1 + 2 + 3 + \ldots + n = n(n + 1)/2$$

Now add $n + 1$ to both sides, as Theorem 4.7 allows:

$$(1 + 2 + 3 + \ldots + n) + n + 1 = n(n + 1)/2 + n + 1$$

The left side of this is just $1 + 2 + 3 + \ldots + (n + 1)$, while the right side is easily simplified (by a couple of high school algebra steps) to $(n + 1)[(n + 1) + 1]/2$. ■

To prove the next theorem, we need to assume some basic properties of exponents, which are proved in Chapter 7.

Theorem 4.12: $1 + 2 + 4 + \ldots + 2^{n-1} = 2^n - 1$

Proof: We proceed by induction on n. For $n = 1$, the statement says $1 = 1$, so it's true. For the induction step, assume $1 + 2 + 4 + \ldots + 2^{n-1} = 2^n - 1$. Now add 2^n to both sides, and obtain

$$\begin{aligned} 1 + 2 + 4 + \ldots + 2^n &= 2^n - 1 + 2^n \\ &= 2(2^n) - 1 \\ &= 2^{n+1} - 1 \end{aligned}$$

as desired. ■

Theorems 4.11 and 4.12 are particular cases of the formulas for the sum of **arithmetic** and **geometric** series. Here are the general formulas, with the proofs left as exercises.

Theorem 4.13 (arithmetic series formula): For any real numbers a and d and any natural number n,

$$a + (a + d) + (a + 2d) + \ldots + (a + (n - 1)d) = n[2a + (n - 1)d]/2$$

(Note that Theorem 4.11 is this formula with $a = d = 1$.)

Proof: See Exercise 4. ■

Theorem 4.14 (geometric series formula): For any real numbers a and r (provided $r \neq 1$) and any natural number n,

$$a + ar + ar^2 + \ldots + ar^{n-1} = a(1 - r^n)/(1 - r)$$

(Note that Theorem 4.12 is this formula with $a = 1$ and $r = 2$.)

Proof: See Exercise 8. ■

In Theorems 4.13 and 4.14, d stands for "difference" and r stands for "ratio." An arithmetic sequence is one in which the difference between successive terms is constant, whereas a geometric sequence has a constant ratio between successive terms. Note that in both theorems, the formula given is for the sum of n terms of a sequence. (The word "series" always means a sum of terms.)

We turn now to a special case of the **division algorithm**, one of the most important basic results in number theory. We prove the general form of the division algorithm in Section 8.2. (We have occasionally referred to odd and even *integers*. When natural numbers are being discussed, it is best to define n to be even iff it is of the form $2m$ and odd iff it is of the form $2m - 1$, where m must also be a natural number in both cases. Do you see why this is preferable to saying an odd number is of the form $2m + 1$?)

Theorem 4.15: Every natural number is either even or odd.

Proof: By induction on n: for $n = 1$, note that $1 = 2(1) - 1$, so 1 is odd, by EG. Of course, this makes it even or odd. Now assume that n is even or odd. We use proof by cases. Case 1: Assume n is even. That means $n = 2m$, for some $m \in \mathbb{N}$. Then $n + 1 = 2m + 1 = 2(m + 1) - 1$, so $n + 1$ is odd. Case 2: Assume n is odd. That means $n = 2m - 1$, for some $m \in \mathbb{N}$. Then $n + 1 = 2m$, so $n + 1$ is even. ■

Compare this proof to Proof Preview 1 (Section 2.2). Exercise 11 asks you to prove that the "or" in this theorem is actually exclusive.

Theorem 4.16 proves a few important facts about natural numbers. It is tempting to just assume these "obvious" facts without proof, but that would be sloppy mathematical practice.

Theorem 4.16: (a) Every natural number is equal to or greater than 1.

(b) If $n > 1$, then $n - 1$ is in $\mathbb{N}$ (and so $n \geq 2$).

(c) There is no natural number (strictly) between n and $n + 1$.

Proof: (a) and (b): These straightforward inductions are left for Exercise 12.

(c) We proceed by induction on n. For $n = 1$, we must prove that there is no natural number between 1 and 2. But part (b) says that any natural number greater than 1 is at least 2, so we are done.

For the induction step, assume there is no natural number between n and $n + 1$; we must show there is none between $n + 1$ and $n + 2$. We proceed by cases on an arbitrary natural number m. If $m = 1$, then m is not between $n + 1$ and $n + 2$, because part (a) guarantees that $n + 1$ is at least 2. On the other hand, if $m > 1$ and we assume that $n + 1 < m < n + 2$, then $m - 1$ would be strictly between n and $n + 1$, and $m - 1$ is a natural number by part (b). This would violate the induction hypothesis, and so it is impossible. ■

It takes a while to learn just how powerful and versatile mathematical induction is and how often it can be used. As a rule of thumb, when you are asked to prove a statement for all values of a variable and that variable can *only* be a natural number, you should consider using induction. Note that Theorems 4.11 through 4.15 all fall into this category, in that n has to be a positive integer for these statements to make sense. For example, an expression like $1 + 2 + 3 + ... + 8.32$ is meaningless.

But there are also many theorems in mathematics for which it's much harder to see that induction must be used. There are also several variants of mathematical induction, and it is not always obvious that one of these is required. Numerous such situations are encountered later in this book, and it seems appropriate to list some of them here, for future reference:

(1) The well-ordering property of $\mathbb{N}$ (Theorem 5.6) and results whose proofs use this property instead of ordinary induction (for example, Theorems 8.14 and 10.10).

(2) Theorems that seem to be about objects other than natural numbers, such as sets or polynomials, but that can somehow be classified in terms of a natural number

variable, and proved by induction (for example, Theorems 5.8 and 6.1). This important method is discussed further before Theorem 5.8.

(3) Induction proofs that begin at 0 or some other integer, instead of at 1. The induction proofs for the theorems mentioned in item 2 begin at 0. See also Exercises 12 and 13 of Section 5.3.

(4) Definitions by induction (Section 7.4).

(5) Double induction (Exercise 20 of Section 8.2).

(6) Complete induction (Lemma 8.20 and Theorem 8.21).

Mathematical Discovery Revisited

In Section 1.2 we discussed the important ideas of discovery and conjecture in mathematics. It was mentioned that in some situations the discovery process and the proof process are very closely linked, and in other situations they are totally separate. Induction is an excellent example of the latter situation. Because of the way induction proofs must be structured, it is impossible to begin an induction proof without already knowing what it is that you are trying to prove. Therefore, in most situations induction is not helpful as a discovery tool.

With this in mind, it is reasonable to ask what sorts of methods are used to discover new information about natural numbers. This is too complex a question to tackle in depth, but it is worth considering briefly. Note that Theorems 4.11 through 4.14 state formulas for the sums of various series. Mathematicians have many techniques, some of which are viewed as tricks (in a positive sense), for discovering such formulas.

Probably the most famous such trick relates to Theorems 4.11 and 4.13. Supposedly, when Carl Friedrich Gauss was about nine years old, his teacher was annoyed with the class and ordered everyone to add up all the numbers from 1 to 100. While the rest of the students toiled away, Gauss found the answer in a few seconds, without knowing Theorem 4.11. How? Quite simply (but ingeniously), he regrouped the numbers in the series, writing them in pairs as follows:

$$(1 + 100) + (2 + 99) + (3 + 98) + \dots + (49 + 52) + (50 + 51)$$

It then becomes a simple matter to compute the sum of these hundred numbers, and the same trick can be used to prove Theorem 4.13, the formula for the sum of an arbitrary arithmetic series (see Exercise 6). Other exercises in this section guide you through the discovery process for various sum formulas, including Theorem 4.14.

Does it seem to you that Gauss's computation should be considered a proof of Theorem 4.11? Most mathematicians would agree that it is a somewhat informal but essentially correct proof. However, they still usually prefer to write induction proofs for these sorts of formulas, even when the formula has already been derived by manipulating the terms of the series.

Carl Friedrich Gauss (1777–1855) was one of the greatest mathematicians of all time and also one of the most brilliant child prodigies in the history of the subject. One story has it that he found a mistake in his father's ledger book when he was three. Gauss's genius was recognized early; this enabled him to accelerate his education and be sponsored by the Duke of Brunswick.

Gauss's first important mathematical result, at the age of nineteen, was a proof that a regular 17-sided polygon can be constructed with straightedge and compass. This problem had evaded solution for over two thousand years. Five years later he completed his doctoral dissertation, the "Fundamental Theorem of Algebra." During his career he contributed to many branches of mathematics, notably differential geometry, number theory, and probability theory. His name is attached to many important concepts, such as gaussian curvature and the gaussian distribution.

Gauss chose to publish only the most significant of his results and only when they were quite complete, rigorous, and polished. The motto on his seal was "Pauca sed matura" ("Few but ripe"). But he left behind an enormous amount of valuable mathematical writing, in twelve volumes of diaries.

Gauss did important work in fields other than mathematics, notably astronomy and physics. He was one of the founders of the modern theory of electromagnetism, and the standard (metric) unit of magnetic strength bears his name. Many people would name Gauss as the last person to reach the very highest ranks of research in both mathematics and physics.

Exercises 4.5

(1) Evaluate by the formulas given in Theorems 4.13 and 4.14:

(a) $1 + 3 + 5 + 7 + \ldots + 399$

(b) $2 + 5 + 8 + 11 + \ldots + 200$

(c) $1 + 2 + 4 + 8 + \ldots + 1024$

(d) $27 + 9 + 3 + 1 + \ldots + 1/81$

(2) Write an induction proof of Theorem 4.10 in which $P(n)$ is $\forall m\ (m + n \in \mathbb{N})$ instead of just $m + n \in \mathbb{N}$.

(3) Prove that $\mathbb{N}$ is closed under multiplication, that is, $\forall m, n\ (mn \in \mathbb{N})$. You may assume Theorem 4.10.

(4) (a) Prove Theorem 4.13 by induction.

(b) The content of Theorem 4.13 can be expressed simply in words, as "The sum of an arithmetic series equals ... ," where the "..." is an expression built up from the first term of the series, the last term, and the number of terms. Complete this statement.

(5) Restate Theorems 4.12 through 4.14 in sigma notation.

(6) (a) Complete Gauss's derivation of the formula in Theorem 4.11, as discussed at the end of the section.

(b) Generalize part (a) to derive Theorem 4.13. Make sure to include the possibility that the series has an odd number of terms.

(7) (a) Noting that the first positive odd number is 1, and odd numbers differ by 2, find a formula for the nth positive odd number.

(b) Use induction to prove this formula.

(c) Derive a formula for the sum of the first n positive odd numbers.

(8) Prove Theorem 4.14 by induction.

(9) Here is the classic trick that can be used to derive Theorem 4.14 without induction:

(a) Start by writing $S = a + ar + ar^2 + \ldots + ar^{n-1}$. We want to find a concise expression for S. What quantity could be multiplied by both sides of this equation so that all terms on the right side of the new equation, except one, would be the same as in the original equation?

(b) Now carry out the step determined in part (a), subtract the new equation from the original one, and solve for S.

(10) In the geometric series formula of Theorem 4.14, if $|r| < 1$ and n gets large, what happens to the term r^n? From this, make a conjecture about the formula for the sum of an *infinite* geometric series with $|r| < 1$.

(11) Prove that the "or" in Theorem 4.15 is exclusive; that is, no natural number is both odd and even. (Don't just assume that 1 is not even or that 1/2 is not in $\mathbb{N}$; prove these things)

(12) Prove parts (a) and (b) of Theorem 4.16.

(13) Consider the series $1/2 + 1/6 + 1/12 + \ldots + 1/n(n + 1)$.

(a) Directly evaluate the sum of this series for several small values of n, and use these results to form a conjecture for the sum of this series in general.

(b) Using the fact that $1/n(n + 1) = 1/n - 1/(n + 1)$, find a trick for deriving the sum of this series without induction.

(c) Use induction to prove the formula for the sum of this series.

(14) Prove the surprising result (predicted in Exercise 6 of Section 1.2) that the sum of the cubes of the first n natural numbers equals the square of the sum of these numbers; that is,

$$1^3 + 2^3 + \ldots + n^3 = (1 + 2 + \ldots + n)^2 \quad \left(\text{that is, } \sum_{i=1}^{n} i^3 = \left(\sum_{i=1}^{n} i\right)^2\right)$$

You may use Theorem 4.11.

(15) (a) Consider Theorem 4.11. If the right side of this formula is expanded, what is its leading (highest power) term?

(b) Repeat part (a) for the formula in Exercise 14, after using Theorem 4.11 to write the right side of that formula in closed form.

(c) On the basis of parts (a) and (b), complete this conjecture: For any natural numbers k and n, the sum $1^k + 2^k + \ldots + n^k$ equals a polynomial in n whose leading term is ____ .

*(16) The goal of this problem is to derive the formula for $1^2 + 2^2 + 3^2 + \ldots + n^2$.

(a) Applying Exercise 15(c) to this series, write a polynomial with unknown coefficients for its sum. How many unknown coefficients must be included?

(b) Now substitute four small values of n into the result of part (a), to get a system of linear equations whose variables are the unknown coefficients. Surprisingly, it's OK and in fact advisable to use 0 as one of your values for n.

(c) Solve this system of equations to determine the exact polynomial that represents the sum of this series.

(d) Show that this polynomial equals $n(n + 1)(2n + 1)/6$.

(e) Now use induction to prove that $1^2 + 2^2 + 3^2 + \ldots + n^2 = n(n + 1)(2n + 1)/6$.

(17) Prove that $n < 2^n$, for any natural number n. You may assume basic facts about the algebra of exponents, as was done in the proof of Theorem 4.12.

(18) Prove that $1 + 1/4 + 1/9 + \ldots + 1/n^2 < 2 - 1/n$.

The next two problems provide examples of the **fundamental counting principle**, which is discussed further in Sections 6.1 and 7.6

(19) (a) At meetings of the Oxnard Pataphysics Club, every person present is required to say hello to every other person there, exactly once. Use trial and error and/or common sense to arrive at a conjecture about how many hellos are spoken at a meeting with n people present.

(b) Use induction to prove this conjecture.

(20) The English alphabet has 26 letters. Prove by induction that, for any n, there are 26^n n-letter words, where a "word" just means any sequence of letters.

(21) Prove the following calculus formulas, where n is any natural number. Use induction and the indicated formulas for each one:

(a) $\dfrac{d}{dx}(x^n) = nx^{n-1}$ Use the product rule and the derivative of x.

(b) $\dfrac{d^n}{dx^n}(e^{kx}) = k^n e^{kx}$ Use the chain rule, the derivative of e^x, and the derivative formula for a constant multiple of a function.

(c) $\dfrac{d^n}{dx^n}(xe^x) = (x+n)e^{kx}$ Use the product rule and the derivative of e^x.

(22) Critique the following proof of Theorem 4.10. (If necessary, review the instructions for this type of problem in Exercises 4.2.)

Proof: First we must prove the theorem for 1. To do this, note that $1 + 1 \in \mathbb{N}$, because $1 \in \mathbb{N}$ (by axiom VI-1), and thus, so is $1 + 1$ (by axiom VI-2). Now assume the theorem is true for n; that is, $n + n \in \mathbb{N}$. Then we must prove that $(n + 1) + (n + 1) \in \mathbb{N}$. But $(n + 1) + (n + 1) = [(n + n) + 1] + 1$, by the commutative and associative laws of addition. So the desired result follows by two applications of axiom VI-2.

(23) Critique the following proof.

Theorem: For every nonnegative integer n, $\sin n + \cos n = 2^n$.

Proof: For $n = 0$, the statement $\sin 0 + \cos 0 = 2^0$, which is true. For the induction step, assume $\sin n + \cos n = 2^n$. By substituting (that is, specifying) the expression $n + 1$ for the variable n, we get $\sin(n + 1) + \cos(n + 1) = 2^{n+1}$, as desired.

4.6 Hints for Finding Proofs

With the exception of Section 4.1, which dealt with different styles of proofs, the purpose of this chapter has been to explain the axioms and rules of inference that are commonly used in mathematical proofs. But we haven't said much about how to *find* proofs of statements. This section discusses how mathematicians go about proving things. It is as if we have just explained the rules of some game, like chess; this section starts to explain how to play the game competently.

Section 4.1 urged you to try to write proofs that are outlines or summaries of formal proofs. But most mathematicians write clear, logical proofs without ever consciously considering formal proofs. How is that done?

Here is another rule of thumb, based on a somewhat different (but not conflicting) perspective from that presented in Section 4.1: *A good proof of a statement should be a clear explanation of why the statement must follow from what you already know*. In other words, if you have a clear understanding of why a statement must be true, then you should be able to convert that understanding into a good proof of that statement. But to make that conversion requires careful analysis of your own understanding, and the ability to explain the sources of that understanding. Understanding why something is true entails more than merely seeing *that* it's true.

Is your understanding based in part on logic and common sense? Then the corresponding part of your written proof will use logical axioms and/or rules of inference. Is your understanding based in part on things you know about the subject matter of the statement? Then the corresponding part of your written proof will use proper axioms and/or previous theorems. Is your understanding based in part on knowing what certain words or symbols mean? Then your written proof will probably need to use the definitions of those words or symbols. Is your understanding of why the statement is true based on some reason why it *couldn't* be *false*? Then you will want to use indirect proof to prove the statement. And so on.

It is not always easy to analyze your own understanding in this way. But although mathematicians sometimes come up with proofs of difficult theorems without being consciously aware of how they did it, most would agree that proofs are based on understanding, and with enough analysis you can usually turn your understanding into a proof.

Gaining Insight into a Proof

We've been discussing how to convert your understanding of why a statement is true into a proof of that statement. But what if you are trying to prove something and you don't see why it's true? In fact, you may barely understand what the statement is *saying*, let alone that it's a true statement, let alone *why* it's true. Believe it or not, this happens frequently to all mathematicians, even the best. How are you supposed to prove something when you don't see what's going on?

Writing proofs when you have insight into the problem can already be hard; it is not something anyone can learn completely from one book or one course or in one year. Learning how to find proofs when you don't have that insight or understanding is that much harder, and it would be absurd to pretend that it's possible to learn this skill quickly. Nonetheless, it's possible to give some hints or guidelines for tackling proofs. Here are some of the more useful ones:

(1) *Analysis of the Structure of the Statement:* What is the logical structure (in terms of connectives and quantifiers) of the statement you are trying to prove? Answering this is not something that usually provides much mathematical understanding, but it may help you choose the right proof technique. Is the statement an implication? Then you almost certainly want to try conditional proof. Is it a negation? Then indirect proof might be a good try. Does the statement begin with one or more quantifiers? Then you probably need universal and/or existential generalization, remembering that existentially quantified variables are to be chosen as functions of the outer universal quantifiers. Tables 4.1 and 4.3 are meant to help you with this process.

(2) *Forward Reasoning:* This term refers to attempting to write a forward proof, perhaps more or less by trial and error. First you must start somewhere. How to start? If you are trying to prove an implication, conditional proof provides you with an assumption to start with. If not, you can try indirect proof, which also provides you with a starting assumption. If that seems inappropriate, you need to start with an axiom or

theorem. Which one? Obviously, you need to find an axiom or theorem that somehow is relevant to what you're trying to prove, but it can take some luck to find the most appropriate one.

Once you have one or more steps to start with, you have to go forward. How? One procedure is to look for an axiom or theorem that says that what you already have implies some other statement. Then you can use modus ponens to get a new step. Another procedure is to look for a rule of inference that would use one or more of the steps you have so far to get a new step. Of course, it is important not to be too random in generating new steps; rather, you need to constantly remember what you are trying to prove and try to keep getting closer to it.

(3) *Reverse Reasoning:* We discussed reverse proofs in Section 4.1, and this is another term for the same technique. If you can't see how to start your proof forward, start at the end and work in reverse. But, as was emphasized in our earlier discussion, you have to be careful when you do this. It's useless to find a new statement that is *implied by* the statement you want to prove; you need to find a new statement that *implies* the statement you are trying to prove. In other words, when trying a proof in reverse, you're always looking for statements that are *sufficient* for something you've already stated. Except for this one important difference, reverse reasoning is similar to forward reasoning.

Sometimes you can attempt a proof by a combination of forward and reverse reasoning. If you are fortunate, your two partial proofs will meet in the middle, and then you have a complete proof.

By the way, do not confuse the idea of reverse proofs with the method of indirect proof. Indirect proof is a valid rule of inference in which you assume the negation of what you're trying to prove. In a reverse proof, you start with the statement you're trying to prove, but you certainly are *not* assuming this statement!

(4) *Definition Unraveling:* This is a fairly simple process that can sometimes make a difficult-looking proof very easy. When the statement you are trying to prove involves defined words or symbols, it's always legitimate to replace them with whatever they are defined to mean. This usually makes the statement longer but may make it easier to understand, since you have replaced some words or symbols with simpler ones. Sometimes, you may be able to repeat this process two or more times, until the original statement has been unraveled into one involving only very basic symbols. At that point, you may see that the statement is one that is easy to prove; it may even be a tautology. Note that unraveling the statement you want to *prove* is a type of reverse reasoning. Unraveling statements that you are *using* in a proof—assumptions, axioms, previous theorems, and so on—is a type of forward reasoning. Both processes are quite legal.

As we see in Chapter 5, this procedure is very useful in basic set theory, a subject in which definitions tend to be used more than axioms. Often, in order to prove a biconditional involving sets, all you need to do is unravel all the defined symbols on both sides of the biconditional, and it then turns out that the two sides are logically equivalent.

(5) *Trying Special Cases:* This is an extremely important technique used by *all* successful mathematicians when they get stuck on a proof or problem of any sort. Most students do not realize how important it is; the sooner you learn to appreciate it and use it fully, the better off you will be. The idea is that most statements that are to be proved begin with at least one universal quantifier (even if it's not explicitly stated). So if you don't see how to prove the statement for an arbitrary value of whatever variable, first try to do it for one or more particular values. Although this can't constitute a complete proof, it's amazing how often doing some simple special cases provides enough insight to enable you to do the problem in full generality.

Example 1: Suppose you are asked to prove a statement of the form "Given any point in the domain of a real-valued function of several variables, ..." This statement contains at least three variables (f for a function, n for the number of variables of f, and c for a point in the domain of f), which are understood to be universally quantified. If you don't see how to prove the statement, it may be because it's hard to visualize things in higher dimensions. So perhaps you should first try to prove it when $n = 1$, the simplest possible special case. If that works, you might then try $n = 2$, and if that works also, perhaps you can see how to do it for arbitrary n. What if you can't do it for $n = 1$? Don't panic; there are other variables to specify. Within the special case $n = 1$, you might try the proof for a simple particular function like $f(x) = x^2$, $f(x) = x$, or even $f(x) = c$. You might even want to choose a specific number for c.

Example 2: Suppose you are asked to prove some geometric statement involving an arbitrary triangle. Geometry problems can be very difficult. Instead of beating your brains out over the full problem, why not try it for a few very special triangles, like equilateral ones and isosceles right triangles. If you succeed with those, you might then try it for all right triangles and/or all isosceles triangles, or you might tackle the general problem.

Typically, when you succeed with one special case, you then try a more general or a harder special case (or the whole problem). But when you can't do one, you then try a more specific or a simpler special case.

No serious mathematician ever gives up on a problem or proof without trying at least one or two special cases, where applicable. The limitation "where applicable" is necessary because some statements have no variables from which to form special cases; but these statements are a minority. A mathematics instructor can do his students a service by refusing to help them with proofs until they have at least looked at a couple of special cases. But why lay this responsibility on your teacher? Establish this practice on your own!

(6) *Trying a Simpler Problem:* This is related to the trying special cases, but can be quite different. Suppose you are asked to prove something about any 3 by 3 matrix. If you first try it for some particular 3 by 3 matrix, that's a special case. But what if you first try it for some particular 2 by 2 matrix or perhaps all 2 by 2 matrices? This could not be called a special case of the problem, because 2 is not in any sense a special case

of 3. (That is, 3 is not a variable.) It's just a different problem. What you are asked to prove may not be true for or even apply to 2 by 2 matrices. But if it does, you might find it's much easier to see what's going on with 2 by 2 matrices, and solve the problem for those. Then the insight gained from that may help you do the 3 by 3 version. This technique is not quite as powerful as the use of special cases, but it's still useful.

Suggestions for Further Reading: The references listed at the end of Chapter 2 all include some coverage of the major ideas of this chapter: formal systems, proofs, rules of inference, and so on. So do Stoll (1979) and Wilder (1965). For an in-depth treatment of the method of mathematical induction, see Sominskii (1961). For elaboration of the discussion in Section 4.6, you are encouraged to read the lucid observations on mathematical insight in the classic works by Polya (1945, 1954, and 1965).

Unit 2

Sets, Relations, and Functions

Chapter 5

Sets

5.1 Naive Set Theory and Russell's Paradox

It can be a challenge to convince people that set theory is a profound and important branch of mathematics. That's because most students get a taste of sets in high school or even earlier, and at that level what's done can seem simpleminded and pointless. Don't be fooled. Set theory (like algebra, another subject that many people think is limited to the high school level) is full of fascinating and deep problems that have stumped many of the world's greatest mathematicians.

Naive Set Theory

The concept of sets in mathematics is quite recent, dating back only to about 1870. The essential idea of set theory is that any collection of objects of any sort that you can list or clearly describe may be considered to form a set. This principle is usually called the **axiom of comprehension.**

The other elementary principle of set theory is that a set has no other characteristics than being a collection of things. In other words, if two sets have exactly the same members, then they must be equal because there's nothing else that could distinguish them. This is called the **axiom of extensionality.**

Notation: You have probably already seen most of the basic notation for describing sets, but let's go over it anyway.

If an object x is in a set A, we say that x is an **element** or a **member** of A, which is written $x \in A$. (The symbol for set membership is a modified Greek epsilon.)

We use capital letters, usually from the beginning of the alphabet, to denote sets. Note that the expression on the right side of a membership statement must always denote a set, but the expression on the left side can denote any type of object.

The standard way of denoting a set is to show its members in braces: $\{ \ldots \}$. If we want to define a set that consists of a small, finite number of members, we can just list them inside the braces. This is called the **roster method** of denoting a set.

Example 1: $\{1, 3, 5, 7, 9\}$ denotes the set of all odd natural numbers less than 10.

A variant of the roster method can be used when a set has many elements. If we wanted to denote the set of all odd natural numbers up to 999, it would be absurd to list all those numbers. But if we write $\{1, 3, 5, 7, \ldots, 999\}$, the meaning is clear enough. This type of notation is acceptable. Furthermore, we could write $\{1, 3, 5, 7, \ldots\}$ as a perfectly clear notation for the set of *all* odd natural numbers. Thus even infinite sets can be shown with the roster method, *provided* that the meaning of the ellipsis is clear.

Note the reference to infinite sets. A rigorous definition of the words "finite" and "infinite" must wait until Section 7.5. In the meantime, a working definition may be helpful. Informally, we can say that a finite set is one that can be defined by the roster method, with no ellipsis. Alternatively, a finite set is one with no elements, or with n elements for some natural number n. "Infinite" simply means "not finite."

In the other method of denoting a set with braces, called **set-builder notation**, the elements are described instead of listed. The usual way to do this is to show a variable followed by a vertical line or a colon and then a proposition that shows what has to be true for something to be in the set.

Example 2: To denote the set of all odd natural numbers by the set-builder method, we can write

$$\{x \mid x \text{ is an odd natural number}\} \quad \text{or}$$

$$\{x \mid \exists n \in \mathbb{N}\ (x = 2n - 1)\}$$

We read the first notation "the set of all x such that x is an odd natural number." The second can be read "the set of all x such that $x = 2n - 1$, for some n in $\mathbb{N}$."

Set-builder notation can sometimes be shortened by restricting the variable to a set. So another way to write this same set would be

$$\{x \in \mathbb{N} \mid x \text{ is odd}\}$$

Of course, if we maintain our convention that the letter n is restricted to natural numbers, then the same set can be written simply as $\{n \mid n \text{ is odd}\}$.

Also, it's sometimes very convenient to use set-builder notation with the vertical line or colon preceded by an expression, rather than just a single letter. For example, the shortest and "neatest" notation for the set of all squares of natural numbers is

$$\{n^2 \mid n \in \mathbb{N}\}$$

Technically, this is an abbreviation for the more cumbersome notation

$$\{x \mid \exists n \in \mathbb{N}\ (x = n^2)\}$$

Every finite set can be described by either the roster method or set-builder notation. But for infinite sets, set-builder notation is more useful than the roster method. By the way, it's *only* in set-builder notation that a vertical line or colon is used as an abbreviation for "such that." Some mathematicians use the symbol $\ni$ as an abbreviation for these words in other contexts.

☞ Here is a *crucial* thing to learn about set-builder notation: suppose that a set has been defined by set-builder notation, say $A = \{x \mid P(x)\}$. Then for any x, it follows that $x \in A$ iff $P(x)$. This is simply the *definition* of this notation, and you don't need any theorems to justify this biconditional. Similarly, if $A = \{x \in B \mid P(x)\}$, then it's understood that $x \in A$ iff $x \in B$ and $P(x)$, for any x.

Example 3: Suppose that $A = \{x \in \mathbb{R} \mid \tan x > 5\}$, and we know that some number u is in A. Of course, we can then write $u \in \{x \in \mathbb{R} \mid \tan x > 5\}$, but this is rather cumbersome and can be stated much more simply by saying $\tan u > 5$. You should learn to make this translation *automatically*.

In the notation $\{x \mid P(x)\}$, the variable x should be considered bound, not free. That means it can be replaced by any other letter, so $\{x \mid P(x)\} = \{u \mid P(u)\}$, and so on. Such a replacement is often necessary to avoid having the same variable be free and bound at the same time.

Example 4: Suppose that for any real number y, we define A_y to be $\{x \mid x < y - 2\}$. So $A_3 = \{x \mid x < 1\}$, and so on. But what if we are considering a real number x? We can't say that $A_x = \{x \mid x < x - 2\}$. Rather, we have to change the dummy variable x; for example, we could write $A_x = \{z \mid z < x - 2\}$.

When set theory was invented, it was based completely on the axioms of comprehension and extensionality, and it's still based intuitively on the same two principles. However, as we soon see, this simple approach had some severe problems. For this reason, this early form of set theory is now called **naive set theory**. Here are the axioms of naive set theory, in symbols.

Extensionality: $A = B \leftrightarrow \forall x\,(x \in A \leftrightarrow x \in B)$

Comprehension: For any proposition $P(x)$, the set $\{x \mid P(x)\}$ exists.

Remarks: (1) In these axioms, and in this unit generally, the variables x, y, and z are not assumed to be restricted to real numbers; they can have any domain. In the extensionality axiom, x should be considered an unrestricted variable, ranging over *all* possible objects.

(2) Some authors call our extensionality axiom the *definition* of set equality. There's nothing wrong with this approach, and it does not differ from ours in any essential way.

(3) The comprehension axiom establishes the existence of any set of the form $\{x \in A \mid P(x)\}$, since that's the same as $\{x \mid x \in A \text{ and } P(x)\}$. This limited version of the comprehension axiom, in which the variable x must be restricted to a set, is called the **axiom of separation**.

Here are three definitions that illustrate the use of set-builder notation. The first two define important number systems that are intermediate between the systems $\mathbb{N}$ and $\mathbb{R}$. Note that we have *not* defined $\mathbb{N}$ and $\mathbb{R}$. Instead, we have taken them to be primitive and included axioms for them; this automatically means they must be undefined. Other approaches to the development of number systems are considered in Chapter 9.

Definition: A real number is called an **integer** iff it can be written as the difference of two natural numbers. The set of all integers is denoted $\mathbb{Z}$. In symbols,

$$\mathbb{Z} = \{n - m \mid m, n \in \mathbb{N}\}$$

An alternate way of describing $\mathbb{Z}$ is given in Exercise 13.

Definition: A real number is called **rational** if it can be written as a quotient of two integers. The set of all rational numbers is denoted $\mathbb{Q}$. In symbols,

$$\mathbb{Q} = \{a/b \mid a, b \in \mathbb{Z} \text{ and } b \neq 0\}$$

Note that the last two definitions illustrate the use of set-builder notation with an expression other than a variable before the vertical line, as discussed earlier in this section.

Definitions (intervals): In what follows, the letters a and b denote specific real numbers (normally with $a < b$), and x also stands for a real number:

$$(a, b) = \{x \mid a < x < b\}$$
$$[a, b] = \{x \mid a \leq x \leq b\}$$
$$(a, b] = \{x \mid a < x \leq b\}$$
$$[a, b) = \{x \mid a \leq x < b\}$$

The first type of set is called an **open interval**, the second a **closed interval**, and the last two **half-open intervals** (less frequently, **half-closed intervals**). When it is important to make a distinction between intervals of the forms $(a, b]$ and $[a, b)$, the terms **open-closed interval** and **closed-open interval** are sometimes used.

The following notations are used to describe what are called **unbounded intervals** or **rays** (open or closed):

$$(a, \infty) = \{x \mid x > a\}$$

$$[a, \infty) = \{x \mid x \geq a\}$$
$$(-\infty, b) = \{x \mid x < b\}$$
$$(-\infty, b] = \{x \mid x \leq b\}$$

When using this notation, it is important to bear in mind that ∞ and $-\infty$ do *not* denote real numbers. The fact that these symbols always appear next to a parenthesis, never a square bracket, is intended to emphasize this fact. These symbols simply indicate that a certain set of real numbers has no end, either in the positive or the negative direction. The notation $(-\infty, \infty)$ is also used occasionally. But since this is just another way of denoting $\mathbb{R}$, it is not particularly useful.

Usually, intervals and interval notation refer to sets of real numbers. But they can be considered in any context where the inequality symbols have meaning.

The Paradoxes of Set Theory

What went wrong with naive set theory? You might guess that such a simpleminded theory, with these two very trivial-looking axioms, might have the drawback of not being powerful enough in the sense that it might not be possible to prove any interesting theorems from them. Surprisingly, the opposite is true. Naive set theory is *too* powerful; in fact it's **inconsistent**, meaning that it leads to contradictions. Several people discovered this about 1900, when the new subject was only a couple of decades old. The various forms of this contradiction in set theory are called paradoxes, but this word doesn't really convey the severity of the situation. A **paradox** usually refers to an *apparent* contradiction that can be straightened out with careful thought. The paradoxes of naive set theory have no solution, except to change the theory substantially.

Here is the simplest and most blatant paradox of set theory, usually credited to the great English philosopher and logician Bertrand Russell.

Theorem 5.1 (Russell's Paradox): Naive set theory is inconsistent; that is, it leads to a contradiction.

Proof: The proof is amazingly short; the core of it is a single use of the comprehension axiom, to form the set of all sets that are not members of themselves. In symbols, let

$$A = \{B \mid B \notin B\}$$

Then we simply ask whether A is a member of *itself!* If $A \in A$, then by the definition of A, $A \notin A$. On the other hand, if $A \notin A$, then since A is a set, we must have $A \in A$. So we have proved $A \in A \leftrightarrow A \notin A$, which is a contradiction. ∎

A popularized version of Russell's paradox is known as the **Barber's paradox**. In a certain town there is a single barber, who is a man. The barber shaves all men in the town who do not shave themselves, and only those men. This sounds plausible, but the question is: Does the barber shave himself? There is no consistent answer to this question. If he does, he doesn't, and if he doesn't, he does.

The discovery of the paradoxes of naive set theory threw the foundations of mathematics for a loop. A mathematical theory that leads to contradictions is of no use. If a subject as simple looking as this could be inconsistent, what assurance is there that other branches of mathematics are consistent? And if there is no such assurance, who can guarantee the soundness of conclusions made in science and engineering on the basis of mathematics?

As a reaction to this development, many scholars in the early part of this century attempted to fix set theory, reformulating its axioms to achieve a consistent theory that would still be productive. A careful examination of the paradoxes of set theory leads to

Bertrand Russell (1872–1970) was not primarily a mathematician but continued an ancient tradition of philosophers making important contributions to the foundations of mathematics. He was born into a wealthy liberal family, orphaned by the age of four, and then raised by his grandmother and tutored privately.

There are two mathematical contributions for which Russell is remembered. One was his discovery of the inconsistency of naive set theory, as discussed in this section. The other was a monumental task that occupied him for over a decade: the three-volume *Principia Mathematica*, written with his former professor, Alfred North Whitehead, and completed in 1913. This work was the manifesto of the logicist school of thought, one response to the crisis in the foundations of mathematics that was precipitated by the paradoxes of set theory.

As a philosopher, Russell is considered one of the major figures in the modern analytical school. He was a prolific writer and wrote many books intended for the general public, notably the best-selling *A History of Western Philosophy* (1950), for which he won the Nobel Prize in Literature.

Outside academic circles, Russell is probably best known for his political and social activism. During World War I, his pacifism led to his dismissal from Trinity College and a six-month prison sentence. Many decades later, he vehemently opposed nuclear weapons, racial segregation, and the U. S. involvement in Vietnam. He advocated trial marriages and sexual freedom as early as the 1930s, a position that caused a court of law to nullify a faculty position that he had been offered by the City College of New York in 1940.

the conclusion that the full comprehension axiom is the culprit because it is just too general. Therefore, the revised set theory that is currently used keeps the original extensionality axiom, but replaces the comprehension axiom with about a half dozen more specific rules postulating the existence of various sets. The most widely used version of modern set theory was developed by Ernst Zermelo (1871–1953) and Abraham Fraenkel (1891–1965) and is called **Zermelo-Fraenkel (ZF) set theory**. We will not be discussing ZF set theory in this book. However, our set axioms (group IV in Appendix 1) are essentially the axioms of ZFC set theory (Zermelo-Fraenkel set theory plus the important axiom of choice, which is discussed in Section 7.7).

Did Zermelo and Fraenkel achieve their goal of creating a consistent version of set theory? Surprisingly, the answer to this is not known, and in a certain sense can *never* be known for sure! One of the reasons that set theory is important is that, with some esoteric possible exceptions, all of current mathematics can be carried out within the framework of ZFC set theory. Therefore, knowing the consistency of set theory would essentially be the same as knowing the consistency of mathematics. However, one of the most amazing and significant discoveries in the history of mathematics, known as **Gödel's incompleteness theorem**, states that the consistency of a "reasonable" mathematical theory cannot be proved without using postulates that go beyond that theory. Therefore, the consistency of set theory simply can't be proved using standard mathematical principles. The best that can be said is that nearly a century of experience with ZFC set theory has not produced any contradictions, and there is every reason to believe that it provides a consistent framework for mathematics.

Exercises 5.1

(1) Rewrite the following sets using the roster method.
   (a) $\{n \in \mathbb{N} \mid n^2 < 36\}$
   (b) $\{n^2 \mid n \in \mathbb{N} \text{ and } n < 6\}$
   (c) $\{x \in \mathbb{R} \mid \sin x = x\}$
   (d) $\{s \mid s \text{ is a New England state}\}$
   (e) $\{x \in \mathbb{Z} \mid |x| \text{ is prime and even}\}$

(2) Rewrite the following sets using set-builder notation.
   (a) $\{1, 4, 9, 16, 25, \ldots, 10{,}000\}$
   (b) $\{1, 4, 9, 16, 25, \ldots\}$
   (c) $\{-2, 4, -8, 16, \ldots\}$
   (d) {a, e, i, o, u, y}
   (e) $\{6, 17, 92\}$

(3) Which of the following sets are equal to each other?
   (a) $\{1, 2, 3\}$
   (b) $\{3, 2, 1\}$
   (c) $\{1, 2, 3, 1.0\}$
   (d) $\{x \in \mathbb{R} \mid 1 \le x \le 3\}$

(e) $\{n \in \mathbb{Z} \mid n^2 = 1 \text{ or } n^2 = 4 \text{ or } n^2 = 9\}$
(f) $\{n \in \mathbb{N} \mid n + 7 < 11\}$
(g) $\{\sqrt{1}, \sqrt{4}, \sqrt{9}\}$

(4) Let $A = \{2^x \mid x \in \mathbb{R} \text{ and } x^3 - x = 17\}$.
(a) Rewrite A in the form $\{y \mid ...\}$.
(b) Rewrite A in the form $\{x \mid ...\}$.
(c) Rewrite $\{2^x \mid x \in \mathbb{R} \text{ and } x^3 - x = 0\}$ using the roster method.

(5) Let $A = \{x + 3 \mid x$ is a real number that equals its tangent$\}$. Which of the following statements are true, if any? Explain your assertion.
(a) For any x in $\mathbb{R}$, $x \in A$ iff $x = \tan x$.
(b) For any x in $\mathbb{R}$, $x \in A$ iff $x + 3 = \tan(x + 3)$.
(c) For any x in $\mathbb{R}$, $x \in A$ iff $x - 3 = \tan(x - 3)$.

(6) Rewrite the following sets in interval notation:
(a) $\{x \in \mathbb{R} \mid x \leq 17\}$
(b) $\{x \in \mathbb{R} \mid x > -2 \text{ and } x \leq -1\}$
(c) $\{x \in \mathbb{R} \mid x \geq 5 \text{ or } -x < -2\}$
(d) $\{x \in \mathbb{R} \mid x^2 \leq 9\}$
(e) $\{x \in \mathbb{R} \mid |x + 3| < 4\}$

(7) Rewrite the following sets using set-builder notation *or* the roster method:
(a) $(-1, 7]$ (b) $(-\infty, 0)$ (c) $[5, 5]$

(8) True or false (with brief explanation):
(a) $3 \in [3, 7]$ (b) $3 \in (3, 7)$
(c) $3 \in [5, 2]$ (d) $3 \in [3, 3)$
(e) $-\infty \in (-\infty, 127)$ (f) $-\infty \in [-\infty, 127]$

(9) For each of the following statements, determine whether it's true or false in (i) $\mathbb{N}$, (ii) $\mathbb{Z}$, (iii) $\mathbb{Q}$, and (iv) $\mathbb{R}$. Give brief explanations. (All in all, this exercise has twelve true/false questions. For some guidelines for this type of problem, refer back to Example 6 of Section 3.3.)
(a) $\forall x,y\ \exists z\ (x - y = y^2 - z)$
(b) $\forall x,y\ [x \geq y \text{ or } \exists z\ (x < z < y)]$
(c) $\exists x,y\ (x^2 - y^2 = 2)$

(10) Using the extensionality axiom, prove that set equality satisfies axioms III-1, III-2, and III-3 (with the variables in those three axioms changed to set variables).

(11) Prove that the set form and the statement form of mathematical induction are equivalent to each other. You may use the axioms of naive set theory as well as all logic and equality axioms.

*(12) Here is an example of a "semantic paradox" known as Berry's paradox. Let A be the set of natural numbers that can be defined by English phrases less than sixty syllables long. (Examples of such phrases that define natural numbers are "the fifth smallest prime number," "the number of days in a week," and so on.) Since there are only a finite number of such phrases in English, A is finite. Therefore there are natural numbers that are not in A. Let n be the smallest natural number not in A.

Now consider the phrase "the smallest number that cannot be defined by an English phrase of fewer than sixty syllables." This phrase has fewer than sixty syllables and defines the number n. Therefore, $n \in A$, which is a contradiction.

Try to explain this paradox. That is, try to explain the flaw in the argument; there should be one since it should not be possible to prove a contradiction from scratch. The explanation is based on subtle philosophical considerations, rather than a technical point or trick. By the way, the problem does *not* lie with the last sentence of the first paragraph. As we will soon see (Theorem 5.6), if there is a natural number with a certain property, then there is a least one.

*(13) (a) Define the set $\mathbb{Z}'$ to consist of all natural numbers, negatives of natural numbers, and zero. In symbols,

$$\mathbb{Z}' = \{x \in \mathbb{R} \mid x \in \mathbb{N} \text{ or } (-x) \in \mathbb{N} \text{ or } x = 0\}$$

Prove that $\mathbb{Z}' = \mathbb{Z}$. Therefore, this is a correct alternate way of defining integers.

(b) Using part (a), deduce that $\mathbb{N}$ is the set of all positive integers.

(14) Let $P(n)$ be a statement with a free *integer* variable n. Suppose that we are able to prove $P(0)$ and $\forall n\, [P(n)$ implies $P(n + 1)$ and $P(n - 1)]$. What would be the logical thing to conclude from this? Prove your claim. You may use the result of the previous exercise.

Critique the proofs in the remaining exercises. (If necessary, review the instructions for this type of problem in Exercises 4.2.)

(15) **Theorem:** If a, b, c and d are real numbers with $a < b$ and $c < d$, then $[a, b] = [c, d]$ iff $a = c$ and $b = d$.

Proof: Assume $a, b, c, d \in \mathbb{R}$, $a < b$, and $c < d$.

For the forward direction, assume $[a, b] = [c, d]$. Since $c \le a \le b$, the definition of intervals tells us that $a \in [a, b]$. So, by extensionality, $a \in [c, d]$. By definition of intervals, this implies that $c \le a$. Similarly, we know that $c \in [c, d]$, hence $c \in [a, b]$, and therefore $a \le c$. From the inequalities $c \le a$ and $a \le c$, it follows (as in Exercise 26, Section 4.4) that $a = c$.

A nearly identical argument shows that $b = d$.

For the reverse direction, assume $a = c$ and $b = d$. By Theorem 4.7 applied to $a = c$, we get $[a, b] = [c, b]$. The same theorem applied to $b = d$ yields $[c, b] = [c, d]$. Therefore, by transitivity of equality (axiom III-3), $[a, b] = [c, d]$.

(16) **Theorem:** Let A and B be any sets, P(x) any proposition, $C = \{x \in A \mid P(x)\}$ and $D = \{x \in B \mid P(x)\}$. Then, $A = B$ iff $C = D$.

Proof: Assume $A = B$. Then Theorem 4.7 immediately implies that $C = D$. By the same reasoning, if $C = D$, then $A = B$.

5.2 Basic Set Operations

Despite the fact that naive set theory is inconsistent, it turns out that as long as you avoid defining sets that are too big, naive set theory works quite well and does not seem to lead to any contradictions. And the more correct modern axiomatic set theory is much more complicated. So most mathematicians use naive set theory unless they are trying to be extremely careful and/or formal; they learn through experience how to use it safely. We take this approach, and you should feel free to do the same for most of your dealings with set theory.

It's impossible to give an ironclad set of guidelines for what to avoid when using naive set theory, but here is the most important one.

Rule of Thumb (When Using Naive Set Theory): Do not try to define the set of *all* sets or a set that involves all sets in its definition (such as the set defined in the proof of Russell's paradox). Any such definition will probably lead to a contradiction.

At times it is convenient to talk about the "class" of all sets. As long as classes of this sort are not allowed to be members of sets, paradoxes do not seem to arise. In contrast, there is no difficulty in defining the set of all real numbers, the set of all *sets* of real numbers, the set of all people who have ever lived, the set of all particles in the universe, and many other sets that might seem very big. If this restriction seems strange and esoteric, don't be concerned; it's not something you have to worry about very often. Instead, let's spend the rest of this section on some simple and familiar operations involving sets, which are shown in standard **Venn diagram** form in Figure 5.1.

Definitions: (a) The **union** of any two sets A and B, denoted $A \cup B$, is the set

$$\{x \mid x \in A \text{ or } x \in B\}$$

(b) The **intersection** of any two sets A and B, denoted $A \cap B$, is the set

$$\{x \mid x \in A \text{ and } x \in B\}$$

(c) The **relative complement of A in B** (or **the complement of A relative to B**), denoted $B - A$, is the set

$$\{x \mid x \in B \text{ and } x \notin A\}$$

(d) The **empty set** or **null set**, denoted $\varnothing$, is the set with no members.

(e) Two sets are called **disjoint** if their intersection is empty.

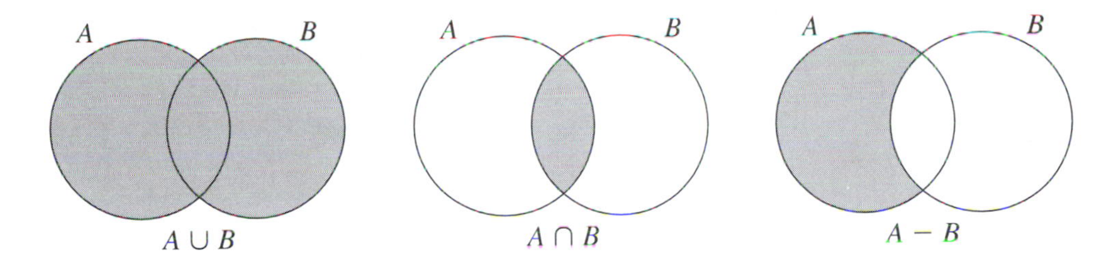

Figure 5.1 Venn diagrams illustrating basic set operations

Remarks: (1) The idea of a set with no members may seem odd at first, but it's a harmless and very useful concept. Also, unless we want to restrict the axioms further, there must be such a set. For example, $\{n \in \mathbb{N} \mid n < n\}$ is empty. Note that extensionality guarantees that any two empty sets are equal. In other words, there's only one of them, so mathematicians normally refer to *the* empty set.

(2) Notice that we've defined *relative* complement as opposed to just complement. That's because when people talk about the complement of a set, they always mean a relative complement—relative to some set that's understood. For example, suppose $A = [1, 3] = \{x \mid 1 \leq x \leq 3\}$, which is an interval on the real number line. If you then see a reference to the complement of A, it almost certainly means the complement of A relative to $\mathbb{R}$, which is $\{x \mid x < 1 \text{ or } x > 3\}$.

But now suppose there were a reference to the complement of the set $\{2, 17, 984\}$. Would that mean the complement of this set relative to $\mathbb{N}$? To $\mathbb{Z}$? To $\mathbb{R}$? Or perhaps relative to some other set? Unless the context made things clear, the reference would be quite ambiguous. Worse yet, suppose we were considering a set like {6, −2.7, Shakespeare, Canada}. Would it make sense to talk about the complement of this set without saying what it is relative to? Not at all.

You might wonder why we can't define the absolute complement of a set, meaning simply the set of all objects (with no restriction) that are not in the set. The reason is that doing so quickly leads to a contradiction similar to Russell's paradox. In particular, the absolute complement of the null set would have to contain all sets. With these considerations in mind, we establish the following convention.

Convention: Suppose that, during a certain discussion, it is understood that all sets being considered are contained in some particular set U. Then it's permissible to write A' or $\overline{A}$ (called the **complement of A** or **A complement**) as an abbreviation for $U - A$ (see Figure 5.2). The set U may be called the **universal set** for the purposes of the discussion. But remember that the idea of a universal set is just a temporary convenience. There is no such thing as *the* universal set.

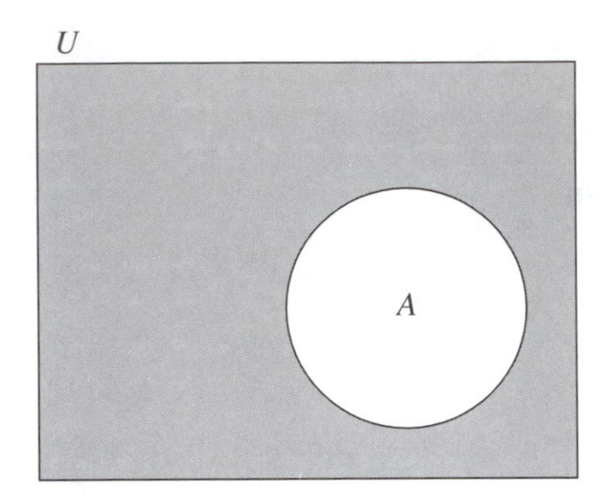

Figure 5.2 Venn diagram illustrating the complement of a set A

Here is a sample of the many elementary results that hold for these basic operations on sets. This subject is sometimes called the **algebra of sets.**

Theorem 5.2: For any sets A, B, C, and D,
(a) $A \cup B = B \cup A$
(b) $A \cap B = B \cap A$
(c) $A \cup (B \cup C) = (A \cup B) \cup C$
(d) $A \cap (B \cap C) = (A \cap B) \cap C$
(e) $A \cup (B \cap C) = (A \cup B) \cap (A \cup C)$
(f) $A \cap (B \cup C) = (A \cap B) \cup (A \cap C)$
(g) $A - (B \cap C) = (A - B) \cup (A - C)$
(h) $A - (B \cup C) = (A - B) \cap (A - C)$
(i) $A \cap (B - C) = (A \cap B) - (A \cap C)$

Proof: We just prove a couple of parts here. The rest are very similar and are left for the exercises.

(a) The usual way to prove two sets are equal is via the extensionality axiom. So we want to show

$$\forall x\,(x \in A \cup B \;\leftrightarrow\; x \in B \cup A)$$

To start, assume $x \in A \cup B$. By the definition of union, that means $x \in A$ or $x \in B$. But this is equivalent to $x \in B$ or $x \in A$. So, by the definition of union, $x \in B \cup A$. We have thus proved $x \in A \cup B \rightarrow x \in B \cup A$. The proof of the converse is similar. Since x is arbitrary, extensionality yields that $A \cup B = B \cup A$.

(e) As with part (a), the main step here is to prove a certain biconditional. But instead of doing that as two separate implications, let's do it as a single proof, a shortcut that was mentioned in the discussion of the biconditional rule:

$$\begin{array}{lll} x \in A \cup (B \cap C) & \leftrightarrow x \in A \text{ or } x \in B \cap C & \text{Definition of } \cup \\ & \leftrightarrow x \in A \text{ or } (x \in B \text{ and } x \in C) & \text{Definition of } \cap \\ & \leftrightarrow (x \in A \text{ or } x \in B) \text{ and } (x \in A \text{ or } x \in C) & \text{Tautology 30} \\ & \leftrightarrow x \in A \cup B \text{ and } x \in A \cup C & \text{Definition of } \cup \\ & \leftrightarrow x \in (A \cup B) \cap (A \cup C) & \text{Definition of } \cap \end{array}$$

So $x \in A \cup (B \cap C) \leftrightarrow x \in (A \cup B) \cap (A \cup C)$. Again applying UG and extensionality, the two sets must be equal. ■

The result of Theorem 5.2(e) is also illustrated in Figure 5.3. Although a picture can never constitute a rigorous proof, a careful Venn diagram can be a pretty reliable way to determine whether a statement of elementary set algebra is necessarily true.

The first six parts of Theorem 5.2 strongly resemble various field axioms (group V in our axiom system). Parts (a) and (b) say that $\cup$ and $\cap$ are commutative, (c) and (d) say that these set operations are associative, and (e) and (f) are distributive laws. Although there is some connection between set algebra and real number algebra, set algebra is more closely connected with propositional logic. The following theorem illustrates this connection further.

Theorem 5.3: Assume that A and B are both contained in some particular set U, and let A' and B' be abbreviations for $U - A$ and $U - B$. Then

(a) $A \cap A' = \varnothing$
(b) $A \cup A' = U$
(c) $(A \cup B)' = A' \cap B'$ (De Morgan's law for sets)
(d) $(A \cap B)' = A' \cup B'$ (De Morgan's law for sets)
(e) $A - B = A \cap B'$
(f) $A' - B' = B - A$
(g) $(A')' = A$

Proof: We just prove part (c), leaving the rest for the exercises. Our proof is very similar to the proof of Theorem 5.2(e), except that we now let the variable x have the set U as its domain.

$$\begin{array}{lll} x \in (A \cup B)' & \leftrightarrow x \notin A \cup B & \text{Definition of } ' \\ & \leftrightarrow \sim (x \in A \text{ or } x \in B) & \text{Definition of } \cup \\ & \leftrightarrow x \notin A \text{ and } x \notin B & \text{De Morgan's law} \\ & \leftrightarrow x \in A' \text{ and } x \in B' & \text{Definition of } ' \\ & \leftrightarrow x \in A' \cap B' & \text{Definition of } \cap \end{array}$$

Applying UG and extensionality to this yields $(A \cup B)' = A' \cap B'$. ■

Remarks: (1) By now, you should be getting the idea of how to prove two sets are equal. In general, if you want to prove $A = B$ using the extensionality axiom, you must prove $x \in A \leftrightarrow x \in B$ (where x is arbitrary). In simple cases, this biconditional can often be proved without splitting it up, as we've done in the previous two cases (and

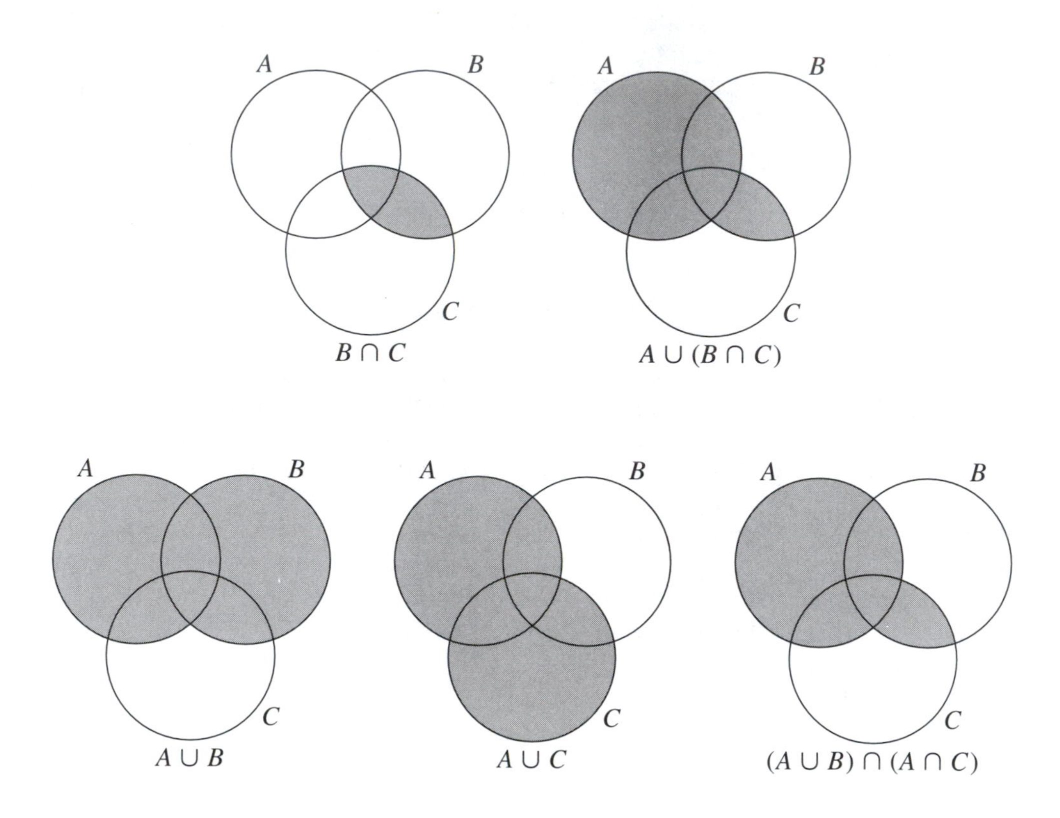

Figure 5.3 Venn diagrams illustrating Theorem 5.2(e)

could have done in Theorem 5.2(a)). In more complex cases, it's often necessary to split the biconditional up into two implications and prove each one separately.

(2) The proofs in this section don't have much content; they could be described as mostly definition unraveling, as discussed in Section 4.6. That is, if you look at the proofs so far in this section, each one simply rewrites the symbols $\cup$, $\cap$, $-$, and $'$ in terms of their definitions, and then uses one tautology to establish the desired biconditional. Of course, the extensionality axiom is also used, and in an indirect sense so is the comprehension axiom. But it's typical of basic set algebra that the proofs involve mostly definitions and propositional logic. The connection between set algebra and propositional logic can be made a bit more precise, as follows: every theorem of set algebra stating that two sets must be equal is, after unraveling the definitions of the set

symbols, equivalent to some tautology. This connection is the basis of a branch of mathematics called **boolean algebra** (see Exercise 16).

(3) Also, you may have noticed that many parts of Theorems 5.2 and 5.3 occur in pairs of similar-looking statements. This **duality**, as it's called, is also a basic aspect of boolean algebra (see Exercise 17).

Subsets, Proper and Otherwise

Definitions: We say that A is a **subset** of B (in symbols, $A \subseteq B$) iff every element of A is in B. Also, A is a **proper subset** of B ($A \subset B$) iff $A \subseteq B$ and $A \neq B$.

It is permissible to write $B \supseteq A$ and $B \supset A$ to mean $A \subseteq B$ and $A \subset B$, respectively. Normally, these reversed symbols and the associated word "superset" are used only when there is some specific reason to do so. For instance, it is simpler to say "every superset of A" than "every set of which A is a subset."

Here are a few simple results involving these concepts.

Theorem 5.4: (a) $A \subseteq A$

(b) $A \not\subset A$

(c) $\varnothing \subseteq A$

(d) $\varnothing \subset A \leftrightarrow A \neq \varnothing$

(e) $A \subseteq B$ and $B \subseteq C \rightarrow A \subseteq C$

(f) $A = B \leftrightarrow A \subseteq B$ and $B \subseteq A$

(g) $A \neq B \leftrightarrow (A - B) \cup (B - A) \neq \varnothing$

(h) $A \subset B \leftrightarrow A \subseteq B$ and $B \not\subseteq A$

Proof: Again, we prove only two parts and leave the rest for the exercises.

(c) By definition, $\varnothing \subseteq A$ means $\forall x\ (x \in \varnothing \rightarrow x \in A)$. Let x be arbitrary. By definition of $\varnothing$, $x \in \varnothing$ is automatically false, and therefore the conditional $x \in \varnothing \rightarrow x \in A$ is automatically true.

(e) This has already been proved in Proof Preview 2 (Section 2.3), so we do not repeat the argument. ■

The word "subset" might give the impression that a set would not be a subset of itself. But as you can see, the definition is written in such a way that *every* set is a subset of itself.On the other hand, *no* set is a *proper* subset of itself. For emphasis, mathematicians may say that every set is an *improper* subset of itself.

Be careful to keep the grammar of these symbols straight. The symbols $\in$ and $\subseteq$ are predicate symbols; that means that $x \in A$ and $A \subseteq B$ are complete statements that can stand alone or be combined using connectives and quantifiers. On the other hand, $\cup$, $\cap$, $-$, and $'$ are operator symbols. So expressions like $A \cup B$, $A \cap (B - C)$, and so on, are terms denoting sets, *not* statements. For example, it is *grammatically impossible* to have a line in a proof that just says $A \cup B$. Exercise 1 tests your understanding of the grammar of set theory. Also, proving some of the parts of Theorem 5.5 should provide good practice with these symbols.

Phrases like "contains" and "is contained in" are used ambiguously by mathematicians. We have already used the latter phrase to mean $\subseteq$, and this is probably how it is used most frequently. But it can also be used to mean $\in$. Fortunately, the ambiguity is not too serious because the context almost always makes it clear which meaning is intended.

Theorem 5.5: (a) $A \subseteq A \cup B$
(b) $A = A \cup B \leftrightarrow B \subseteq A$
(c) $A \cap B \subseteq A$
(d) $A = A \cap B \leftrightarrow A \subseteq B$
(e) $A \cap B = A \cup B \leftrightarrow A = B$
(f) $A \subseteq B \cap C \leftrightarrow A \subseteq B$ and $A \subseteq C$
(g) $A \cup B \subseteq C \leftrightarrow A \subseteq C$ and $B \subseteq C$

Proof: (b) For the forward direction, assume $A = A \cup B$. We want $B \subseteq A$, so assume $x \in B$. (Note we're doing a conditional proof within a conditional proof.) Therefore, $x \in A$ or $x \in B$, by propositional logic. So $x \in A \cup B$, by definition of $\cup$. Thus $x \in A$, by the first assumption. Since x was arbitrary, we've shown that $B \subseteq A$, as desired.

For the reverse direction, assume $B \subseteq A$. We want $A = A \cup B$, which by Theorem 5.4(f) says $A \subseteq A \cup B$ and $A \cup B \subseteq A$. We have just shown $A \subseteq A \cup B$ in part (a). For the other part, assume $x \in A \cup B$. So $x \in A$ or $x \in B$. But since $B \subseteq A$, if $x \in B$, then $x \in A$. Therefore we can conclude that $x \in A$. Since x was arbitrary, we've shown $A \cup B \subseteq A$, and so we obtain $A = A \cup B$.

The proofs of the other parts are left for the exercises. ■

Now that we know more about sets, we can prove an important consequence of mathematical induction.

Theorem 5.6 : Every nonempty set of natural numbers has a *least* element.

Proof: We prove this by induction, but we must carefully phrase what we are proving. Let P(n) be the statement "Every set that contains a natural number less than or equal to n contains a least natural number." Note that P has a quantified set variable, but the natural number variable n is free in it. We wish to prove $\forall n$ P(n), by induction.

Let $n = 1$. By Theorem 4.16(a), 1 is the least natural number. So if a set contains a natural number less than or equal to 1, it contains 1, which is the least natural number in the set.

Now assume P(n), and let A be any set that contains a natural number less than or equal to $n + 1$. We must show that A contains a least natural number. In the case that $n + 1$ is the least number in A, we are of course done. If not, let $B = \{m \in A \mid m < n + 1\}$. The set B contains natural numbers that are less than $n + 1$, and therefore, by Theorem 4.16(c), equal to or less than n. Thus we can apply the induction hypothesis to B; so B contains a least natural number, which is also the least natural number in A.

This completes the proof of $\forall n$ P(n). The theorem easily follows: if A is a nonempty subset of $\mathbb{N}$, then there is some natural number in A. Let n be such a number (by ES). Using P(n), we conclude that A contains a least natural number. ■

Theorem 5.6 can be used in conjunction with indirect proof to prove statements by what is sometimes called the "no least counterexample" method (see Exercise 8).

Theorem 5.6 states an important property of $\mathbb{N}$ that fails for many other number systems and ordered structures. In particular, it fails with the word "natural" replaced by "real." (For instance, $\mathbb{Z}$ is a nonempty subset of $\mathbb{R}$ with no least member; so is $\mathbb{R}$ itself.) An ordering relation with the property described in this theorem is called a **well-ordering**. In other words, the theorem asserts that $\mathbb{N}$ is **well ordered** and is sometimes called the **well-ordering property of** $\mathbb{N}$ (see Exercises 16 through 18 of Section 6.3). Assuming a few basic properties of $\mathbb{N}$, the well-ordering property of $\mathbb{N}$ is equivalent to mathematical induction, and is sometimes used as an axiom instead of induction.

The Sum Rule for Counting

Set theory can be extremely abstract, but it also deals with many problems that are very concrete. In particular, the study of finite sets is closely related to *counting problems*, problems whose goal is to determine the number of members in some finite set. The following counting formula is quite simple to understand (though not to prove rigorously), and yet is surprisingly useful.

Theorem 5.7 (Sum rule for counting): Let A and B be finite sets, with m and n members, respectively.

(a) If A and B are disjoint, then $A \cup B$ has $m + n$ members.

(b) More generally, if $A \cap B$ has k members, then $A \cup B$ has $m + n - k$ members.

Proof: Rather than providing an extremely informal proof of this result now, we prove it rigorously in Section 7.6, where we examine counting problems in depth. ■

Clearly, part (b) of this theorem makes part (a) superfluous. We state part (a) separately because it is such an important special case. Part (a) easily generalizes to three or more sets. It is more complicated to generalize part (b), as the next example illustrates.

Example 1: Mudville High has three varsity teams. The table tennis team has 13 members, the Ultimate Frisbee team has 21 members, and the boomerang team has 16 members. How many varsity athletes are there? The obvious answer is 50, but this is wrong if there is duplication on the teams. Suppose there are 11 students who are on more than one team. Can we conclude that there are 39 athletes? This would also be wrong if there are people who play on all three teams. Exercise 12 asks you to investigate this further and find the correct formula.

We conclude this section with an overview of the most direct ways to prove basic relationships between sets.

Table 5.1 Summary of How to Prove Statements about Sets

(1) To prove a statement of the form $A \subseteq B$, assume that $x \in A$ (where x is arbitrary) and show that $x \in B$.

(2) To prove a statement of the form $A = B$, prove both $A \subseteq B$ and $B \subseteq A$.

(3) To prove a statement of the form $A \not\subseteq B$, find a member of A that is not in B.

(4) To prove a statement of the form $A \subset B$, prove $A \subseteq B$ and $B \not\subseteq A$.

(5) To prove a statement of the form $A \neq B$, prove $A \not\subseteq B$ or $B \not\subseteq A$. That is, find an element of either set that is not in the other one.

Exercises 5.2

(1) Classify each of the following expressions as either (i) a grammatically correct *statement*, (ii) a grammatically correct *expression* denoting a set, or (iii) grammatically incorrect and therefore meaningless. Assume that A, B, and C are set variables and P and Q are propositional variables.

(a) $A \cup B \subseteq C$
(b) $A \cup (B \subseteq C)$
(c) $A \leftrightarrow B$
(d) $(A \cup B) = \mathrm{P}$
(e) $\mathrm{P} \cup \mathrm{Q} \rightarrow \mathrm{P}$
(f) $(x \in A) \cup (x \in B)$
(g) $\mathrm{P} \wedge A \cup B \subseteq C$
(h) $\{x \in A \mid B \cup \{x\} \subseteq C\} \cap C$
(i) $A \cup B \in A \cap C$

(2) Prove any two parts of Theorem 5.2 that were not proved in the text.

(3) Prove any two parts of Theorem 5.3 that were not proved in the text.

(4) Prove any two parts of Theorem 5.4 that were not proved in the text.

(5) Prove any two parts of Theorem 5.5 that were not proved in the text.

(6) For each of the following statements, either prove that it is true for all sets or find a counterexample to show that it is not. Also, in parts (a) and (b), if the statement is not always true, at least try to prove that one side must be a subset of the other side.

(a) $A \cup (B - C) = (A \cup B) - (A \cup C)$ (Note this resembles Theorem 5.2(i)).
(b) $(A - B) \cup (A - C) = A - (B \cup C)$
(c) $A \subseteq B$ iff $(A - B) = \varnothing$
(d) $A \subseteq (B \cup C)$ iff $A \subseteq B$ or $A \subseteq C$

(7) True or false (with brief explanations):
(a) $\mathbb{R} \cup [3, 7) \subseteq \mathbb{R}$
(b) $[1, 4] \cup (3, 6] = [1, 9) \cap [2, 6]$
(c) $[1, 6] - [2, 5] = [1, 2] \cup [5, 6]$
(d) $[3, 6] \cup [6, 8] = [3, 8]$
(e) $(3, 6) \cup (6, 8) = (3, 8)$
(f) $\mathbb{R} - (\mathbb{Q} - \mathbb{N}) = (\mathbb{R} \cup \mathbb{N}) - \mathbb{Q}$

(8) Suppose we want to prove a statement of the form $\forall n\ P(n)$. If we want to use indirect proof, what do we assume? From that assumption, what can we assert to exist? Then, using Theorem 5.6, how can we strengthen this assertion? Often, this last assertion easily leads to a contradiction.

(9) Reprove Theorem 4.9 by the method outlined in Exercise 8. You may use Theorem 4.16, which does not require Theorem 4.9 in its proof.

(10) (a) Prove that $\mathbb{N} \subset \mathbb{Z}$.
*(b) Prove that $\mathbb{Z} \subset \mathbb{Q}$. ***Hint:*** Show that 1/2 is not in $\mathbb{Z}$.

(11) At a meeting of the Swampscott Phrenology Club, 37 members are present. Of these, 13 are wearing glasses, 8 are wearing sandals, and 20 are wearing *neither* glasses nor sandals. According to the sum rule, how many must be wearing *both* glasses and sandals?

(12) (a) Investigate the situation described in Example 1. If you wish, stick to the numbers in that example, but try various possibilities for the number of athletes on each pair of teams and the number on all three teams. On the basis of your results, conjecture a formula for the total number of athletes in terms of the number on the individual teams, the number who are on more than one team, and the number who are on all three teams. You might find Venn diagrams helpful for your investigation.

(b) Assuming your formula from part (a) is correct, derive a formula for the number of elements in the union of any three finite sets A, B, and C, in terms of the number of members in $A, B, C, A \cap B, A \cap C, B \cap C$, and $A \cap B \cap C$.

*(c) Carefully compare Theorem 5.7(b) and the formula you found in part (b) of this problem. Then try to describe (in words) how to calculate the number of elements in the union of four or more finite sets, in terms of the number of elements in the individual sets and the number of elements in the intersections of combinations of those sets.

Critique the proofs in Exercises 13 and 14. (If necessary, review the instructions for this type of problem in Exercises 4.2.)

(13) **Theorem:** Let a, b, and c be real numbers with $a < b < c$. Then $(a, b) \cup (b, c) = (a, c)$.

Proof: For the forward direction, assume x is any member of $(a, b) \cup (b, c)$. Then x is either in (a, b) or in (b, c), and we may proceed by cases. Case 1: Assume $x \in (a, b)$. That means $a < x < b$. But from $x < b$ and $b < c$ we obtain $x < c$, by transitivity. Therefore, $a < x < c$; this says that $x \in (a, c)$. Case 2: Assume $x \in (b, c)$. That means $b < x < c$. But from $b < x$ and $a < b$, we obtain $a < x$. Thus, $a < x < c$.

For the reverse direction, assume x is any member of (a, c). We again proceed by cases. Case 1: Assume $x < b$. Since $x \in (a, c)$, we also have $a < x$. Thus $a < x < b$; so $x \in (a, b)$. This implies that $x \in (a, b) \cup (b, c)$. Case 2: Assume $x > b$. We also have $x < c$. Thus $b < x < c$; so $x \in (b, c)$. This implies $x \in (a, b) \cup (b, c)$.

(14) **Theorem:** If $A \subset B$ and $B \subseteq C$, then $A \subset C$.

Proof: Assume $A \subset B$ and $B \subseteq C$. By the definition of $\subset$, $A \subseteq B$ and $A \neq B$. From $A \subseteq B$ and $B \subseteq C$, we have $A \subseteq C$, by Theorem 5.4(e). We must also show $A \neq C$. So assume, on the contrary, that $A = C$. Then $B \subseteq C$ becomes $B \subseteq A$. So we have $A \subseteq B$ and $B \subseteq A$, which yield $A = B$, by Theorem 5.4(f). This contradicts the fact that $A \neq B$.

(15) For any sets A and B, define $A \Delta B$, the **symmetric difference** of A and B, to be the set $(A - B) \cup (B - A)$. Prove.

(a) Commutativity of Δ: $A \Delta B = B \Delta A$
(b) Associativity of Δ: $A \Delta (B \Delta C) = (A \Delta B) \Delta C$
(c) The empty set is an identity for Δ: $A \Delta \varnothing = A$
(d) Each set is its own inverse for Δ: $A \Delta A = \varnothing$
(e) $\cap$ distributes over Δ: $A \cap (B \Delta C) = (A \cap B) \Delta (A \cap C)$

*(16) Remark 2 after Theorem 5.3 mentions a strong connection between set algebra and propositional logic. We now make this more precise. Consider any statement of set algebra that does not contain quantifiers or the symbols $\in$, $-$, or $\subset$. (Most of the results in Theorems 5.2 through 5.5 fit this description.) Turn the statement into a statement of pure logic by making the following replacements: change $\cup$ to $\vee$, $\cap$ to $\wedge$, $=$ to $\leftrightarrow$, $\subseteq$ to $\rightarrow$, $\varnothing$ to any contradiction, U (the universal set if one is being used) to any tautology, $'$ to $\sim$, and finally, every set variable to a propositional variable. You might also need to put in some parentheses to keep things grouped the way they were originally. Then, the original statement is a valid theorem of set algebra iff the new statement is a tautology.

(a) Transform each of the following results in the manner just described, and verify that the new statement is a tautology: parts (a) and (f) of Theorem 5.2; parts (a), (b), and (d) of Theorem 5.3; parts (c), (e), and (f) of Theorem 5.4; and parts (b), (c), and (f) of Theorem 5.5.

(b) Using our list of tautologies (Appendix 3) and the transformation process of part (a) *in reverse*, find at least two correct theorems of set algebra that have not been given in Theorems 5.2 through 5.5. Be careful, because not every tautology corresponds to a grammatically meaningful statement of set algebra; for example, tautology 12 does not.

(c) Figure out how to extend the above transformation process to include statements of set algebra that also contain the symbols $-$ and $\subset$.

*(17) Remark 3 after Theorem 5.3 mentions the duality principle of Boolean Algebra. For set algebra, this may be stated as follows. Consider any statement of set algebra of the type described in Exercise 16, except that it may also contain the symbol $\subset$. Form a new statement of set algebra in this way: change every $\cup$ to $\cap$ and vice versa; change every $\varnothing$ to U and vice versa; and wherever a $\subseteq$ or $\subset$ occurs, switch the *expressions* on the left and right sides of the symbol. The new statement is called the **dual** of the original. Then the original is a valid theorem iff its dual is valid.

(a) Explain why the dual of the dual of any statement is the original statement. In other words, if Q is the dual of P, then P is the dual of Q.

(b) Identify at least five dual pairs of statements in Theorems 5.2 through 5.5.

(c) It is possible for a statement to be its own dual. Find three such statements.

(d) Using the ideas of this exercise and the previous one, describe how to define the dual of any statement of propositional logic.

5.3 More Advanced Set Operations

Set theory gets much more complicated when it starts dealing with sets of sets. A set of sets is often called a **collection** or a **family** of sets. Collections of sets are the main topic of this section. One simple way to define a set of sets is to start with any set and then consider the set of all the subsets of the original set. This process is so important that it deserves a name:

Definition: The **power set** of any set A is the set of all subsets of A, denoted $\wp(A)$. In symbols,

$$\wp(A) = \{B \mid B \subseteq A\} .$$

Note that $\wp(A)$ is automatically a set of sets, no matter what kind of set A is. Working with power sets can take some care, as the following examples illustrate.

Example 1: Let's figure out the members of $\wp(A)$, where $A = \{4, 7\}$. Clearly, A has one subset with two elements (itself), two subsets with one element, and one subset with no elements. Therefore, $\wp(A) = \{\varnothing, \{4\}, \{7\}, \{4, 7\}\}$ (see Figure 5.4).

Braces within braces are tricky at first, but with some practice you will find them familiar and easy to work with.

Example 2: What would $\wp(\varnothing)$ be? Well, what are the subsets of $\varnothing$? Does it have any subsets? Yes, it has one—itself. Therefore, $\wp(\varnothing) = \{\varnothing\}$. It's important to see that $\{\varnothing\}$ is not the same as $\varnothing$. The set $\varnothing$ has no members, while $\{\varnothing\}$ has one member.

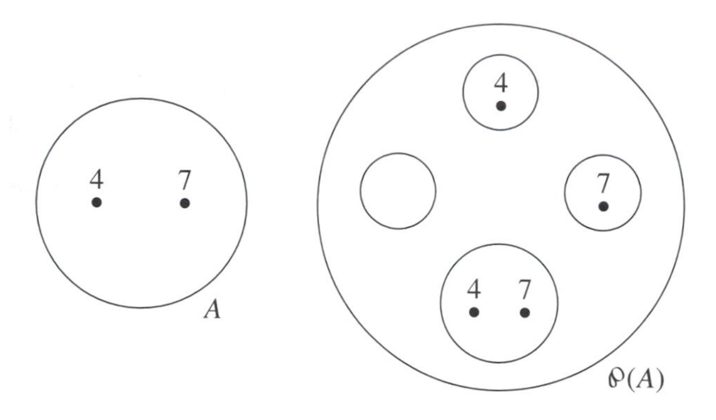

Figure 5.4 A simple set and its power set

Example 3: To carry this a step further, let's find the elements of $\wp(\wp(\varnothing))$. By Example 2, that has to be $\wp(\{\varnothing\})$. So what are the subsets of $\{\varnothing\}$? Well, this is a set with one element, so it must have two subsets: $\varnothing$ and the whole set. In other words, $\wp(\wp(\varnothing)) = \{\varnothing, \{\varnothing\}\}$.

The following theorem is helpful for keeping track of the number of elements in a power set and explains the origin of this term. It is also the first instance in this book of an induction proof that begins at $n = 0$ instead of $n = 1$. We mentioned in Section 4.5 that this is allowed, and Exercise 14 asks you to justify it.

Another complicating feature of this induction proof is that the theorem seems to be about sets more than it is about integers or natural numbers. Integers are mentioned in the statement of the theorem only to measure the number of elements in a set. It is fine to try to prove such a statement by induction, but it is usually necessary to have P(n) be the statement that the theorem is true for *all* sets with n elements. That will be our approach here. To put it another way, these theorems are basically about all *finite* sets. But rather than try to prove them for all finite sets at once, we prove them by induction on the number of elements in the set.

Theorem 5.8: If a set A has n elements, where n is any nonnegative integer, then $\wp(A)$ has 2^n elements.

Proof: We prove this by induction on n, starting at 0. For $n = 0$, the only set with 0 elements is $\varnothing$. We know that $\wp(\varnothing) = \{\varnothing\}$, which has one element. And since $2^0 = 1$, the theorem holds for $n = 0$.

For the induction step, assume the theorem holds for every set with n elements, and let A be any set with $n + 1$ elements. Pick any particular element of A, and call it c. *[This step is justified by ES.]* Let $B = A - \{c\}$. Note that B has n elements. Let's count the subsets of A. A subset of A either contains c as a member or it doesn't. The subsets of A that don't contain c are precisely the subsets of B, and so by the induction hypothesis there are 2^n of them. Furthermore, if D is any subset of B, then $D \cup \{c\}$ is a subset of A that contains c. It is also easy to see that every subset of A that contains c is of this form. Therefore, there are also 2^n subsets of A that contain c.

So the total number of subsets of A is $2^n + 2^n$, which equals 2^{n+1}, as desired. ■

Remarks: (1) This is probably the least formal proof in this book up to this point. You could try to prove this theorem more formally, but it is difficult to do so without using material from Chapter 7. Specifically, the argument in the third paragraph of the proof, and in fact the rigorous notion of what it means for a set to have n elements, are based on the concept of a one-to-one correspondence. Also, the definition of exponents is an inductive definition. But for most purposes, our proof is fine; it certainly conveys the main idea of why the theorem is true. A different approach to this proof is given in Theorem 7-17(d).

(2) Note that Examples 1 through 3 are of course consistent with Theorem 5.8: a set with 0, 1, or 2 elements must have (respectively) 1, 2, or 4 subsets.

(3) From Exercise 11 of Section 4.6, plus the fact that $0 < 2^0$, we know that $n < 2^n$ for any nonnegative whole number n. It follows that for every finite set A, $\wp(A)$ has more elements than A does. Theorem 7.26 shows, by a famous and ingenious argument, that this fact also holds for all infinite sets.

Here are a few basic results involving power sets:

Theorem 5.9: For any sets A and B:

(a) $\wp(A \cap B) = \wp(A) \cap \wp(B)$

(b) $A = B \leftrightarrow \wp(A) = \wp(B)$

(c) $A \subseteq B \leftrightarrow \wp(A) \subseteq \wp(B)$

(d) $A \subset B \leftrightarrow \wp(A) \subset \wp(B)$

Proof: (a) $C \in \wp(A \cap B) \leftrightarrow C \subseteq A \cap B$

$\leftrightarrow C \subseteq A$ and $C \subseteq B$ By Theorem 5.5(f)

$\leftrightarrow C \in \wp(A)$ and $C \in \wp(B)$

$\leftrightarrow C \in \wp(A) \cap \wp(B)$

(b) If $A = B$, then $\wp(A) = \wp(B)$ by Theorem 4.7. For the reverse direction, assume $\wp(A) = \wp(B)$. Since $A \subseteq A$ and thus $A \in \wp(A)$, it follows by axiom III-4 that $A \in \wp(B)$. Therefore $A \subseteq B$. Similar reasoning shows $B \subseteq A$. By Theorem 5.4(f), $A = B$.

(c) and (d) See Exercise 6. ■

Indexed Families of Sets

Now let's consider more general sets of sets than just power sets. In theory, no special notation is needed to describe sets of sets. We could just begin a discussion or a proof with a statement such as "Let A be a collection of sets." However, certain notation for sets of sets has come into general use. For one thing, to distinguish them from ordinary sets, sets of sets are usually denoted by capital script letters. For the most part, we follow this practice.

It is also common, when describing a collection of sets, to denote the individual sets in the collection with a subscripted variable called an **index**. So a mathematician might define sets A_n for each natural number n, and then define the collection of all these sets A_n. In this type of situation, the set $\mathbb{N}$ (that is, the set over which the *subscript* ranges) is referred to as an **index set** for the collection of sets, and the collection itself is called an **indexed family of sets**.

Example 4: For each n in $\mathbb{N}$, let A_n be the closed interval $[n, n + 1/n]$. So $A_1 = [1, 2]$, $A_2 = [2, 2.5]$, and so on. Then we can define an indexed family of sets $\mathscr{A}$ by

$$\mathscr{A} = \{A_n \mid n \in \mathbb{N}\}$$

It is important to see that $\mathscr{A}$ is *not* a set of real numbers. It is a set of sets of real numbers. That is, $\mathscr{A}$ is not a subset of $\mathbb{R}$ or a member of $\wp(\mathbb{R})$; rather it's a subset of $\wp(\mathbb{R})$ and an element of $\wp(\wp(\mathbb{R}))$. You might be tempted to think of the collection $\mathscr{A}$ as an *infinite sequence* of sets rather than a set of sets. This is a plausible alternative view of the situation, and it becomes the preferable way to view indexed families when the *order* of the sets in the family is important or it is desirable to allow *repetition* of sets in the family. See Example 6 of Section 7.4.

$\mathbb{N}$ is not the only possible index set. Any set can be one. Here's an example where the set of real numbers is an index set.

Example 5: For each real number c, let

$$L_c = \{(x, y) \mid x \in \mathbb{R} \text{ and } y \in \mathbb{R} \text{ and } y = cx\}$$

Then we can define $\mathscr{A} = \{L_c \mid c \in \mathbb{R}\}$. Graphically, you can see that L_c is a straight line of slope c through the origin, in the xy plane. So $\mathscr{A}$ can be thought of as a set of lines in the plane (see Figure 5.5).

When an unspecified set is used as an index set, the letter I or J is usually used for the index set, and i or j (respectively) is usually used as the subscript. By the way, even though we are restricting our attention to indexed families of sets, it is permissible to define indexed families of any kind of mathematical object.

Unions and Intersections of Collections of Sets

We've discussed the familiar operations of forming the union and intersection of two sets. By repeating these, it's a simple matter to form the union and intersection of any

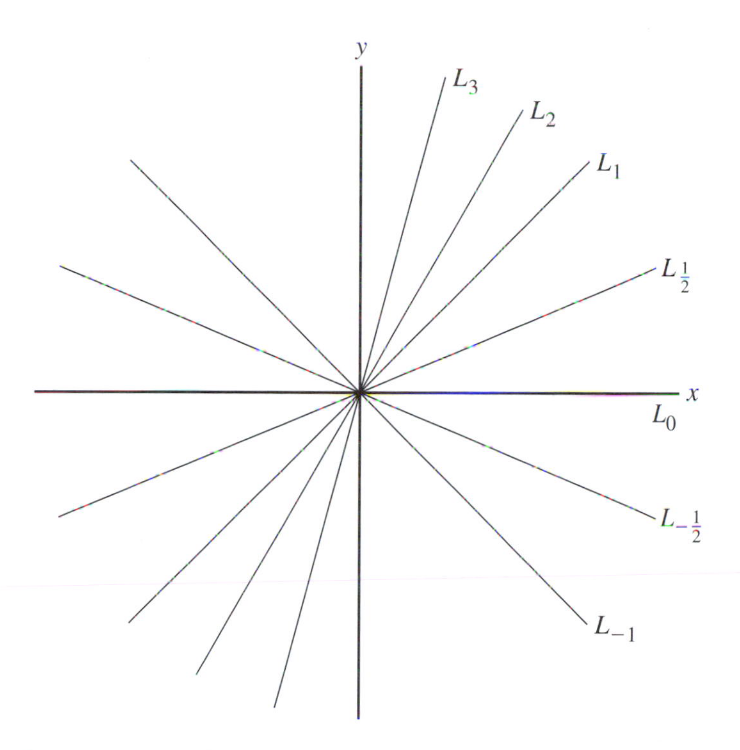

Figure 5.5 Illustration of Example 5: a family of sets indexed by $\mathbb{R}$

finite number of sets. We now define the important notions of the union and intersection of an arbitrary (and so, possibly infinite) collection of sets.

Definitions: Let $\mathscr{A}$ be any set of sets. Then the **union of** $\mathscr{A}$ or the **union over** $\mathscr{A}$, denoted $\bigcup\limits_{A \in \mathscr{A}} A$ or simply $\cup\mathscr{A}$, is the set

$$\{x \mid \exists A \in \mathscr{A}(x \in A)\}$$

The **intersection of** $\mathscr{A}$ or the **intersection over** $\mathscr{A}$, denoted $\bigcap\limits_{A \in \mathscr{A}} A$ or simply $\cap\mathscr{A}$, is the set

$$\{x \mid \forall A \in \mathscr{A}(x \in A)\}$$

Unless there is a universal set in the discussion, $\mathscr{A}$ must be nonempty for the second definition to make sense. The simple notations $\cup\mathscr{A}$ and $\cap\mathscr{A}$ introduced here are found in most set theory books but for some reason are rarely used by mathematicians other than set theorists. The longer notations are more common.

Notation: When working with an indexed family of sets, yet another notation is used for its union or intersection. If $\mathscr{A} = \{A_i \mid i \in I\}$, then the most common notation for the union of $\mathscr{A}$ is $\bigcup_{i \in I} A_i$, and similarly for the intersection of $\mathscr{A}$.

Take the time to see that these definitions say what they ought to. The union of $\mathscr{A}$ consists of all things that are in at least one of the sets in the collection $\mathscr{A}$, so it consists of all the sets in $\mathscr{A}$ "put together." Similarly, the intersection of $\mathscr{A}$ consist of all things that are in all the sets in the collection $\mathscr{A}$.

It was mentioned earlier that the quantifiers $\exists$ and $\forall$ are related to the connectives "or" and "and," respectively. The definitions of union and intersection of $\mathscr{A}$, together with the definitions of ordinary unions and intersections, are a good illustration.

Example 6: Let $\mathscr{A}$ be as defined in Example 4. Then the set $\bigcup\mathscr{A}$ could also be denoted $\bigcup_{n \in \mathbf{N}} A_n$. While $\mathscr{A}$ is not a set of real numbers, $\bigcup\mathscr{A}$ is; it's a set consisting of an infinite number of intervals. The set $\bigcap\mathscr{A}$ is also a set of real numbers, namely $\varnothing$.

Example 7: Let $\mathscr{A}$ be the collection defined in Example 5. Then $\bigcap\mathscr{A} = \{(0,0)\}$, since the origin is the one point common to all the lines L_c. Exercise 5 asks you to describe the set $\bigcup\mathscr{A}$.

Example 8: Here is the definition of the **Cantor set**, also known as **Cantor's discontinuum**. This set of real numbers is important in higher mathematics, and you will almost certainly encounter it again. To define the Cantor set, start with the closed unit interval [0, 1]. Then define sets A_n as follows:

Let A_1 be the open interval (1/3, 2/3).

Let $A_2 = (1/9, 2/9) \cup (7/9, 8/9)$.

Let $A_3 = (1/27, 2/27) \cup (7/27, 8/27) \cup (19/27, 20/27) \cup (25/27, 26/27)$.

To see the pattern here, note that A_1 is the middle third of the original interval, A_2 consists of the two middle thirds of what's left of the original interval after removing A_1, and so on (see Figure 5.6). Continuing in this way, we define sets A_n for every natural number n. (This definition can be made more algebraic and precise, if desired.) The Cantor set is then defined to be

$$[0, 1] - \bigcup_{n \in \mathbf{N}} A_n$$

Exercises 18 through 20 deal with some of the interesting features of the Cantor set.

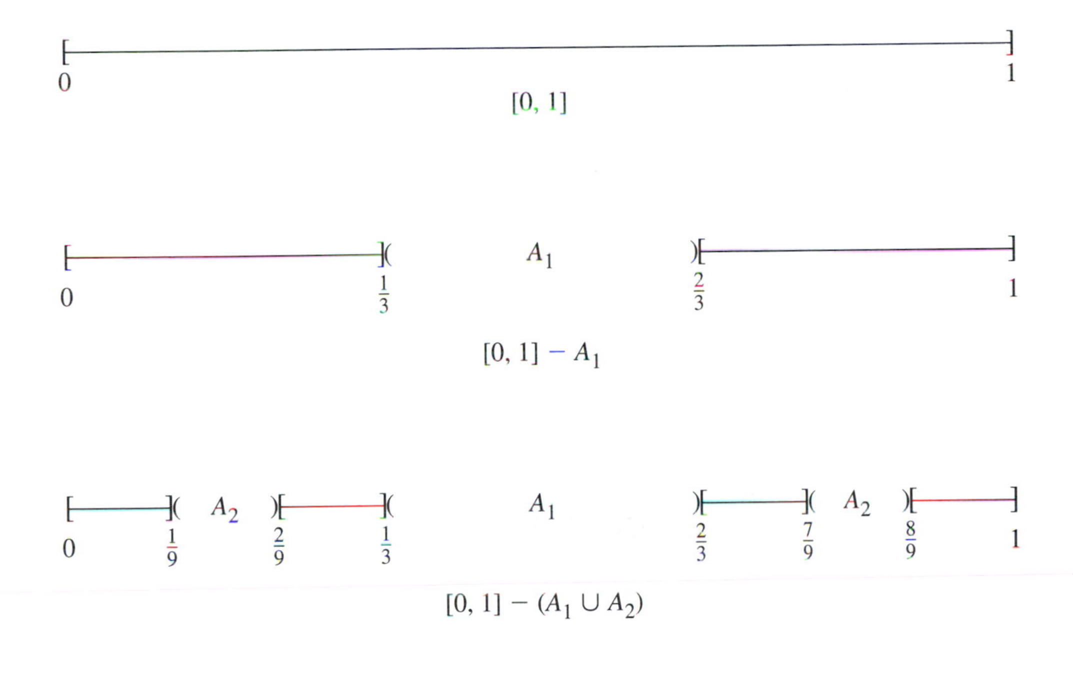

Figure 5.6 The first three or four stages in the construction of the Cantor set

Unions and intersections of infinite families of sets are sometimes called **infinitary** operations, as opposed to ordinary **finitary** operations. Most results about the finitary operations have infinitary analogs. Here are a few of these: see if you can figure out which parts of Theorems 5.2 and 5.3 the results of the following theorem are related to.

Theorem 5.10: Let B be any set, and $\mathscr{A} = \{A_i \mid i \in I\}$ any family of sets. As usual, when we use the symbol $'$, it means complement relative to some specified universal set.

(a) $B \cup \left(\bigcap_{i \in I} A_i\right) = \bigcap_{i \in I} (B \cup A_i)$

(b) $B \cap \left(\bigcup_{i \in I} A_i\right) = \bigcup_{i \in I} (B \cap A_i)$

(c) $(\bigcup_{i \in I} A_i)' = \bigcap_{i \in I} (A_i')$

(d) $(\bigcap_{i \in I} A_i)' = \bigcup_{i \in I} (A_i')$

(e) For each $i \in I$, $A_i \subseteq \bigcup_{i \in I} A_i$

(f) For each $i \in I$, $\bigcap_{i \in I} A_i \subseteq A_i$

Proof: (a) $x \in B \cup \left(\bigcap_{i \in I} A_i\right) \leftrightarrow x \in B$ or $x \in \left(\bigcap_{i \in I} A_i\right)$

$\leftrightarrow x \in B$ or $\forall i \in I\,(x \in A_i)$

$\leftrightarrow \forall i \in I\,(x \in B$ or $x \in A_i)$ By law of logic 17 (Figure 4.2)

$\leftrightarrow \forall i \in I\,(x \in B \cup A_i)$

$\leftrightarrow x \in \bigcap_{i \in I} (B \cup A_i)$

(e) Let $i \in I$ be arbitrary. If $x \in A_i$, then clearly $\exists i \in I\ (x \in A_i)$, so $x \in \bigcup_{i \in I} A_i$.

Therefore, $A_i \subseteq \bigcup_{i \in I} A_i$.

The proofs of the other parts are left for the exercises. ■

We have called parts (c) and (d) of Theorem 5.3 De Morgan's laws for sets. Similarly, parts (c) and (d) of Theorem 5.10 may be viewed as set versions of De Morgan's laws for quantifiers.

Exercises 5.3

(1) List all the members of:
(a) $\wp(\{1, 2, 3\})$
(b) $\wp(\{2, \{2\}\})$
(c) $\wp(\wp(\{\varnothing\}))$
(d) $\wp(\{1, 3, 5\}) \cap \wp(\{5, 6, 7\})$
(e) $\wp(\{1, 2, 3\}) - \wp(\{1, 3\})$

(2) True or false, with brief explanation:
(a) $3 \in \wp(\mathbb{N})$
(b) $\{3\} \in \wp(\mathbb{N})$
(c) $\{3\} \subseteq \wp(\mathbb{N})$
(d) $\{\varnothing\} \in \wp(\{\{\varnothing\}\})$
(e) $\wp(\mathbb{Z} \cap (2, 4)) = \{\varnothing, \{3\}\}$

(3) Characterize each of the following statements as always true, always false, or sometimes true and sometimes false. Explain briefly.

(a) $A \in \wp(A)$ (b) $A \subseteq \wp(A)$

(c) $A \cap B \in \wp(A \cup B)$ (d) $\wp(A) - \wp(B) = B - A$

(e) $\wp(\wp(A)) \in \wp(\wp(\wp(A)))$ (f) $\wp(A - B) \cap \wp(B - A) = \{\varnothing\}$

(4) Prove whichever of the following equations are true for all sets. For each one that's *not* always true, try to prove that one side is a subset of the other, and give a counterexample to the other direction. If neither side must be a subset of the other, give a counterexample to both directions.

(a) $\wp(A \cup B) = \wp(A) \cup \wp(B)$ (Compare with Theorem 5.9(a).)

(b) $\wp(A - B) = \wp(A) - \wp(B)$

(c) $\bigcup(\wp(A)) = A$

(d) $\wp(\bigcup\mathscr{A}) = \mathscr{A}$

(5) Referring to Examples 5 and 7, give a simple *geometric* description of $\bigcup\mathscr{A}$.

(6) Prove parts (c) and (d) of Theorem 5.9.

(7) Prove Theorem 5.10(b).

(8) Prove Theorem 5.10(c).

(9) Prove Theorem 5.10(d).

(10) Prove Theorem 5.10(f).

(11) For each $n \in \mathbf{N}$, let A_n be the interval $[2^{-n}, 2^{1-n})$. Give a simple description of the set $[0, 1] - \bigcup_{n \in \mathbf{N}} A_n$. Explain briefly.

(12) Let $B = \{(x, y) \mid y = x\}$, where x and y are real variables. For each real number r, let $A_r = \{(x, y) \mid x^2 + y^2 = r^2\}$. Then $\{A_r \mid r \in \mathbf{R}\}$ is a family of circles indexed by $\mathbf{R}$.

(a) Note that using $\mathbf{R}$ as the index set causes most of the circles in this indexed family to be repeated. This is allowed but may cause needless confusion. Name two smaller index sets that can be used to define the same collection of circles without any repetition. On the other hand, it is occasionally useful to allow repetition of sets in an indexed family.

(b) What is the union of all the A_rs?

(c) Describe $B \cap A_r$.

(d) Verify that Theorem 5.10(b) holds in this case.

(13) Let $A = (0, 1)$ and $B = (0, 1]$. For any $C \subseteq \mathbf{R}$ and $y \in \mathbf{R}$, let $C + y$ be the set $\{x + y \mid x \in C\}$. That is, $C + y$ is the set C shifted or **translated** y units. Describe the following sets in words or in more concise mathematical notation.

(a) $\bigcup_{n \in \mathbf{N}} (A + n)$ (b) $\bigcup_{n \in \mathbf{N}} (B + n)$

(c) $\bigcup_{x \in A} (\mathbb{Z} + x)$ (d) $\bigcup_{x \in B} (\mathbb{Z} + x)$

(14) (a) Justify the use of induction proofs that start at 0 instead of at 1, as in Theorem 5.8. That is, prove

$$[P(0) \land \forall n \geq 0\, (P(n) \rightarrow P(n+1))] \rightarrow \forall n \geq 0\, P(n)$$

where n is an integer variable.

*(b) Generalizing part (a), show that an induction proof can start at any integer k. That is, prove

$$[P(k) \land \forall n \geq k\, (P(n) \rightarrow P(n+1))] \rightarrow \forall n \geq k\, P(n)$$

where n and k are both integer variables. ***Hint:*** Let Q(n) be the statement P($n + 1 - k$). Also, you may use the result of Exercise 13 of Section 5.1.

(15) This problem illustrates a natural situation where you might do an induction proof beginning at a number other than 0 or 1: suppose that you want to prove that some property holds for all *polygons*. It might be natural to attempt this by induction on the number of sides.

(a) In such a proof, what would be the initial value of n (represented by k in Exercise 12(b))?

(b) For what class of polygons would you first have to show the result?

(c) State carefully what the induction step would be.

(16) Mathematicians often say things like "Let $\mathscr{A}$ be a collection of nonempty disjoint sets." From this, which of the following can you conclude?

(a) Each set in $\mathscr{A}$ is nonempty.

(b) Each set in $\mathscr{A}$ is disjoint.

Carefully explain the difference between parts (a) and (b). To avoid this subtle linguistic confusion, it is more precise to say, "Let $\mathscr{A}$ be a collection of nonempty, *pairwise* disjoint sets."

(17) Critique the following well-known and entertaining proof. It is included here because it is similar to the proof of Theorem 5.8. (If necessary, review the instructions for this type of problem in Exercises 4.2.)

Theorem: All horses are the same color.

Proof: By induction on n, we prove that, given any set of n horses, they are all the same color. Clearly, this implies all horses are the same color. For $n = 1$ (or $n = 0$ if we choose to start there), the statement is trivial. So assume the statement holds for n, and let A be any set of $n + 1$ horses. Let c be any horse in A. By the induction hypothesis, all the horses in $A - \{c\}$ are the same color. Let k be any other horse in A. Again, all the horses in $A - \{k\}$ are the same color. So c and k are both the same color as all the other horses in A. Therefore, all the horses in A are the same color, as desired.

*(18) Prove that the Cantor set (Example 8) consists of all numbers in [0,1] that have a base 3 expansion with no 1's.

*(19) Prove that the Cantor set contains no intervals. You may use the result of Exercise 18.

*(20) Prove that the Cantor set contains no isolated points. (A number x is an **isolated point** of a set A of real numbers if $x \in A$ and for some $c > 0$, no other number between $x - c$ and $x + c$ is in A.)

Suggestions for Further Reading: For a more complete treatment of basic set theory, see Stoll (1979), Suppes (1960), or Vaught (1995). Devlin (1993) is a good text at a somewhat higher level. Most logic and set theory books discuss paradoxes, including the paradoxes of set theory; Kline (1982) does so in more detail than most. Many authors have attempted to capture the brilliant essence of Gödel's incompleteness theorem, including Nagel and Newman (1958), Hofstadter (1989), and Smullyan (1992). For more information about the subject of boolean algebra, see Pfleeger and Straight (1985), Rueff and Jeger (1970), or Stoll (1979). The first two of these cover important applications of boolean algebra such as switching theory.

Chapter 6

Relations

6.1 Ordered Pairs, Cartesian Products, and Relations

In this chapter, we study the important subject of binary relations. This section is devoted primarily to definitions of basic concepts, and Sections 6.2 and 6.3 discuss two useful types of relations. The single most important type of relation, functions, is covered in Chapter 7.

In Chapters 6 and 7, several concepts are defined in two different ways. The first version is always a rather nontechnical or intuitive one, and the second is a more rigorous one involving sets. Some books give only one of the two definitions in each case, but it's more educational to see both approaches.

Section 5.1 mentions that virtually all mathematical concepts can be defined and developed in terms of set theory. Even numbers (including natural numbers, rational numbers, real numbers, and so on) can be defined as special types of sets. All mathematicians are aware of this set-theoretic approach to mathematics but generally find it artificial and prefer not to think of relations and functions (let alone numbers!) as types of sets. Also, remember that set theory is only a hundred years old, whereas the study of numbers and functions is much older. So most of the fundamental ideas of mathematics are not based on set theory. On the other hand, the set-theoretic approach is very useful for proofs and theoretical work. So you can see why you should learn both approaches to these basic concepts.

Definition (intuitive): For any two objects a and b, the **ordered pair** (a, b) is a notation specifying the two objects a and b, in that order.

Definition (set-theoretic): For any two objects a and b, the **ordered pair** (a, b) is defined to be the set $\{\{a\}, \{a, b\}\}$.

Perhaps you can see why neither definition of ordered pairs is totally satisfactory. The intuitive definition is not a rigorous mathematical definition, any more than it would be to define a natural number as a sequence of digits. That sounds fine at first, but a number is definitely not the same thing as the *numeral* used to denote it. On the other hand, the set-theoretic definition is very strange looking and conveys none of the intuitive meaning of what an ordered pair is. For these reasons, many mathematicians view the concept of ordered pairs as an undefined, primitive notion.

There are just two important properties of ordered pairs. The first is that you can form the ordered pair of any two objects whatsoever. The second is the familiar condition for equality of ordered pairs:

$$(a, b) = (c, d) \text{ iff } a = c \text{ and } b = d$$

These properties appear as set axioms IV-4 and IV-5 in Appendix 1. However, if the set-theoretic definition of ordered pairs is used, these axioms are provable and therefore superfluous (see Exercise 7).

Definition: For any two sets A and B, their **cartesian product** is the set of all ordered pairs whose first member is in A and whose second member is in B; in symbols,

$$A \times B = \{(x, y) \mid x \in A \text{ and } y \in B\}$$

René Descartes (1596–1650), from whose name the word "cartesian" is derived, was an extremely important figure in the development of modern mathematics and philosophy. As a child his health was poor, and he developed a lifelong habit of spending his mornings in bed, thinking and writing. At the age of eighteen his life entered a less intellectual phase, including a short period of heavy gambling and several years of intermittent military service. Fortunately, inspired in part by three vivid dreams in 1619, Descartes quit the military and devoted the rest of his life to academic pursuits.

Descartes's major mathematical achievement was the invention of analytic geometry: the system whereby equations can be graphed and, conversely, geometric figures can be analyzed algebraically. The importance of this contribution to mathematics—a two-way link between symbolic entities (equations and inequalities) and pictorial entities (straight lines, circles, parabolas, and so on)—would be difficult to overestimate. This achievement also strongly influenced his philosophical views, notably that "pure reason," of a mathematical sort, was the correct path to truth and knowledge. His famous conclusion, "Cogito, ergo sum" ("I think, therefore I am"), also emphasized the importance of the individual and rationality.

Descartes was deeply religious, but his emphasis on the individual and reason was not consistent with the views of the Catholic Church. For this and other reasons, he spent the second half of his life away from his native France, mostly in Holland.

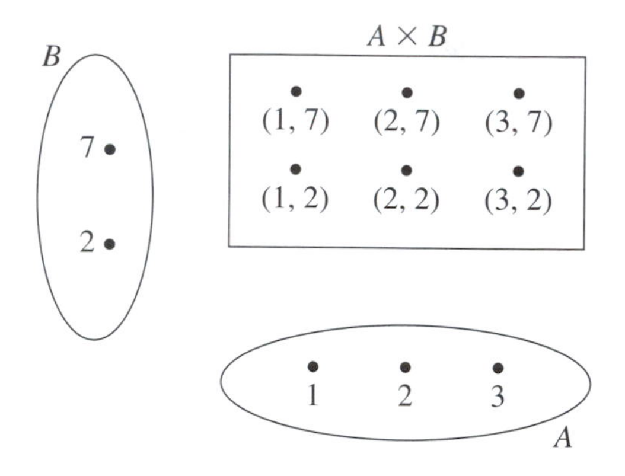

Figure 6.1 A simple cartesian product

You have been using cartesian products since eighth or ninth grade, whether or not you have called them that. Every time you draw a graph in an ordinary two-dimensional rectangular coordinate system, the coordinate system allows you to represent ordered pairs of real numbers on paper. In other words, the coordinate system turns your piece of paper into a picture of the cartesian product $\mathbb{R} \times \mathbb{R}$.

In the notation $A \times B$, the $\times$ is normally pronounced "cross" (not "times" or "ex"), and a cartesian product may also be called a cross product. But there is absolutely no connection between this notion and vector cross product. However, there is a definite connection between cartesian products and ordinary multiplication, as we now see.

Theorem 6.1: If A and B are sets, with m and n members respectively ($m, n \geq 0$), then $A \times B$ has mn members.

Proof: This proof can be done by a straightforward induction on either m or n (starting at 0), and we leave it for Exercise 9. ■

Example 1: Let $A = \{1, 2, 3\}$ and $B = \{7, 2\}$. Then $A \times B$ is the set $\{(1, 7), (1, 2), (2, 7), (2, 2), (3, 7), (3, 2)\}$. Note that $A \times B$ has six ordered pairs and that we have listed them in a systematic manner (see Figure 6.1).

Example 2: Suppose a certain T-shirt comes in sizes small, medium, large, and extra large and in colors red, blue, green, orange, and purple. If we write $A = \{\text{S, M, L, X}\}$ and $B = \{\text{R, B, G, O, P}\}$, then $A \times B$ consists of 20 ordered pairs. These ordered pairs may be viewed as representing all the possible choices of size and color for this type of shirt. Figure 6.2 is a tree diagram illustrating this situation.

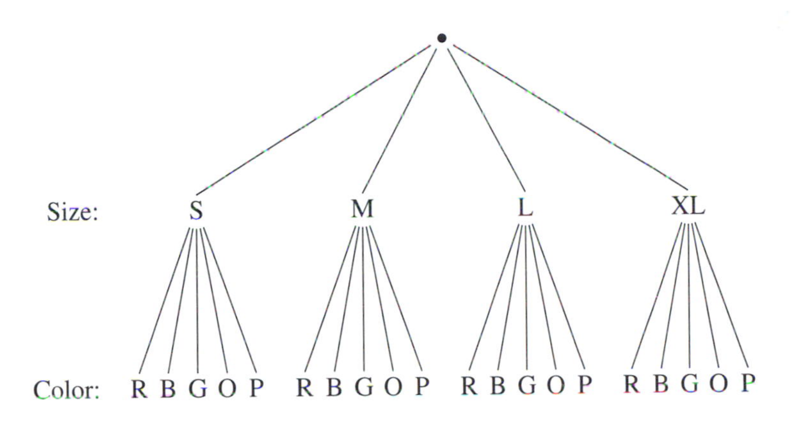

Figure 6.2 Tree diagram for Example 2

Is cartesian product a commutative operation? In other words, does $A \times B = B \times A$ in general? It's pretty clear that the answer is no. For instance, in Example 1, $A \times B$ contains (2, 7) but not (7, 2), whereas $B \times A$ contains (7, 2) but not (2, 7). And we know that $(2, 7) \neq (7, 2)$. So $A \times B$ and $B \times A$ are different as sets of ordered pairs. On the other hand, Theorem 6.1 at least guarantees that if A and B are finite sets, then $A \times B$ and $B \times A$ must have the same number of elements.

Is cartesian product an associative operation? In other words, does $(A \times B) \times C = A \times (B \times C)$, in general? Once again, the answer is no, but in this case the difference between the two cartesian products is often insignificant. For instance, let $A = B = C = \mathbb{N}$. Then $(A \times B) \times C$ contains ((2, 7), 9), while $A \times (B \times C)$ contains (2, (7, 9)). Does ((2, 7), 9) = (2, (7, 9)) ? Technically, no, but in many situations, it might as well. This discussion leads us to the concepts of ordered triples, ordered n-tuples, and cartesian products of more than two sets.

Definitions: For any three objects a, b, and c, the **ordered triple** (a, b, c) is defined to be $((a, b), c)$.

For any sets A, B, and C, $A \times B \times C = \{(a, b, c) \mid a \in A,\ b \in B, \text{ and } c \in C\}$.

Remarks: (1) These definitions represent a somewhat arbitrary choice: technically, (a, b, c) means $((a, b), c)$, not $(a, (b, c))$. And $A \times B \times C$ means $(A \times B) \times C$, not $A \times (B \times C)$. But as the previous discussion indicates, this distinction can often be ignored.

(2) These definitions can be extended to define **ordered n-tuples** and cartesian products of any number of sets. These would be so-called **inductive** definitions. In other words, if we have already defined what $(a_1, a_2, ..., a_n)$ means, then we can define

$(a_1, a_2, ..., a_{n+1})$ to be $((a_1, a_2, ..., a_n), a_{n+1})$. And if we have already defined what $A_1 \times A_2 \times ... \times A_n$ means, we can then define the cartesian product $A_1 \times A_2 \times ... \times A_{n+1}$ to be $(A_1 \times A_2 \times ... \times A_n) \times A_{n+1}$. Inductive definitions are studied in Chapter 7. (One often writes A^2 for $A \times A$, A^3 for $A \times A \times A$, and so on.) Theorem 6.1 generalizes to cartesian products of three or more sets (see Exercise 10).

(3) Theorem 6.1 is the basis for an extremely useful formula called the **fundamental counting principle** or the **product rule for counting**. Recall Example 2. In that situation, there are four possibilities for the size of a shirt and five possibilities for the color. To select a shirt, a person must decide on both size and color. Theorem 6.1 says that the number of possible ways to make this *sequence* of decisions is 4 × 5, or 20. Like Theorem 6.1, this formula can be extended to a sequence of three or more decisions. In its general form, the product rule may be stated as follows:

☞ *Suppose that it is required to make a sequence of k decisions. If there are n_1 possible choices for the first decision, n_2 possible choices for the second decision, and so on, then the number of possible ways to make the whole sequence of decisions is $n_1 n_2 ... n_k$.*

Relations

We're now ready to define the major concept of this chapter. Once again, we give two definitions, one intuitive and one set-theoretic. Neither one is particularly complicated.

Definition (intuitive): A **binary relation** is a statement with two free variables (which are usually assumed to have specific sets as their domains, although the variables may be unrestricted).

Definition (set-theoretic): A **binary relation** is a set of ordered pairs.

The word "binary" means "pertaining to the number two"; in the term "binary relation" it refers to the fact that there are two free variables. Similarly, a **ternary relation** is a statement with three free variables, or a set of ordered triples. In general, an ***n*-ary relation** is a statement with n free variables or a set of ordered n-tuples. Also, the term **unary relation** is occasionally used, but this is essentially just a fancy term for a set or a subset of a specified universal set. Our primary interest is in binary relations, and so we normally omit the word "binary" when discussing them. Another (equivalent) way to give the set-theoretic definition of a relation is to say that it's a subset of some cartesian product.

Example 3: The intuitive definition of a binary relation describes what most people would call a relationship between two things. For example, the statement that one number is less than another is a binary relation. The statement that two numbers add up to 74 is a binary relation. The statement that one person is the father of another is a binary relation.

You can see that both forms of the definition of binary relations are quite simple. But it's not so easy to see the connection between the two or why they are two definitions of the same concept. Here is an example of how the two are connected.

Example 4: Let x and y be real variables. Then one simple type of relation (in the intuitive sense) would be an equation, like $x^2 + y^2 = 25$. The set-theoretic counterpart of this relation is then the set of all ordered pairs that satisfy the equation, or $\{(x, y) \mid x^2 + y^2 = 25\}$. So it would include ordered pairs such as (0, 5), (−5, 0), (4, −3), and so on. So when you *graph* an equation, you are really drawing a picture of the set of ordered pairs corresponding to that equation. Some mathematicians prefer to say that the equation is the relation, and the set of ordered pairs you get is not the relation but rather the graph of the relation.

Given any statement with two free variables, it's easy to form a set of ordered pairs in this manner. (You might have to make an arbitrary choice of which variable will be first and which second in the ordered pairs.)

Definitions: A subset of $A \times B$ is called a **relation between A and B**. Also, a subset of $A \times A$ is called a **relation on A**.

Notation: The letters R, S, and T denote binary relations. We sometimes write xRy as a shorthand for $(x, y) \in R$.

Definitions: Given a relation R, the **domain** (respectively, **range**) of R is the set of all objects that occur as a first (respectively, second) member of some ordered pair in R. In symbols,

$$\mathrm{Dom}(R) = \{x \mid \exists y\, (xRy)\}$$

$$\mathrm{Rng}(R) = \{y \mid \exists x\, (xRy)\}$$

Note that the concept of the domain of a *relation* is somewhat different from the concept of the domain of a *variable*, which we have been using since Chapter 3.

Example 5: Let $R = \{(3, 6), (8, 2), (1, 2), (0, 0), (-5, 3)\}$. Then R is clearly a relation, that is, a set of ordered pairs. By inspection, $\mathrm{Dom}(R) = \{3, 8, 1, 0, -5\}$, while $\mathrm{Rng}(R) = \{6, 2, 0, 3\}$.

Example 6: Again consider the relation on $\mathbb{R}$ defined by the equation $x^2 + y^2 = 25$. Whether we consider this relation to be an equation or a set of ordered pairs, its domain and range are both the interval $[-5, 5]$. How do we know this? The standard algebraic method is to solve the equation for each of the variables. If this equation is solved for y, it becomes $y = \pm\sqrt{25 - x^2}$. Since negative numbers do not have square roots, this is possible if and only if $x^2 \leq 25$, which in turn means $-5 \leq x \leq 5$, or $x \in [-5, 5]$. (Here we are assuming that every nonnegative number has a square root.) Exercise 12 asks you to show this more rigorously.

Example 7: Consider the "less than" relation on $\mathbb{R}$, that is, $\{(x, y) \mid x, y \in \mathbb{R}$ and $x < y\}$. Its domain and range are both $\mathbb{R}$. What does the graph of this relation look like?

Example 8: Let A be the set of all people who have ever lived, and let R be the "parenthood" relation on A; that is, $R = \{(u, v) \mid u, v \in A$ and u is a parent of $v\}$. Then Dom(R) is the set of all people who have ever had a child, while Rng(R) might be A—might be because whether one believes in evolution or in divine creation, there is some question about whether the first humans are in Rng(R).

Inverse Relations

We conclude this section with a brief discussion of one of the most important ways of forming new relations. This concept is of primary importance in Chapter 7.

Definition: If R is any binary relation, the **inverse** of R, denoted R^{-1} and read R inverse, is the relation obtained by reversing the order of all the ordered pairs in R. In symbols,

$$R^{-1} = \{(y, x) \mid (x, y) \in R\}$$

This definition is given in terms of ordered pairs. If we are thinking of relations as propositions, we can pretty much say that "inverse" means "reverse," as the following examples illustrate.

Example 9: Recalling Example 7, the inverse of the "less than" relation is $\{(y, x) \mid y, x \in \mathbb{R}$ and $x < y\}$, which is the same as $\{(x, y) \mid x, y \in \mathbb{R}$ and $x > y\}$. In other words, the inverse of the "less than" relation is the "greater than" relation.

Example 10: Similarly, the inverse of the relation in Example 8 is $\{(u, v) \mid u, v \in A$ and u is a child of $v\}$. Thus, the inverse of the "parent of" relation is the "child of" relation.

Example 11: Figure 6.3 shows several pairs of graphs of relations and their inverses, involving real variables. Note that in each case the equation or other algebraic statement for the inverse relation is obtained simply by switching the variables. However, people often like to see graphs of equations solved for y, not x. For instance, it might be preferable to transform the first inverse equation in Figure 6.3 into the equivalent form $y = \pm\sqrt{x + 5}$, using elementary algebra.

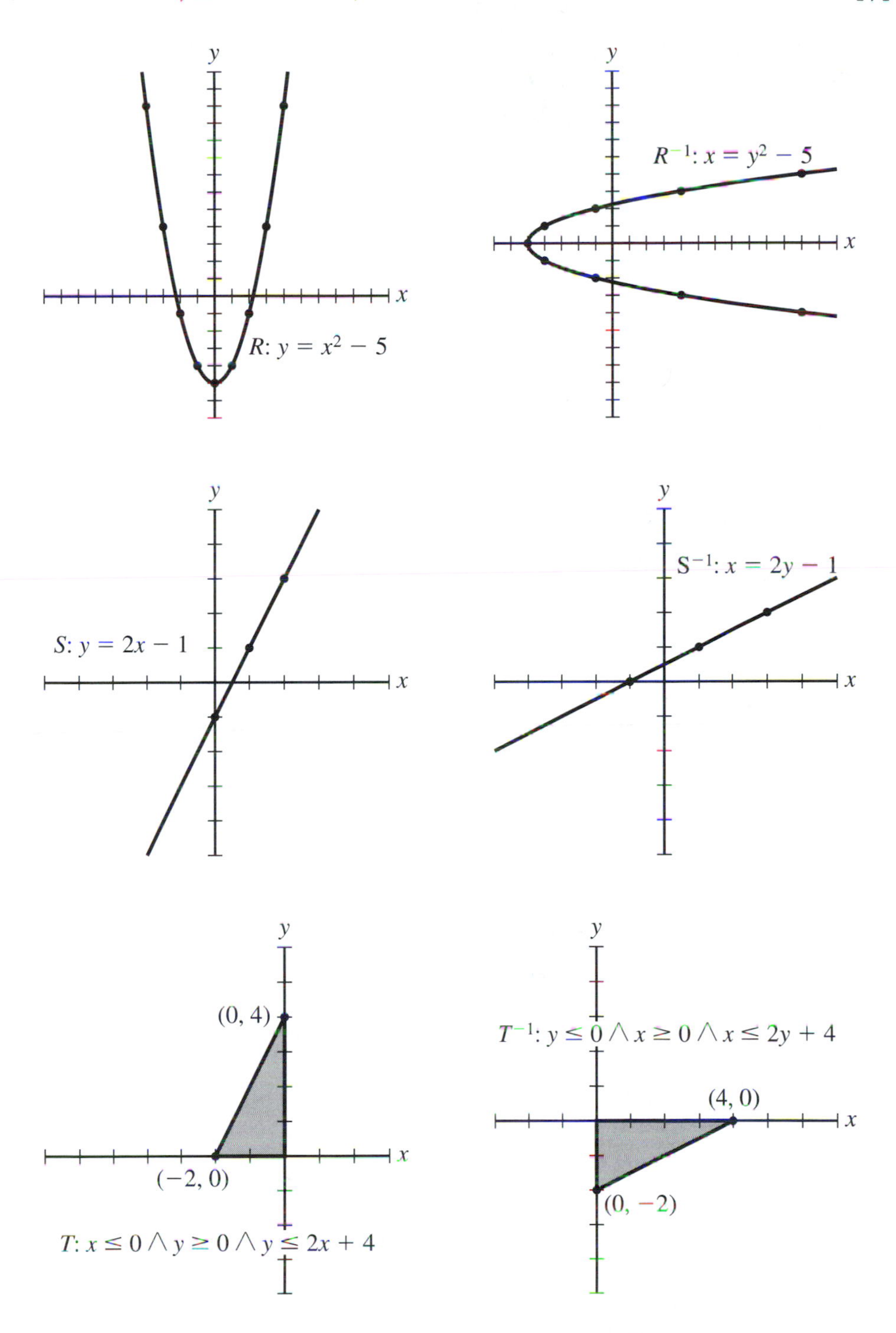

Figure 6.3 Graphs of several relations and their inverses

Observe (from Figure 6.3) that there is a basic geometric similarity between the graphs of R and R^{-1}: *For any relation R on* ℝ, *the graph of* R^{-1} *is obtained from the graph of R by reflecting it across the line* $y = x$. Note that we don't get the inverse graph from the original by a rotation; an actual reflection, or flip, is required.

Example 12: A relation can be its own inverse. Refer to Example 4.

We conclude this section with some simple facts:

Theorem 6.2: For any relation R:
(a) $(R^{-1})^{-1} = R$
(b) $\text{Dom}(R^{-1}) = \text{Rng}(R)$
(c) $\text{Rng}(R^{-1}) = \text{Dom}(R)$
Proof: See Exercise 11. ∎

Another way of putting part (a) of Theorem 6.2 is that if S is the inverse of R, then R is also the inverse of S. In other words, the property that one relation is the inverse of the other is symmetric. So it is common to say simply, "R and S are inverses."

Exercises 6.1

(1) Let $R = \{(8, 4), (-3, 4), (2, 1)\}$ and $S = \{(8, 2), (2, 3), (8, 4), (\pi, \pi)\}$. Find:
(a) $\text{Dom}(R)$
(b) $\text{Rng}(R)$
(c) $\text{Dom}(S)$
(d) $\text{Rng}(S)$
(e) R^{-1}
(f) $R \cup S^{-1}$
(g) $\text{Dom}(R \cap S)$
(h) $\text{Dom}(R) \cap \text{Dom}(S)$
(i) $\text{Dom}(R \cup S)$
(j) $\text{Dom}(R) \cup \text{Dom}(S)$

(2) Graph the following subsets of $\mathbb{R} \times \mathbb{R}$.
(a) The relation R of Exercise 1
(b) The relation S of Exercise 1
(c) $\{(x, y) \mid y = x^2 \text{ or } x = y^2\}$
(d) $\{(x, y) \mid y = x^2 \text{ and } x = y^2\}$
(e) $\{(x, y) \mid y \geq 2x \text{ and } x \leq 3y + 2\}$
(f) $\{(x, y) \mid y \geq 2x \text{ or } x \leq 3y + 2\}$
(g) $\{(x, y) \mid y = x^3 \text{ and } y \neq x\}$
(h) $\{(x, y) \mid |x| - |y| = 5\}$
(i) $\ln y = 3 \ln x$
(j) $\{(x, y) \mid x^2 + y^2 = 4 \text{ or } x^2 + y^2 = 0 \text{ or } (x - 1/2)^2 + (y - 1)^2 = 0 \text{ or } (x + 1/2)^2 + (y - 1)^2 = 0 \text{ or } y = -\sqrt{1 - x^2}\}$. ***Hint:*** Have a nice day!

(3) Find the domain and range of each of the following relations on ℝ. (The variables x and y denote real numbers). As much as possible, use the method shown in

Example 6. But you may need to resort to graphing or calculus. Justify your answers with more than a graph.

(a) $x^2 + y^3 = 7$ (b) $xy = 1$
(c) $|x| + |y| = 5$ (d) $|x + y| = 5$
(e) $y = \sin x + \cos x$ (f) $e^x + e^y = 0$

(4) Let $A = \{1, 2, 3\}$ and $B = \{2, 5\}$. How many elements are there in these sets?

(a) $\wp(A \times B)$ (b) $\wp(A) \times \wp(B)$
(c) $(A \times B) - (A \cup B)$ (d) $(A \cup B) - \wp(A \cup B)$
(e) $B \times B \times B \times B$

(5) In the country of Tannu Tuva, a valid license plate consists of any digit except 0, followed by any two letters of the alphabet, followed by any two digits.

(a) Let D be the set of all digits and L the set of all letters. With this notation, write the set of all possible license plates as a cartesian product.

(b) How many possible license plates are there?

(6) The Fishskill Numismatics Club has 10 members. The club must choose a slate of officers: president, treasurer, and secretary. How many possible slates are there, given that:

(a) Two or more positions may be filled by the same person.
(b) The officers must be three different people.

(7) Using the set-theoretic definition of ordered pairs (and naive set theory), derive axiom IV-5, the condition for the equality of ordered pairs.

(8) (a) Using the definition of ordered triples and the condition for the equality of ordered pairs, prove that $(x, y, z) = (u, v, w)$ iff $x = u$, $y = v$, and $z = w$.

(b) By induction, extend part (a) to the obvious condition for equality of two ordered n-tuples $(a_1, a_2, ..., a_n)$ and $(b_1, b_2, ..., b_n)$.

(9) Prove Theorem 6.1. Do *not* try to make the proof of the induction step very rigorous; just make sure you do a careful count. ***Hint:*** You may want to refer to the proof of Theorem 5.8, which has a similar flavor.

(10) Using Theorem 6.1 and induction, state and prove a generalization of Theorem 6.1 to cartesian products of any number of sets.

(11) Prove Theorem 6.2.

(12) Prove rigorously that the range of the relation $\{(x, y) \mid x, y \in \mathbb{R} \text{ and } x^2 + y^2 = 25\}$ is $[-5, 5]$. You may use the discussion in Example 6 as a starting point, and you may assume without proof that every nonnegative real number has a square root. You may also use Theorem A-11 of Appendix 2.

(13) Prove that for any relations R and S:
 (a) $\text{Dom}(R \cup S) = \text{Dom}(R) \cup \text{Dom}(S)$
 (b) $\text{Rng}(R \cup S) = \text{Rng}(R) \cup \text{Rng}(S)$

(14) Prove or disprove: for all relations R and S,
 (a) $\text{Dom}(R \cap S) = \text{Dom}(R) \cap \text{Dom}(S)$
 (b) $\text{Rng}(R \cap S) = \text{Rng}(R) \cap \text{Rng}(S)$
 (c) $\text{Dom}(R - S) = \text{Dom}(R) - \text{Dom}(S)$
 (d) $\text{Rng}(R - S) = \text{Rng}(R) - \text{Rng}(S)$
For any part that isn't true, try to prove that one side is a subset of the other.

(15) Prove or disprove: $\forall A, B, C\ (A \times B = A \times C \text{ iff } B = C)$

Critique the proofs in the remaining exercises in this section. (If necessary, refer to Exercises 4.2 for the instructions for this type of problem).

(16) **Theorem:** For any sets A and B, $\text{Dom}(A \times B) = A$ and $\text{Rng}(A \times B) = B$.
Proof: Let's first show $\text{Dom}(A \times B) = A$. To do this, we must show each side is a subset of the other. So first assume $x \in \text{Dom}(A \times B)$. Then, for some y, (x, y) is in $A \times B$. Thus $x \in A$. Conversely, assume $x \in A$. Then, for any $y \in B$, we have that (x, y) is in $A \times B$. Therefore, $x \in \text{Dom}(A \times B)$. This completes the proof that $\text{Dom}(A \times B) = A$. The proof that $\text{Rng}(A \times B) = B$ is almost identical.

(17) **Theorem:** For any relations R and S:
 (a) $(R \cup S)^{-1} = R^{-1} \cup S^{-1}$
 (b) $(R \cap S)^{-1} = R^{-1} \cap S^{-1}$
 (c) $(R - S)^{-1} = R^{-1} - S^{-1}$
Proof: (a) For any objects x and y,

$$\begin{aligned}(x, y) \in (R \cup S)^{-1} &\text{ iff } (y, x) \in (R \cup S)\\ &\text{ iff } (y, x) \in R \text{ or } (y, x) \in S\\ &\text{ iff } (x, y) \in R^{-1} \text{ or } (x, y) \in S^{-1}\\ &\text{ iff } (x, y) \in R^{-1} \cup S^{-1}\end{aligned}$$

By extensionality, we are done.
 (b) and (c) These arguments are almost identical to the argument for part (a).

6.2 Equivalence Relations

Definitions: A relation R is called

- **reflexive on A** iff $\forall x \in A, xRx$.
- **symmetric** iff $\forall x, y\ (xRy \text{ implies } yRx)$.
- **transitive** iff $\forall x, y, z\ (xRy \text{ and } yRz \text{ implies } xRz)$.

An **equivalence relation on** A is a relation on A that is reflexive on A, symmetric, and transitive.

It would be a good idea for you to take note of exactly where the phrase "on A" appears in these definitions and to try to see *why* it appears where it does. None of the instances of this phrase can be casually deleted. On the other hand, mathematicians sometimes call a relation R **reflexive**, with no reference to a set A. Technically, this means that R is reflexive on the set $\text{Dom}(R) \cup \text{Rng}(R)$. Similarly, if we simply say that R is an **equivalence relation**, we mean that R is an equivalence relation on $\text{Dom}(R)$. Exercises 5 and 6 are intended to clarify some subtleties in the above terminology, as is the following:

Theorem 6.3: (a) R is reflexive iff, whenever (x, y) is in R, so are (x, x) and (y, y).

(b) If R is symmetric, then $\text{Dom}(R) = \text{Rng}(R)$.

(c) If R is symmetric and transitive, then R is an equivalence relation (on the set $\text{Dom}(R)$).

Proof: (a) and (b) See Exercise 10.

(c) Assume R is symmetric and transitive. By part (b), R is a relation on $\text{Dom}(R)$. We also must show that R is reflexive. By part (a), we just need to show that xRy implies xRx and yRy. This is also left for Exercise 10. ■

Theorem 6.3(c) can be a time-saver. If we just want to show that R is an equivalence relation, with no reference to a domain, then we don't need to prove it's reflexive. In practice, the more usual situation is that we know R is a relation on some set A, and we want to show it's an equivalence relation on A. Then, besides proving symmetry and transitivity, we must also show that $\text{Dom}(R) = A$.

Notation: If R is known to be an equivalence relation, it is fairly common to write $x \equiv y$, $x \approx y$, or $x \sim y$ instead of xRy. In fact, mathematicians sometimes use these symbols to name equivalence relations, for example, "Let $\equiv$ be an equivalence relation."

Example 1: There's one equivalence relation that we've been working with since Chapter 3, namely the **identity, or equality, relation**. The first three of our equality axioms say that the relation $x = y$ is reflexive, symmetric, and transitive. Hopefully, it makes sense to you that equality is one way to define how things can be equivalent; in fact it can be argued that equality is the simplest possible equivalence relation.

The one thing that's a bit unusual about equality as a relation is that there's no particular domain for it. It can be thought of as a relation whose domain consists of all objects. Another approach is to say that given any set A, the equality relation, if restricted to $A \times A$, is an equivalence relation on A.

Notation: The identity relation on a set A is denoted id_A. In symbols,

$$\text{id}_A = \{(x, x) \mid x \in A\}$$

Example 2: The relation $\{(1, 1), (2, 2), (2, 3), (3, 2), (3, 3), (4, 4)\}$ is an equivalence relation on $\{1, 2, 3, 4\}$, as is easily verified (see Exercise 7).

Example 3: Let's find all possible equivalence relations on the set $A = \{1, 2\}$. This is simpler than it sounds. To be reflexive on A, such a relation must include (1, 1) and (2, 2). The only other ordered pairs it could contain are (1, 2) and (2, 1). And, by symmetry, if it contains one of these, then it contains the other. So there are only two equivalence relations on $\{1, 2\}$: one is id_A and the other is $A \times A$. Exercise 19 asks you to carry out more investigations of this sort.

Example 4: The "siblinghood" relation on the set P of all people who have ever lived is an equivalence relation, provided that we agree that each person is to be considered his or her own sibling. That is, let $S = \{(x, y) \mid x \text{ and } y \text{ have the same parents}\}$, where x and y are people variables. It is then quite simple to verify that S is reflexive on P, symmetric, and transitive. Similarly, let $F = \{(x, y) \mid x \text{ and } y \text{ have the same father}\}$ and $M = \{(x, y) \mid x \text{ and } y \text{ have the same mother}\}$. Then F and M are also equivalence relations on P.

Example 5: On the other hand, consider the "half-siblinghood" relation, that is, $\{(x, y) \mid x \text{ and } y \text{ have at least one parent in common}\}$. This is not an equivalence relation. Can you see why?

Example 6: **Similarity** is an equivalence relation on the set of all triangles. Recall that two triangles are **similar** iff they have equal angles. Alternatively, two triangles are similar iff one is a scale model of the other (with a reflection, or flip, also allowed). Similarity is also an equivalence relation on the set of all polygons. For polygons with more than three sides, similarity is defined by the scale model idea, not by angles. The definition using angles would still yield an equivalence relation, but it's not called similarity.

Example 7: Another equivalence relation on the set of all triangles or on the set of all polygons is **congruence.** Recall that two polygons are **congruent** iff they have the same angles and sides, in the same order (except that one can be clockwise and the other counterclockwise). This is the same as saying that if one of the polygons were rigidly constructed out of sticks, you could turn it into the other one just by moving the first polygon and/or flipping it over.

Example 8: On the set $\mathbb{R}$, the relation $\{(x, y) \mid x - y \text{ is an integer}\}$ is an equivalence relation. More generally, let c be any fixed real number, and replace the words "is an integer" in this definition with the words "is an integer multiple of c." Then for each c, this is an equivalence relation, called **congruence modulo c**. The usual notation for congruence modulo c is $x \equiv y \pmod{c}$. This type of equivalence relation, with c being an integer, is extremely important in number theory. It is discussed further in Section 8.3. By the way, congruence modulo c has nothing to do with the geometric notion of congruence discussed in Example 7. But in general, when mathematicians refer to any type of congruence or similarity, it always denotes an equivalence relation.

Example 9: There are many other ways to define an equivalence relation on $\mathbb{R}$. One is $|x| = |y|$, which is the same as saying $x = \pm y$. A somewhat more complicated one would be $\sin x = \sin y$. These equivalence relations are examples of a very general sort, and even though we have not studied functions yet, here is the general way such equivalence relations are formed.

Theorem 6.4: If A is any set, and f is any function that is defined on all the elements of A, then the relation $\{(x, y) \mid f(x) = f(y)\}$ is an equivalence relation on A.

Proof: This theorem doesn't take much proving. Since $f(x) = f(x)$ in all cases, the relation is reflexive. If $f(x) = f(y)$, then $f(y) = f(x)$, so the relation is symmetric. Transitivity is equally simple to show. ■

Note that Theorem 6.4 is derived directly from the fact that equality is an equivalence relation. It says that any function defined on a set creates an equivalence relation on that set. It turns out that a sort of converse holds: given any equivalence relation R on a set A, there's a function f defined on A such that, for any $x, y \in A$, xRy holds if and only if $f(x) = f(y)$. (See Exercise 9 of Section 7.1.)

For instance, in Example 9, the first relation corresponds to the function $f(x) = |x|$ and the second to the function $f(x) = \sin x$. In Example 4, F corresponds to the function $f(x) = x$'s father, M corresponds to the function $f(x) = x$'s mother, while for the relation S, $f(x)$ could be the ordered pair (x's father, x's mother).

In Example 8 it's somewhat harder to find the appropriate functions, but it can be done. For the first equivalence relation, we could use $f(x) = x - \lfloor x \rfloor$, where $\lfloor x \rfloor$ means the greatest integer that is equal to or less than x. (So $\lfloor 5 \rfloor = 5$, $\lfloor 3.27 \rfloor = 3$, $\lfloor -3.27 \rfloor = -4$, and so on. Therefore, $f(5) = 0$, $f(3.27) = 0.27$, $f(-3.27) = 0.73$, and so on. For positive numbers, $f(x)$ is the decimal part of x, but for negative numbers, it's a bit different.)

The point of this discussion and these examples is the following rule of thumb:

☞ *An equivalence relation almost always expresses some way in which two things are the same or alike.*

This follows from Theorem 6.4: To say that an equivalence relation must be definable by a statement of the form $f(x) = f(y)$ is to say that the relation must express that some characteristic or property of x and y is the same. A binary relation that does not express some kind of alikeness is probably not an equivalence relation.

Example 10: Consider the relation on $\mathbb{R}$ defined by $R = \{(x, y) \mid x - y < 1\}$. This is not an equivalence relation; for one thing it's not symmetric. If we instead define $S = \{(x, y) \mid |x - y| < 1\}$, then S is reflexive and symmetric, but not transitive. If you try to find a function that corresponds to R or S, in the sense of Theorem 6.2, you won't be able to.

Example 11: Referring to Examples 4 and 5, note that "x and y have the same mother *and* father" defines an equivalence relation, whereas "x and y have the same mother *or* father" does not. We just saw how to define a function that corresponds to the former relation; there's no way to define one for the latter relation.

This is part of a general phenomenon. The intersection of two equivalence relations is always an equivalence relation, but their union is usually not (see Exercises 12 and 13). Here is another example of this.

Example 12: On the set of all people, let

$$R = \{(x, y) \mid x \text{ and } y \text{ have the same hair color}\}$$

$$S = \{(x, y) \mid x \text{ and } y \text{ have the same eye color}\}$$

Let's assume (perhaps unrealistically) that hair and eye color are defined precisely enough so that R and S are equivalence relations. Then $R \cap S = \{(x, y) \mid x$ and y have the same hair and eye color$\}$, another equivalence relation. But $R \cup S$ is the same relation with "and" replaced by "or," which is not transitive.

We've been discussing the connection between equivalence relations and functions. There is an even more direct connection between equivalence relations and **partitions**.

Definition: Let A be a set. A **partition** of A is a collection of nonempty sets such that any two of them are disjoint, and the union of all of them is A. (See Exercise 21 for an alternative definition of partitions.)

Example 13: Let A = the set of all male people and B = the set of all female people. Then $\{A, B\}$ is a partition of the set of all people (assuming every person is male or female, exclusively).

Example 14: For each integer i, let A_i be the half-open interval $[i, i + 1)$. Then the collection of intervals $\{A_i \mid i \in \mathbb{Z}\}$ is a partition of $\mathbb{R}$. Note that this only works with half-open intervals. If we used closed intervals, they would not be disjoint from each other. If we instead used open intervals, the union of all the A_is would not be all of $\mathbb{R}$ (although these open intervals would form a partition of $\mathbb{R} - \mathbb{Z}$). Of course, we would still get a partition of $\mathbb{R}$ by letting $A_i = (i, i + 1]$.

Definitions: If R is an equivalence relation on a set A and $x \in A$, the **equivalence class** of x, denoted $[x]_R$, is the set $\{y \in A \mid xRy\}$. The collection of all these equivalence classes is called A **modulo** R, denoted A/R.

If there is no possibility of confusion (that is, if only one equivalence relation is being discussed), we just write $[x]$ instead of $[x]_R$.

Example 15: Let A be the set of all people, and let the equivalence relation R on A be defined by "x and y are the same age (in years)." If Lucian is a nine-year-old, then $[\text{Lucian}]_R$ is the set of all nine-year-olds. Any person's equivalence class (with respect to R) is the "club" consisting of all people of the same age as that person. Clearly, there is no overlap between different clubs, and each person is in exactly one of these clubs. The following lemma and theorem generalize these observations.

Lemma 6.5: Let R be an equivalence relation on a set A, and let $x, y \in A$. Then

(a) $x \in [x]$

(b) xRy iff $[x] = [y]$

(c) $\sim(xRy)$ iff $[x]$ and $[y]$ are disjoint.

Proof: We prove part (b) and leave the other parts for Exercise 11. For the forward direction, assume xRy. To show $[x] = [y]$, we need to show that, given any z, $z \in x$ iff $z \in y$, or equivalently, xRz iff yRz. *[Note how we must introduce the new variable z here.]* So now assume xRz. Since xRy we also have yRx, by symmetry. So yRz by transitivity. The converse is analogous. For the reverse direction, assume $[x] = [y]$. Then, since $y \in [y]$ by part (a), we have $y \in [x]$. So xRy by definition of equivalence classes. ■

We can now prove what is probably the most important single result about equivalence relations.

Theorem 6.6: For any equivalence relation, its equivalence classes form a partition of its domain.

Proof: Say R is an equivalence relation on A. By Lemma 6.5(a), each element of A is in some equivalence class. So each equivalence class is a nonempty subset of A, and the union of all the equivalence classes is A. And, by Lemma 6.5(b) and (c), any two distinct equivalence classes are disjoint. ■

Theorem 6.6 provides yet another way of understanding equivalence relations. An equivalence relation on a set partitions, or breaks up, the set into disjoint subsets (the equivalence classes). Each class is formed as a set of things that are alike in whatever sense corresponds to that equivalence relation. A converse to Theorem 6.6 holds, rather trivially: if $\mathscr{A}$ is any partition of a set B, then the relation $\{(x, y) \mid x \text{ and } y \text{ are in the same member of } \mathscr{A}\}$ is an equivalence relation on B (see Exercise 20).

Example 16: Consider the equality relation on any set A. For each x, $[x] = \{x\}$. So id_A partitions A into one-element sets. This is called the **finest** possible partition on A.

Example 17: For any set A, let $R = A \times A$. So xRy holds for all $x, y \in A$. It follows that R is an equivalence relation; the only equivalence class is A itself. This is called the **coarsest** possible partition on A.

Example 18: Consider the "siblinghood" relation of Example 4. Then for any person x, $[x]$ consists of x and all his or her siblings. If x is an only child, then $[x] = \{x\}$. So this relation partitions the set of all people who have ever lived into sibling classes.

Example 19: Referring to Example 8, consider congruence modulo 2 on the set $\mathbb{Z}$. Then if n is even, $[n]$ consists of all the even integers, whereas if n is odd, $[n]$ consists of all the odd integers. So $\mathbb{Z}$ is partitioned into two subsets by this relation.

Exercises 6.2

(1) Let A be the set of all people who have ever lived. For each of the following, state whether it's an equivalence relation on A and justify your assertion. If it is an equivalence relation, describe the equivalence classes and give an example of an equivalence class.

(a) x and y were born in the same year.
(b) x and y were born less than a week apart.
(c) x and y have the same maternal grandfather.
(d) x and y have the same four grandparents.
(e) x and y are first cousins or $x = y$.
(f) The set of all of x's children equals the set of all of y's children.

(2) For each of the following, state whether it's an equivalence relation on the specified set and justify your assertion.

(a) The relation $A \subseteq B$, on $\mathcal{P}(\mathbb{N})$
(b) The relation m and n have the same digit in the 100's place, on $\mathbb{N}$
(c) The relation x and y differ by a rational number, on $\mathbb{R}$
(d) The relation A and B are not disjoint, on $\mathcal{P}(\mathbb{N})$
(e) The relation A and B have the same smallest member, on $\mathcal{P}(\mathbb{N}) - \{\varnothing\}$
(f) The relation A and B have the same smallest member, on $\mathcal{P}(\mathbb{R}) - \{\varnothing\}$

(3) For each of the following relations on $\mathbb{R}$, state whether it's an equivalence relation (on whatever its domain is) and justify your assertion. If it is, describe the equivalence classes. In particular, describe $[3]$ and $[\pi]$, provided that these numbers are in the domain of the relation.

(a) $x^2 - 5 = y^2 - 5$
(b) $\sin x = \sin y$
(c) $\tan x = \tan y$
(d) $x + y$ is an integer.
(e) $|x| - |y|$ is an integer.
(f) $x - y$ is irrational.
(g) x/y is an integer.
*(h) $x/y = 2^i$, for some $i \in \mathbb{Z}$
*(i) $x - y = a + b\pi$, for some $a, b \in \mathbb{Q}$.

(4) Give an example of a relation R on $\mathbb{N}$ satisfying each of the following or explain why it is not possible to have one.

(a) R is reflexive on $\mathbb{N}$ and symmetric but not transitive.
(b) R is reflexive on $\mathbb{N}$ and transitive but not symmetric.

(c) R is symmetric and transitive but not reflexive on $\mathbb{N}$.
(d) R is reflexive on $\mathbb{N}$ but neither symmetric nor transitive.
(e) R is symmetric but not reflexive on $\mathbb{N}$ or transitive.
(f) R is transitive but not reflexive on $\mathbb{N}$ or symmetric.
(g) R is not reflexive on $\mathbb{N}$, symmetric, or transitive.

(5) Let R be the relation $\{(1, 1), (2, 2), (3, 3)\}$.
(a) Is R reflexive?
(b) Is R an equivalence relation?
(c) Is R a relation on $\mathbb{N}$?
(d) Is R reflexive on $\mathbb{N}$?
(e) Is R an equivalence relation on $\mathbb{N}$?
(f) Explain why your answers to parts (a) through (e) are not contradictory.

(6) Are the following statements true or false? If true, explain why; if not, find a counterexample.
(a) Whenever R is reflexive on A, then R is a relation on A.
(b) Whenever R is a relation on A and R is reflexive, then R is reflexive on A.
(c) An equivalence relation on A is precisely an equivalence relation whose domain is A.

(7) Show that the relation defined in Example 2 is an equivalence relation.

(8) Let $A = \mathbb{Z} \times (\mathbb{Z} - \{0\})$. In other words, A is the set of all ordered pairs of integers in which the second integer is nonzero. Define a relation $\sim$ on A by

$$(a, b) \sim (c, d) \text{ iff } ad = bc$$

Prove that $\sim$ is an equivalence relation on A. (The idea behind this is that the ordered pair (a, b) may be used to represent the fraction a/b, in which case $\sim$ becomes the standard cross-multiplication condition for the equality of fractions. This equivalence relation is important in the construction of the rationals from the integers; see Sections 8.3 and 9.5.)

(9) For any $A, B \subseteq \mathbb{R}$, A is said to be a **translate** of B iff there is a real number k such that $B = \{x + k \mid x \in A\}$.
(a) Prove that this relation is an equivalence relation on $\wp(\mathbb{R})$.
(b) Describe the equivalence class of the set of negative real numbers.
(c) Prove that every translate of a closed interval is also a closed interval.
(d) Find two other phrases that could replace the words "closed interval" in part (b) and still yield a true statement. Prove these statements.
(e) Find two other phrases that could replace the words "closed interval" in part (b) and yield a *false* statement. Give counterexamples to verify that the statements are false.

(10) (a) Prove Theorem 6.3(a).
(b) Prove Theorem 6.3(b).
(c) Complete the proof of Theorem 6.3(c).

(11) Prove parts (a) and (c) of Lemma 6.5.

(12) (a) Prove that the intersection of any two equivalence relations on A is an equivalence relation on A.
*(b) More generally, prove that the intersection of *any collection* of equivalence relations on A is an equivalence relation on A.

(13) Give two examples to show that the union of two equivalence relations is not necessarily an equivalence relation.

(14) This exercise continues ideas introduced in Examples 16 and 17. If $\mathscr{B}$ and $\mathscr{C}$ are two partitions of a set A, we say $\mathscr{B}$ is a **refinement** of $\mathscr{C}$ or a **finer partition** than $\mathscr{C}$ iff every set in $\mathscr{B}$ is a subset of some set in $\mathscr{C}$. Prove that if R and S are equivalence relations on A, then the partition created by R is a refinement of the partition created by S if and only if $\forall x, y\ (xRy \rightarrow xSy)$.

(15) Refer to Example 8 and Exercise 14. Let A be any one of the sets $\mathbb{R}$, $\mathbb{Q}$, or $\mathbb{Z}$. If m and n are natural numbers, under what condition is the partition created by congruence modulo m a refinement of the partition created by congruence modulo n? Prove your assertion.

*(16) This exercise generalizes the ideas introduced in Example 8 and Exercise 2(c). Let A be as in Exercise 15. Let B be a nonempty subset of A that is closed under subtraction; that is, $\forall x, y \in B\ (x - y \in B)$. Such a set B is called a **subgroup of A under addition**. Now define a relation R on A by xRy iff $x - y \in B$. Prove that R is an equivalence relation on A, and describe the equivalence classes. (R is called **congruence modulo B**.)

(17) How many equivalence classes are there under the congruence modulo 3 relation on $\mathbb{Z}$? Describe them.

(18) Let $R = \{(m, n) \mid m, n \in \mathbb{Z}$ and $3m + 4n$ is a multiple of $7\}$.
(a) Is R an equivalence relation on $\mathbb{Z}$? Prove your claim.
*(b) R is a relation of the type discussed in Example 8. By experimentation, determine exactly which relation R is. You needn't prove your conclusion.

(19) (a) Find all equivalence relations on the set $\{1, 2, 3\}$.
*(b) Find all equivalence relations on the set $\{1, 2, 3, 4\}$.

(20) Prove the converse of Theorem 6.6 mentioned in the text.

(21) Suppose A is a set and $\mathscr{B}$ is a collection of sets. Prove that $\mathscr{B}$ is a partition of A iff $\mathscr{B}$ is a collection of nonempty subsets of A and every member of A is in exactly one member of $\mathscr{B}$.

*6.3 Ordering Relations

Ordering relations are just about as important in mathematics as equivalence relations. They are also a good deal more familiar to most students than equivalence relations, and so you will probably find this section conceptually simpler than the previous one. On the other hand, there are more details and minor variations involved with ordering relations than with equivalence relations, so you have to be careful to keep things straight.

There are two general types of orderings. This section is primarily devoted to reflexive orderings. At the end of the section we briefly discuss irreflexive orderings and show the simple, close connection between these two types of orderings. Irreflexive orderings are also used in Appendices 1 and 2.

Definitions: A relation R is **antisymmetric** iff whenever xRy and yRx, then $x = y$.

A **partial ordering** is a relation that is reflexive, antisymmetric, and transitive.

A **total ordering** is a partial ordering R in which xRy or yRx holds for every x and y in the domain of R.

As with equivalence relations, when we refer to an ordering (partial or total) on a set A, we mean an ordering whose domain is A.

Note that the only difference between the definitions of equivalence relations and partial orderings is symmetry versus antisymmetry. But this is a crucial difference. Whereas an equivalence relation expresses some sort of alikeness of elements and groups them together (in equivalence classes), an ordering sets elements apart by putting them in a hierarchy, or order.

If R is an ordering on A, we also say that A is **partially** (or **totally**) **ordered** by R. Total orderings may also be called **linear orderings** or **simple orderings**.

Notation: If R is an ordering, we may use the more common notation $x \le y$ or $x \ge y$ to mean xRy. When this is done, standard abbreviations are automatically assumed: $x \ge y$ means the same as $y \le x$, and $x < y$ means $x \le y$ and $x \ne y$ (as does $y > x$).

Example 1: The set of real numbers $\mathbb{R}$ has a standard ordering on it. Both $\le$ and $\ge$ are total orderings on $\mathbb{R}$ (and therefore also on any subset of $\mathbb{R}$, such as $\mathbb{N}$, $\mathbb{Z}$, or $\mathbb{Q}$). The relations $<$ and $>$ on $\mathbb{R}$ are not total orderings in the sense defined here, since they are not reflexive. They are *irreflexive* total orderings.

Example 2: It's instructive to consider some orderings on small sets. Let $A = \{1, 2, 3\}$. As a subset of $\mathbb{R}$, A inherits a total ordering, as mentioned in the previous example. If we use $\le$ as the basis of this relation, what ordered pairs would it contain?

It can't contain just (1, 2) and (2, 3), because this set of ordered pairs is neither reflexive nor transitive. Exercises 1 and 9 ask you to answer this question and some similar ones. There are many ways to totally order this set or almost any set.

Definition: If R is a partial ordering, x and y are called **comparable** (with respect to R) iff xRy or yRx holds. If x and y are in Dom(R) but are not comparable, then they are called **incomparable**.

So a partial ordering is total iff it has no incomparable pairs of elements.

Example 3: Many important orderings are not total. Let $\mathscr{A}$ be any collection of sets. It is simple to show that the relation $A \subseteq B$ is a partial ordering on $\mathscr{A}$. In fact, all three of the conditions for this are proved in Chapter 5. This relation is often considered with $\mathscr{A} = \wp(C)$, for some set C. Choose some small set for C. Is the relation $\subseteq$ on $\mathscr{A}$ total? (See Exercise 5.)

Example 4: Let $C = \mathbb{R} \times \mathbb{R}$. We can use the total ordering on $\mathbb{R}$ to define a *partial* ordering on C, as follows: let's say that one ordered pair is related to another if the first pair $\leq$ the second in *both* coordinates. That is, define the relation S on C by

$$(a, b)\, S\, (c, d) \quad \text{iff} \quad a \leq c \ \text{ and } \ b \leq d$$

It's not hard to show that S is a partial ordering on C. It's also clear that S is not total. For example, consider (3, 7) and (9, 4). You can see that these ordered pairs are incomparable.

This example is a specific case of the following definition.

Definition: Let R and S be partial orderings on sets A and B, respectively. Then the **product ordering** $R \times S$ is the relation on $A \times B$ defined by

$$R \times S = \{((a, b), (a', b')) \mid aRa' \text{ and } bSb'\}$$

This definition may seem confusing at first because it involves ordered pairs of ordered pairs. But if you understand Example 4, you should see that this definition just formalizes the idea of that example. By the way, this notation is somewhat sloppy, in that $R \times S$ is not literally the cartesian product of R and S.

Theorem 6.7: Let R and S be partial orderings on sets A and B, respectively. Then the product ordering $R \times S$ is a partial ordering on $A \times B$. However, if both A and B have more than one element, then $R \times S$ is not total.

Proof: The proof that $R \times S$ is a partial ordering is left for Exercise 3(a). To prove the second claim, assume that a and a' are distinct elements of A, and b and b' are distinct elements of B. By antisymmetry, we can't have both aRa' and $a'Ra$, so without loss of generality let's say $\sim (aRa')$. (So either $a'Ra$, or a and a' are incomparable—it doesn't matter.) Similarly, without loss of generality let's say $\sim (bRb')$.

Now consider the ordered pairs (a, b') and (a', b). It's a simple matter to show that these are incomparable in the ordering $R \times S$ (see Exercise 3(b)). ■

Example 5: Here's another example of product orderings. Suppose we want to rank mixed-doubles tennis teams (that is, teams of one male and one female player). Let M be the set of all male tennis players and F the set of all female ones. On each of these sets, we have the ordering defined by one player's being at least as good as another. For simplicity, let's assume these orderings are total; this means that for any two players, it's possible to say which one is better. (This assumption is somewhat unrealistic in tennis and most other sports, but let's not worry about that.)

We can then define the product ordering on $M \times F$, the set of all possible mixed-doubles teams. Under the product ordering, one team is at least as good as another iff its male and female players are both at least as good as the corresponding players of the other team. But this is not a total ordering. For example, since Pete Sampras is at least as good as Bob Wolf and Martina Hingis is at least as good as Roseanne Barr, the product ordering would say the team of Sampras and Hingis is at least as good as the team of Wolf and Barr. But the product ordering would not settle which is better, Sampras-Barr or Wolf-Hingis. These two teams would be incomparable under the product ordering.

Product orderings have some practical value. But, as you can imagine, they're not likely to be completely accurate in a situation like the ranking of sports teams.

Orderings (especially nontotal ones) on small sets can often be clearly shown using **lattice diagrams**. In a lattice diagram for a relation R, each point of the domain is represented as a dot, and the fact that aRb is represented by an upward (but not necessarily vertical) path from a to b. Such a path from a to b may pass through other points; transitivity requires us to interpret things this way. A lattice diagram for a total ordering is a single line, or chain, so it isn't very interesting. Figure 6.4 shows some lattice diagrams, and Exercise 4 asks you to construct several others.

Example 6: This example is not of a product ordering, but it illustrates a related idea. Let A be the set of all people. For simplicity, let's assume that it's possible to accurately measure everyone's height and weight and that no two people have exactly the same height or weight. Then we get two different *total* orderings on A, defined by height and weight, respectively. That is, let

$$R = \{(a, b) \mid a, b \in A \text{ and } a \text{ is at least as tall as } b\}$$

$$S = \{(a, b) \mid a, b \in A \text{ and } a \text{ is at least as heavy as } b\}$$

A simple way to use these two orderings to form a new one is to take their intersection $R \cap S$ as sets of ordered pairs. Since intersection is defined by the word "and," it's easy to understand $R \cap S$: it consists of all ordered pairs (a, b) such that a is taller *and* heavier

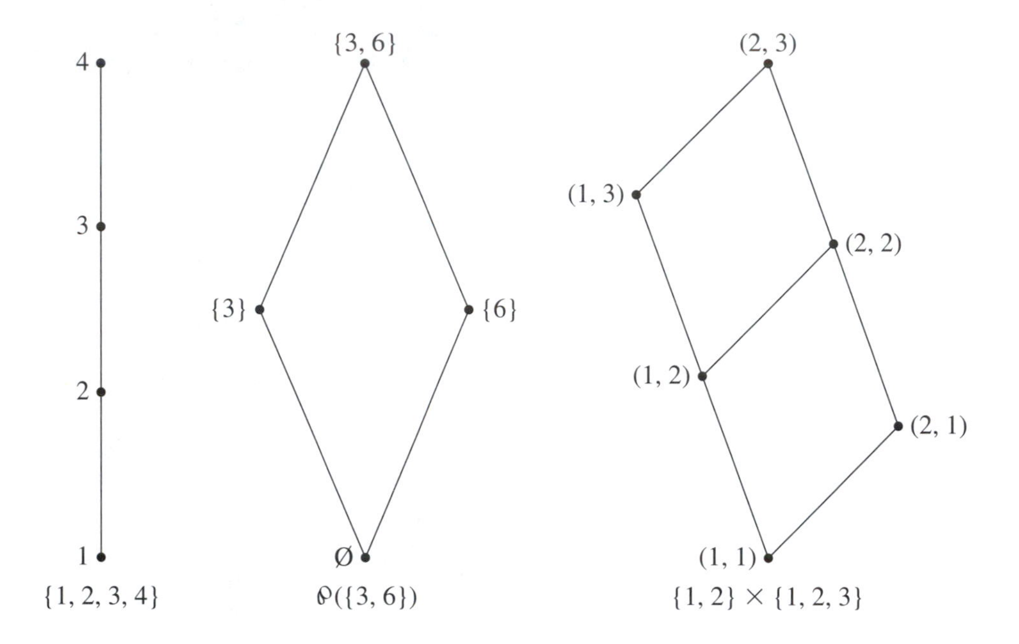

Figure 6.4 Lattice diagrams for three partial orderings

than b. This new relation must still be a partial ordering (see Exercise 6), but it's not total, since one person could be shorter and heavier than another. Two such people would be incomparable in the intersection ordering.

What makes intersections of orderings similar to product orderings? The simplest way to answer that is to say that both are defined with the word "and." This simple logical similarity is the reason why both products and intersections of partial orderings must be partial orderings, but they are rarely total.

Example 7: In the previous example, what would happen if we looked at $R \cup S$ instead of $R \cap S$? This union would consist of all ordered pairs (a, b) of people such that a is at least as tall *or* at least as heavy as b. This is not even a partial ordering (see Exercise 7). Compare this to Exercises 12 and 13 of Section 6.2.

Example 8: Let's reconsider Example 5. Recall that the product ordering on mixed-doubles teams is not total. Is there a way to set up a *total* ordering on $M \times F$, defined directly in terms of the separate orderings on M and F?

There are actually several ways to do this. Perhaps the simplest way is to give one sex or the other precedence for ranking teams. For example, it could be decided to give precedence to male players; this would mean that if one team had a better male player than another, that team would be automatically ranked higher than the other, no matter how the two female players stacked up. Only if two teams had the same male player (or, in a more practical setting, male players of equal ability) would the relative abilities of the teams' female players be taken into account.

Like some of our previous examples, this one is not completely realistic. If one team has a slightly better male player than another but a vastly inferior female player, it is unlikely that the first team should be considered the better one. But the way we are defining this ordering of teams, it would be ranked as the better one.

Let's now formalize this example and give it a name.

Definition: Let R and S be partial orderings on sets A and B, respectively. Then the **lexicographic ordering** on $A \times B$ (associated with R and S) is

$$\{((a, b), (a', b')) \mid (aRa' \text{ and } a \neq a') \text{ or } (a = a' \text{and } bSb')\}$$

That is, the first coordinates take precedence for ordering ordered pairs. Only in the case that the first coordinates are equal are the second coordinates used.

It would also be possible to define a type of lexicographic ordering that favors the second coordinate instead of the first, and this would give a very different ordering on $A \times B$ from the one defined above (see Exercise 10(c)). In contrast, there's only one way to define a product ordering $R \times S$, since that notion does not involve choosing which coordinate to give precedence.

Theorem 6.8: (a) If R and S are partial orderings on sets A and B, then the associated lexicographic ordering is a partial ordering on $A \times B$.

(b) If, in addition, R and S are total, then the associated lexicographic ordering is also total.

Proof: (a) Let L be the lexicographic ordering on $A \times B$. We must show L is reflexive on $A \times B$, antisymmetric, and transitive. To prove reflexivity, let $(a, b) \in A \times B$. We know that $a = a$, and bSb because S is reflexive. Therefore, $((a, b), (a, b))$ is in L, because it satisfies the second disjunct in the definition of lexicographic order.

To prove antisymmetry, assume $(a, b)\, L\, (a', b')$ and $(a', b')\, L\, (a, b)$. Both these assumptions are disjunctions, so there are four possible ways to make them both true. Three of these directly say that $a = a'$, and the fourth requires that aRa', $a \neq a'$, and $a'Ra$, which violates the antisymmetry of R. So $a = a'$. But in that case, the two assumptions require, respectively, that bSb' and $b'Sb$. But then $b = b'$, by the antisymmetry of S. So we have shown that $(a, b) = (a', b')$, as desired.

Exercise 11 asks you to prove the transitivity of L, as well as part (b). ■

Preorderings

We have seen several examples of orderings that are not total because they have incomparable pairs of elements. The next two examples illustrate another phenomenon that prevents some relations from being even partial orderings.

Example 9: For integers m and n, we say m **divides** n, denoted $m|n$, iff for some integer k, $n = km$. It is not difficult to show that | is a partial ordering on $\mathbb{N}$ (see Exercise 12). (Specifically, this means that m and n are restricted to be positive; it doesn't matter whether k is similarly restricted.) This ordering is certainly not total; for instance, 2 and 3 are incomparable.

Does | define a partial ordering on $\mathbb{Z}$? Perhaps surprisingly, it does not, because it's not antisymmetric. For example, $5|(-5)$ and $(-5)|5$, but $5 \neq -5$. Note that this is very different from saying that 5 and -5 are incomparable. Relations like this fall into a different category called preorderings. The relation | is discussed further in Section 8.2.

Definition: A **preordering** is a relation that is reflexive and transitive.

Obviously, every partial ordering is a preordering. Example 9 shows that the converse of this does not hold. However, every preordering naturally gives rise to both an equivalence relation and a partial ordering (see Exercise 13). Here is a simple relation from real life that can be viewed as a preordering.

Example 10: As in Example 6, let R be the height relation on the set of all people. Again, if we assume that no two people are exactly the same height, then R is a total ordering. But what if there are people of the same height? (Or we could guarantee this by letting xRy mean that y is at least as tall as x, with height measured to the nearest centimeter.) If x and y are two people of the same height, then xRy and yRx, so R is not antisymmetric. It would make sense to say that such people are tied in this relation. Note that all people are comparable in height, so R may be called a **total preordering**. In contrast, the relation | on $\mathbb{Z}$ has both incomparable elements (such as 2 and 3) and tied elements (such as 5 and -5).

Irreflexive Orderings

We conclude this section by showing that the difference between reflexive orderings and irreflexive orderings is not very profound. The definition and some simple results involving irreflexive orderings appear in Appendices 1 and 2, because the order axioms of $\mathbb{R}$ are usually stated in terms of $<$ rather than $\leq$.

Recall that $\text{id}_A = \{(x, x) \mid x \in A\}$.

Definitions: A relation R is called **irreflexive** iff xRx is *not* true for any x.

A relation that is irreflexive and transitive is called an **irreflexive partial ordering**.

An **irreflexive total ordering** is an irreflexive partial ordering that also satisfies **trichotomy:** for every x and y in the domain of R, xRy, yRx, or $x = y$.

Note that "irreflexive" means more than merely "not reflexive," just as "antisymmetric" means more than merely "not symmetric." Also note that, in the context of irreflexivity, we must add the disjunct $x = y$ to the condition for being total.

Antisymmetry does not need to be included in the definition of an irreflexive partial ordering because it follows from the other two conditions. Furthermore, if R is irreflexive, then xRy, yRx, and $x = y$ cannot all be true. Therefore, an irreflexive partial ordering must satisfy **strong antisymmetry**: (xRy and yRx) is always false. This conclusion is proved as Theorem A-8 of Appendix 2.

The standard symbols $<$ and $>$ are typically used to denote irreflexive orderings. If $<$ is an irreflexive ordering, it would seem sensible that we could obtain a reflexive ordering $\leq$ by defining $x \leq y$ to mean $x < y$ or $x = y$, as usual. Conversely, if $\leq$ is a reflexive ordering, it would seem sensible that we could obtain an irreflexive ordering $<$ by defining $x < y$ to mean $x \leq y$ and $x \neq y$. This is indeed the simple link between the two types of orderings and is stated more precisely in the following result.

Theorem 6.9: (a) Let R be an irreflexive partial ordering on a subset of A. Then $R \cup \mathrm{id}_A$ is a reflexive partial ordering on A. The ordering R is total on A iff $R \cup \mathrm{id}_A$ is.

(b) Let S be a reflexive partial ordering on a set A. Then $S - \mathrm{id}_A$ is an irreflexive partial ordering on a subset of A. The ordering S is total on A iff $S - \mathrm{id}_A$ is.

Proof: (a) Assume R is as described, and let $S = R \cup \mathrm{id}_A$. Since the domain and range of R are subsets of A, the domain and range of S are both equal to A. Since id_A is reflexive on A, so is S. To show that S is transitive, assume aSb and bSc. Then either aRb or $a = b$; similarly, either bRc or $b = c$. We get a total of four possible cases, all of which imply that aSc. We also need to show that S is antisymmetric. So assume aSb and bSa. We need to show $a = b$. But $a \neq b$ would imply aRb and bRa, which would contradict the strong antisymmetry of R.

To prove the second statement, note that

R is total on A iff (xRy or yRx or $x = y$), for all x, y in A
iff [(xRy or $x = y$) or (yRx or $y = x$)], for all x, y in A
iff (xSy or ySx), for all x, y in A
iff S is total on A

The proof of part (b) is similar and is left for Exercise 14. ■

Exercises 6.3

(1) (a) As discussed in Example 2, list all the ordered pairs in the ordering $\leq$ on the set $\{1, 2, 3\}$. How many ordered pairs are in this relation?

(b) Repeat part (a) for the sets $\{1\}$, $\{1, 2\}$ and $\{1, 2, 3, 4\}$.

(2) Draw the cartesian plane $\mathbb{R} \times \mathbb{R}$, and choose an arbitrary point (a, b). Now consider the product ordering on $\mathbb{R} \times \mathbb{R}$ discussed in Example 4.

(a) Relative to (a, b), where are the points that are greater than it?
(b) Relative to (a, b), where are the points that are less than it?
(c) Relative to (a, b), where are the points that are incomparable to it?
(d) Repeat parts (a) and (b) for the lexicographic ordering on $\mathbb{R} \times \mathbb{R}$.

(3) (a) Prove the first part of Theorem 6.7.
(b) Complete the proof of the second part of Theorem 6.7.

(4) (a) Draw a lattice diagram for the product ordering on $\{1, 2, 3\} \times \{1, 2, 3\}$.
(b) Draw a lattice diagram for the ordering $\subseteq$ on $\wp(\{1, 3, 8\})$. (This can not be done without having paths cross each other.)
(c) Pick at least a half dozen people you know, and for this set of people, draw a lattice diagram for the ordering of Example 6.

(5) Find and prove a necessary and sufficient condition on a set A for the ordering $\subseteq$ on $\wp(A)$, discussed in Example 3, to be total.

(6) Prove that the intersection of two partial orderings on A is again a partial ordering on A, as stated in Example 6.

(7) Show that the relation $R \cup S$ defined in Example 7 is not a partial ordering.

(8) Prove that R is a partial (respectively, total) ordering on A iff R^{-1} is.

(9) (a) On the basis of Exercise 1, make a conjecture of the form "The number of ordered pairs in a total ordering on an $(n + 1)$ element set is ____ more than the number of ordered pairs in a total ordering on an n element set."
(b) By induction (and not too formally), prove your conjecture from part (a).
*(c) Find and prove a formula for the number of ordered pairs in any total ordering on a set with n elements.

(10) (a) Look up "lexicographic" in a dictionary, and then explain our use of the term "lexicographic ordering."
(b) Draw a graph showing the elements of $\mathbb{N} \times \mathbb{N}$, as a subset of $\mathbb{R} \times \mathbb{R}$. Then give a simple pictorial description of the lexicographic ordering on $\mathbb{N} \times \mathbb{N}$ that is based on the standard ordering on $\mathbb{N}$.
(c) Give a precise definition of the lexicographic ordering on $A \times B$, associated with R and S, in which the second coordinate takes precedence.
(d) Give a precise definition of the lexicographic ordering on $A \times B \times C$, associated with partial orderings R, S, and T (on A, B, and C, respectively).
*(e) A language such as English does not consist of just two-letter words or three-letter words. The ordering of words in a dictionary must allow words of any length. In analogy to this, give an informal but clear definition of the lexicographic ordering on the set of *all finite sequences* of elements of a set A, based on an ordering R on A.

(11) (a) Complete the proof of Theorem 6.9(a).
(b) Prove Theorem 6.9(b).

(12) Prove that the relation | , defined in Example 9, defines a partial ordering on $\mathbb{N}$.

(13) (a) Prove that if R is a preordering on A, then $R \cap R^{-1}$ is an equivalence relation on A.

(b) Applying part (a) to Example 10, describe the equivalence classes.

(c) A corollary to part (a) states that the original preordering R defines a "natural" partial ordering whose domain is the set of *equivalence classes* of $R \cap R^{-1}$. Explain this in the context of Example 10.

(14) Prove Theorem 6.9(b).

(15) Let A be any set with one element, perhaps $\{5\}$.
(a) How many reflexive orderings are there on A? Describe all of them.
(b) Are there any irreflexive orderings whose domain is A?
(c) Does this situation contradict Theorem 6.9? Explain.

The next three exercises concern topics introduced after Theorem 5.6.

(16) Say we have a partially ordered set A. On the basis of the last part of this section, it doesn't matter whether this ordering is reflexive or irreflexive. For any $x \in A$, we say x is:

Minimal iff $\sim \exists y \in A\ (y < x)$

Maximal iff $\sim \exists y \in A\ (y > x)$

A **least element** iff $\forall y \in A\ (x \le y)$

A **greatest element** iff $\forall y \in A\ (x \ge y)$

Prove: (a) There cannot be more than one least element or more than one greatest element. (Hence, one usually refers to *the* least and *the* greatest element, if they exist.)

(b) A least (respectively, greatest) element must be minimal (respectively, maximal).

(c) There can be more than one minimal (respectively, maximal) element. Therefore, the converses of part (b) fail.

(17) A partial ordering on A is called a **well-ordering** on A iff every nonempty subset of A has a least element. A set with a well-ordering defined on it is said to be **well ordered**. Prove:
(a) A well-ordering on A must be total.
(b) Every subset of a well-ordered set is well ordered.
(c) Every subset of $\mathbb{N}$ is well ordered by $\le$. (Recall Theorem 5.6.)
(d) The sets $\mathbb{Z}$, $\mathbb{Q}$, and $\{x \in \mathbb{R} \mid x \ge 0\}$ are *not* well ordered by $\le$.

(18) By induction on n, prove that every subset of $\mathbb{R}$ with n members is well ordered. (You may begin this proof with either $n = 1$ or $n = 0$, as you wish. Essentially, what you are proving is that every *finite* subset of $\mathbb{R}$ is well ordered.)

(19) Prove or find a counterexample: if R is a partial ordering on A and S is a partial ordering on B, then $R \cup S$ is a partial ordering on $A \cup B$.

Critique the proofs in the remaining exercises in this section. (If necessary, refer to Exercises 4.2 for the instructions for this type of problem.)

(20) **Theorem:** If R is a partial ordering on A and S is a partial ordering on B, where A and B are *disjoint* sets, then $R \cup S$ is a partial ordering on $A \cup B$.

Proof: First, assume $x \in A \cup B$. If $x \in A$, then xRx. If $x \in B$, then xSx. In either case, $(x, x) \in R \cup S$, so $R \cup S$ is reflexive. To prove antisymmetry, assume (x, y) and (y, x) are both in $R \cup S$. If they are both in R, then $x = y$, by the antisymmetry of R. If they are both in S, we similarly have $x = y$. And it's not possible that $(x, y) \in R$ and $(y, x) \in S$, because A and B are disjoint. To prove transitivity, assume (x, y) and (y, z) are both in $R \cup S$. Again using the disjointness of A and B, the only way this can occur is that (xRy and yRz) or (xSy and ySz). This implies xRz or xSz, as desired.

(21) **Theorem:** If R is a total ordering on A and S is a total ordering on B, where A and B are *disjoint* sets, then $R \cup S$ is a total ordering on $A \cup B$.

Proof: Assume the givens. The proof that $R \cup S$ is a partial ordering on $A \cup B$ is as in the previous problem. And since R is total on A and S is total on B, $R \cup S$ is total on $A \cup B$.

Suggestions for Further Reading: The material in this chapter is covered in most books on set theory, such as the first four references given at the end of Chapter 5. See also Feferman (1989), Pfleeger and Straight (1985), or Ross and Wright (1985).

Chapter 7

Functions

7.1 Functions and Function Notation

All mathematicians would agree that functions and their use are vital to the understanding of mathematics. It is fairly safe to say that the three most important concepts in mathematics are the concepts of numbers, functions, and sets. Let's get right to the definition of functions. As in Chapter 6, some concepts are given both an intuitive definition and a set-theoretic one.

Definitions (intuitive): A **function from *A* to *B*** is a *rule* (or *procedure* or *set of instructions*) that assigns, to each member of set A, *exactly one* member of set B.

A is called the **domain** of the function, and B is called its **codomain**. The set of all members of B that are actually assigned to some member of A is called the **range** of the function.

Definitions (set-theoretic): A **function from *A* to *B*** is a relation between A and B (that is, a subset of $A \times B$) that pairs each member of A with *exactly one* member of B.

Again, B is called the **codomain** of the function. The domain and range are defined as in Chapter 6; this guarantees that the domain is A and the range is a *subset* of B.

A function from A to itself is called a **function on *A***. If we simply say that f is a function, we mean that it's a function from some set A to some set B, or, equivalently, that it's a function from $\text{Dom}(f)$ to $\text{Rng}(f)$.

Functions are sometimes called **mappings**, **maps**, **transformations**, **operators**, or **operations**. These words can have more specialized meanings in particular branches of mathematics. But, for the most part, they are synonyms for the word "function."

Notation: (a) The letters f, g, F, and G almost always denote functions.

(b) The statement that f is a function from A to B is abbreviated $f: A \to B$. (The arrow in this notation has no connection with implications.)

It is worthwhile to compare the intuitive definitions of the concepts "relation" (Section 6.1) and "function." A relation is simply a statement with two free variables. This definition has a deliberately passive feel and makes no distinction between the variables. A function, on the other hand, is a rule that assigns some things to other things. These words connote a certain activity; a function is supposed to *do* something. We think of the domain of a function as the set of *inputs*, and the range as the set of *outputs*. In fact, it is often helpful to think of a function as a sort of computer or input-output machine. As long as the computer is fed a legal input, it performs some sort of calculation and spits out exactly one answer or output (see Figure 7.1). This image may seem too simple to be useful, but it can be an instructive way to illustrate concepts involving combinations of functions.

In the intuitive definition, we call a function a type of rule, procedure, or set of instructions. Other words that are sometimes used here include "algorithm" and "correspondence." These words are all deliberately imprecise, because it is important not to be too restrictive about what constitutes a function. Later in this section we give several examples to show the wide variety allowed in functions. On the other hand, for the vast majority of functions used in *mathematics*, the rules used to define them are *equations*.

Example 1: You have undoubtedly seen statements like "Consider the function $y = x^2 + 5$." In high school algebra and calculus, the usual understanding is that x and y are real variables, x is the input variable or **independent** variable, and y is the output variable or **dependent** variable. Since the variable y in this equation appears alone on the left side, and the right side is an expression in x that can only have one value for any given value of x, this equation certainly defines a function. Also, since the right side is defined for every real number x, the domain of this function is all of $\mathbb{R}$. (It could be specified to be a smaller set or even a bigger set like all complex numbers. But, as indicated, the normal unwritten convention is that the domain of any function that is defined by an equation in x and y is the set of all real values of x for which a value of y can be found to make the equation true.) If there were any values of x for which the right side were not defined, they would have to be excluded from the domain.

Since the domain consists of all real numbers and the output y also must be a real number, we would probably view this as a function from $\mathbb{R}$ to $\mathbb{R}$. How about the range of this function? Is it all of $\mathbb{R}$? Just because we've specified the codomain to be $\mathbb{R}$ doesn't mean the range must be $\mathbb{R}$. In fact, you can see that y cannot be less than 5. Some thought should also convince you that any value equal to or greater than 5 can be

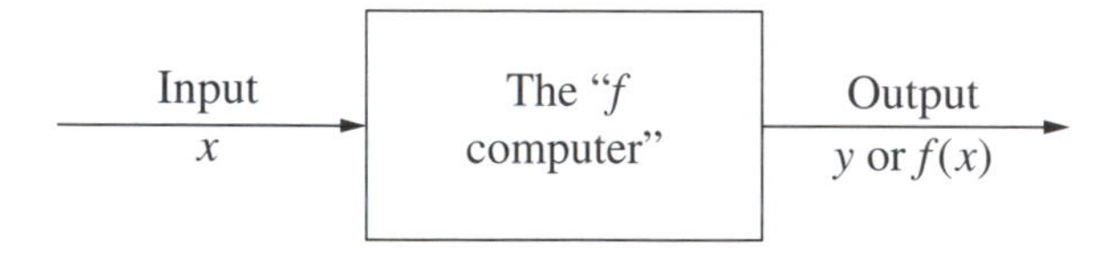

Figure 7.1 A function as an input-output machine

obtained for y. So the range is the set of all real numbers that are at least 5, in symbols $\{y \mid y \geq 5\}$ or $[5, \infty)$. Exercise 3 asks you to do this more rigorously. (The equations discussed in Examples 1 through 3 are graphed in Figure 7.2.)

As with relations, it takes some thought to see the connection between the two definitions of functions. And as with relations, it's not hard to explain how to go from the intuitive definition to the set-theoretic one (and we won't worry about the other direction). Under the intuitive definition, a function is a rule that tells you how to take any input from some set and get exactly one output value or object. We've just seen how an equation like $y = x^2 + 5$ can be viewed as establishing a rule that fits the intuitive definition of a function. But this equation is also a statement with two free variables, so it's a relation (in the intuitive sense). And if we take the set of ordered pairs of real numbers that satisfy this equation, we get a relation in the set-theoretic sense. Furthermore, it should be clear to you that this set of ordered pairs satisfies the set-theoretic definition of a function. Naturally, when you plot this set of ordered pairs of real numbers, you get the *graph* of this function.

In the next three examples, we continue to assume that x and y are real variables and that the given relation includes all possible ordered pairs of real numbers that work for that relation.

Example 2: Consider the equation $y = 5/(x^2 - 1)$. On the basis of the reasoning in Example 1, this defines a function. Since a denominator can't be zero, the domain of this function is not all of $\mathbb{R}$. The values $x = 1$ and $x = -1$ must be excluded, since then the expression is undefined. What's the range? We could determine this by solving the equation for x, but here's another way: we know that $x^2 - 1$ can take on all values ≥ -1. Since $x^2 - 1$ takes on all possible positive values, so does $5/(x^2 - 1)$. And since $x^2 - 1$ takes on all negative values ≥ -1, $5/(x^2 - 1)$ takes on all negative values ≤ -5. In interval notation, the range is $(-\infty, -5] \cup (0, \infty)$. Exercise 4 asks you to do this more rigorously.

Example 3: Consider the relation $x^2 + y^2 = 25$, discussed in Section 6.1. Note that this equation doesn't have y by itself on one side, so we can't be sure it's a function. If we solve it for y, we get $y = \pm\sqrt{25 - x^2}$. So this relation is definitely not a function, because for a single value of x, we can get two values of y. For instance, (3, 4) and (3, −4) are both in this relation.

The graphs in Figure 7.2 illustrate a familiar fact, the **vertical line test:** *A relation on* $\mathbb{R}$ *is a function if and only if no vertical line intersects its graph more than once.*

Function Notation

Notation: If f is any function and x is any member of the domain of f, then the expression $f(x)$ denotes the unique output that f assigns to x. More concretely, an equation of the form $f(x) = y$ means the same thing as $(x, y) \in f$.

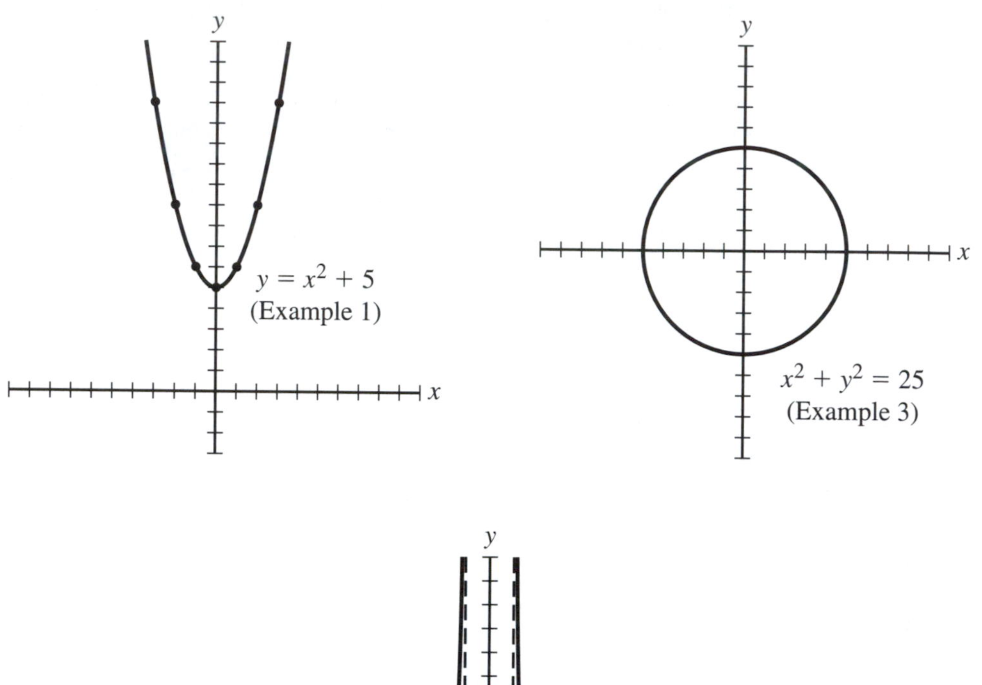

y

x

$$y = \frac{5}{x^2 - 1}$$

(Example 2)

Figure 7.2 Graphs of three relations

If $f(x) = y$, we say that y is *the* **image of x under f**, and x is *a* **preimage of y under f**. We also say that y is the **value of f at x**, and **f maps x to y**.

Notice the italicized words in the preceding sentence; their intent should be clear. By definition, a domain member can have only one image. In contrast, a range member can have many preimages.

As you know, the usefulness of function notation is that once a function has been defined using the notation $f(x)$, we can then plug in any number or expression for the variable x. Technically, this is an application of universal specification, since the defining equation is normally understood to hold for all x in the domain.

Example 4: If we define a function f by $f(x) = 3x - 2$, then we can say things like $f(5) = 13$, $f(-1) = -5$, $f(2b) = 6b - 2$, $f(\cos 2y) = 3(\cos 2y) - 2$, and even $f(f(x)) = 9x - 8$. If we also define $g(x) = 5 \sin x$, then we have $f(g(x)) = 15 \sin x - 2$, $g(f(x)) = 5 \sin(3x - 2)$, $g(f(4)) = 5 \sin 10$, and so on. Note that $f(g(x))$ and $g(f(x))$ are not the same.

Example 5: You have probably seen function notation used for so-called functions of two (or more) variables. For example, we can consider the function $f(x, y) = 2xy - \sin y$, where x and y are real variables. For this function, two real numbers have to be specified in order to compute an answer; and the graph of f is viewed as a set of ordered triples (x, y, z), where z is the output for any given x and y. But this seems to violate the definition of a function, which specifies that an input to a function consists of one object, not two; and the graph of a function consists of ordered pairs, not ordered triples.

The technical solution to this difficulty is to say that the domain of f consists of ordered pairs (x, y); that is, $\text{Dom}(f) = \mathbb{R} \times \mathbb{R}$. So the graph of f is a subset of $(\mathbb{R} \times \mathbb{R}) \times \mathbb{R}$, which, according to Chapter 6, is exactly what is meant by $\mathbb{R} \times \mathbb{R} \times \mathbb{R}$. So it is valid to think of f as being made up of either ordered pairs or ordered triples!

Similar points hold for functions of three or more variables. For instance, a function of five variables can be thought of as a set of ordered 6-tuples or as a set of ordered pairs in which the first members are 5-tuples. Of course, the graph of a function of three or more real variables cannot be drawn, but mathematicians still talk about the graphs of such functions.

Why Codomains?

We know that every relation has a domain and a range, and a function is a type of relation. So why is the concept of a codomain introduced in connection with functions? If the statement $f: A \to B$ requires that $\text{Dom}(f) = A$, why does it require only that $\text{Rng}(f) \subseteq B$, rather than $\text{Rng}(f) = B$? There are several answers to these questions, but the simplest has to do with convenience. When we define a function, we usually have a precise domain in mind, but it may not be easy to determine the exact range. It is easier to specify a codomain, since that can be any set that contains all the outputs. Examples 1 and 2 illustrate this. In Example 1, we have an equation in which each real input x yields exactly one output y, which is also a real number. This allows us to say that we have a function from $\mathbb{R}$ to $\mathbb{R}$, without needing to calculate the exact range.

There are also more theoretical (but no less important!) reasons for using codomains. Generally speaking, the set of all functions from A to B is easier and more fruitful to work with than the set of all functions with domain A and range B. On the other hand, it is sometimes easier to work with functions without specifying codomains. To that end, it is possible to define the word "function" without ever mentioning the words "from A to B": a function is a relation in which no domain member is paired with more than one range member. We can state an even simpler definition.

Alternative Definition: A function is a relation in which no two ordered pairs have the same first member. In other words, a function is a relation in which, whenever (x, y) and (x, z) are in it, then $y = z$.

This definition is equivalent to the set-theoretic one given at the beginning of the section (see Exercise 10). Under this definition, a function does not have a predefined codomain, although any set that contains the range may be designated as the codomain if desired.

Most mathematicians prefer not to use the words "mapping" or "map" to describe a function unless its domain and codomain are specified. The next three examples illustrate the subtleties inherent in the term "codomain" and in the different ways of viewing functions.

Example 6: Let $f = \{(3, 6), (8, 2), (1, 2), (0, 0), (-5, 3)\}$. Then f is clearly a relation in the set-theoretic sense; in fact, it is the same relation defined in Example 5 of Section 6.1. Is f a function? Unlike all of the previous examples in this section, there is no apparent rule relating the first and second members of the ordered pairs in f. But f is definitely a function because it satisfies the simple condition in the above definition: no two ordered pairs in f have the same first member. (The ordered pairs (8, 2) and (1, 2) are in f and have the same *second* member, but that doesn't matter.)

$\text{Dom}(f) = \{3, 8, 1, 0, -5\}$ and $\text{Rng}(f) = \{6, 2, 0, 3\}$. If desired, we can give f a codomain; any set that contains the range is allowed. So we can say $f\colon \{3, 8, 1, 0, -5\} \to \{6, 2, 0, 3\}$, $f\colon \{3, 8, 1, 0, -5\} \to \mathbb{R}$, $f\colon \{3, 8, 1, 0, -5\} \to \mathbb{Z}$, and so on.

Example 7: Consider the function $f(x) = x^2 + 1$, where x is a real variable. The range of f is the set $B = \{y \in \mathbb{R} \mid y \geq 1\}$. Therefore, we can say $f\colon \mathbb{R} \to \mathbb{R}$. But we could also say $f\colon \mathbb{R} \to \mathbb{C}$ or $f\colon \mathbb{R} \to B$. If we decided to restrict the domain of f to integers (which would technically make it a different function), we could say $f\colon \mathbb{Z} \to \mathbb{Z}$, $f\colon \mathbb{Z} \to \mathbb{R}$, $f\colon \mathbb{Z} \to \mathbb{C}$, $f\colon \mathbb{Z} \to \mathbb{N}$, and so on.

Example 8: Let A and B be sets with $A \subseteq B$. Then there is a unique function $f\colon A \to B$ with $f(x) = x$ for every x in A. This function is called the **inclusion map** from

A to B. However, as a set of ordered pairs, f is simply id_A. Thus, the idea of inclusion maps makes sense only when codomains are specified.

Example 8 indicates that even the basic notion of equality of functions changes depending on whether or not codomains are considered part of what constitutes a function.

Definition: Two *mappings* f and g are called **equal** iff they have the same domain and codomain and, for every x in this domain, $f(x) = g(x)$.

This definition gives us the option of saying that id_A and the inclusion map from A to B are not the same mapping, provided that $A \subset B$. If we don't care about codomains, then the above definition remains valid if we simply drop the reference to codomains. In fact, if functions are viewed as sets of ordered pairs, there is no need to call this a definition because it is easily derivable (see Exercise 11).

Other Ways of Defining Functions

Possibly the most important way of defining a function in mathematics, other than by a single equation, is by cases. The justification for defining functions by cases is given in Theorem 7.6.

Example 9: An important function defined by cases is the absolute value function, defined near the end of Appendix 2. Another typical definition of this sort, which is the subject of Exercise 12, is:

$$g(x) = \begin{cases} x^2 & \text{if } x \geq 0 \\ -x^2 & \text{if } x < 0 \end{cases}$$

Example 10: Let A and B be sets with $B \subseteq A$. The **characteristic function of B** (with domain understood to be A) is the function χ_B: $A \to \{0, 1\}$ defined by cases as

$$\chi_B(x) = \begin{cases} 1 & \text{if } x \in B \\ 0 & \text{if } x \notin B \end{cases}$$

(The symbol χ is the Greek letter chi, so this function is called "chi sub B.") In order to talk about characteristic functions, the set A must be understood. Often it is $\mathbb{R}$.

Almost all our examples of functions so far have been very mathematical. Specifically, they have been defined by algebraic relationships involving real numbers. In reality, there are many other types of situations that give rise to functions. The next several examples illustrate just a few of the many possibilities.

Example 11: Before a passenger plane takes off, each passenger is assigned a seat. This assignment can be viewed as a function from the set of passengers on the plane to the set of seats on the plane, for each passenger is sent, by some procedure, to a unique seat, even if there is no obvious formula for who gets which seat.

Example 12: Many physical situations give rise to functions; some of these can best be understood and/or graphed using some sort of instrument. For instance, suppose that a thermometer is placed at a particular location, from noon to 4 p.m. on a certain day. We can then say that the thermometer is evaluating a function; the inputs of this function are moments of time in a specific interval, while the outputs are values of temperature. If the thermometer is hooked up to a graphing device, then the graph of this function is supplied physically, without a mathematical formula. In the real world, time is probably the most common independent variable for functions.

In many situations of this sort, it can be convenient to view the inputs and outputs of the function as real numbers. In this case, we could take the domain to be the interval $[0, 4]$ and the range to be the numerical values of temperature (in some agreed-on units such as Celsius or Fahrenheit).

Example 13: There are also physical situations that give rise to functions defined by mathematical formulas. One famous example is the **universal law of gravitation**, discovered by Isaac Newton. It states that the gravitational force between two objects is given by the equation $F = Gm_1m_2/r^2$, where m_1 and m_2 are the masses of the objects, r is the distance between them, and G is an important constant, the **gravitational constant**. One way to view this formula is that it gives F as a function of three variables. But if we have two specific objects in mind (for example, the sun and Mars), then F is a function of the single variable r. This formula is an example of an **inverse square law**, so named because of the way r appears in it.

As in Example 12, we have a choice here: we can view the inputs and outputs of this function as physical quantities (distances and forces), or we can view them as unitless (positive real numbers).

Example 14: Functions are also important in social sciences such as economics. For instance, consider a factory that produces widgets (a very popular product in the world of economics). If n is any nonnegative integer, it would be useful to know the total cost for that factory to produce n widgets on a given day. It is common to denote this cost as $C(n)$ and say that C is the **cost function** for this factory.

Cost functions provide useful information. For example, $C(0)$, the cost to produce no widgets, is probably not zero. Rather, its value represents the daily **overhead**, or **fixed costs**, at that factory. Also, the behavior of the cost function as n increases gives important information about the efficiency of the factory.

Example 15: Computer programs often define functions, some of which may be hard to fit into mathematical formulas. One amusing category is screen-saver programs. You probably have one of these on your computer—perhaps "Flying Toasters." There are all sorts of functions defined by such a program. You could start the screen saver and let $f(x)$ be the number of toasters on the screen x minutes after the program starts or the distance from the top of the screen to the nearest toaster at that time. If you do this for k minutes, then f should be a function with domain $[0, k]$. Is there a mathematical formula for f? Screen savers use a type of program called a **random-number generator**, but most of these are still deterministic programs. So if the computer's clock is accurate, the programmer who wrote the screen saver or someone else who understands it very well might be able to write a mathematical formula for the function.

We conclude this section with a picture that is frequently used to illustrate concepts connected with functions. Figure 7.3 illustrates the image of a function as a collection of arrows. If $f: A \to B$, we know that f assigns exactly one element of B to each element of A. If we use an arrow going from x to y whenever $f(x) = y$, we get a picture like the one shown. The important thing to see in this picture is the difference between the appearance of A and that of B. Since f is a function, no point in A can have more than one arrow coming from it. And since $\mathrm{Dom}(f) = A$, each point in A has ***exactly*** one arrow coming from it. But in B, a point can have more than one arrow going to it. And since the range of f need not be all of B, it is also possible that a point in B can have no arrows going to it. By the way, Figure 7.3 gives the impression that A and B are disjoint, but they need not be.

Exercises 7.1

(1) Find the domain and range of each of the following, and determine whether it's a function. For some of the solutions, you may need to use methods not discussed in this book, such as techniques involving calculus.

(a) $\{(1, 2), (2, 2), (4, -2)\}$
(b) $\{(1, 2), (2, 2), (2, 1)\}$
(c) $\{(n, 2n - 1) \mid n \in \mathbb{N}\}$
(d) $\{(2n - 1, n) \mid n \in \mathbb{N}\}$
(e) $\varnothing$
(f) $\{(x, y) \mid x, y \in \mathbb{R} \text{ and } y = x^2 - 2x + 7\}$
(g) $\{(x, y) \mid x, y \in \mathbb{R} \text{ and } y = 3/(x^2 + 1)\}$
(h) $\{(x, y) \mid x, y \in \mathbb{R} \text{ and } x = y^3 - 4\}$
(i) $\{(x, y) \mid x, y \in \mathbb{R} \text{ and } x = y^3 - y\}$
*(j) $\{(x, y) \mid x, y \in \mathbb{R} \text{ and } x = y^3 + y\}$

(2) If $f(x) = 3 \sin x$, $g(x) = 1/x$, and $h(x) = x^2 - 1$, compute expressions for:

(a) $f(h(x))$
(b) $g(g(x))$
(c) $h(x^3)$
(d) $g(2/f(x))$
(e) $[g(x + z) - g(x)]/z$
(f) $h(h(4))$

(3) Use the method described in Example 6 of Section 6.1 (solving for x), to determine the ranges of the functions in Example 1 of this section.

(4) Repeat the previous exercise for Example 2 of this section.

(5) Define five functions whose domains and ranges consist of objects other than numbers.

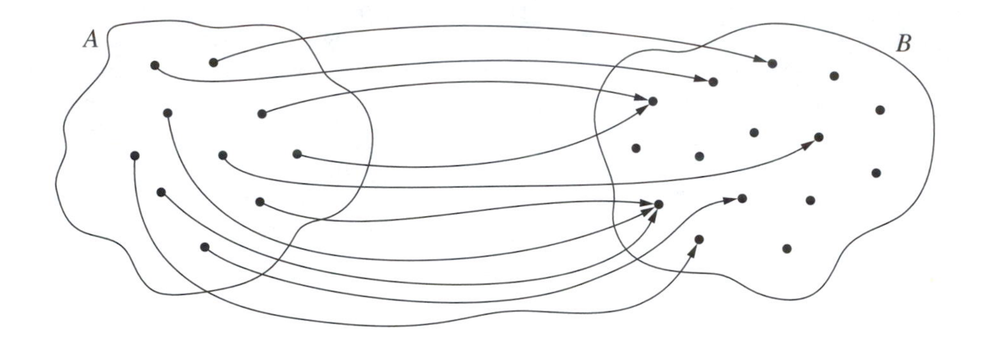

Figure 7.3 A function $f: A \to B$ shown by arrows

(6) Let $A = \{1, 2\}$ and $B = \{2, 4, 5\}$. Give an example of each of the following, or explain briefly why none can exist. For each example you provide, write it three ways: as a set of ordered pairs, as a function defined with function notation, and as a diagram like Figure 7.3.

(a) $f: A \to B$
(b) f is a function with domain A and range B.
(c) $f: B \to A$
(d) $f: B \to A$ and each member of $\text{Rng}(f)$ has only one preimage.

(7) Determine whether each of these rules defines a function whose domain is the specified set of inputs. Justify your answers.

(a) For each person who has ever lived, $f(x) = x$'s mother.
(b) For each person who has ever lived, $f(x) = x$'s brother.
(c) For each person who has ever lived, $f(x) = x$'s youngest brother.
(d) For each natural number, $f(n) = n + 1$.
(e) For each natural number, $f(n) = n - 1$.
(f) For each real number, $f(x) = \tan x$.
(g) For each real number, $f(x) = e^x$ if $x \geq 0$, and $f(x) = \cos x$ if $x \leq 0$.
(h) For each *set* of real numbers, $f(A) =$ the smallest number in A.
(i) For each *set* of *natural* numbers, $f(A) =$ the smallest number in A.
(j) For each polynomial, $F(g) =$ the derivative of g.
(k) For each polynomial, $F(g) =$ the maximum value of g.
(l) For each polynomial, $F(g) =$ the maximum value of g on the interval [2, 17].

(8) Consider $\chi_{\mathbb{Z}}$, the characteristic function of $\mathbb{Z}$ (with domain $\mathbb{R}$), as discussed in Example 10.

(a) What is the image of 1.4 under $\chi_{\mathbb{Z}}$?

(b) List three preimages of 1 under χ_Z.
(c) Describe the set of all preimages of 0 under χ_Z.

(9) Let A be any set, R an equivalence relation on A, and $B = A/R$, the collection of all equivalence classes under R. Now define $f: A \rightarrow B$ by $f(x) = [x]$. Prove:

(a) For any x and y in A, $f(x) = f(y)$ iff xRy. (Therefore, f is a function that establishes the converse of Theorem 6.4 discussed after that theorem.)

(b) For any x and y in A, y is a preimage of $[x]$ under f iff xRy.

(c) For any b in B, the set of preimages of b under f is b.

The function f is called the **mapping to the equivalence classes** defined by R.

(10) This exercise provides the verification that the alternative definition of functions is equivalent to the set-theoretic definition given at the beginning of the section. Prove that for any set of ordered pairs R, the following are equivalent:

(a) R is a function in the sense defined at the beginning of the section.

(b) No member of Dom(R) is paired in R with more than one member of Rng(R).

(c) No two ordered pairs in R have the same first member.

(d) Whenever (x, y) and (x, z) are in R, then $y = z$.

(11) Prove that for any functions f and g (viewed as sets of ordered pairs), $f = g$ iff Dom(f) = Dom(g) and, for all x in this domain, $f(x) = g(x)$.

(12) This exercise investigates the subtleties inherent in finding the derivative of a function defined by cases. Do not try to give rigorous justifications for your solutions to this exercise.

(a) Graph the absolute-value function and the function $g(x)$ defined in Example 9. Describe these graphs in words.

(b) Determine the derivative of the function $|x|$. You have to consider *three* cases: $x > 0$, $x < 0$, and $x = 0$. Graph your result.

(c) Repeat part (b) for the function $g(x)$. Pay special attention to the case $x = 0$. What do you notice about your answer?

(13) Let A be any set, and let $\mathscr{F}$ be the set of all functions from A to $\mathbb{R}$. For any f and g in $\mathscr{F}$, define two new functions $f + g$ and fg, also in $\mathscr{F}$, by the rules $(f + g)(x) = f(x) + g(x)$, and $(fg)(x) = f(x)g(x)$, for every x in A. These are the standard definitions of **function addition** and **function multiplication**. The purpose of this exercise is to see that these operations make $\mathscr{F}$ into a well-behaved algebraic structure.

(a) Prove that $f + g = g + f$, for all f, g in $\mathscr{F}$ (function addition is commutative).

(b) What would be the sensible definitions of $f - g$ and $-f$?

(c) Which functions would serve the role of the additive identity and the multiplicative identity in $\mathscr{F}$, as defined in group V of the axioms in Appendix 1?

(d) Under what circumstances, if any, does axiom V-11 hold in $\mathscr{F}$?

(14) Give an example of each of the following, or explain briefly why none can exist. (Here, f and g are functions, viewed as sets of ordered pairs with no specified codomain.)

(a) $f \cup g$ is a function.
(b) $f \cup g$ is not a function.
(c) f^{-1} is a function.
(d) f^{-1} is not a function.
(e) $f \cap f^{-1}$ is a function.
(f) $f \cap f^{-1}$ is not a function.

(15) This exercise introduces some basic ideas from **linear algebra**. Recall that $\mathbb{R}^n$ denotes the set of all n-tuples of real numbers, which may also be thought of as ***n*-dimensional real vectors**. You probably know that addition and scalar multiplication of vectors are defined coordinate by coordinate.

A function $f: \mathbb{R}^m \to \mathbb{R}^n$ is called **linear** iff, for all u and v in its domain and all reals c, $f(u + v) = f(u) + f(v)$ and $f(cu) = cf(u)$. (This is a more restrictive definition of linearity than the one usually used for functions from $\mathbb{R}$ to $\mathbb{R}$.) Prove:

(a) If f is linear, then $f(u - v) = f(u) - f(v)$, for all u and v in the domain.

(b) If f is linear, then $f(\mathbf{0}) = \mathbf{0}$, where $\mathbf{0}$ denotes a vector that is all zeros.

(c) Now let $f: \mathbb{R}^2 \to \mathbb{R}^2$. Prove that f is linear iff there are constants a, b, c, and d such that $f(x, y) = (ax + by, cx + dy)$, for all reals x and y. ***Hint:*** If f is linear, use $f(1, 0)$ and $f(0, 1)$ to determine the constants.

(16) Prove or find a counterexample:

(a) For any function f, $\text{Dom}(f) = \varnothing$ iff $\text{Rng}(f) = \varnothing$.

(b) Whenever $f: A \to B$, $A = \varnothing$ iff $B = \varnothing$.

7.2 One-to-one and "Onto" Functions; Inverse Functions and Compositions

In this section we define and study some important types of functions and ways of forming new functions. We have already discussed inverse relations; now we take up the more specific and important concept of inverse functions. Similarly, most of the ideas we encounter in this section pertain to relations in general, not just to functions, and our definitions reflect that generality. The reason that some of these concepts are not defined in Chapter 6 is that they are infrequently used with relations that are not functions.

Recall the discussion of inverses in Section 6.1. Since every relation has an inverse, every function has an inverse. But the inverse of a function is not necessarily a function. This should be apparent from the first pair of graphs in Figure 6.2. The original parabola is a function but its inverse is not, as is easily seen using the vertical line test. It is not hard to see how this can occur. When we change from R to R^{-1}, the roles of the axes are reversed; this means that the roles of horizontal and vertical lines are reversed. So there is no reason to expect R^{-1} to be a function, unless the graph of R passes an analogous *horizontal* line test. The next definition makes this notion rigorous.

For the rest of this chapter, definitions are usually phrased in terms of the set-theoretic (set of ordered pairs) view of relations and functions.

Definition: A relation is **one-to-one** iff no two ordered pairs in it have the same second member. In other words, a relation is one-to-one iff, whenever (x, z) and (y, z) are both in it, then $x = y$.

The main thing to see about this definition is that it is *exactly equivalent to the definition of a function, except that the roles of the first and second members of ordered pairs are switched.* If you refer to the alternative definition of functions in Section 7.1, you see this immediately. Here is another way to think of this: a relation is a function iff no *input* is matched with more than one *output*. A relation is one-to-one iff no *output* is matched with more than one *input*.

Perhaps you have noticed that our definitions of functions use the words "exactly one" instead of the more mathematical word "unique." The reason for this is that the way that "unique" is used in ordinary life can muddle the distinction between the concepts of function and one-to-one. For instance, what would you conclude if your bank sent you a letter saying that "we have given every customer a unique ID number"? You would conclude that each customer has exactly one ID number, so there is a *function* from customers to ID numbers. But you would also probably conclude that no two customers have the same ID number, that is, that this function is *one-to-one*. In mathematics it is important *not* to attach the second meaning to the word unique.

☞ In the case of functions, the definition of a one-to-one relation becomes easier to state: f is one-to-one iff whenever $f(x) = f(y)$, then $x = y$.

The next result is a continuation of Theorem 6.2 and is just as simple.

Theorem 7.1: (a) If R and S are inverses, then one of them is a function if and only if the other is one-to-one.

(b) If f is a one-to-one function, then so is f^{-1}.

Proof: See Exercise 1. ■

Definition: A one-to-one function is called **injective** or an **injection**.

Notation: The notation $f\colon A \xrightarrow{1-1} B$ or $f\colon A \xrightarrow{\text{inj}} B$ may be used to mean that f is a one-to-one function from A to B.

The notation $f\colon A \xrightarrow{\text{onto}} B$ or $f\colon A \xrightarrow{\text{sur}} B$, read "$f$ is a function from A **onto** B," means that f is a function, $\text{Dom}(f) = A$, and $\text{Rng}(f) = B$.

Equivalently, $f\colon A \xrightarrow{\text{onto}} B$ means that $f\colon A \to B$ and $\forall y \in B\ \exists x \in A\ f(x) = y$.

Definitions: If the codomain of f has been specified and the range of f is all of its codomain, f may be called **onto** or **surjective** or a **surjection**.

If f is a one-to-one function from A onto B, f is called a **bijection** or a **one-to-one correspondence** between A and B. If the sets A and B are understood, we may simply say that f is **bijective**. The notations $f\colon A \xrightarrow{\text{bij}} B$ and $f\colon A \xrightarrow[\text{onto}]{1-1} B$ are used.

Remarks: (1) Mathematicians are often somewhat imprecise when they call a function onto. After all, every function is onto some set, namely its range. In other words, it makes absolutely no sense to call a function onto, or surjective, unless some set has been specified as the codomain. In practice, this may not be done explicitly. For example, mathematicians often say things like "Is the function $f(x) = x^2 + 1$ onto?" Without any other information given, you are probably supposed to assume that the domain of f is $\mathbb{R}$ (since x is usually a real variable) *and* that the intended codomain is also $\mathbb{R}$. Under this interpretation, the answer to the question is no.

Similarly, the terms "bijection" and "one-to-one correspondence" are sometimes used imprecisely. Every one-to-one function is automatically a one-to-one correspondence between its domain and its range. But when a function is called a bijection, that means it not only is an injection but also is onto some intended codomain.

This situation is similar to the one involving the complement of a set. Recall from Section 5.2 that mathematicians often refer to the complement of a set, but that term is meaningless unless some universal set is understood.

(2) A good way to understand the terms "injection," "surjection," and "bijection" is to refer to Figure 7.3. Remember that every point in A must have exactly one arrow from it, but a point in B may have more than one arrow or no arrow going to it. But if f happens to be surjective (that is, onto B), then every point in B must have at least one arrow going to it.

If f is injective, then no point in B has more than one arrow going to it. This means that if we reverse all the arrows to form f^{-1}, we get a function (in fact a bijection) from Rng(f) to A. This is a consequence of Theorems 6.2 and 7.1. If f is a bijection, then every point in B has exactly one arrow going to it. In this case, if we reverse all the arrows, we see that f^{-1} is a bijection from B to A.

(3) With Figure 7.3 still in mind, let's think a bit more about the idea of a one-to-one correspondence between A and B. Note that in this case there's exactly one arrow leaving each point in A, and exactly one arrow arriving at each point in B. In other words, a bijection is sort of a perfect match-up between two sets, with no points omitted and no duplication. To test your intuition of this concept, consider the following three assertions:

(a) Given two finite sets, there's a bijection between them iff they have the same number of elements.

(b) There can't be a bijection between a finite set and an infinite set.

(c) Given any two infinite sets, there's a bijection between them.

Do these assertions seem to be clearly correct? Interestingly, the first two are correct, but the third is not. Our consideration of these issues is resumed in Section 7.5.

Compositions

You've definitely had experience with compositions of functions, since they are probably the most important way of forming new functions from old ones. The composition of two functions is the new function obtained by performing one followed by the other.

Example 1: If $f(x) = 2x + 5$ and $g(x) = \sin x$, we can set up the composition $f(g(x)) = 2 \sin x + 5$, which denotes "g followed by f" on any input value x. We can also set up the composition $g(f(x)) = \sin(2x + 5)$, which denotes "f followed by g."

The idea of compositions is implicit in function notation, as we can see in Example 1 here, and also in Example 4 of Section 7.1. To make it explicit, a composition of functions is denoted with a small circle, which is read "circle." So in the above example, $f \circ g$ is the function defined by $f \circ g(x) = 2 \sin x + 5$. Note that since $f \circ g(x)$ means "f of g of x," $f \circ g$ means "g then f," *not* "f then g"!

If all we intended to do was to define the composition of *functions*, the above would be plenty of introduction. However, it's sometimes useful to talk about the composition of arbitrary *relations*, and for that a more careful analysis is required. To that end, see Figure 7.4. This shows the composition $f \circ g$ as two computers or boxes in sequence. When a composition is being discussed, it's often helpful to use three variables for the inputs and outputs at various stages. Here we've used x for the input of $f \circ g$, which is actually used as the input for g. We then use y to represent $g(x)$, the output of g, which then becomes the input for f. Finally, we use z for the output of the whole composition, so $z = f(y) = f(g(x)) = f \circ g(x)$.

Now, let's ask what is required for an input x to be paired with an output z under the composition $f \circ g$. Of course, we can say that what's required is that $z = f(g(x))$. But if we want to say this without function notation, we can say that what's required is that there is some y such that x is paired with y by g, and y is paired with z by f. In other words, (x, z) is in $f \circ g$ iff for some y, (x, y) is in g and (y, z) is in f. With this in mind, we are led to the following.

Definition: For any relations R and S, their **composition** $R \circ S$ is defined to be

$$\{(x, z) \mid \exists y\ [(x, y) \in S \text{ and } (y, z) \in R]\}$$

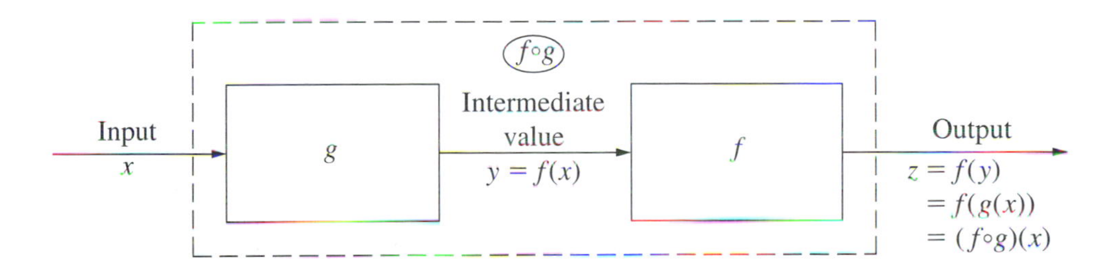

Figure 7.4 Composition of functions viewed as linked boxes

Example 2: Here is a concrete example to show that this definition makes sense. Let f and g be the functions with domain ℝ defined by $f(x) = x + 3$ and $g(x) = 4x$. Then $f \circ g$ should also have domain ℝ and satisfy $f \circ g(x) = f(g(x)) = f(4x) = 4x + 3$. According to the above definition,

$$\begin{aligned} f \circ g &= \{(x, z) \mid \exists y\, ((x, y) \in g \text{ and } (y, z) \in f)\} \\ &= \{(x, z) \mid \exists y\, (x \in \mathbb{R} \text{ and } y = 4x \text{ and } y \in \mathbb{R} \text{ and } z = y + 3)\} \\ &= \{(x, z) \mid \exists y\, (x \in \mathbb{R} \text{ and } y = 4x \text{ and } y \in \mathbb{R} \text{ and } z = 4x + 3)\} \\ & \qquad\qquad \text{By Theorem 4.7} \\ &= \{(x, z) \mid x \in \mathbb{R} \text{ and } z = 4x + 3 \text{ and } \exists y\, (y = 4x \text{ and } y \in \mathbb{R})\} \\ & \qquad\qquad \text{By law of predicate logic 18} \\ &= \{(x, z) \mid x \in \mathbb{R} \text{ and } z = 4x + 3\} \\ & \qquad\qquad \text{Since the existence of such a } y \text{ is automatic} \end{aligned}$$

Thus the rigorous definition of $f \circ g$ matches the informal one.

Example 3: Suppose that $R = \{(1, 3), (2, 3), (5, 7), (3, 4)\}$ and $S = \{(5, 7), (3, 3), (3, 8), (4, 1)\}$. Then, by going through all the possibilities, we find that $S \circ R = \{(1, 3), (1, 8), (2, 3), (2, 8), (3, 1)\}$. Exercise 8 asks you to find $R \circ S$.

Example 4: On the set of all people who have ever lived, let P be the relation "y is a parent of x," and let B be "y is a brother of x." Then $B \circ P$ is the "unclehood" relation. What is $P \circ B$? (See Exercise 9.)

Inverse Functions

Theorem 7.1 tells us that only *one-to-one* functions have inverse relations that are also functions. The following theorem provides more information about inverse functions, characterizing them in terms of compositions.

Theorem 7.2: Suppose $f: A \to B$. Then

(a) If f is a bijection, then f^{-1} is the unique function from B to A such that $f^{-1} \circ f = \text{id}_A$ and $f \circ f^{-1} = \text{id}_B$.

(b) If f is not a bijection, then there is no function g such that $g \circ f = \text{id}_A$ and $f \circ g = \text{id}_B$.

Proof: (a) Assume $f: A \xrightarrow{\text{bij}} B$. We want to show $f^{-1} \circ f = \text{id}_A$. By definition, $\text{Dom}(\text{id}_A) = A$. Since $\text{Dom}(f) = A$, it is clear that $\text{Dom}(f^{-1} \circ f) \subseteq A$ (see Exercise 16). Given any x in A, $f(x)$ is defined. Also, if $f(x) = y$, then $f^{-1}(y) = x$. *[The fact that f is one-to-one justifies using function notation for f^{-1}.]* Therefore $f^{-1} \circ f(x) = f^{-1}(f(x)) = f^{-1}(y) = x$. Therefore, $\text{Dom}(f^{-1} \circ f) = A = \text{Dom}(\text{id}_A)$, and for every x in A, $f^{-1} \circ f(x) = x = \text{id}_A(x)$. By Exercise 11 of the previous section, we are done.

The proof that $f \circ f^{-1} = \text{id}_B$ is similar and we omit it. Next we need to prove uniqueness. Assume $g: B \to A$, $g \circ f = \text{id}_A$, and $f \circ g = \text{id}_B$. Let $y \in B$. Since f is onto B, there's an x such that $f(x) = y$. Then $g(y) = g(f(x)) = g \circ f(x) = \text{id}_A(x) = x = f^{-1}(y)$. So we have that $g(y) = f^{-1}(y)$ for every y in B, and this implies that $g = f^{-1}$.

(b) If $f\colon A \to B$ but f is not a bijection, then either f is not onto B or f is not one-to-one. If f is not onto B, let $y \in B - \text{Rng}(f)$. Since y is not an output of f, it can't be an output of any composition $f \circ g$. Therefore, there is no g such that $f \circ g = \text{id}_B$.

If f is not one-to-one, assume $f(u) = f(x) = y$ and $u \neq x$. If there were a function g such that $g \circ f = \text{id}_A$, we would have $g(y) = g(f(u)) = u$ and $g(y) = g(f(x)) = x$. Since g is a function, this implies $u = x$, a contradiction. ∎

Theorem 7.2 says that an inverse function really reverses or undoes the action of the original function, in the sense that if you compose an injective function with its inverse function (in either order), you get your original input back. With this in mind, it is sometimes easy to figure out the inverse of a function without doing any computation.

Example 5: For instance, the inverse of $f(x) = x - 5$ (subtracting 5 from any input) must be $f^{-1}(x) = x + 5$ (adding 5). The inverse of $f(x) = 2x$ (doubling) must be $f^{-1}(x) = x/2$ (halving). The inverse of $f(x) = -x$ happens to be itself. The inverse of $f(x) = 2x - 5$ (doubling and then subtracting 5) must be $f^{-1}(x) = (x + 5)/2$ (adding 5 and then halving). Note the reversed order of the operations in the inverse; this phenomenon is proved in Theorem 7.3(a).

Example 6: Various important functions in mathematics are *defined* as inverse functions. For instance, taking the cube root of a real number may be defined as the inverse of cubing a number (see Figure 7.5). Note that

$$\sqrt[3]{x^3} = x \quad \text{and} \quad (\sqrt[3]{x})^3 = x \quad \text{for all } x$$

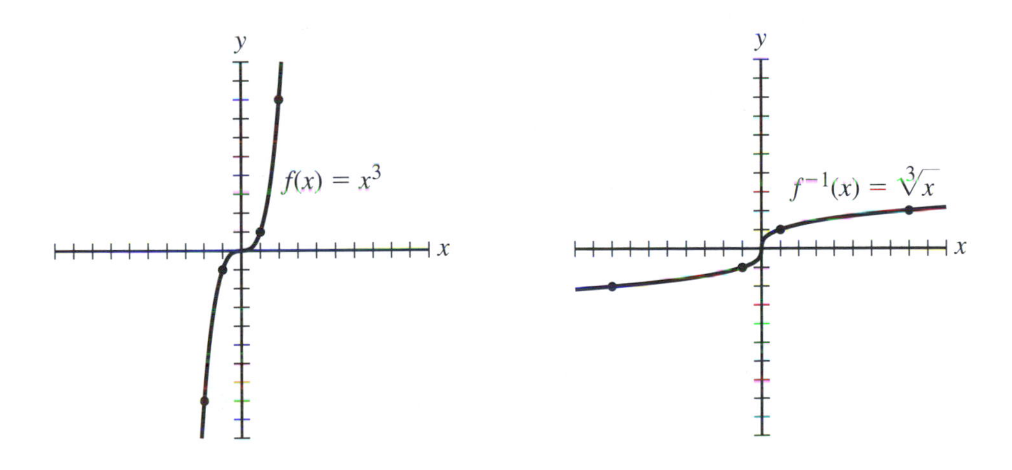

Figure 7.5 The cube root function defined as an inverse

Example 7: Another example of this is logarithms: the function $\log_b x$ is usually defined as the inverse of the exponential function b^x, where b is any positive constant (the base) (see Figure 7.6). Some calculus textbooks do this in the other order: they define the natural logarithm function as an integral, then define the function e^x as its inverse, and then finally define logarithm and exponential functions with other bases. Whichever way they're defined, we get two important identities from Theorem 7.2:

$$\log_b(b^x) = x, \text{ for all } x \in \mathbb{R}, \quad \text{and} \quad b^{(\log_b x)} = x, \text{ for all } x > 0$$

Restricting the Domain of Functions

Having an inverse function is so useful in mathematics that often a sort of partial inverse function is defined even for functions that are not one-to-one. This is done by restricting the domain of the original function, a concept that is important enough to deserve its own notation:

Definitions: Let f be any function and $A \subseteq \text{Dom}(f)$. Then the **restriction** of f to A, denoted $f|_A$, is the set of ordered pairs $\{(x, y) \mid x \in A \text{ and } f(x) = y\}$. If $g = f|_A$ for some A, then g may be called a **restriction** of f, and f may be called an **extension** of g. Note that these statements simply say that $g \subseteq f$.

It is clear that the domain of $f|_A$ is A and, for any x in A, $f|_A(x) = f(x)$ (see Exercise 10). Mathematicians are sometimes a bit careless about the requirement that $A \subseteq \text{Dom}(f)$ in this definition. For instance, they might refer to "the restriction of the tangent function to the positive real numbers," even though the domain of the tangent function does not include all positive real numbers. What is meant, of course, is the restriction of the tangent function to the positive real numbers *that are in the domain of the tangent function.*

The existence of partial inverse functions is based on the following principle: for any function f, there is a subset A of the domain of f such that $f|_A$ is one-to-one and $\text{Rng}(f|_A) = \text{Rng}(f)$. In other words, every function can be restricted to a one-to-one function without losing any members of the range. Don't bother trying to prove this principle; it's equivalent to the powerful axiom of choice (Section 7.7).

However, in most applications, it is possible not only to find a partial inverse function but also to have the restricted domain A be a relatively simple set. In the two important examples following, A is an interval.

Example 8: Consider the squaring function $f(x) = x^2$, defined on $\mathbb{R}$ (see Figure 7.7). This is not one-to-one, so its inverse is not a function. But if we let $g = f|_A$, where A is the set of all nonnegative real numbers, then g is one-to-one. The function g^{-1} is an important function called the **principal square root** function, $\sqrt{x}$. Note that the conditions described in Theorem 7.2, $(\sqrt{x})^2 = x$ and $\sqrt{x^2} = x$ hold true for all nonnegative numbers. However, they fail if $x < 0$. Thus g is not a true inverse function of the original function f.

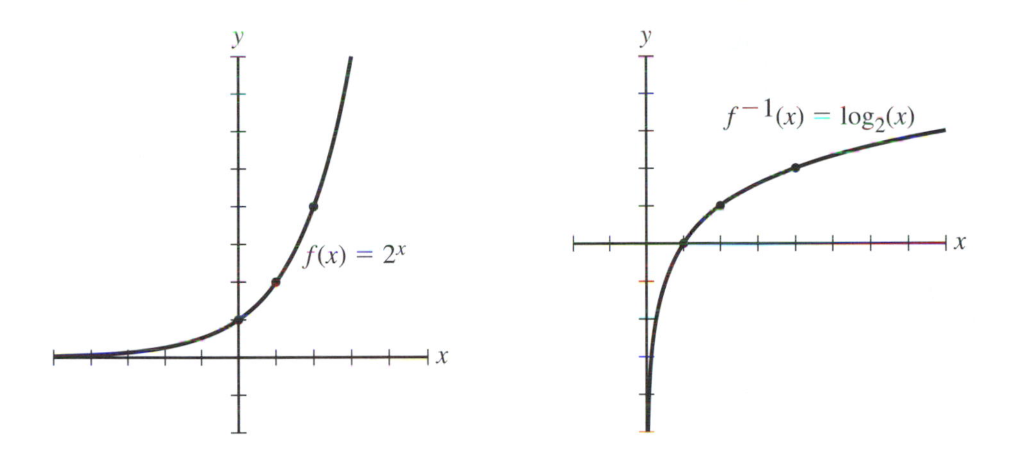

Figure 7.6 A logarithm function defined as an inverse

Example 9: As a final example of this, consider the **inverse trigonometric functions** (Figure 7.8). None of the regular trigonometric functions is one-to-one. Therefore their inverses are relations but not functions. Some mathematicians denote these inverse relations using lowercase designations like arc sin x, $\sin^{-1} x$, and so on. If we want inverse trigonometric *functions,* we must restrict the domains of the original trigonometric functions. This must be done with some care: it is important that the restricted function still hits every number in the original range but just once. For the trigonometric functions, this is standardly done as follows: the sine function is restricted to $[-\pi/2, \pi/2]$, the cosine function to $[0, \pi]$, the tangent function to $(-\pi/2, \pi/2)$, and so on. Exercise 11 asks you to find appropriate restricted domains for the three other trigonometric functions.

Exercises 7.2

(1) Prove Theorem 7.1.

(2) Let $A = \{1, 2, 3\}$ and $B = \{2, 3, 6, 8\}$. Try to find functions from A to B that are:

(a) One-to-one and onto
(b) Neither one-to-one nor onto
(c) One-to-one but not onto
(d) Onto but not one-to-one

If any of these functions is impossible to find, explain why (nonrigorously). You may use this result, which is proved in Section 7.6: if f is a function whose domain is a finite set A, then Rng(f) cannot have more elements than A.

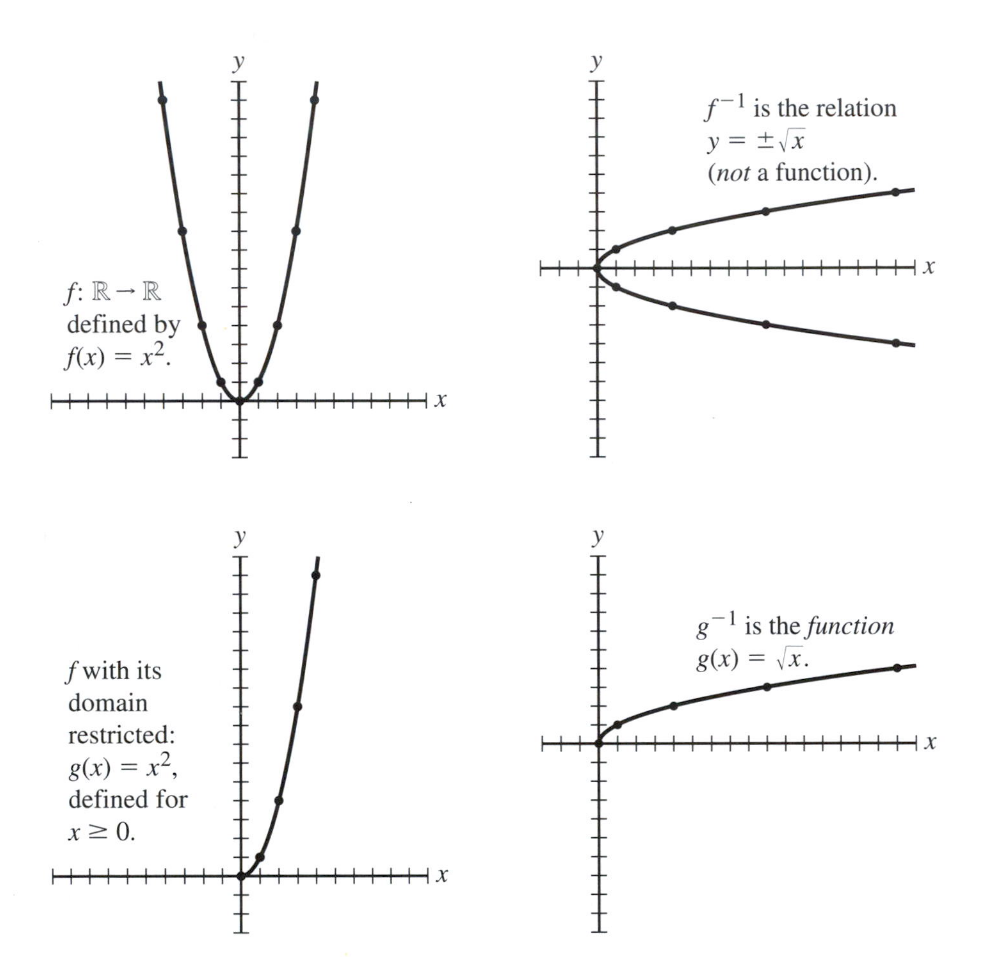

Figure 7.7 Construction of a partial inverse function to the squaring function

(3) Repeat Exercise 2 for functions from B to A.

(4) Repeat Exercise 2 for functions from A to A.

(5) Repeat Exercise 2 for functions from $\mathbb{R}$ to $\mathbb{R}$.

(6) Repeat Exercise 2 for functions from $\mathbb{N}$ to $\mathbb{N}$.

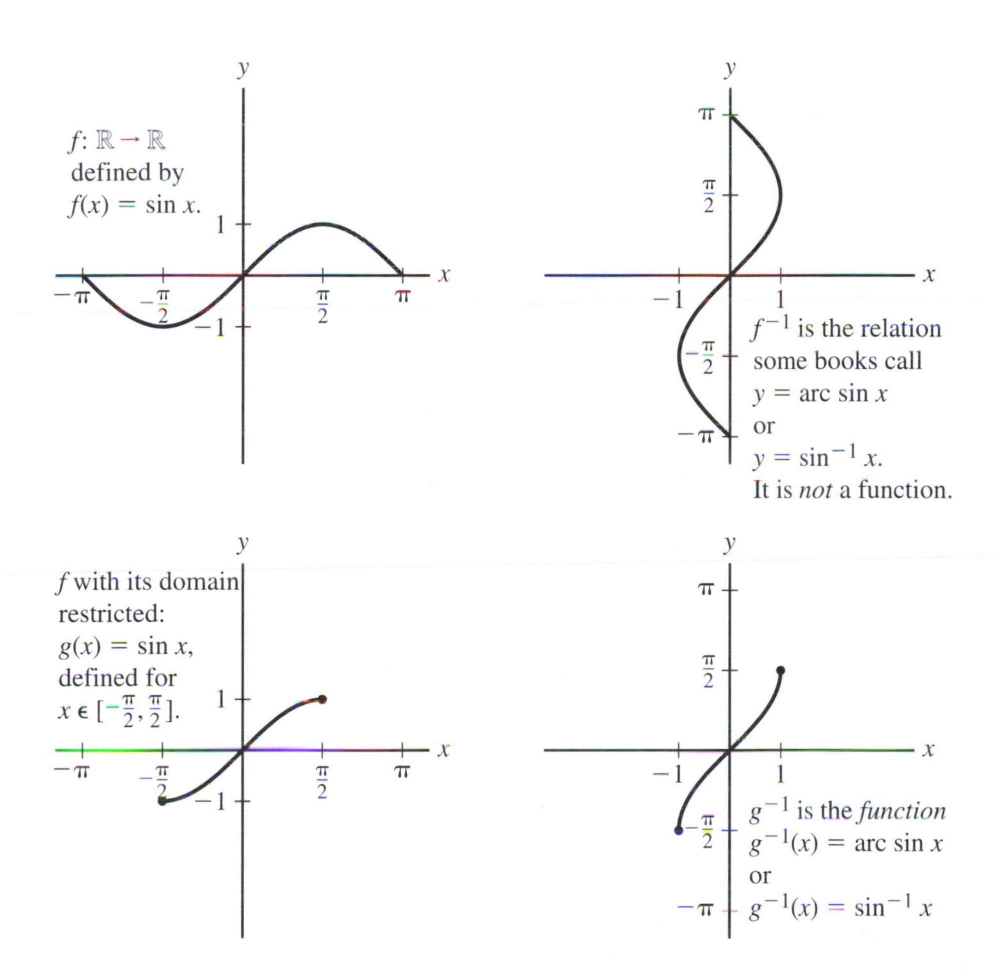

Figure 7.8 Construction of a partial inverse function to the sine function

(7) In a certain town, every person lives in exactly one house. (So nobody is homeless, and nobody has two residences.) Thus we have a function from people in the town to houses in the town.

(a) What would it mean to say this function is one-to-one?

(b) What would it mean to say this function is onto?

(8) With R and S as defined in Example 3, find $R \circ S$.

(9) Let P and B be as in Example 4. Also, let M be the "motherhood" relation. Describe these relations:

(a) $P \circ B$ (b) $P \circ P$
(c) $M \circ P$ (d) $P \circ M$
(e) $P^{-1} \circ P$ (f) $P \circ P^{-1}$

(10) Let f be any function and $A \subseteq \mathrm{Dom}(f)$. Prove that the domain of $f|_A$ is A and, for any x in A, $f|_A(x) = f(x)$.

(11) Find appropriate restricted domains for the secant, cosecant, and cotangent functions to define partial inverse functions for them (refer to Example 9). There are many reasonable answers in each case; you need to find just one.

(12) Refer to Example 7. For fixed positive b, the function b^x is a bijection between $\mathbb{R}$ and the set of all positive reals and satisfies the familiar properties of exponents: $b^x b^y = b^{x+y}$, $b^x/b^y = b^{x-y}$, and $(b^x)^y = b^{xy}$. From these and the definition of logarithms, derive the corresponding properties of logarithms, for any positive x and y and any z:

(a) $\log_b(xy) = \log_b x + \log_b y$
(b) $\log_b(x/y) = \log_b x - \log_b y$
(c) $\log_b(x^z) = z \log_b x$

(13) Determine whether the following functions are one-to-one. For each one that is, find a formula for f^{-1}, and find the domain and range of f^{-1} (x and y are real variables).

(a) $f(x) = (2 - 5x)/3$ (b) $f(x) = (2 - 5x)/(3x + 4)$
(c) $f(x) = \sqrt{x}$ (d) $f(x) = x^3 - x$

(14) Let $g(x) = x^3 + x$, defined on $\mathbb{R}$.

(a) Use calculus methods to show that g is one-to-one.

Now, without finding a formula for g^{-1} (which is not easy to do), evaluate:

(b) $g^{-1}(0)$ (c) $g^{-1}(10)$
(d) $g^{-1}(g(5))$ (e) $g(g^{-1}(-7))$

(15) Let $f: A \rightarrow B$. Prove:

(a) f is one-to-one iff no member of B has more than one preimage under f.
(b) f is onto iff every member of B has a preimage under f.

(16) (a) For any relations R, S, prove that $\mathrm{Dom}(R \circ S) \subseteq \mathrm{Dom}(S)$ and that $\mathrm{Rng}(R \circ S) \subseteq \mathrm{Rng}(R)$. Show by counterexamples that $\subseteq$ can not be strengthened to $=$ in either statement.

(b) Fill in the blanks to make a correct statement, and then prove it:
For any functions f and g, $\mathrm{Dom}(f \circ g) = \{x \in \mathrm{Dom}(g) \mid \_\_\_ \in \mathrm{Dom}(\_\_)\}$.

(17) If R is a relation on a set A, prove :

(a) R is reflexive on A iff $\mathrm{id}_A \subseteq R$.

(b) R is symmetric iff $R = R^{-1}$.
(c) R is transitive iff $R \circ R \subseteq R$.
(d) R is irreflexive iff $\mathrm{id}_A \cap R = \varnothing$.
(f) R is antisymmetric iff $R \cap R^{-1} \subseteq \mathrm{id}_A$.
(e) R is strongly antisymmetric iff $R \cap R^{-1} = \varnothing$.

The point of this exercise is that it provides an alternative definition or characterization of equivalence relations and partial orderings, based on the identity relation, inverses, and compositions.

(18) Let f be a linear map on $\mathbb{R}^2$. According to Exercise 15 of Section 7.1, that means f has the form $f(x, y) = (ax + by, cx + dy)$, for all reals x and y. Prove that the following are equivalent:
(a) f is one-to-one.
(b) f is onto.
(c) $ad - bc \neq 0$. (By the way, you may recall that $ad - bc$ is the value of the determinant $\begin{vmatrix} a & b \\ c & d \end{vmatrix}$.)

Hint: Let $f(x, y) = (ax + by, cx + dy) = (u, v)$, and try to solve the second equation for (x, y) to get a formula for f^{-1}.

Critique the proofs in the remaining exercises. (If necessary, refer to the instructions for this type of problem in Exercises 4.2.)

(19) **Theorem:** For all f, A, and B, $(f|_A)|_B = f|_{A \cap B}$.
Proof:

$$\begin{aligned} (f|_A)|_B &= \{(x, y) \mid x \in B \wedge (x, y) \in f|_A\} && \text{By definition of } | \\ &= \{(x, y) \mid x \in B \wedge (x \in A \wedge y = f(x))\} && \text{By the same reason} \\ &= \{(x, y) \mid x \in A \cap B \wedge y = f(x)\} && \text{By definition of } \cap \\ &= f|_{A \cap B} && \text{By definition of } | \end{aligned}$$

(20) **Theorem:** If $f(x) = 1/x$, where x is a real variable, then $f \circ f = \mathrm{id}_{\mathbb{R}}$.
Proof: For any x, $f \circ f(x) = f(f(x)) = f(1/x) = 1/(1/x) = x = \mathrm{id}_{\mathbb{R}}(x)$.

7.3 Proofs Involving Functions

In this section, we begin with some general guidelines for proving things about functions (and about relations, to a lesser extent) and then prove a number of basic results using these methods. Unless stated otherwise, we do not require that functions have specified codomains.

Guidelines for Proving Things about Functions

(1) *To prove that a relation R is a function*, the most common approach is to assume that (x, y) and (x, z) are both in R and show that $y = z$.

(2) *To prove that a relation R is one-to-one*, assume that (x, z) and (y, z) are both in R and derive that $x = y$. In the case that the relation is a function f, this says: assume $f(x) = z$ and $f(y) = z$, and show that $x = y$.

(3) *To prove that* $f: A \to B$, prove that f is a function, $\text{Dom}(f) = A$, and $\text{Rng}(f) \subseteq B$. The first part of this is normally proved as in guideline 1. The second and third parts may be proved as discussed in Section 5.2.

(4) *To prove that* $f: A \xrightarrow{\text{onto}} B$, prove guideline 3 but with the stronger condition $\text{Rng}(f) = B$. That is, also show that if $y \in B$, then there's some $x \in A$ such that $f(x) = y$.

(5) *To prove that* $f: A \xrightarrow{1-1} B$, combine guidelines 2 and 3.

(6) *To prove that* $f: A \xrightarrow{\text{bij}} B$, combine guidelines 4 and 2.

(7) *To prove that two relations are equal*, prove that an ordered pair is in one if and only if it's in the other.

(8) *To prove that two functions f and g are equal*, one approach is to use guideline 7. A more common method is to prove that $\text{Dom}(f) = \text{Dom}(g)$ and that for every x in that domain, $f(x) = g(x)$ (as in Exercise 11 of Section 7.1).

Theorem 7.3: For any relations R, S, and T:
(a) $(R \circ S)^{-1} = S^{-1} \circ R^{-1}$
(b) $(R \circ S) \circ T = R \circ (S \circ T)$

Proof: (a) $(x, y) \in (R \circ S)^{-1}$ iff $(y, x) \in R \circ S$
iff $\exists z\,[ySz \text{ and } zRx]$
iff $\exists z\,[(z, y) \in S^{-1} \text{ and } (x, z) \in R^{-1}]$
iff $(x, y) \in S^{-1} \circ R^{-1}$.

(Formally, we should have used ES and then EG here, but it is common to leave in existential quantifiers in this manner.)
(b) See Exercise 1. ■

Both parts of Theorem 7.3 can be easily expressed in words. Part (a) says that to invert a sequence of steps, you invert them one by one but in the *opposite order*. This is easy to illustrate with examples from everyday life. For instance, in the morning a person puts on socks and then shoes. In what order are these steps reversed or undone?

Part (b) says that composition is associative. This is worth noting because we have already seen that it is not commutative. Because of this associativity, expressions like $f \circ g \circ h$ are unambiguous, analogously to ones like $x + y + z$ (see Exercise 2).

Theorem 7.4: (a) The composition of two functions is a function.
(b) The composition of two one-to-one relations is one-to-one.
(c) If $f: A \to B$ and $g: B \to C$, then $g \circ f: A \to C$.
(d) If $f: A \xrightarrow{1-1} B$ and $g: B \xrightarrow{1-1} C$, then $g \circ f: A \xrightarrow{1-1} C$.

(e) If $f\colon A \xrightarrow{\text{onto}} B$ and $g\colon B \xrightarrow{\text{onto}} C$, then $g \circ f\colon A \xrightarrow{\text{onto}} C$.
(f) If $f\colon A \xrightarrow{\text{bij}} B$ and $g\colon B \xrightarrow{\text{bij}} C$, then $g \circ f\colon A \xrightarrow{\text{bij}} C$.

Proof: (a) Assume (x, y) and (x, z) are both in $f \circ g$. Then by the definition of composition, there exist u and v such that $g(x) = u$ and $f(u) = y$, and $g(x) = v$ and $f(v) = y$. But since g is a function, from $g(x) = u$ and $g(x) = v$ we get $u = v$. But then we have $f(u) = y$ and $f(u) = z$, so since f is a function, $y = z$.

(b) This proof is similar to that of part (a) and is left for Exercise 3.

(c) Assume $f\colon A \to B$ and $g\colon B \to C$. By part (a), $g \circ f$ is a function. We need to show $\text{Dom}(g \circ f) = A$. So assume $x \in A$. Then there's a $y \in B$ such that $f(x) = y$. In turn, there's a $z \in C$ such that $g(y) = z$. So $g \circ f(x) = z$, and therefore $x \in \text{Dom}(g \circ f)$. Conversely, assume $x \in \text{Dom}(g \circ f)$. Then for some z, $g \circ f(x) = z$. By the definition of composition, there must be a y such that $f(x) = y$ and $g(y) = z$. But from $f(x) = y$, we have $x \in \text{Dom}(f) = A$. This completes the proof that $\text{Dom}(g \circ f) = A$. Showing that $\text{Rng}(g \circ f) \subseteq C$ is also left for Exercise 3.

(d) This follows immediately from parts (b) and (c).

(e) Assume $f\colon A \xrightarrow{\text{onto}} B$ and $g\colon B \xrightarrow{\text{onto}} C$. By part (c), all that's left to prove is that the range of $g \circ f$ is all of C. So assume $z \in C$. Since g is onto, there is a y in B such that $g(y) = z$. But then since f is onto, there's an x in A such that $f(x) = y$. So $g(f(x)) = z$, and thus $z \in \text{Rng}(g \circ f)$.

(f) This follows immediately from parts (d) and (e). ∎

Theorem 7.4 can be summarized as saying that many properties are **preserved under composition**. Specifically, being a function, being one-to-one, being onto, and being a bijection are preserved under composition.

Theorem 7.5: If $f\colon A \to B$, then $f \circ \text{id}_A = \text{id}_B \circ f = f$.

Proof: By Theorem 7.4(c), $f \circ \text{id}_A\colon A \to B$. So this composition has the same domain as f does. Also, if $x \in A$, then $f \circ \text{id}_A(x) = f(\text{id}_A(x)) = f(x)$. By guideline 8, this establishes $f \circ \text{id}_A = f$.

The proof that $\text{id}_B \circ f = f$ is left for Exercise 4. ∎

Definition: A bijection from a set A to itself is called a **permutation on** A. This term is used mostly when A is a finite set but may be used even when it is infinite.

Example 1: Here is an example from **group theory**, which is a major part of the subject of abstract algebra: let A be any fixed set, and then let S be the set of all permutations on A. For any two permutations f and g in S, define their product fg to be $f \circ g$. Also, define 1 in S to be id_A. Now refer to the field axioms (axioms V-1 to V-12), and note the following.

By Theorem 7.4(f), the composition of two permutations on A is again a permutation on A, so axiom V-2 is true if we replace $\mathbb{R}$ by S and let x and y have S as their domain.

By Theorem 7.3(c), this multiplication is associative; that is, axiom V-4 is true (again, with the variables having S as their domain).

By Theorem 7.5, axiom V-9 holds in S. And by Theorem 7.2(a), a version of axiom V-11 holds in S. This version is $\forall f(f \circ f^{-1} = f^{-1} \circ f = 1)$. There is no 0 in S that must be excluded from this axiom, as there is in a field.

In summary, we have defined a sort of multiplication operation on S that satisfies the properties of closure, associativity, identity, and inverses. This is what is meant by saying that S, with this multiplication operator, is a **group**. This multiplication is not commutative (as long as A has more than two elements), but that's permissible in a group. In fact, such **permutation groups** are the main example of groups that are not commutative (see Exercise 13).

Note that in a field there are by definition two basic binary operations (called addition and multiplication), whereas in a group there's only one.

Next we present a simple result that justifies defining functions by cases.

Theorem 7.6: If f and g are functions whose domains are disjoint, then $f \cup g$ is a function.

Proof: See Exercise 5. ■

What does this theorem have to do with definitions by cases? Well, a typical such definition has the form

$$f(x) = \begin{cases} g(x) & \text{if } P(x) \\ h(x) & \text{if } Q(x) \end{cases}$$

Here it is understood that x is restricted to some set A; that $\{x \in A \mid P(x)\} \subseteq \text{Dom}(g)$; that $\{x \in A \mid Q(x)\} \subseteq \text{Dom}(h)$; and finally, that there is no x in A such that P(x) and Q(x) are both true. The last assumption is necessary to guarantee that Dom(g) and Dom(h) are disjoint. Then the function f is simply $g \cup h$.

The next two theorems are more sophisticated than the ones in this section so far. The proof of the next theorem might be easier to follow with a diagram like Figure 7.3.

Theorem 7.7: Say $f: A \to B$. Then f is one-to-one iff for all functions g and h with codomain A, ($f \circ g = f \circ h$ implies $g = h$).

Proof: For the forward direction, assume $f: A \xrightarrow{1-1} B$, and let g and h be functions with codomain A. We must prove the conditional in parentheses. So assume $f \circ g = f \circ h$. To show $g = h$, we follow guideline 8. First we show Dom(g) = Dom(h). Let Dom(g) = C, and assume $x \in C$. We have $g: C \to A$ and $f: A \to B$, and so by Theorem 7.4(c), $f \circ g: C \to B$. So $x \in \text{Dom}(f \circ g)$, and by Theorem 4.7 again, $x \in \text{Dom}(f \circ h)$. But if $f(h(x))$ is defined, we must have $h(x)$ defined, so $x \in \text{Dom}(h)$. We have just shown that Dom(g) $\subseteq$ Dom(h). The proof of the other direction is similar and we omit it. Thus we have Dom(g) = Dom(h). Finally, let $x \in \text{Dom}(g)$. We want $g(x) = h(x)$. Since $f \circ g = f \circ h$, we know that $f(g(x)) = f(h(x))$. But f is one-to-one. Therefore, $g(x) = h(x)$.

For the reverse direction, we use indirect proof. Assume that the conditional in parentheses holds for all g and h with codomain A, and assume that f is *not* one-to-one. Then we have some x and y in A with $f(x) = f(y)$. Now define functions g and h from $\{1\}$ to A by $g = \{(1, x)\}$ and $h = \{(1, y)\}$. (Since the domain of these two functions has only one member, each function consists of just one ordered pair.) Note that $g \neq h$, but since $f(g(1)) = f(x) = f(y) = f(h(1))$, we have that $f \circ g = f \circ h$. This contradicts the assumption. ■

Theorem 7.8: If $f: A \to B$, then f is onto if and only if for all g and h with domain B, $g \circ f = h \circ f$ implies $g = h$.

Proof: This proof is similar to that of Theorem 7.7 (see Exercise 11). ■

Induced Set Operations

The rest of this section is devoted to some concepts and notation that provide an important tool for understanding and working with functions.

Definition: Let $f: A \to B$ be a function. Then, for any $C \subseteq A$, the set $\{f(x) \mid x \in C\}$ is called the **image of C under f**, denoted $f(C)$.

Note that, for any $C \subseteq A, f(C) \subseteq B$. In other words, we have defined a new function from $\wp(A)$ to $\wp(B)$, which is called the **forward set operation induced by f**. Since it is a new function, it should technically be named with a different symbol. Some authors use F or f^* to denote this function. But the most common practice is to use the same letter for the new function as for the original. So we write both $f(x)$ and $f(C)$, x being an *element* of the domain, and C a *subset*. The confusion can be compounded by the fact that the word "image" may be used in both cases. But the idea is simple enough, as shown in Figure 7.9, and the context usually makes it clear which type of image is meant.

Example 2: Let $f(x) = x^2$, on the domain $\mathbb{R}$. Then we can write $f(3) = 9$ and say that the image of 3 under f is 9. But we can also write things like $f(\mathbb{N}) = \{1, 4, 9, ...\}$, $f(\{-3, 3\}) = \{9\}$, and even $f(\{3\}) = \{9\}$, with corresponding uses of the word "image."

Here are some simple facts involving this new notion.

Theorem 7.9: Let $f: A \to B$, and $C, D \subseteq A$. Then

(a) $f(C) = \text{Rng}(f|_C)$
(b) f is onto iff $f(A) = B$.
(c) $\forall x \in A, f(\{x\}) = \{f(x)\}$
(d) $f(C \cup D) = f(C) \cup f(D)$
(e) $f(C \cap D) \subseteq f(C) \cap f(D)$
(f) f is one-to-one iff $\forall C\, [f(A - C) = f(A) - f(C)]$.

Proof: Parts (a) through (c) are extremely simple, so we omit their proofs. We prove part (d) and leave parts (e) and (f) for Exercise 15.

(d) Assume $y \in f(C \cup D)$. By definition, that says $y = f(x)$, for some $x \in C \cup D$. So $x \in C$ or $x \in D$; this means $y \in f(C)$ or $y \in f(D)$. Therefore, $y \in f(C) \cup f(D)$. For the converse, just reverse these steps. ■

Inverse Images

Inverse images are just like forward images but in reverse. The idea is no more difficult, but the notation is even more ambiguous than in the forward case. As before, let $f: A \to B$. Recall that f^{-1} automatically exists as a *relation* and that it's a function if and only if f is one-to-one. In that case the notation $f^{-1}(b)$, where $b \in B$, would have its usual meaning based on function notation. But whether or not f is one-to-one, we have the following.

Definitions: Let $f: A \to B$. For any $C \subseteq B$, the set $\{x \in A \mid f(x) \in C\}$ is called the **inverse image of C under f** and is denoted $f^{-1}(C)$.

Note that $f^{-1}(C) \subseteq A$, so we have defined a new function from $\wp(B)$ to $\wp(A)$, called the **inverse set operation induced by f**. As before, there should technically be a new symbol for this function, but it is almost always simply denoted f^{-1}.

Whenever f denotes a function from A to B, the same letter denotes the induced forward set operation from $\wp(A)$ to $\wp(B)$, so it is a bit overworked. But now we see that the symbol f^{-1} is even more overworked. It always denotes a relation, it may denote a function from Rng(f) to A, and it is used to denote the inverse set operation just defined. Most of the time, the context makes the meaning clear; but *you* have to do your part by paying careful attention. It is also probably helpful that different words are used; we refer to *preimages* of a member of B, as opposed to *the inverse image*. Figure 7.9 illustrates inverse images.

Example 3: With f as in Example 2, we have $f^{-1}(\{4, 9\}) = \{2, -2, 3, -3\}$, $f^{-1}([9, 16]) = [3, 4] \cup [-4, -3]$, and $f^{-1}(\mathbb{R}) = \mathbb{R}$. Also, $f^{-1}(\{9\}) = \{3, -3\}$, $f^{-1}(\{-4\}) = \varnothing$, and $f^{-1}(\{0\}) = \{0\}$. Note that the last equation is not saying that $f^{-1}(0) = 0$, which would be incorrect in this situation since f is not one-to-one.

Example 4: Define $g: \mathbb{R} \to \mathbb{R}$ by $g(x) = e^x$. This function is one-to-one, so we can write equations like $g^{-1}(1) = 0$ and $g^{-1}(e) = 1$. Here g^{-1} is the usual inverse function. But now we can also use inverse image notation and say $g^{-1}(\{1\}) = \{0\}$ and $g^{-1}(\{e\}) = \{1\}$, as well as $g^{-1}(\{0\}) = \varnothing$, $g^{-1}(\{1, e\}) = \{0, 1\}$, $g^{-1}([1, e^3]) = [0, 3]$, and so on.

Theorem 7.10: Suppose $f: A \to B$, $y \in B$, and $C, D \subseteq B$. Then

(a) f is onto iff, for every y, $f^{-1}(\{y\}) \neq \varnothing$.
(b) f is one-to-one iff, for every y, $f^{-1}(\{y\})$ has at most one member.
(c) $f^{-1}(C \cup D) = f^{-1}(C) \cup f^{-1}(D)$
(d) $f^{-1}(C \cap D) = f^{-1}(C) \cap f^{-1}(D)$
(e) $f(f^{-1}(C)) \subseteq C$
(f) For any $E \subseteq A$, $E \subseteq f^{-1}(f(E))$

Proof: We prove part (c) only and leave some of the other parts for Exercise 16.

(c) $x \in f^{-1}(C \cup D)$ iff $f(x) \in C \cup D$
iff $f(x) \in C$ or $f(x) \in D$

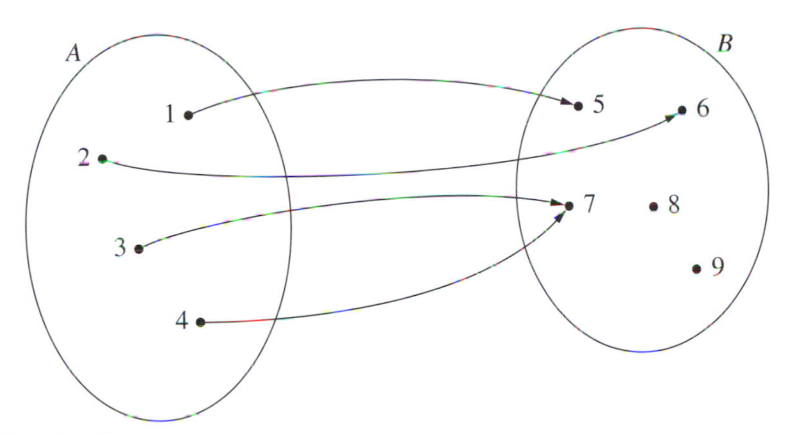

For the function f:

$f(1) = 5; f(\{1\}) = \{5\}; f(\{2, 3\}) = \{6, 7\}; f(\{3, 4\}) = \{7\}; f(A) = \{5, 6, 7\} = \text{Rng}(f)$

Also,

$f^{-1}(\{5\}) = \{1\}$

$f^{-1}(\{8\}) = \emptyset$

$f^{-1}(\{7, 9\}) = f^{-1}(\{7\}) = \{3, 4\}$

$f^{-1}(B) = f^{-1}(\{5, 6, 7\}) = A$

Figure 7.9 Some simple examples of forward and inverse images

iff $x \in f^{-1}(C)$ or $x \in f^{-1}(D)$
iff $x \in f^{-1}(C) \cup f^{-1}(D)$. ■

Exercises 7.3

(1) Prove Theorem 7.3(b).

(2) In the following, verify the associativity of composition (Theorem 7.3(b)), by computing $f \circ g$, $(f \circ g) \circ h$, $g \circ h$ and $f \circ (g \circ h)$. Make sure to specify the domain of each function.

(a) $f(x) = x + 3$, $g(x) = x^2 - 5$, $h(x) = 1 - 2x$
(b) $f(x) = x^2 - 2x$, $g(x) = 2 + 1/x$, $h(x) = x - 3/x$

(3) (a) Prove Theorem 7.4(b).
(b) Complete the proof of Theorem 7.4(c).

(4) Complete the proof of Theorem 7.5.

(5) Prove Theorem 7.6.

(6) On the basis of the discussion after Theorem 7.6, describe the general setting necessary to define a function by *three* cases.

(7) (a) Is it possible for the union of two functions to be a function even if their domains are not disjoint?

(b) State a more general version of Theorem 7.6 that begins, "For any functions f and g, $f \cup g$ is a function iff ... ."

(c) Prove this new version of Theorem 7.6.

(8) Define three functions by cases in which the domains do not consist of numbers.

(9) This problem illustrates part of Theorem 7.7. Let $f: \mathbb{R} \to \mathbb{R}$ be defined by $f(x) = x^2$. Note that f is not one-to-one. Find functions g and h with codomain $\mathbb{R}$ such that $f \circ g = f \circ h$ but $g \neq h$.

(10) This problem illustrates part of Theorem 7.8. Let f be as in Exercise 9. Note that f is not onto. Find functions g and h with domain $\mathbb{R}$ such that $g \circ f = h \circ f$ but $g \neq h$.

*(11) Prove Theorem 7.8.

(12) (a) List all the permutations on each of these sets.

(i) $\varnothing$ (ii) $\{1\}$ (iii) $\{1, 2\}$ (iv) $\{1, 2, 3\}$

(b) Using your answers to part (a) and the fundamental counting principle, conjecture a formula for the number of permutations on a set with n elements, and then show that your conjecture is correct.

(13) Referring to Example 1, show that multiplication (composition) *is* commutative in the group of all permutations on $\{1, 2\}$ but *not* in the group of all permutations on $\{1, 2, 3\}$.

(14) Given any two functions $f: A \to B$ and $g: C \to D$, define a new function $h: A \times C \to B \times D$ by $h(x, y) = (f(x), g(y))$. (This new function may be denoted $f \times g$, although it is not literally the cartesian product of f and g.)

(a) Prove that if both f and g are one-to-one, so is h.

(b) Prove that if both f and g are onto, so is h.

(c) Show by counterexamples that the converses of parts (a) and (b) do not hold in general. What additional assumptions are needed to make these converses true?

(15) Prove parts (e) and (f) of Theorem 7.9.

(16) Prove parts (d), (e), and (f) of Theorem 7.10.

(17) Theorems 7.9(e) and 7.10(e) and (f) make assertions using the symbol $\subseteq$. Show by counterexamples that these cannot be strengthened to equalities.

(18) For each of the three assertions mentioned in Exercise 17, find a simple additional condition on the function f that allows it to be strengthened to an equality. Prove these results.

(19) Let $f: \mathbb{R} \to \mathbb{R}$ be the absolute value function. Determine

(a) $f([-3, 2])$
(b) $f(\mathbb{Z})$
(c) $f(\{2\})$
(d) $f^{-1}(\mathbb{N})$
(e) $f^{-1}(\{-5\})$
(f) $f^{-1}(\{3, -4, 5, \pi\})$

(20) Let $g: \mathbb{R} \to \mathbb{R}$ be the cosine function. Determine

(a) $g(\mathbb{R})$
(b) $g([\pi/2, 5\pi/4))$
(c) $g^{-1}([1, 5])$
(d) $g^{-1}(\{-0.5\})$

(21) Consider these rectangles in $\mathbb{R}^2$: $A = [0, 1] \times [0, 1]$, $B = [1, 2] \times [0, 2]$, and $C = [-1, 0] \times [1, 2]$. Find the image and inverse image of these rectangles under the following mappings on $\mathbb{R}^2$. You need not prove your results.

(a) $f(x, y) = (2x, 2y)$
(b) $f(x, y) = (x + y, y)$
(c) $f(x, y) = (-y, x)$
(d) $f(x, y) = (-x, y)$
(e) $f(x, y) = (0, y)$
(f) $f(x, y) = (x/3, y/3)$
(g) $f(x, y) = (x + 3, y - 4)$
(h) $f(x, y) = (-x, -y)$
(i) All but one of these mappings is linear. Which is not?
(j) Match the mappings in parts (a) through (h) with the following descriptions: translation, reflection about a line, contraction, expansion, shear, 90° rotation, projection onto the Y axis, and reflection about a point.

(22) Let $f: A \to B$ and $g: B \to C$. Prove or find a counterexample:

(a) For any $D \subseteq A$, $g(f(D)) = (g \circ f)(D)$.
(b) For any $E \subseteq C$, $f^{-1}(g^{-1}(E)) = (f \circ g)^{-1}(E)$.

7.4 Sequences and Inductive Definitions

Definition: An **infinite sequence** is a function whose domain is $\mathbb{N}$. Finite sequences are not discussed much in this text, so we usually drop the word "infinite."

We include 0 in the domain of a sequence whenever it's convenient to do so. This should not be viewed as a matter of much significance. Most of our examples are sequences of real numbers; that is, the range of the sequence is a subset of $\mathbb{R}$. But this is certainly not a requirement.

Notation: Even though a sequence is technically a function, the usual function notation is rarely used for sequences. First of all, the letters f, g, and h are not usually used to denote sequences; instead, the letters a, b, and c are typically used. Furthermore, instead of denoting the value of the sequence a at the input n by $a(n)$, we denote it by a_n and call it the ***n*th term** of the sequence. Finally, instead of writing "the sequence a," mathematicians usually write "the sequence (a_n)" or "the sequence $(a_1, a_2, a_3, ...)$." The idea behind this is that we usually don't think of a sequence as a set of ordered pairs; rather, we think of it as an infinite list, consisting of all the outputs written in order.

At the same time, it's important not to confuse the *sequence* $(a_1, a_2, a_3, ...)$ with the *set* $\{a_1, a_2, a_3, ...\}$. Remember that order does not matter in a set, whereas it certainly matters in a sequence. Also, repetition is allowed in a sequence.

Example 1: Consider the sequence a defined by $a_n = 3n + 1$. Technically, a is the set of ordered pairs $\{(1, 4), (2, 7), (3, 10), ...\}$. But it is more natural to think of this sequence as the infinite list $(4, 7, 10, 13, ...)$.

Example 2: The formula $a_n = 6$ defines the sequence $(6, 6, 6, ...)$. This simple type of sequence is called a **constant sequence**.

Example 3: The formula $a_n = (-1)^n$ defines the sequence $(-1, 1, -1, 1, ...)$. This is an example of an **oscillating sequence.**

Note that the ranges of the sequences in the last two examples are, respectively, $\{6\}$ and $\{-1, 1\}$. This should highlight the fact that an infinite sequence is not the same as an infinite set.

Example 4: Let a_n be the digit in the nth decimal place of the number π. Since π is irrational, it is known that the terms of this sequence never go into a permanent repeating pattern.

Example 5: Let a_n be the nth derivative of the function e^{2x}. The terms of this sequence are functions rather than numbers. In this case, it's not hard to write down a general formula for a_n. If we let b_n be the nth derivative of the tangent function, it is no longer simple to find a formula for the nth term of the sequence (see Exercise 6).

Example 6: Let (A_n) be the sequence of closed intervals defined by $A_n = [n, n + 1/n]$. We have encountered this sequence before in the guise of an indexed family of intervals, as Example 4 of Section 5.3. There are two ways to interpret an indexed family of sets. Most of the time, it's just viewed as a convenient notation for a *set* of sets. But in other situations, such as when the sets are supposed to be in a certain *order*, a family of sets indexed by I may be viewed as a *function* that maps each i in I to the set A_i. Note that if the index set is $\mathbb{N}$, this is precisely what is meant by an infinite sequence of sets.

Definitions by Induction

We have talked quite a bit about mathematical induction, but only as a way of *proving* things. Certainly, induction is primarily a principle of proof. But induction is also very important as a means of defining sequences.

Example 7: A typical example of a sequence defined by induction (or by **recursion**, a term commonly used in computer science) is the **factorial** function, $f(n) = n!$. The standard definition of this sequence is

$$0! = 1$$

$$(n+1)! = (n!)(n+1)$$

What makes this an inductive definition? The main feature is that $f(n+1)$ is defined in terms of $f(n)$. In fact, the structure of this definition is completely parallel to a standard proof by induction. First, $f(0)$ is defined directly, as a specific number. (As with proofs by induction, definitions by induction need not begin at $n = 1$.) Then f of any larger number is defined in terms of f of the previous number, just as in a proof by induction $P(n+1)$ is proved using $P(n)$ as an assumption.

You might think that the sequence $n!$ does not need to be defined by induction, because it could be defined instead by the rule $n! = 1 \times 2 \times \ldots \times n$ (except that $0!$ must be defined separately). However, use of an ellipsis in a mathematical definition is not rigorous. In fact, using an ellipsis to define the output of a function is always an abbreviated way of describing an inductive definition.

Example 8: The intuitive definition of the exponential expression x^n is $x \cdot x \cdot x \ldots x$, with n x's being multiplied. But the rigorous definition requires induction:

$$x^0 = 1$$

$$x^{n+1} = x \cdot x^n$$

As in the previous example, the inductive definition has the advantage that it can start at $n = 0$, but the intuitive definition makes no sense with $n = 0$.

Both x and n are variables in this expression, so we are defining a function of two variables here. But the induction is on n only. The other variable x need not even be a natural number. Recall the similar situation for proofs by induction, in which a proof typically uses induction on only one variable even though a statement with two or more free variables is being proved.

Example 9: The inductive definition $a_1 = 1$ and $a_{n+1} = 2a_n + 1$, defines the sequence 1, 3, 7, 15, … . Exercise 3 asks you to show that this sequence can also be defined by the direct (noninductive) rule $a_n = 2^n - 1$. The latter equation may also be referred to as a **closed formula** for this sequence. Closed formulas have an obvious advantage over inductive definitions: if you want to evaluate the thousandth term of a sequence, a closed formula lets you do so without first evaluating terms 1 through 999.

Of course, since the previous example shows that exponents are technically defined inductively, this closed formula is only *apparently* noninductive. This is often the case with closed formulas for sequences.

Justification of Inductive Definitions (Optional Material)

We now prove the important result that justifies the most common type of definition by induction. For the rest of this section, we usually use function notation rather than sequence notation for the sequences under discussion.

To motivate this theorem, consider the most straightforward way of inductively defining a sequence of elements of some set A. The first term in the sequence is directly defined to be some specific member of A. Then, after m terms have been defined, the next term is usually defined in terms of the mth term *and* the number m. That means that there must be a *function* from $\mathbb{N} \times A$ to A that is used to determine the next term.

Theorem 7.11: Let A be a set, $c \in A$, and $g: \mathbb{N} \times A \to A$. Then there is a unique function $f: \mathbb{N} \to A$ such that $f(1) = c$ and $f(m + 1) = g(m, f(m))$ for every m in $\mathbb{N}$.

Proof: We give the proof but not every detail. We need to prove the existence (and uniqueness) of a certain function with domain $\mathbb{N}$. To do this, we first show that there exist similar functions with finite domains. Specifically, we show that for every n, there's a unique function f_n from $\{1, 2, \dots, n\}$ to A, such that $f_n(1) = c$ and $f_n(m + 1) = g(m, f_n(m))$ for every $m < n$. We do this by induction on n.

For $n = 1$, we want a function f_1 from $\{1\}$ to A such that $f_1(1) = c$. (The second condition is automatically true since the condition $m < 1$ cannot occur.) So let $f_1 = \{(1, c)\}$. It is clear that this is the unique possibility for f_1.

For the induction step, assume there is a unique function f_n with the properties described above. To define f_{n+1}, all we do is extend f_n to one more domain member; that is, let $f_{n+1} = f_n \cup \{(n + 1, g(n, f_n(n))\}$. This is clearly a function with domain $\{1, 2, \dots, n, n + 1\}$, satisfying the required properties. We must also show it's unique. For this, assume h is any function with domain $\{1, 2, \dots, n, n + 1\}$, satisfying the same properties required of f_{n+1}. By the uniqueness of f_n, h must equal f_{n+1} on all inputs from 1 to n, or else h restricted to $\{1, 2, \dots, n\}$ would be another function satisfying the definition of f_n. But then we also have $h(n + 1) = g(n, h(n)) = g(n, f_n(n)) = f_n(n)$. So this yields that $h = f_{n+1}$, as desired.

This completes the proof of the existence and uniqueness of the f_n's. The proof also shows that, for all n, f_{n+1} is an extension of f_n (that is, $f_n \subseteq f_{n+1}$). From this and the transitivity of $\subseteq$ (Theorem 5.4(e)), it follows easily (see Exercise 12) that $f_m \subseteq f_n$ whenever $m < n$. Therefore, for any m, n, and k, if $f_m(k)$ and $f_n(k)$ are both defined, they are equal. We can think of this as saying that the f_n's are all consistent with each other.

We are now ready to define f: simply let it be the union of all the f_n's defined above. It is obvious that $\text{Dom}(f) = \mathbb{N}$ and $\text{Rng}(f) \subseteq A$. The fact that f is a function follows immediately from the last sentence of the previous paragraph. And since f is constructed directly from the f_n's, we certainly have $f(1) = c$, and the other desired condition. So we have shown the existence of the f claimed by the theorem.

Finally, we need to show the uniqueness of f. So assume h is another function meeting the same requirements. If $f \neq h$, then there must be an n such that $f(n) \neq h(n)$. But then restricting both f and h to the domain $\{1, 2, \dots, n\}$ would give us two different functions that fit the definition of f_n. This contradicts what we just proved. This indirect proof establishes the uniqueness of f. ■

This theorem does not deal with inductive definitions in their most general form. There are at least three ways to make the theorem more general, all of which have already been encountered. One way is to have the initial domain value be a number other than 1, as in the definition of $n!$.

A second generalization is to allow additional variables in the function being defined, as in Example 8. There, we defined a function $f(n, x)$ or x^n by induction on n. In this situation, the extra variable x may have any domain whatsoever, and the function g mentioned in Theorem 7.11 should also be allowed to depend on x. That is, it has the general form $g(n, f(n), x)$ instead of just $g(n, f(n))$. (In Example 8, $g(n, f(n), x) = xf(n)$, so x is involved although n is not.) Furthermore, the constant c should also be allowed to depend on x; that is, it becomes a function too, which we might denote $h(x)$. (In Example 8, the definition starts with $f(1, x) = x$, so we could say that $h(x) = x$. If we started the definition at $n = 0$, we'd have $f(0, x) = x^0 = 1$, and so the initial values would be constant, independent of x. But normally the initial values can depend on the extra variable(s).)

The third important way to generalize the theorem is to note that, in the inductive definition of a function, the formula for $f(n + 1)$ may involve one or more values of $f(k)$ with $k < n$. Intuitively, this is quite legitimate, since if we are defining the values of f sequentially, at the point where we define $f(n + 1)$ we must have already defined not just $f(n)$ but also all values $f(k)$ for $k < n$. In Chapter 8, we discuss proofs by **complete induction**, which are based on the same idea.

Exercise 13 asks you to state (but not prove) the first two generalizations of Theorem 7.11 just discussed. Here is a rigorous statement of the third one.

Corollary 7.12: Let A be a set and $h: B \to A$, where B is the collection of all functions from a set of the form $\{k \mid k < n\}$ into A. Then there is a unique function $f: \mathbb{N} \to A$ such that, for every n, $f(n) = h(f|_{\{k \mid k < n\}})$.

Proof: This proof does not differ significantly from that of Theorem 7.11. We leave it for Exercise 9. ■

It takes some thought to see that Corollary 7.12 says what it should. The point is that whenever f has been defined for all natural numbers less than n, we can plug that partial function into h to get a new member of A to be used for $f(n)$. Note that this even applies to the initial value of f: when $n = 1$, we obtain $f(1) = h(\varnothing)$. Thus, in contrast to ordinary inductive definitions, we don't need to specify $f(1)$ separately.

Example 10: The **Fibonacci sequence** is defined by a more complicated induction than the ones given so far, in that terms are calculated from *two* previous terms, not just one: $a_1 = a_2 = 1$ and, for subsequent terms, $a_{n+1} = a_n + a_{n-1}$. Let's see how Corollary 7.12 can be used to define the Fibonacci sequence. First of all, we let $A = \mathbb{N}$. Since we want $f(1) = 1$, define $h(\varnothing)$ to be 1. Similarly, to obtain $f(2) = 1$, we must define $h(\{(1, 1)\})$ to be 1. Finally, for any $g \in B$ whose domain is of the form $\{1, 2, \ldots, k\}$ with $k \geq 2$, let $h(g) = g(k) + g(k - 1)$.

Note that we have not specified the values of h on all of B. But it does not matter how h is defined on functions of the form $\{(1, m)\}$, with $m \neq 1$.

Exercises 7.4

(1) Write the first six terms of the sequences defined by each of the following formulas for a_n.

(a) $n^2 - n + \sin \pi n$
(b) $(-2)^{-n}$
(c) n^{n-4}
(d) $[(n + 1)! - n!]/n!$

(2) For each of the following sequences, the first five terms are given; find a closed formula for a_n that fits the terms.

(a) 1, 5, 9, 13, 17, ...
(b) −1, 5, −9, 13, −17, ...
(c) 1, 3, 6, 10, 15, ...
(d) 1, 0, −1, 0, 1, ...

(3) Use induction to prove the claim made in Example 9.

(4) For each of the following inductively defined sequences, write out the first five terms, find a closed formula for a_n (as in Example 9), and prove that your formula defines the same sequence as the given one.

(a) $a_1 = 2; a_{n+1} = a_n - 3$
(b) $a_1 = 1; a_{n+1} = 2a_n$
(c) $a_1 = 4; a_{n+1} = -a_n$
(d) $a_1 = 2; a_{n+1} = (a_n)^2$
(e) $a_1 = 1; a_{n+1} = 1 + a_1 + a_2 + \ldots + a_n$
(f) $a_1 = 1; a_{n+1} = a_n + 1/(n^2 + n)$.

(5) (a) Write the first ten terms of the sequence defined inductively by $a_1 = 0$, $a_2 = 1$, and thereafter $a_{n+2} = a_{n+1} - a_n$.

(b) Evaluate a_{1000}. Explain.

(6) (a) Find a formula for a_n in Example 5.

(b) Find b_1, b_2, and b_3 in Example 5.

(7) For each of the following functions, find a formula for its nth derivative $f^{(n)}$. In most cases, your answer will need to involve cases and/or factorials.

(a) $f(x) = \sin x$
(b) $f(x) = \sin 3x$
(c) $f(x) = 1/x$
*(d) $f(x) = \sqrt{x}$

(8) Using the inductive definition in Example 8, prove these well-known laws of exponents (assuming $m, n \in \mathbb{N} \cup \{0\}$):

(a) $x^m x^n = x^{m+n}$ (b) $(x^m)^n = x^{mn}$

*(9) If $x \neq 0$, the definition of x^n can be extended to all integer values of n, using a definition by cases: if $n \geq 0$, x^n has already been defined, and if $n < 0$, x^n is $1/(x^{-n})$. (Note that in this definition, the value of the function in the second case is based on some value from the first case. This is fine.) Prove the same laws of exponents as in the previous exercise, but now assuming m and $n \in \mathbb{Z}$.

(c) Also prove the law: $x^m/x^n = x^{m-n}$, provided $x \neq 0$

Hint: Since the definition of integer exponents involves both induction and cases, your proofs should use both these methods.

(10) (a) Write out the first ten terms of the Fibonacci sequence (Example 10).

(b) Prove that $a_n < a_{n+1} < 2a_n$, for all $n \geq 3$.

(11) Prove by induction:

(a) For every $n \geq 0$, $(n + 1)! - n! = n(n!)$

(b) For every $n \geq 4$, $n! > 2^n$

(c) For every $n \geq 7$, $n! > 3^n$

*(d) For every $n \geq 4$, $n! > n^2$

(e) For every $n \geq 1$, $\sum_{k=1}^{n} k(k!) = (n + 1)! - 1$

The remaining problems pertain to the optional material at the end of this section.

(12) This exercise fills in some steps required in the proof of Theorem 7.11.

(a) If (A_n) is a sequence of sets and $A_n \subseteq A_{n+1}$ for every n, prove that $A_m \subseteq A_n$ whenever $m < n$.

(b) Using part (a), prove that if (f_n) is a sequence of functions and $f_n \subseteq f_{n+1}$ for every n, then $\bigcup_{n \in \mathbb{N}} f_n$ is a function.

(c) State and prove an equation of the form $\mathrm{Dom}(\bigcup_{n \in \mathbb{N}} f_n) = \ldots$.

(13) State rigorously the first two generalizations of Theorem 7.11 mentioned in the discussion after it.

*(14) State a version of Theorem 7.11 encompassing all three of the ways mentioned to generalize it.

(15) Prove Corollary 7.12. For any part of your proof that's the same as the corresponding part of the proof of Theorem 7.11, say so rather than repeating the argument.

(16) Show that Corollary 7.12 really does provide a generalization of Theorem 7.11, in the sense that any function f that can be defined via the theorem could also be defined via the corollary.

(17) Apply Theorem 7.11, using $c = 3$, and $g(n, k) = 3k$. What function f is obtained? Prove your answer.

*(18) Repeat Exercise 17 using $c = -2$ and $g(n, k) = n + k$. ***Hint:*** In this problem, it is much harder to determine the formula for $f(n)$. Here is one approach: compute at least a half-dozen values of f, and notice the simple formula that apparently fits the value of $f(n + 1) - f(n)$. It can be shown that $f(n)$ must be a polynomial in n, of degree one higher than that of $f(n + 1) - f(n)$. Now use some of the specific values of f to solve for the coefficients of this polynomial.

(19) If R is any finite set of ordered pairs of real numbers, define $h(R)$ to be the sum of all the individual first and second members of the ordered pairs in R. For instance, $h(\{(3, 2), (3, -4), (1, 3)\}) = 8$. Apply Corollary 7.12 with this h.

(a) Write out the first six values of the function f obtained.
(b) Prove by induction that $f(n + 1) - f(n) = 2^n - 1$, for all n.
(c) Prove by induction that $f(n) = 2^n - n - 1$, for all n.

7.5 Cardinality

As opposed to algebraic structures like fields and groups, which by definition have algebraic operations defined on them, a set does not have any structure in and of itself. It is just a collection of objects. Aside from the question of *what* particular objects are in a set, the most obvious question to ask about a set is *how many* elements are in it.

At first glance, it might seem that this question could not be very fruitful. A set either has some finite number of elements, like 7 or 354, or else it is infinite, in which case there might seem to be nothing more worth saying. But it turns out there are different sizes of infinity among sets, and learning how to compare them is a vital part of higher mathematics.

Definition: For any sets A and B, $A \sim B$ means there is a bijection between A and B.

The statement $A \sim B$ is usually read "A and B are **equivalent**," "A and B are **in one-to-one correspondence**," or "A and B have the same **cardinality**." The intended meaning of this is that A and B have the same *size* or the same *number of elements*. Essentially, "cardinality" is the technical term for the size of a set.

Example 1: Let $A = \{5, 7, 18\}$ and $B = \{\text{Rome}, 19, \text{Jodie Foster}\}$. Note that each of these sets has three elements, so they certainly should have the same size. In fact they do. For example, $\{(5, \text{Rome}), (7, \text{Jodie Foster}), (18, 19)\}$ is a bijection between A and B. Therefore $A \sim B$. Note that it's not necessary to give an explicit rule for the bijection.

Example 2: Let $A = \{1, 2\}$ and $B = \{3, 4, 5\}$. Since A has fewer members than B, it would seem that these sets should not have the same cardinality, and in fact they don't. A bit of trial and error will convince you that there is no bijection between A and B. Soon we prove this.

Example 3: Is the set $\{1, 2, 3, \ldots, 1000\}$ the same size as $\mathbb{N}$? Informally, we know that the first set is finite and the second one is infinite, so presumably the answer is no. This is also proved later in this section.

Example 4: Is the set $E = \{2, 4, 6, 8, \ldots\}$ the same size as $\mathbb{N}$? Since E is a proper subset of $\mathbb{N}$, it might appear that the answer must be no. But the simple function $f(n) = n/2$ defines a bijection from E to $\mathbb{N}$, so $E \sim \mathbb{N}$! As we soon see, this strange situation is characteristic of infinite sets.

It is natural to think of $\sim$ as a relation between sets. Technically, its domain is the class of all sets, as mentioned in Section 5.2.

Theorem 7.13: The relation $\sim$ is an equivalence relation.
Proof: For any set A, id_A is a bijection from A to itself. So $\sim$ is reflexive. To show it's symmetric, assume $A \sim B$. That means there's a bijection f from A to B. Then, by Theorem 7.1, f^{-1} is a bijection from B to A, so $B \sim A$. For transitivity, assume $A \sim B$ and $B \sim C$. Then there are bijections $f: A \to B$ and $g: B \to C$. It follows by Theorem 7.4(f) that $A \sim C$. ■

Theorem 7.13 should come as no surprise, since $A \sim B$ is supposed to mean that A and B have the same size, and a binary relation that expresses some way in which two things are alike is generally an equivalence relation. An equivalence class of sets under $\sim$ is sometimes called a **cardinal**. This usage is a bit imprecise, for the reason mentioned before Theorem 7.13. Still, it is a useful concept.

Besides talking about sets having the same size, we also need to compare their sizes. This is done using the following relations.

Definitions: We write $A \preceq B$ (or $B \succeq A$) to mean that there is a one-to-one function from A to B. And $A \prec B$ (or $B \succ A$) means that $A \preceq B$ but $A \nsim B$.

The definition of $\prec$ is a bit complex, and it's important to give it some thought. One good way to think of what $A \prec B$ means is that there are one-to-one functions from A to B but none of these is onto B. On the other hand, it is easier to come up with good words for $\prec$ than for $\preceq$. We read $A \prec B$ as "A has **smaller cardinality** than B" or simply "A is smaller than B."

Proposition 7.14: (a) $A \preceq B$ iff A has the same cardinality as some *subset* of B.
(b) $A \subseteq B$ implies $A \preceq B$.
Proof: See Exercise 1. ■

Theorem 7.15: The relation $\preceq$ is reflexive and transitive.
Proof: This proof is similar to that of Theorem 7.13 and is left for Exercise 2. ■

In the terminology of Section 6.3, Theorem 7.15 says that $\preceq$ is a preordering, again on the class of all sets. We now turn our attention to the classification of sizes of sets.

Finite and Infinite Sets

Notation: For each nonnegative integer k, $\mathbb{N}_k = \{n \in \mathbb{N} \mid n \leq k\}$.
So $\mathbb{N}_0 = \varnothing$, $\mathbb{N}_1 = \{1\}$, $\mathbb{N}_2 = \{1, 2\}$, and so on.

Theorem 7.16: (a) If $k < m$, then $\mathbb{N}_k \prec \mathbb{N}_m$.
(b) For any k, $\mathbb{N}_k \prec \mathbb{N}$.
(Note that in the statement of this theorem, the first < is ordinary "less than," whereas the symbol $\prec$ pertains to cardinality.)

Proof: (a) By induction on k, we prove $\forall k$ P(k), where P(k) is the statement $\forall m > k\ (\mathbb{N}_k \prec \mathbb{N}_m)$. First we must prove P(0). For any nonempty set A, there's exactly one function from $\varnothing$ to A, namely the empty function (with no ordered pairs). This function is one-to-one but not onto A, so $\varnothing \prec A$. If $m > 0$, then $\mathbb{N}_m$ is nonempty, so $\varnothing \prec \mathbb{N}_m$. For the induction step, assume P(k) and let $m > k + 1$. We must show $\mathbb{N}_{k+1} \prec \mathbb{N}_m$. First note that $\mathbb{N}_{k+1} \subseteq \mathbb{N}_m$ and so $\mathbb{N}_{k+1} \preceq \mathbb{N}_m$ by the function $f(x) = x$. It remains to show $\mathbb{N}_{k+1} \nsim \mathbb{N}_m$, which we do indirectly: assume $g\colon \mathbb{N}_{k+1} \xrightarrow{\text{bij}} \mathbb{N}_m$. We use g to construct a bijection h from $\mathbb{N}_k$ to $\mathbb{N}_{m-1}$. If $g(k + 1) = m$, simply let $h = g - \{(k + 1, \mathrm{m})\}$. Otherwise, say $g(k + 1) = j$. Since g is onto $\mathbb{N}_m$, there must also be an n such that $g(n) = m$. Then let $h = g - \{(k + 1, j), (n, m)\} \cup \{(n, j)\}$. In either case, $h\colon \mathbb{N}_k \xrightarrow{\text{bij}} \mathbb{N}_{m-1}$. (Exercise 5 asks you to prove this in more detail.) But since $m > k + 1$, $m - 1 > k$. Therefore, having a bijection from $\mathbb{N}_k$ to $\mathbb{N}_{m-1}$ violates the induction hypothesis.
(b) See Exercise 5. ■

If $m > k$ then, by Theorem 7.16(a), there is no one-to-one function from $\mathbb{N}_m$ to $\mathbb{N}_k$. This result is known as the **pigeonhole principle**, often stated in the form: If $k + 1$ (or more) letters are put into k mailboxes, then at least one box must receive more than one letter. For example, the pigeonhole principle guarantees that in any group of eight people, there must be two people born on the same day of the week. In spite of its simplicity, the pigeonhole principle is quite useful in higher mathematics.

Definitions: A set A is called **finite** iff $A \sim \mathbb{N}_k$, for some k.
A set is called **infinite** iff it's not finite.

Notation: By Theorem 7.16, if A is finite, then there's a *unique* k such that $A \sim \mathbb{N}_k$; this number k is denoted Card(A), $|A|$, $\overline{\overline{A}}$, or #A and is called the **cardinality** of A. This use of the term "cardinality" is consistent with its use earlier in this section.

If Card(A) = k, we simply say that A has k members. Note that there is nothing to prove here, because it's a definition.

The next two theorems are perfect examples of "completely obvious" statements that are tempting to just assume. As we have mentioned from time to time, it is *imperative* to prove such things. Better safe than sorry!

Theorem 7.17: Every subset of a finite set is finite.

Proof: Assume B is finite and $A \subseteq B$. So there's a bijection f between B and some N_k. Then $f|_A$ is a bijection between A and some subset of N_k. Therefore, since $\sim$ is transitive, we are done if we show that any subset of any N_k is finite. We do this by induction on k.

$\mathrm{N}_0 = \varnothing$, whose only subset is itself, which is finite by definition. Therefore, every subset of N_0 is finite.

For the induction step, assume every subset of N_k is finite, and let C be any subset of N_{k+1}. If $k+1 \notin C$, then $C \subseteq \mathrm{N}_k$ by Theorem 4.16(c), so C is finite by the induction hypothesis. On the other hand, if $k+1 \in C$, let $D = C - \{k+1\}$. Again, the induction hypothesis tells us that D is finite. So there's a bijection g between D and some N_j. Let $h = g \cup \{k+1, j+1\}$. It is then easy to show that h is a bijection between C and N_{j+1}, and so C is finite (see Exercise 6). ■

Before we continue our study of finite and infinite sets, it is helpful to prove an important theorem that shows that this whole idea of cardinality is "reasonable." What do we mean by such a subjective-sounding word? For one thing, consider the notation $A \preceq B$. We might read this as "A is the same size as B or smaller." For this to make sense, the conjunction $A \preceq B$ and $B \preceq A$ should imply that A and B have the same size. But if you try to prove this obvious-looking implication, you will find that it's quite difficult. The proof of this result was one of the first significant achievements in the development of set theory.

Theorem 7.18 (Cantor-Schröder-Bernstein (CSB) theorem): If $A \preceq B$ and $B \preceq A$, then $A \sim B$.

Proof: Assume $A \preceq B$ and $B \preceq A$; this means that we have one-to-one functions $f: A \to B$ and $g: B \to A$. Let $C = \mathrm{Rng}(f)$ and $D = \mathrm{Rng}(g)$. So we can think of $f: A \to C$, $g: B \to D$, and also $g^{-1}: D \to B$ as being bijections.

We define a function h that is a bijection from A to B. The function h is a combination of f and g^{-1}, in the sense that for every x in A, $h(x)$ is either $f(x)$ or $g^{-1}(x)$. Furthermore, we always let $h(x)$ be $f(x)$ unless there is a compelling reason that it should be $g^{-1}(x)$. We now make this precise by defining the set E of elements x for which $h(x) = g^{-1}(x)$.

Let y be any fixed member of $B - C$. (Note that if $B - C$ is empty, we are done, because f is then onto B). We show that y forces an entire infinite sequence of members of A to be in E. First, let $x_1 = g(y)$ (see Figure 7.10). For the range of h to include y, we must have $h(x_1) = g^{-1}(x_1)$, because y is not in the range of f. So there is no way we can have $h(x) = f(x) = y$, for some x in A. And since g^{-1} is one-to-one, x_1 is the only element of A that g^{-1} sends to y. So $h(x_1) = g^{-1}(x_1) = y$ and $x_1 \in E$.

But now consider $f(x_1) \in B$ and $g(f(x_1)) \in A$, which we call x_2. We claim that x_2 must also be in E, or else $f(x_1)$ would not be in the range of h. Of course, $f(x_1)$ is in the range of f, because f maps x_1 to $f(x_1)$. But we have already established that $h(x_1)$ must be y, which is not $f(x_1)$, since y is not in Rng(f). Therefore, since x_1 is the only object that f sends to $f(x_1)$, and x_2 is the only object that g^{-1} sends to $f(x_1)$, the only way to get $f(x_1)$ to be in the range of h is to let $h(x_2) = g^{-1}(x_1) = f(x_1)$.

Continuing in this manner (still with just one y), we define, by induction, an entire infinite sequence $(x_1, x_2, x_3, \ldots)$, all of which must be in E. Let's call the range of this sequence $S(y)$, since it depends on the choice of y. Figure 7.10 makes it clear what's happening. Since x_1 must be mapped to y by h, all the succeeding elements x_2, x_3, and so on must be mapped "up" according to g^{-1} rather than "down" according to f, in the definition of h. (Technically, this can be proved by induction on the subscript n of x_n.)

We now let E be the union of all the sets $S(y)$, taken over all y's in $B - C$. *[So we are using $B - C$ as the index set for this generalized union.]* E is a subset of A, in fact a subset of D, since all the members of any of the sets $S(y)$ are in the range of g. So we can define the function $h: A \to B$ by:

$$h(x) = \begin{cases} g^{-1}(x) & \text{if } x \in E \\ f(x) & \text{otherwise} \end{cases}$$

It remains to show that h is one-to-one and onto. This is left for Exercise 7. ■

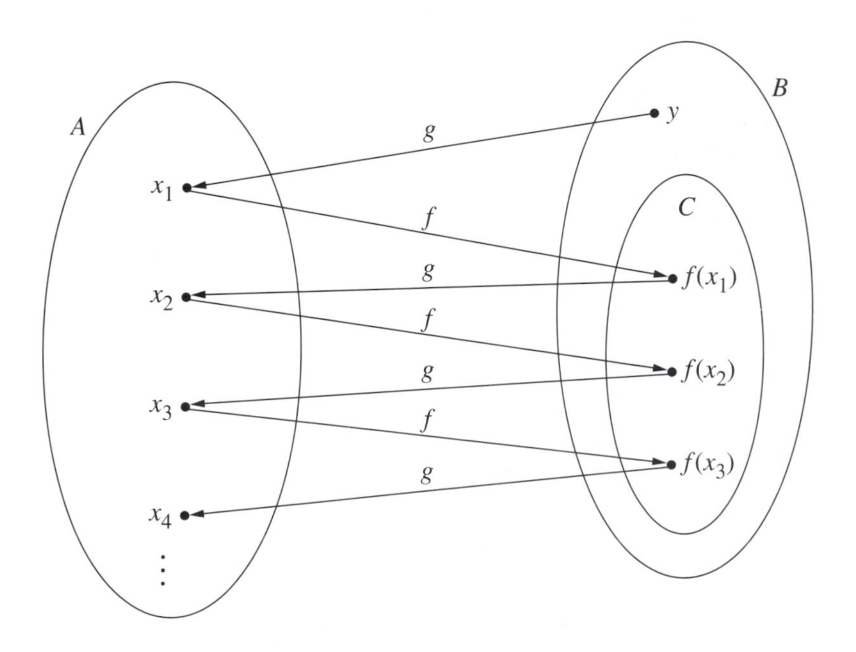

Figure 7.10 Construction of the sequence (x_n) in the proof of Theorem 7.18

Georg Ferdinand Cantor (1845–1918) is generally considered to be the main founder of the subject of set theory. Cantor's father wanted him to study engineering, but Georg was more interested in philosophy, theology and mathematics. Eventually, Cantor decided to concentrate on mathematics and received his doctorate from the University of Berlin in 1867.

Cantor's early research was in the fields of number theory and trigonometric series, but his nature caused him to think more and more about the role of the infinite in mathematics. He was also greatly influenced by the spirit of the times, since it was during the 1860s and 1870s that mathematicians such as Weierstrass and Dedekind (and Cantor himself) showed how to use infinite sets or sequences to develop a rigorous theory of the real numbers. In 1874, Cantor published the first paper that seriously considered infinite sets as actual objects, and he devoted the rest of his career to this subject.

Cantor's work encountered a degree of resistance that, in retrospect, seems unfair and regrettable. Gauss, certainly the most influential mathematician of the first half of the nineteenth century, vehemently shared the ancient Pythagorean "horror of the infinite." The main advocate of this "finitist" school of thought during Cantor's time was Leopold Kronecker, who was often vicious in his criticisms of other mathematicians. Kronecker's attacks on the free use of the infinite angered Weierstrass and Dedekind but had a more profound effect on Cantor. Kronecker used his influence to block Cantor's applications for positions at Germany's most prestigious universities; thus Cantor spent his entire forty-four year career at the relatively minor Halle University. Cantor became exhausted and discouraged by the resistance to his work, and he began to have bouts of severe depression and mental illness in 1884. Cantor did very little new research during the last thirty years of his life and, even though his work finally received proper recognition after the turn of the century, he died in a mental institution in Halle.

Example 5: A nice application of the CSB theorem is the fact that the open interval $(0, 1)$ and the closed interval $[0, 1]$ have the same cardinality. To apply the theorem, we need to show that each interval can be mapped one-to-one to the other. Since $(0, 1) \subseteq [0, 1]$, the inclusion map $f(x) = x$ maps the open interval one-to-one to the closed one. To map $[0, 1]$ one-to-one to $(0, 1)$, the simplest way is to use a linear function that sends both 0 and 1 to numbers in between. For example, we can use $g(x) = (x + 1)/3$. So, by the CSB theorem, there is a bijection between these intervals.

The proof of the CSB theorem is **constructive**, meaning it shows how to define the desired bijection. Let's see how that would work in this example. Following the notation used in that proof, we have $A = C = (0, 1)$, while $B = [0, 1]$. Therefore $C - B$ consists only of the numbers 0 and 1. So the set E consists of two infinite sequences of numbers. For $y = 0$, we get the sequence $\{g(0), g(f(g(0))), g(f(g(f(g(0))))), ...\}$. Since f is the identity, this simplifies to $\{g(0), g(g(0)), g(g(g(0))), ...\}$, for which the numbers are $\{1/3, 4/9, 13/27, ...\}$. For $y = 1$, we similarly find that the numbers $\{2/3, 5/9, 14/27, ...\}$ must all be in E.

For any member of E, h corresponds to g^{-1}. This means that $h(1/3) = 0$, $h(4/9) = 1/3$, $h(13/27) = 4/9$, and so on, and also $h(2/3) = 1$, $h(5/9) = 2/3$, $h(14/27) = 5/9$, and so on. For all numbers in $(0, 1)$ that are not in E, $h(x) = x$. It would be worthwhile to convince yourself that h really is a bijection between $(0, 1)$ and $[0, 1]$; graphing it might help you to see this.

The same type of reasoning can be used to show that all bounded intervals on the real line (open, closed, and half-open) have the same cardinality. Note, however, that h is not a continuous function; that is, its graph has breaks in it. In fact, it is impossible to have a continuous bijection between an open interval and a closed interval.

Corollary 7.19: The relation $\prec$ is irreflexive and transitive. (That is, it is an irreflexive partial ordering.)

Proof: Since $A \sim A$ for any set, it follows that $A \nprec A$. So $\prec$ is irreflexive. To show it's transitive, assume $A \prec B$ and $B \prec C$. This implies that $A \preceq B$ and $B \preceq C$, which yields $A \preceq C$ by Theorem 7.15. We must also show $A \nsim C$. So assume, on the contrary, that $A \sim C$. Then $C \preceq A$. So $B \preceq C \preceq A$; this implies $B \preceq A$. But then $A \sim B$ by the CSB theorem, and this contradicts the assumption that $A \prec B$. ■

Corollary 7.20: Both $A \prec B$ and $B \prec A$ cannot hold. (That is, the relation $\prec$ is strongly antisymmetric.)

Proof: The proof is immediate by Corollary 7.19 and the proof of Theorem A-8 (Appendix 2). Alternatively, note that this is just a thinly disguised way of stating the CSB theorem. ■

Corollary 7.21: (a) If $A \prec \mathrm{B}$ and $\mathrm{B} \preceq \mathrm{C}$, then $A \prec \mathrm{C}$.

(b) If $A \preceq \mathrm{B}$ and $\mathrm{B} \prec \mathrm{C}$, then $A \prec \mathrm{C}$.

Proof: See Exercise 8. ■

Note that all our corollaries to the CSB theorem are obvious looking, based on the symbols and words that they involve. But none of them is easy to prove without the CSB theorem.

We now have more tools for studying finite sets.

Theorem 7.22: (a) If A is finite and $A \subset B$, then $A \prec B$.

(b) If B is finite and $A \subset B$, then $A \prec B$.

(c) If B is finite and $A \preceq B$, then A is finite.

(d) The union of a finite collection of finite sets is finite.

(e) The cartesian product of a finite number of finite sets is finite.

(f) If $\mathscr{A}$ is a collection of sets containing at least one finite set, then $\bigcap \mathscr{A}$ is finite.

Proof: We prove parts (a) and (d) and leave the rest for Exercise 9.

(a) Since A is finite, there is a bijection f from A to some $\mathbb{N}_k$. We also know that $B - A$ is nonempty, so pick any c in it (by ES), and let $D = A \cup \{c\}$. The relation $f \cup \{(c, k + 1)\}$ is clearly a bijection between D and $\mathbb{N}_{k+1}$. Theorem 7.16(a) and a couple of applications of Corollary 7.21 yield $A \prec D$. But we also have $D \subseteq B$, and so $D \preceq B$. Again using Corollary 7.21, we get $A \preceq B$.

(d) We want to show that if $\mathscr{A}$ is any finite collection of sets, all of whose members are also finite, then so is $\bigcup \mathscr{A}$. We prove this by induction on Card($\mathscr{A}$): if Card($\mathscr{A}$) = 0, then $\mathscr{A}$ is empty, and thus so is $\bigcup \mathscr{A}$. And $\varnothing$ is certainly finite. (Of course, it would be no more difficult to begin this proof at 1.)

For the induction step, assume the result holds whenever Card($\mathscr{A}$) = n and let $\mathscr{A}$ be any collection of $n + 1$ finite sets. We can write $\mathscr{A} = \{A_1, A_2, \dots, A_n, A_{n+1}\}$. Then $\bigcup \mathscr{A} = A_1 \cup A_2 \cup \dots \cup A_n \cup A_{n+1} = (A_1 \cup A_2 \cup \dots \cup A_n) \cup A_{n+1}$. But $A_1 \cup A_2 \cup \dots \cup A_n$ is finite by the induction hypothesis, so we just need to show that the union of two finite sets is finite. We could prove this now, but it is more convenient to delay this proof until the next section (Theorem 7.28). ■

Definition: A subset A of $\mathbb{N}$ is called **bounded** iff there is some natural number that is larger than every number in A; in symbols, iff $\exists n \, \forall m \in A \, (n > m)$.

Theorem 7.23: Assume $A \subseteq \mathbb{N}$. Then

(a) A is unbounded iff $A \sim \mathbb{N}$.

(b) A is bounded iff A is finite iff $A \prec \mathbb{N}$.

Proof: (a) If A is an unbounded subset of $\mathbb{N}$, we can inductively define a bijection between $\mathbb{N}$ and A that lists the members of A in increasing order (see Exercise 10). Thus $A \sim \mathbb{N}$. The other direction of part (a) and part (b) are also left for Exercise 10. ■

Countable and Uncountable Sets

Theorems 7.22 and 7.23 show that the behavior of finite sets and subsets of $\mathbb{N}$ is pretty reasonable as far as cardinality is concerned. We now turn to the more challenging and fascinating study of infinite cardinalities.

Definitions: A set A is called:

denumerable if and only if $A \sim \mathbb{N}$

countable if and only if $A \preceq \mathbb{N}$

uncountable if and only if $\mathbb{N} \prec A$

Simply put, a set is denumerable if and only if its member can be arranged in a *single infinite sequence* with no repetitions.

By Theorem 7.23(b), A is finite iff $A \prec \mathbb{N}$. It follows that a set is countable iff it's finite or denumerable. Denumerable sets may also be called **countably infinite.** Some mathematicians use "countable" as a synonym for "denumerable," but it's more efficient to give these two words different meanings.

Theorem 7.24: (a) $\mathbb{N} \times \mathbb{N} \sim \mathbb{N}$

(b) The cartesian product of a finite number of countable sets is countable.

(c) The union of a finite number of countable sets is countable.

Proof: (a) For any m and n in $\mathbb{N}$, let $f(m, n) = 2^{m-1}(2n-1)$. Since m and n are positive integers, so are 2^{m-1} and $2n-1$. Therefore, f is a function from $\mathbb{N} \times \mathbb{N}$ to $\mathbb{N}$.

To prove that f is a bijection, we need a basic result from number theory: every natural number can be written in a *unique* way in the form $2^a b$, where $a \geq 0$ and b is an odd natural number. It follows that f is onto, since every odd natural number is of the form $2n - 1$. But we can use the same result to show that f is one-to-one:

$$\begin{aligned} f(m, n) = f(j, k) \quad &\text{iff} \quad 2^{m-1}(2n-1) = 2^{j-1}(2k-1) \\ &\text{iff} \quad m - 1 = j - 1 \text{ and } 2n - 1 = 2k - 1 \qquad \text{By uniqueness} \\ &\text{iff} \quad m = j \text{ and } n = k. \end{aligned}$$

By the way, this number-theoretic result is proved in Chapter 8.

(b) By induction on n, we prove that every cartesian product of n countable sets is countable. We start the proof at $n = 2$. *[It could be started at $n = 1$, with the convention that the cartesian product of one set is just that set. The $n = 1$ case would then be trivial.]* So we must first prove that if A and B are both countable, then so is $A \times B$. Assume f: $A \to \mathbb{N}$ and g: $B \to \mathbb{N}$ are both injections. Then we can define h: $A \times B \to \mathbb{N} \times \mathbb{N}$ by $h(x, y) = (f(x), g(y))$. By Exercise 14(b) of Section 7.3, h is also one-to-one. Thus $A \times B \preceq \mathbb{N} \times \mathbb{N}$, and it follows from part (a) that $A \times B$ is countable.

For the induction step, assume that every cartesian product of n countable sets is countable and let $B = A_1 \times A_2 \times \ldots \times A_{n+1}$. By definition, this means that $B = (A_1 \times A_2 \times \ldots \times A_n) \times A_{n+1}$. So, by the induction hypothesis, B is the product of just two countable sets and is therefore countable by the argument for $n = 2$.

(c) We outline this proof and leave the details for Exercise 11. Again we use induction, starting with two sets. So assume A and B are countable. Recall Example 4, which showed that there is a bijection between $\mathbb{N}$ and the even natural numbers. Similarly, there is a bijection between $\mathbb{N}$ and the odd natural numbers. Using these functions, we can define a one-to-one function from $A \cup B$ to $\mathbb{N}$ that maps the members of A to even numbers and the members of B to odd numbers. The induction step is similar to the one for part (b). ■

Remarks (1) Parts (b) and (c) of this theorem remain true if "countable" is changed to "denumerable" throughout (see Exercise 12).

(2) Figure 7.11 illustrates an interesting alternative proof of Theorem 7.24(a). This figure shows a path that covers all the members of $\mathbb{N} \times \mathbb{N}$. Following this path arranges

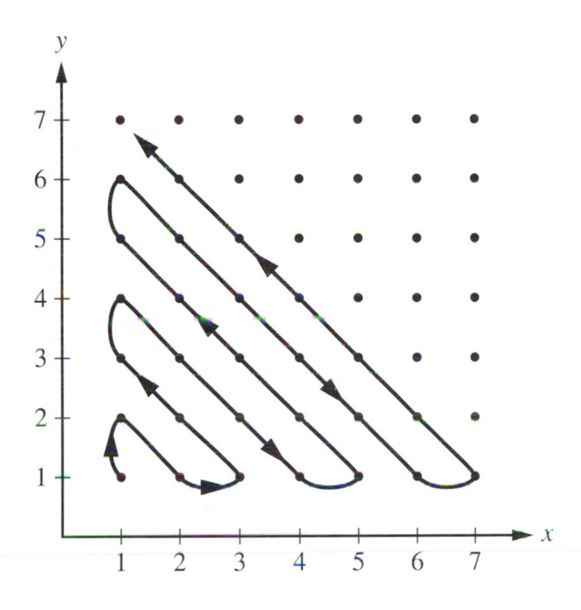

Figure 7.11 Another bijection between $\mathbb{N} \times \mathbb{N}$ and $\mathbb{N}$

the members of $\mathbb{N} \times \mathbb{N}$ in a single infinite sequence; this means that $\mathbb{N} \times \mathbb{N} \sim \mathbb{N}$. There is no simple equation for this bijection, but that doesn't matter.

Figure 7.11 can also be used to clarify the proof of part (a) given in the text. We can think of $\mathbb{N} \times \mathbb{N}$ as the union of an infinite number of copies of $\mathbb{N}$, each vertical column being one copy. The function f maps the first copy of $\mathbb{N}$ to the odd natural numbers, the second copy of $\mathbb{N}$ to the odd multiples of 2, the third copy of $\mathbb{N}$ to the odd multiples of 4, and so on.

(3) The previous remark points out that $\mathbb{N} \times \mathbb{N}$ can be viewed as the union of a denumerable number of denumerable sets. Therefore, it would appear that Theorem 7.24(a) directly implies part (c), and more. In fact, part (a) can be used to prove that a countable union of countable sets is countable, but we need to delay this proof because it requires the axiom of choice.

(4) The methods used in the proof of Theorem 7.24 may be thought of as coding techniques. Part (a) shows how to code, or "blend," a pair of natural numbers into one, without losing any information. (That is, the coding is invertible.) Part (b) shows how to similarly code longer sequences of numbers. In practice, functions like this could be

used to assign a number unambiguously to every possible English sentence, or to every possible computer program in a particular programming language, and so on.

(5) The main theme of Theorem 7.24 is that countability and denumerability are very stable properties. Specifically, two obvious attempts at creating bigger sets—unions and cartesian products—do not succeed in forming an uncountable set from countable ones. For a famous and amusing illustration of this phenomenon, see Exercise 24.

Corollary 7.25: The sets $\mathbb{Z}$ and $\mathbb{Q}$ are denumerable.

Proof: We do the proof for $\mathbb{Q}$ first. We know $\mathbb{N} \subset \mathbb{Q}$, so $\mathbb{N} \preceq \mathbb{Q}$. To show $\mathbb{Q} \preceq \mathbb{N}$, first note that $\mathbb{Q}$ is the union of the positive rationals, the negative rationals, and $\{0\}$. By Theorem 7.24(c), it suffices to show these three sets are all countable. We do the proof for $\mathbb{Q}^+$, the positive rationals. The proof for the negative rationals is almost identical, and $\{0\}$ is clearly countable.

We define an injection f from $\mathbb{Q}^+$ to $\mathbb{N} \times \mathbb{N}$, which by Theorem 7.24(a) is sufficient to prove what we want. For any positive rational number r, there are natural numbers a and b such that $r = a/b$. To be specific, we can take the fraction a/b to be in *lowest terms. [As in the proof of Theorem 7.24(a), we are assuming some number theory results here.]* So let $f(r) = (a, b)$. This f is clearly one-to-one, since $f(r) = f(s)$ implies $r = a/b = s$.

Since $\mathbb{N} \subset \mathbb{Z} \subset \mathbb{Q}$ and $\mathbb{N} \sim \mathbb{Q}$, it follows that $\mathbb{N} \sim \mathbb{Z}$. ∎

We have yet to show that there are any uncountable sets. We conclude this section by describing the two main methods of constructing them.

Theorem 7.26 (Cantor's theorem): For every set A, $A \prec \mathcal{P}(A)$.

Proof: First of all, the function $f: A \to \mathcal{P}(A)$ defined by $f(x) = \{x\}$ is one-to-one, so $A \preceq \mathcal{P}(A)$. It remains to show that $A \nsim \mathcal{P}(A)$. We do this by showing that a function from A to $\mathcal{P}(A)$ can't be onto. So assume $g: A \to \mathcal{P}(A)$. For each $x \in A$, note that $g(x)$ is a subset of A, so we can ask whether $x \in g(x)$. Let $B = \{x \in A \mid x \notin g(x)\}$. We claim that $B \notin \text{Rng}(g)$. That is, $\forall x \in A, B \neq g(x)$. To see this, note that if $x \in g(x)$, then $x \notin B$, so $B \neq g(x)$. But if $x \notin g(x)$, then $x \in B$, so again $B \neq g(x)$. ∎

The proof of Theorem 7.26 should remind you of the reasoning used in Russell's paradox. In this instance, however, we don't get an outright contradiction; we just get the result that a certain function can't exist.

Theorem 7.26 implies that no matter how big a set we have, we can always find one of bigger cardinality. In other words, there's an unbounded, or never-ending, hierarchy of sizes of infinite sets.

As an immediate consequence of Theorem 7.26, $\mathcal{P}(\mathbb{N})$ is uncountable, so this is our first example of an uncountable set. It also turns out that $\mathcal{P}(\mathbb{N}) \sim \mathbb{R}$, so the real line is uncountable. (Even though the proof of this is not difficult, we delay it until Exercises 9.2, where we will have more tools for proving things about $\mathbb{R}$.) Surprisingly, every interval of real numbers has the same cardinality as $\mathbb{R}$. In analogy to Theorem 7.24(b), it can also be proved that $\mathbb{R} \times \mathbb{R}$, $\mathbb{R} \times \mathbb{R} \times \mathbb{R}$, and so on, are all the same size as $\mathbb{R}$ (see Exercises 20 and 21). Give this some thought: *in terms of cardinality, a line segment one millimeter long is just as big as an entire infinite line. In fact, it is just as big as all of three-dimensional space!*

Before proving Theorem 7.26 in general, Cantor first proved the specific result that $\mathbb{N} \prec \mathbb{R}$ or, equivalently, $\mathbb{N} \prec \wp(\mathbb{N})$. The core of the proof is that there is no function from $\mathbb{N}$ *onto* the interval [0, 1]: If $f: \mathbb{N} \rightarrow [0, 1]$, each member of Rng($f$) is a decimal, which is an infinite sequence of digits. Therefore, if we list the decimals $f(1), f(2), f(3)$, and so on, we get an infinite square array of digits. If we then go down the diagonal of this array and change each digit in the diagonal, we get a decimal number in [0, 1] that cannot be in Rng(f). So f is not onto (see Exercise 22). Proofs based on this idea, including our proof of Theorem 7.26, are called **diagonalization arguments**.

Here is the other standard way of creating larger infinite sets from given ones.

Notation: For any sets A and B, the set of all functions from A to B is denoted B^A.

Theorem 7.27: (a) For every set A, $\wp(A) \sim \{0, 1\}^A$.

(b) If A is any set and B has more than one member, then $A \prec B^A$.

Proof: (a) We want to define a bijection f between $\wp(A)$ and $\{0, 1\}^A$. For each $C \subseteq A$, let $f(C)$ be the characteristic function of C with domain A. The simple verification that f is a bijection is left for Exercise 14.

(b) By Cantor's theorem, $A \prec \wp(A)$. And if B has more than one member, then $\wp(A) \preceq B^A$ by an adaptation of the proof of part (a): specifically, we replace 0 and 1 with any two fixed members of B. Since B may have more than two members, we can no longer prove that f is onto, but it is still one-to-one. ■

Theorem 7.27(a) provides a generalization (to all sets, not just finite ones) of Theorem 5.8, and Theorem 7.27(b) generalizes Exercise 17 of Section 4.5. On the basis of Theorem 7.27(a) and the analogy to Theorem 5.8 the set $\wp(A)$ may be denoted 2^A.

Theorem 7.27 is a good example of a theorem that looks difficult because it is very abstract. Part (a) involves a *function* between a *set of sets* and a *set of functions*, which sounds very complicated. Actually, the proof of (a) is quite simple once you get past the abstraction.

Still, the concept of B^A takes some getting used to. Not only does it look strange, but it is one of our first encounters with a set of functions (as opposed to a set of ordered pairs). The next four examples should help clarify this concept.

Example 6: What is B^A if $A = \{1, 2, 3, 4, 5\}$ and $B = \{7\}$? In this case, there is only one way to define a function from A to B, which is to map all members of the domain to 7. Therefore, B^A consists of just one function, namely $\{(1, 7), (2, 7), (3, 7), (4, 7), (5, 7)\}$.

Example 7: With A and B as in Example 6, what is A^B? Now the domain has only one member. So any member of A we choose as the image of 7 defines a function from B to A. So A^B contains five functions: $\{(7, 1)\}$, $\{(7, 2)\}$, $\{(7, 3)\}$, $\{(7, 4)\}$, and $\{(7, 5)\}$.

Example 8: What is B^A if $A = \{7, 12\}$ and $B = \{1, 5\}$? Some trial and error with arrow diagrams like Figure 7.3 makes it apparent that there are four functions from A to B. They are shown in Figure 7.12.

Example 9: The set $\mathbb{R}^{\mathbb{N}}$ is by definition the set of all infinite sequences of real numbers. Geometrically, it can be thought of as one version of infinite-dimensional space. Similarly, $[0, 1]^{\mathbb{N}}$ consists of all infinite sequences from the unit interval. It can be viewed as an infinite-dimensional cube.

Exercises 7.5

(1) Prove Proposition 7.14.

(2) Prove Theorem 7.15

(3) Show that each of the following pairs of sets have the same size, by defining a specific bijection between them:

(a) $\mathbb{N}_k$ and $\mathbb{N}_{2k} - \mathbb{N}_k$, for any nonnegative integer k

(b) $\mathbb{N}_k \times \mathbb{N}_k$ and $\mathbb{N}_{k^2}$, for any nonnegative integer k

(c) $\mathbb{N}$ and the set of odd natural numbers

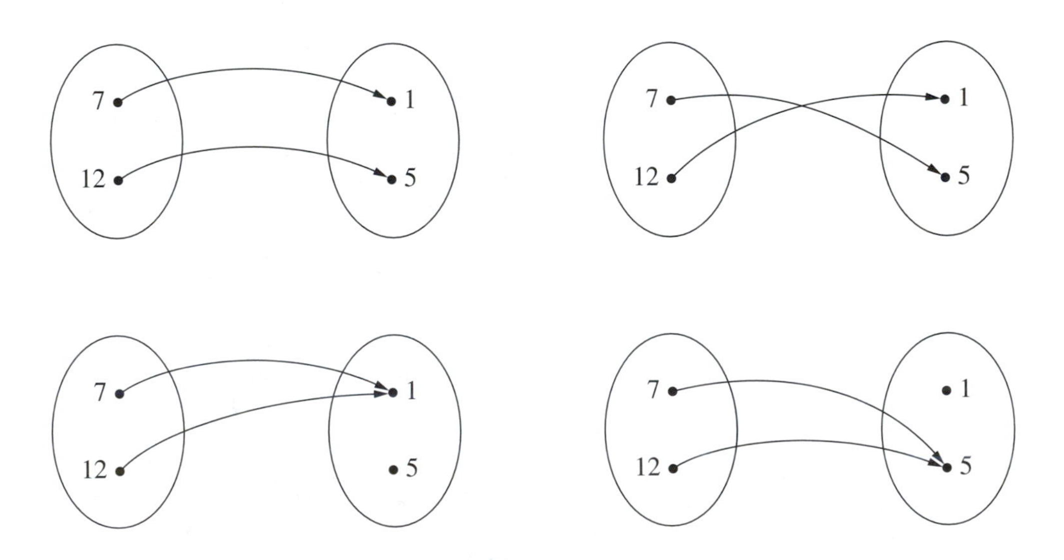

Figure 7.12 Arrow diagrams for the four functions from {7, 12} to {1, 5}

(4) Classify each of the following sets as finite, denumerable, the same size as $\mathbb{R}$, or larger than $\mathbb{R}$. Justify briefly, using any of the results of this section.

(a) $\mathbb{Q} \times \mathbb{Q}$ (b) $\mathbb{Q} \cap [1, 3]$
(c) $\mathbb{N} \cap [1, 3]$ (d) $\mathbb{R} \times \mathbb{N}$
(e) $\mathbb{N}^{\mathbb{R}}$ (f) $\{5\}^{\mathbb{R}}$

(5) (a) Prove the indicated part of Theorem 7.16(a).
(b) Prove Theorem 7.16(b).

(6) Complete the proof of Theorem 7.17.

*(7) Complete the proof of Theorem 7.18 by showing that h is a bijection.

(8) Prove Corollary 7.21.

(9) (a) Prove Theorem 7.22(b).
(b) Prove Theorem 7.22(c).
(c) Prove Theorem 7.22(e).
(d) Prove Theorem 7.22(f).

(10) (a) Prove Theorem 7.23(a).
(b) Prove Theorem 7.23(b).

(11) Fill in the details of the proof of Theorem 7.24(c).

(12) Prove Theorem 7.24(b) and (c) replacing "countable" with "denumerable."

(13) Theorem 7.24 and the remarks after it seem to imply that the set of all finite sequences of natural numbers is denumerable. Describe in words (so not very formally) a bijection between this set and $\mathbb{N}$.

(14) Complete the proof of Theorem 7.27(a).

Critique the proofs in Exercises 15 through 17. (If necessary, refer to Exercises 4.2 for the instructions for this type of problem.)

(15) **Theorem:** If $A \sim B$ and $C \sim D$, then $A \cup C \sim B \cup D$.

Proof: Assume $A \sim B$ and $C \sim D$. Then there are functions $f: A \xrightarrow{\text{bij}} B$ and $g: C \xrightarrow{\text{bij}} D$. Pick such an f and g (by ES), and define $h: A \cup C \to B \cup D$ by

$$h(x) = \begin{cases} f(x) & \text{if } x \in A \\ g(x) & \text{if } x \in C \end{cases}$$

Clearly, h is a bijection.

(16) **Theorem:** If $A \sim B$ and $C \sim D$, then $A \times C \sim B \times D$.

Proof: If $A \sim B$, then, for some m, A and B both have m members. Similarly, if $C \sim D$, then, for some n, C and D both have n members. Thus $A \times C$ and $C \times D$ both have mn members, by Theorem 6.1.

(17) **Theorem:** If $A \sim B$ and $C \sim D$, then $A^C \sim B^D$.

Proof: Assume $f_1: A \xrightarrow{\text{bij}} B$ and $f_2: C \xrightarrow{\text{bij}} D$. For any $h \in A^C$, define $G(h)$ to be $f_2^{-1} \circ h \circ f_1$. By Theorem 7.4(f) applied twice, $G(h) \in B^D$. Thus $G: A^C \rightarrow B^D$. *[Note that we're defining a function between two sets of functions. This may require some thought.]* To show G is onto, given any $h' \in B^D$, let $h = f_2 \circ h' \circ f_1^{-1}$. Again by Theorem 7.4(f), $h \in A^C$. And by Theorems 7.3(b) and 7.5, $G(h) = h'$. To show G is one-to-one, assume $G(h_1) = G(h_2)$, so $f_2^{-1} \circ h_1 \circ f_1 = f_2^{-1} \circ h_2 \circ f_1$. Then $f_2 \circ (f_2^{-1} \circ h_1 \circ f_1) \circ f_1^{-1} = f_2 \circ (f_2^{-1} \circ h_2 \circ f_1) \circ f_1^{-1}$, and again by Theorems 7.3(b) and 7.5, this simplifies to $h_1 = h_2$.

(18) Prove: If $A \sim B$ and $C \sim D$, then $A \times C \sim B \times D$.

(19) Prove: If $A \sim B$, then $\wp(A) \sim \wp(B)$.

(20) The goal of this exercise is to show that all intervals in $\mathbb{R}$ have the same cardinality. We always assume $a < b$ and $c < d$.

(a) Prove that any two closed intervals $[a, b]$ and $[c, d]$ have the same cardinality.

(b) Prove that $[a, b]$, (a, b), $[a, b)$, and $(a, b]$ all have the same cardinality. Use the CSB theorem.

(c) Find a bijection between an open interval and $\mathbb{R}$. You need not prove that the function you find works.

(d) Conclude (with explanation) that all real intervals containing more than one point, including unbounded ones, have the same cardinality as $\mathbb{R}$.

(21) (a) Prove that $\wp(\mathbb{N}) \times \wp(\mathbb{N}) \sim \wp(\mathbb{N})$. ***Hint:*** You need to find a way of coding any two sets of natural numbers into one. The proof of Theorem 7.24(c) might help.

(b) Using part (a) and Exercise 18, show that $[\wp(\mathbb{N})]^m \sim \wp(\mathbb{N})$, for any m in $\mathbb{N}$.

(c) From this exercise and the result (mentioned in the text) that $\wp(\mathbb{N}) \sim \mathbb{R}$, prove that $\mathbb{R} \sim \mathbb{R} \times \mathbb{R} \sim \mathbb{R} \times \mathbb{R} \times \mathbb{R}$, and so on.

(22) Cantor's original diagonalization argument is discussed in the text after Theorem 7.26. Make this proof more rigorous. That is, given any function f from $\mathbb{N}$ to $[0, 1]$, explicitly define a number in $[0, 1] - \text{Rng}(f)$. Because there is duplication in the decimal representation of numbers (for example, $0.2999 \ldots = 0.3$), it is safest not to use any 0's or 9's in the number you define.

(23) Show that, for any set B and any natural number k, $B^k \sim B^{\mathbb{N}_k}$. By definition, the set on the left is a set of k-tuples, and the set on the right is a set of functions.

(24) You are the desk clerk at **Infinity Hotel**, also known as **Hilbert Hotel** in honor of David Hilbert, who thought up this amusing situation. The hotel has an infinite number of rooms (numbered 1, 2, 3, and so on), and is full.

(a) Suddenly a man comes in, desperately wanting a room. At first you tell him that he can't have one because the hotel is full, but then you realize you can give him a room, provided that you are willing to move people around (but *not* force people to share a room who aren't already together). How do you do that?

(b) Later, an even bigger problem occurs. There is another Infinity Hotel across the street, and it burns down. Suddenly a denumerable set of customers arrives, all wanting rooms in your hotel. How can that be done?

(c) Now comes the true disaster. Across town, there is an infinite sequence of Infinity Hotels, all full, and they all burn down. All the customers from all those hotels appear at your desk, wanting rooms. How can you accommodate them?

7.6 Counting and Combinatorics

We now return to the study of finite sets and derive rules for calculating their sizes exactly. Results of this type are called **counting principles**. Counting principles can be very useful in proofs, sometimes providing an appealing alternative to mathematical induction. Furthermore, counting principles have significant practical applications and are the foundation for the subject of **discrete mathematics** or **combinatorics**.

Let's begin by restating the fundamental counting principle, introduced in Section 6.1. We do not prove this principle, in part because it is stated so informally. But it certainly can be stated formally and then proved mathematically, by an argument similar to the proof of Theorem 6.1. Almost all the proofs in this section are based on this formula and are relatively informal.

Fundamental Counting Principle, or **Product Rule for Counting:** Suppose that it is required to make a sequence of k decisions. If there are n_1 possible choices for the first decision, n_2 possible choices for the second decision, and so on, then the number of possible ways to make the whole sequence of decisions is $n_1 n_2 \ldots n_k$.

The next theorem lists some important counting principles, most of which appeared in earlier chapters but are repeated here for easy reference.

Theorem 7.28: For any finite sets A and B:

(a) $\text{Card}(A \cup B) = \text{Card}(A) + \text{Card}(B) - \text{Card}(A \cap B)$

(b) $\text{Card}(A \times B) = \text{Card}(A)\,\text{Card}(B)$

(c) $\text{Card}(\wp(A)) = 2^{\text{Card}(A)}$

(d) $\text{Card}(B^A) = \text{Card}(B)^{\text{Card}(A)}$

Proof: (a) This is a restatement of the sum rule for counting (Theorem 5.7). We first prove it for the special case where A and B are disjoint. Under that assumption, let $\text{Card}(A) = k$ and $\text{Card}(B) = m$. So there are bijections $g: A \to \{1, 2, \ldots, k\}$ and $h: B \to \{1, 2, \ldots, m\}$. Define a function f with domain $A \cup B$ by

$$f(x) = \begin{cases} g(x) & \text{if } x \in A \\ h(x) + k & \text{if } x \in B \end{cases}$$

It is straightforward to verify that f is a bijection from $A \cup B$ to $\mathbb{N}_{k+m}$, and we leave the details for Exercise 6. Therefore, Card$(A \cup B) = k + m$, as desired.

To prove the general case, we apply the special case twice. First of all, note that $B = (A \cap B) \cup (B - A)$, and the sets on the right side of this equation are disjoint. Therefore, Card(B) = Card$(A \cap B)$ + Card$(B - A)$, which becomes Card$(B - A)$ = Card(B) − Card$(A \cap B)$.

Next, note that $A \cup B = A \cup (B - A)$. Again, the sets on the right side are disjoint, so Card$(A \cup B)$ = Card(A) + Card$(B - A)$ = Card(A) + Card(B) − Card$(A \cap B)$.

By the way, this argument requires Theorem 7.17. Can you see where? (See Exercise 6.)

(b) This is a restatement of Theorem 6.1. It also follows immediately from the fundamental counting principle.

(c) This is a restatement of Theorem 5.8. It's also instructive to prove this using the fundamental counting principle: given a set A with (say) k members, what sequence of decisions do we have to make to select a subset of A? Well, for each member of A, we have two possibilities: include it in the subset, or don't. So we have to make a sequence of k decisions, each of which allows two possibilities. Thus there are 2^k subsets.

(d) This also follows from the fundamental counting principle. Let A and B be sets of cardinality k and m, respectively. How many functions from A to B are there? To determine a function from A to B, we have to decide where to send each member of A. *[Figure 7.3 might help you see what's going on here.]* Since A has k members, k decisions have to be made. And since B has m members, there are m possibilities for each decision. Thus there are m^k such functions. ■

Theorem 7.28 shows some of the usefulness of the fundamental counting principle, but we have hardly begun. Simply put, it is one of the most powerful formulas in mathematics. Among other things, notice how much terminology and notation is directly based on counting formulas that follow from the fundamental counting principle, for example, the notion of a cartesian "product," the concept of a "power" set, and the unusual notation B^A.

The next few problems illustrate some standard counting methods, using both the product rule and the sum rule.

Example 1: Recall Exercise 5 of Section 6.1 The solution to that problem is that there are 9 × 26 × 26 × 10 × 10, or 608,400, possible license plates in Tannu Tuva. Now suppose Tannu Tuva needs to plan for its growing population. In addition to the old license plates, there will be new plates consisting of any three letters followed by any three digits. How many cars will this system accommodate?

Solution: By the product rule, there are $26^3 \times 10^3$, or 17,576,000, valid new plates. Since the sets of old and new plates are disjoint, there are 18,184,400 plates in all.

Example 2: The Cucamonga Philately Club has ten members and must elect a slate of three officers: president, vice president, and treasurer. How many possible slates are there, if each of the following hold?

(a) Two or more positions may be filled by the same person.

(b) The officers must be three different people.

(c) The officers must be three different people *and* Jim can't be treasurer because of a conflict of interest.

Solution: This is a direct application of the product rule. The answer to part (a) is 10^3, or 1000, and the answer to part (b) is $10 \times 9 \times 8$, or 720. Part (c) gets quite confusing if we consider the positions in the order given. The time-saving trick is to start with the treasurer: there are 9 possibilities for treasurer, then there are still 9 choices for president, and then 8 choices for vice president. So the answer is $9 \times 9 \times 8$, or 648.

Permutations

It is worthwhile to think about the difference between parts (a) and (b) of Example 2. In part (a), we need to count *sequences* of three club members, in which *repetition is allowed*. The word "sequence" is meant to emphasize that *order matters*; the slate of John, Mary, and Alice is not the same as the slate of Alice, John, and Mary. If the set of club members is C, we are asking for the cardinality of C^3, which is given by Theorem 6.1. In part (b), we need to count *sequences* of three club members, in which *repetition is not allowed*. There is no standard notation for this set, but there is a standard notation for its *cardinality*.

Definition: Let n and r be integers with $0 \leq r \leq n$. The number of sequences of r different elements from a set with n elements is called the **number of permutations of *n* things taken *r* at a time**, denoted $P(n, r)$ or ${}_nP_r$.

It is implicit in this definition (and easy enough to show) that $P(n, r)$ depends only on the values of n and r and not on which particular set with n elements is used. The empty sequence is the only sequence of 0 elements, so $P(n, 0) = 1$.

Theorem 7.29: $P(n, r) = n(n-1)(n-2) \ldots (n-r+1) = n!/(n-r)!$

Proof: Informally, the first equation is immediate by the product rule, and the second can be seen by writing out what the fraction with factorials means and canceling terms. (Both equations could be proved rigorously by induction.) ■

Example 3: A geneticist might need to calculate the number of sequences of 5 distinct genes from a set of 1000 genes. The most succinct expression for this number would be $P(1000, 5)$ or $1000!/995!$. However, if we wanted to evaluate this number, we would not evaluate these huge factorials. The answer is $1000 \times 999 \times 998 \times 997 \times 996$.

In Section 7.3 the word "permutation" is used with what appears to be a very different meaning from the one just defined. Here is the result that links these meanings.

Theorem 7.30: Let A have n members. Then the following sets have $n!$ members:

(a) The set of all bijections on A (permutations in the sense of Section 7.3)
(b) The set of all sequences of (all) the members of A, without repetition
(c) The set of all total orderings on A

Proof: First, note that the cardinality of set (b) is by definition $P(n, n)$, which we know equals $n!/0!$, or simply $n!$. The fact that the three sets have the same size is easy to see informally. Given any sequence as in set (b), the sequence determines a total ordering on A, in which the members are ordered in the same order they appear in the sequence. This mapping from sequences to orderings is one-to-one (different sequences give different orderings) and onto (every total ordering is obtained). So we have a bijection between the sets of (b) and (c).

The set in (b) may be thought of as the set of all bijections between $\mathbb{N}_n$ and A. (We could consider this the definition of this set.) So let g be any *fixed* bijection between $\mathbb{N}_n$ and A. Then, given any bijection f on A, map f to the member $f \circ g$ of set (b). It is straightforward to show that this mapping is a bijection.

Exercise 7 asks you to carry out the details of this proof. ■

Among other things, Theorem 7.30 shows that the number of permutations on a set with n elements, in the sense of Section 7.3, is $P(n, n)$ in our current notation. Loosely speaking, a permutation of a set is thought of as an arrangement of the members of A. Mathematically, the intended meaning is usually (a) or (b). Note that if A happens to be a set of the form $\{1, 2, \ldots, n\}$, then (a) and (b) are the same.

Example 4: Since the word "care" has 4 letters, it has 4!, or 24, permutations. Of these, "race" and "acre" are also words; the other 21 are not.

Combinations

Permutations may always be thought of as sequences, in which order matters by definition. The corresponding notion for situations in which order does not matter is **combinations.**

Example 5: Exercise 19 of Section 4.5 asks for the number of hellos that are spoken if everyone in a group of n people greets everyone else exactly once. This is a permutation problem. We can use the product rule or the formula for $P(n, 2)$ to find the answer. For instance, if there are 20 people, then there are 380 hellos.

Now suppose everyone in this group also shakes hands exactly once with everyone else. How many handshakes occur? At first it might seem that this is the same problem. But note that there are two hellos spoken between each (unordered!) pair of people, but only one handshake. So the answer is 190, half the previous answer. This problem is asking for the number of *sets* of two distinct members of the group, not sequences.

Example 6: The Rego Park Conga Society has ten members. In how many ways could this organization choose a committee of three members? A committee is not the same as a slate of officers. Order does not matter in a committee; this new problem is asking about sets, not sequences. To find the answer, note that each *set* of three members could be arranged to form 3!, or 6, different slates. Therefore, since there are 720 possible slates, there are 720/6, or 120, possible committees.

Definition: Let n and r be integers with $0 \le r \le n$. The number of sets of exactly r elements from a set with n elements is called the **number of combinations of n things taken r at a time** or simply **n choose r**, denoted $C(n, r)$, ${}_nC_r$, or $\binom{n}{r}$.

Theorem 7.31: $C(n, r) = P(n, r)/r! = n!/r!(n - r)!$

Proof: The idea of the proof is exactly as in Example 5. Each set of r elements corresponds to exactly $r!$ sequences of r elements. ■

Example 7: In how many ways can the Rego Park Conga Society choose a slate of three different officers plus a steering committee of three other members?

Solution: By the product rule and the formula for combinations, the answer is $10 \times 9 \times 8 \times C(7, 3) = 720 \times (7 \times 6 \times 5)/3! = 25{,}200$. You should verify that the answer doesn't change if we choose the committee first and then the officers.

Theorem 7.32: (a) $C(n, r) = C(n, n - r)$

(b) $C(n, n) = C(n, 0) = 1$

(c) If $0 < r < n$, then $C(n, r) = C(n - 1, r) + C(n - 1, r - 1)$.

(d) **Binomial theorem**: $(x + y)^n = \sum_{r=0}^{n} C(n, r)x^{n-r}y^r$

Proof: The proofs of (a), (b), and (c) are simple computations (see Exercise 8).

(d) We prove this by induction on n. For $n = 0$, the formula becomes $(x + y)^0 = 1$, which is correct. If we feel safer starting at $n = 1$, we have to show $(x + y)^1 = x + y$, which is also correct. For the induction step, assume the formula for n. Then $(x + y)^{n+1} = (x + y)(x + y)^n = (x + y)\sum_{r=0}^{n} C(n, r)x^{n-r}y^r = \sum_{r=0}^{n} C(n, r)x^{n-r}y^r(x + y)$.

If we examine the terms inside the summation, we see all of them take the form of a number times $x^a y^b$, where $a + b = n + 1$. The only term with x^{n+1} is $C(n, 0)x^{n+1} = 1 \cdot x^{n+1} = C(n + 1, 0)x^{n+1}$. Similarly, the only term with y^{n+1} is $C(n, n)y^{n+1} = 1 \cdot y^{n+1} = C(n, n)y^{n+1}$.

If $0 < r < n + 1$, then there are two terms with $x^{(n+1)-r}y^r$. Their coefficients are $C(n, r)$ and $C(n, r - 1)$. By part (c), these terms add to $C(n + 1, r)x^{(n+1)-r}y^r$. Combining all these terms yields the desired formula for $n + 1$. ■

Because of their role in the binomial theorem, the numbers $C(n, r)$ are also called **binomial coefficients**. The binomial theorem is an extremely useful and versatile formula. The exercises cover some of its other versions and generalizations.

Example 8: It is easily checked that $C(5, 0) = C(5, 5) = 1$, $C(5, 1) = C(5, 4) = 5$, and $C(5, 2) = C(5, 3) = 10$. Therefore, $(x + y)^5 = x^5 + 5x^4y + 10x^3y^2 + 10x^2y^3 + 5xy^4 + y^5$. If we substitute $2y$ for y in this formula, we obtain

$$(x + (2y))^5 = x^5 + 10x^4y + 40x^3y^2 + 80x^2y^3 + 80xy^4 + 32y^5$$

Binomial expansions of this sort can be computed even more quickly with the well-known Pascal's triangle, described in Exercise 9.

Exercises 7.6

(1) The Lodi Jotto Club has ten members.

(a) In how many ways can the club choose a slate of three different officers if Fred can't be the president and Joyce can't be the treasurer?

(b) In how many ways can the club choose a slate of three officers in which any two offices can be held by the same person but all three can't?

(2) The Humptulips Chess Club has ten members.

(a) In how many ways can the club choose two disjoint committees, one with three members and the other with two members?

(b) In how many ways can the club choose two disjoint committees, each with two members? There are two interpretations for this problem. Explain what they are, and give both answers.

(3) Four married couples are to be seated in a row of eight seats at a theater. In how many ways can the seating be done if each of the following hold?

(a) There are no restrictions.
(b) Each person must sit next to his or her spouse.
(c) Sexes must alternate.
(d) The restrictions of parts (b) and (c) both apply.
(e) The two people at the ends must be a couple.

(4) (a) There are 10 contact points on a circuit board. In how many ways can they be connected with 9 wires to form a single sequence of 10 points, with one end designated positive and the other end negative?

(b) What is the answer to (a) without the designation of positive and negative?

(c) In how many ways can the 10 points be connected with 10 wires to form a circuit (with no specified direction)?

(5) This problem is of a type not directly discussed in the text, called **permutations with some objects alike**.

(a) How many permutations are there of the word "gorse"?

(b) How many permutations are there of the word "goose"? The two o's are considered *indistinguishable*. In other words, switching the o's does not constitute a new permutation.

(c) How many permutations are there of the word "geese"?

(d) How many permutations are there of the word "deeds"?

(6) (a) Complete the proof of Theorem 7.28(a), as indicated in the text.

Blaise Pascal (1623–1662) was one of the most brilliant and strange figures in the history of mathematics. Even though his father, Etienne, was a skilled mathematician, he excluded mathematics from his son's education and went so far as to hide all mathematics books from Blaise. He might have feared that mathematics would be too taxing for his sickly and nervous son. However, at the age of 12, Blaise was curious to know what geometry was. Even though he was given only a vague definition of the subject, he gave up his playtime and secretly derived much of basic geometry during the ensuing weeks. When his father discovered this, he immediately realized that Blaise's mathematical talent should no longer be restrained.

Pascal was a generalist who contributed to many branches of mathematics. His first efforts were in geometry and projective geometry, including a remarkably fruitful result known as Pascal's mystic hexagram theorem, presented by him at the age of 16. Soon thereafter, Pascal turned his efforts to inventing the first calculating machine, which worked but was too expensive and cumbersome to be a commercial success. Later, in the 1650s, came the correspondence with Fermat that led to the foundations of probability theory, which in turn stimulated Pascal to discover many interesting properties of the "arithmetical triangle" that now bears Pascal's name even though it was invented much earlier. Pascal came very close to understanding the major principles of calculus and, under different circumstances, might well have been considered its primary inventor.

Pascal's mathematical achievements are all the more remarkable when one considers his short life, his poor health, and his fickle and unstable nature. His mathematical research was confined to three rather short periods of his life. In 1650, his poor health led him to abandon mathematics and concentrate on religious meditation. He resumed mathematics in 1653, only to give it up again in 1654, supposedly because he interpreted his lucky survival of a carriage accident to be a sign from God. His only subsequent mathematical activity took place in 1658, after he noticed that a toothache went away while some mathematical thoughts occurred to him; Pascal took this to be another sign from God.

Pascal's enormous talent was not confined to mathematics. He did important work in physics, and the construction of his calculating machine required substantial engineering ability. Furthermore, he was one of the most important religious philosophers of all time. As if all this weren't enough, Pascal also invented the modern wheelbarrow and devised the plan for the first public bus.

(b) Where in this proof is Theorem 7.17 used?

(7) Fill in the details of the proof of Theorem 7.30.

(8) Prove Theorem 7.32 (a), (b), and (c).

(9) **Pascal's triangle** may be defined as follows: (i) the mth row has m entries; (ii) the rows are arranged in a triangle or pyramid; (iii) each row begins and ends with a 1; and (iv) each entry of any row, except the first and last, is the sum of the two entries diagonally above it.

(a) Construct the first seven rows of the triangle.

(b) Prove that the kth entry of the mth row of the triangle is $C(m - 1, k - 1)$.

(c) Use Pascal's triangle to quickly write out the binomial expansions of $(x + y)^4$, $(x - y)^4$, and $(x + 2y)^5$.

(10) Compare parts (a) and (b) of Theorem 7.28. Note that cartesian product corresponds perfectly to multiplication of cardinalities for finite sets, but union does not correspond perfectly to addition. Here is a set operation that does correspond to addition: the **formal disjoint union** of any sets A and B, denoted $A \sqcup B$, is the set $(A \times \{1\}) \cup (B \times \{2\})$. Prove:

(a) For A and B finite, $\text{Card}(A \sqcup B) = \text{Card}(A) + \text{Card}(B)$.

(b) If $A \sim B$ and $C \sim D$, then $A \sqcup C \sim B \sqcup D$.

(11) Give counterexamples to show that both parts of Exercise 10 fail with $\sqcup$ replaced by $\cup$. All counterexamples for part (b) also use finite sets only.

(12) Let $A = \varnothing$, $B = \{3\}$, $C = \{1, 4\}$, and $D = \{2, 3, 4\}$. List all the members of the following sets. (Note that the members of these sets are *functions*, not just single numbers or ordered pairs.) You should use Theorem 7.28(d) to make sure you have all the members.

(a) A^C (b) C^A (c) B^D (d) D^B (e) C^C

(13) (a) Let B, C, and D be as in Exercise 12. Using Theorem 7.28, compute the cardinality of each of the following sets:

(i) $D^{B \cup C}$ (ii) $D^B \times D^C$ (iii) $(C \times D)^C$

(iv) $C^C \times D^C$ (v) $(C^D)^C$ (vi) $C^{D \times C}$

(b) Which answers to part (a) come out equal? Do these results remind you of any laws of exponents? State conjectures based on these observations.

*(c) Prove these conjectures, *not just for finite sets*. Your only restriction on the sets B, C, and D should be that B and C are disjoint.

*7.7 The Axiom of Choice and the Continuum Hypothesis

Section 7.6 provides a solid introduction to the subject of cardinality. Let's summarize what we have learned, concentrating on the ordering of sets based on size, defined by the relation $\prec$ (or $\preceq$). When discussing this ordering, it is a bit easier to think of the things being ordered as cardinals (equivalence classes of sets under $\sim$) rather than individual sets. The cardinality notation that was introduced in Section 7.6 for finite sets is often extended to infinite sets as well. Thus, for any sets A and B, $|A| = |B|$, $\overline{\overline{A}} = \overline{\overline{B}}$, and Card($A$) = Card($B$) all mean that $A \sim B$. The notation $|A| < |B|$ means that $A \prec B$, whereas $\overline{\overline{A}} \le \overline{\overline{B}}$ means that $A \preceq B$, and so on.

The smallest sets are the finite sets, and the **finite cardinals** are ordered in the obvious way. The smallest cardinal of all is the one including only the empty set, the next smallest cardinal consists of all sets with one element, then comes the cardinal of all two-element sets, and so on.

Then come the infinite sets. It *seems* clear that the smallest infinite cardinal consists of the denumerable sets, sets with the same size as $\mathbb{N}$. This cardinal includes many proper subsets of $\mathbb{N}$, namely the unbounded ones. It also includes many sets that seem much bigger than $\mathbb{N}$, such as $\mathbb{Q}$ and $\mathbb{N} \times \mathbb{Q}$.

We have one primary method of creating larger infinite sets, the power set operator. Thus $\wp(\mathbb{N})$ is uncountable. Its cardinal includes $\mathbb{R}$ as well as all intervals in $\mathbb{R}$. As with the denumerable sets, unions and cartesian products of sets of this cardinality are no bigger. However, we can get bigger and bigger sets by considering $\wp(\mathbb{R})$, $\wp(\wp(\mathbb{R}))$, and so on.

You might think that we have pretty much settled all reasonable questions about cardinality, but this is not true at all. There are many difficult problems in this area that we have not even considered. For one thing, although everything stated above about *finite* sets is fairly routine to prove, it is not so clear that the denumerable sets really are the smallest infinite cardinal. (We certainly haven't proved this.) An even more glaring gap in our exposition is that we have not proved that the ordering on cardinals is *total*. That is, we have not proved that $A \prec B$, $A \sim B$, or $B \prec A$ must be true for all sets A and B. This is another one of those deceptively "obvious" statements. Surprisingly, neither of these simple-sounding conclusions can be proved without a special, powerful axiom.

The Axiom of Choice

In the early part of the twentieth century, mathematicians wanted to find axioms for set theory that would avoid difficulties like Russell's paradox without weakening the subject significantly. The axioms developed by Zermelo and Fraenkel worked well for the most part, but one important method of forming new sets did not seem to follow from those axioms. Since this way of forming sets was useful and seemed intuitively valid, people eventually added a new axiom to set theory, which states that sets defined in this special way do exist.

By now, virtually all mathematicians believe that the axiom of choice (AC for short) is a correct principle that should be accepted as an axiom. Yet it is still generally viewed as special and less obvious than the other axioms. More precisely, AC is typically used to prove the existence of a set that cannot be explicitly defined in set-builder notation. That is, AC allows you to define sets that you can't describe or see directly; this can cause a sort of vagueness which some mathematicians find troublesome. So many mathematicians try to prove theorems without using AC whenever possible and inform their readers when they do use this axiom.

Here's what this axiom says: suppose $\mathscr{A}$ is a collection of nonempty sets. Then, intuitively, we ought to be able to *choose* one element from each set in $\mathscr{A}$. The axiom AC states this more mathematically by saying that there must be a *function* that maps each set in $\mathscr{A}$ to a member of itself.

Definition: The Axiom of Choice (AC) is the following statement:
For every collection $\mathscr{A}$ of nonempty sets, there is a function f such that, for every B in $\mathscr{A}$, $f(B) \in B$. Such a function is called a **choice function** for $\mathscr{A}$. (For definiteness, we can specify $f: \mathscr{A} \to \bigcup\mathscr{A}$ if we wish.)

In abstract mathematics, it is useful to acquire a feel for when AC is required to define a set, and when it isn't. Not every choosing process requires AC. For one thing, AC is not required to prove the existence of a choice function when $\mathscr{A}$ is a *finite* collection of nonempty sets (see Exercise 3). Also, AC is not required if a rule for a choice function can be stated explicitly. The next three examples clarify this subtle point.

Example 1: Suppose we want to define a choice function on $\mathcal{P}(\mathbb{N}) - \{\varnothing\}$, that is, a function that chooses a member of each nonempty subset of $\mathbb{N}$. This is no problem; for each nonempty $B \subseteq \mathbb{N}$, let $f(B)$ = the least number in B. Theorem 5.6 guarantees that this f works as it should. And since f is defined explicitly, AC has not been used.

Example 2: In contrast, suppose we want a choice function on $\mathcal{P}(\mathbb{R}) - \{\varnothing\}$. How can we systematically choose a member of every nonempty set of real numbers? The above rule for f certainly doesn't work, since many subsets of $\mathbb{R}$ (for instance, $\mathbb{R}$ itself) do not have a least element. For fun, spend a few minutes trying to define such a function. Don't try too hard, because it's impossible. Thus, if we need to work with a choice function on this set, we need to use AC to assert that one exists (see Exercise 2).

Example 3: Here is a well-known and cute way of understanding when the axiom of choice is required. Suppose you are a clerk at Galaxy o' Shoes. The store contains an infinite number of pairs of shoes. The boss comes in and says, "Quick—I need a way of picking one shoe out of every pair." "No problem," you say, "Just take the right shoe of each pair." Mathematically, you have explicitly defined a choice function without using AC.

Now suppose the store also contains an infinite number of pairs of socks, and the boss makes a similar request for socks. You are stuck. Since the two socks of a pair are generally indistinguishable, there is no way to describe how to do this other than randomly choosing one sock from each pair. In other words, AC is required to define this choice function.

There are many valid ways of stating (and thinking of) the axiom of choice. If we add the extra hypothesis that the sets in $\mathscr{A}$ are *disjoint* (two at a time, that is), we can replace the conclusion about a choice function with the simpler conclusion that there is a *set* that contains exactly one member of each set in $\mathscr{A}$. This alternative to AC seems more narrow but is equivalent to AC as defined. Still, the form involving choice functions is often more convenient to work with.

Here is a theorem, without proof, which lists some of the many interesting statements that are equivalent to AC. The exercises ask you to prove several parts of it.

Theorem 7.33: The following are equivalent:

(a) The axiom of choice.

(b) For every collection $\mathscr{A}$ of nonempty, *disjoint* sets, there is a set consisting of exactly one member of each set in $\mathscr{A}$.

(c) Every set can be well ordered (the **well-ordering principle**).

(d) For every function f, there is a subset A of Dom(f) such that $f|_A$ is one-to-one and Rng($f|_A$) = Rng(f). As discussed in Section 7.2, this guarantees the existence of partial inverse functions.

(e) The cardinality ordering on sets is total; that is, for any sets A and B, either $A \prec B$, $A \sim B$, or $B \prec A$.

In addition to the fact that it postulates the existence of functions and sets that cannot be described explicitly, another reason that the axiom of choice was originally controversial is that it has some very strange, even wrong-looking consequences. The most famous of these is the theorem that an ordinary solid sphere can be decomposed into five pieces that can be reassembled into *two* solid spheres with the same radius as the original sphere! This sounds impossible, but it's a valid consequence of AC. Many people consider this theorem, called the **Banach-Tarski paradox**, to be the most bizarre result in all of mathematics.

Theorem 7.33(e) answers one of our earlier questions about cardinality. That is, the axiom of choice is required to conclude that all sets are comparable in size. Let us now use AC to settle some other cardinality issues.

Theorem 7.34: Let A be any set.

(a) If $\mathbb{N} \preceq A$, then A is infinite.

(b) Assuming AC, the converse of part (a) also holds.

(c) $\mathbb{N} \preceq A$ if and only if A is the same size as a proper subset of itself.

Proof: (a) If $\mathbb{N} \preceq A$, then the CSB theorem implies that $A \nprec \mathbb{N}$.

(b) Assume AC and that A is infinite. Let g be a choice function on the nonempty subsets of A. We use induction to define a one-to-one function f from $\mathbb{N}$ to A. Informally, let $f(1) = g(A), f(2) = g(A - \{f(1)\}), f(3) = g(A - \{f(1), f(2)\})$. In words, to obtain each output of f (even the first one), simply apply the choice function g to the set of all members of A that have not already been used as outputs of f. This can be turned into a rigorous definition; it is the type of inductive definition discussed at the end of Section 7.4, in which the value of the function at a number n depends on *all* values of the function for smaller inputs. The fact that A is infinite guarantees that we never run out of elements of A to use for the next output of f. Therefore, $f: \mathbb{N} \to A$. And the fact that $f(n)$ is always chosen from the members of A not previously used by f implies that f is one-to-one.

(c) We outline the proof and leave the details for Exercise 7. For the forward direction, assume $g: \mathbb{N} \to A$ is one-to-one. So g defines a bijection between $\mathbb{N}$ and some subset C of A. We know there are bijections from $\mathbb{N}$ to a proper subset of itself. Using g, we can identify C with $\mathbb{N}$ and thereby get a bijection from C to a proper subset of itself. Then simply extend this function to all of A by mapping all members of $A - C$ to themselves.

For the reverse direction, assume g is a bijection from A to a proper subset of itself. Let c be any member of A - Rng(g). Consider the infinite sequence $(c, g(c), g(g(c)), \ldots)$. An infinite sequence can always be viewed as a function with domain $\mathbb{N}$. In this case, all the outputs are in A, so we have a function from $\mathbb{N}$ to A. And it is not hard to show that this sequence has no repetition; that is, the function is one-to-one. ■

So, if we accept the axiom of choice, there are three equivalent ways to define infinity of sets. The idea of defining an infinite set to be one which can be put in one-to-one correspondence with a proper subset of itself is rather appealing.

Here is the strengthened version of Theorem 7.24(c) mentioned in the remarks following that theorem.

Theorem 7.35: Assuming AC:

(a) The union of a countable number of countable sets is countable.

(b) The union of a denumerable number of denumerable sets is denumerable.

Proof: We prove part (b); the proof of (a) is very similar. Let $B = \bigcup_{n \in \mathbb{N}} A_n$, where each A_n is denumerable. First note that $\mathbb{N} \preceq A_1 \subseteq B$, so $\mathbb{N} \preceq B$. By the CSB theorem, it suffices to also prove $B \preceq \mathbb{N}$. By definition, there is a bijection from $\mathbb{N}$ to A_n, for every n. By AC, there is an infinite sequence of functions $(f_1, f_2, f_3, \ldots)$ such that each f_n is a bijection from $\mathbb{N}$ to A_n. *[This is a typical use of AC, that is, to make an infinite set of choices. You might find it instructive to write down an acceptable codomain for the choice function in this case.]* Using this sequence, we define $g: \mathbb{N} \times \mathbb{N} \to B$ by $g(k, m) = f_k(m)$. The function g needn't be one-to-one (since the A_i's aren't necessarily disjoint), but it is clearly onto. Therefore, by Theorem 7.28(b), $B \preceq \mathbb{N} \times \mathbb{N}$. Since $\mathbb{N} \times \mathbb{N} \sim \mathbb{N}$ (Theorem7.24(a)), we obtain $B \preceq \mathbb{N}$, as desired. ■

Our final application of the axiom of choice to cardinality is a generalization of Theorem 7.24 to all infinite cardinalities. That is, unions and cartesian products can never produce a larger infinite set than the ones you start with. The proof is too technical to include.

Theorem 7.36: Assume AC. Let A be infinite and $B \preceq A$. Then:

(a) $A \cup B \sim A$

(b) If $B \neq \varnothing$, then $A \times B \sim A$. (In particular, $A \times A \sim A$.)

This theorem can be generalized to unions and cartesian products of any finite number of sets.

The Continuum Hypothesis

By Cantor's theorem (Theorem 7.26), we know that $A \prec \wp(A)$, for every set A. If A is a finite set with at least two members, then there is a set B such that $A \prec B \prec \wp(A)$; that is, the cardinality of $\wp(A)$ is not the *next* cardinality after that of A. But if A is infinite, the situation is less obvious. For example, is there a set B such that $\mathbb{N} \prec B \prec \wp(\mathbb{N})$, or equivalently, an uncountable subset of $\mathbb{R}$ that has smaller cardinality than $\mathbb{R}$? The conjecture that there is *no* such set, made by Cantor, is known as the **continuum hypothesis**. The same statement, generalized from $\mathbb{N}$ to all infinite sets A, is called the **generalized continuum hypothesis**. (The word "continuum" refers to the set $\mathbb{R}$.)

For many decades, mathematicians tried in vain to prove or disprove either of these conjectures. Finally, in the late 1930s, Kurt Gödel showed they could not be *disproved* from the axioms of set theory, even using the axiom of choice. Then, in 1963, Paul Cohen showed they could not be *proved* either. Taken together, these results establish that the continuum hypothesis and the generalized continuum hypothesis are **independent** from the usual axioms of mathematics; that is, they are **undecidable** questions, ones that cannot be answered. Gödel and Cohen also showed that the axiom of choice is independent from the rest of the axioms of mathematics. These results of Gödel and Cohen were among the most surprising, deep, and far-reaching achievements in the history of mathematics.

Again, almost all current mathematicians believe the axiom of choice and accept it as an axiom. There is less of a consensus about the continuum hypothesis. There are some very compelling reasons for accepting it, notably, the fact that no one has ever produced a set that is between $\mathbb{N}$ and $\mathbb{R}$ in size or even found much evidence for the existence of such a set (see Exercise 8). On the other hand, people are reluctant to accept such a specific statement about cardinalities as a fact, without at least some compelling informal rationale for it. The continuum hypothesis is probably the most well-known example of a mathematical conjecture whose status is genuinely controversial.

Exercises 7.7

(1) (a) Define a choice function on the collection of all nonempty subsets of $\mathbb{Z}$.
(b) Define a choice function on the collection of all nonempty subsets of $\mathbb{Q}$.

(2) Make at least three attempts to explicitly define a choice function on the collection of all nonempty subsets of $\mathbb{R}$. Explain how each of your attempts fails. As the text explains, it is not possible to do this.

(3) It is claimed in the text that AC is not needed to prove the existence of a choice function on a *finite* collection of nonempty sets. Carry out this proof.

(4) In Theorem 7.33, prove that (a) implies (b).

(5) In Theorem 7.33, prove that (c) implies (a).

(6) In Theorem 7.33, prove the equivalence of (a) and (d).

(7) Prove Theorem 7.34(c) in more detail. In particular, write the explicit definition of the bijection needed for the forward direction.

*(8) Here are three attempts to define a set that is between $\mathbb{N}$ and $\mathbb{R}$ in size. Show that each of them is actually the same size as $\mathbb{N}$ or $\mathbb{R}$. This is one sort of investigation that has led mathematicians to consider the continuum hypothesis plausible.

(a) $\{a + b\sqrt{n} \mid a, b \in \mathbb{Q} \text{ and } n \in \mathbb{N}\}$
(b) $\mathbb{R} - \mathbb{Q}$, the set of irrational numbers
(c) $(\mathbb{R} - \mathbb{Q}) \cap [0, 1]$

Critique the proofs in the remaining exercises. If necessary, refer back to the instructions for this type of problem in Exercises 4.2. Also, if a proof is correct but requires AC, point out where AC is needed.

(9) **Theorem:** If there is a surjection from A to B, then $B \preceq A$.

Proof: Since f is onto B, every member of B has at least one preimage under f. For each y in B, let $g(y)$ be any one of these preimages. It is clear that g is a one-to-one function from B to A.

(10) **Theorem:** If A is an infinite set, then $A \times A \sim A$.

Proof: Assume A is infinite. Then $A \sim \mathbb{N}$ and $A \times A \sim \mathbb{N} \times \mathbb{N}$. By Theorem 7.24(a) and the transitivity of $\sim$, we are done.

(11) **Theorem:** Given well-orderings on sets A and B, the associated lexicographic ordering on $A \times B$ is also a well-ordering.

Proof: Let R be any nonempty subset of $A \times B$. Then Dom(R) must be a nonempty subset of A. With respect to the given well-ordering on A, there is a least element of Dom(R). Call it a. By the definition of a domain, $\{y \mid (a, y) \in R\}$ is a nonempty subset of B. Call its least element b. Then (a, b) is the least element of R in the lexicographic ordering.

Suggestions for Further Reading: All the references mentioned in Chapter 6 also apply to this chapter. Stoll (1979) has a very clear treatment of cardinality and the axiom of choice. For more on combinatorics, see Grimaldi (1994) or Tucker (1994). For a good combined treatment of cardinality and combinatorics, see Pfleeger and Straight (1985).

Unit 3

Number Systems

Chapter 8

The Integers and the Rational Numbers

8.1 The Ring $\mathbb{Z}$ and the Field $\mathbb{Q}$

Having spent Unit II discussing concepts that are rather abstract and general, we spend the rest of this book considering a much more concrete and familiar topic: numbers. We have been discussing some of the most important number systems of mathematics since Chapter 4. The purpose of this unit is to study these and other number systems in more depth. Before moving on, you might want to review the basic algebraic properties of $\mathbb{R}$ (the ordered field axioms) presented in Appendix 2, the material on $\mathbb{N}$ in Section 4.5, and the brief introduction to $\mathbb{Z}$ and $\mathbb{Q}$ in Section 5.1.

Notation: Throughout this chapter, the letters a, b, c, i, j, k, m, and n always stand for integers. If we wish to restrict them further (for example, to $\mathbb{N}$), we say so.

Definitions: A **ring** is a number system that satisfies (at least) field axioms V-1 through V-5, V-7, V-8, and V-10, of Appendix 1. A ring that also satisfies axiom V-6 is called **commutative**. A ring that also satisfies axioms V-9 and V-12 is called a **ring with unity**. A ring that also has an ordering relation defined on it, satisfying axioms V-13 through V-17, is called an **ordered ring**.

Remarks: (1) The term "number system" is being used here rather loosely, in that a number system in this context might be made up of objects that would not normally be considered numbers. In the same vein, the operations called addition and multiplication in a ring need not be anything like the standard operations of arithmetic. It would be more accurate to say that a ring is a type of **algebraic structure**. In general, an algebraic structure is a set with one or more operations (and possibly one or more relations) defined on it, satisfying certain properties. The branch of mathematics that studies algebraic structures is called **abstract algebra**.

(2) Rings and fields are two of the most important types of algebraic structures. They are rather similar, in that both rings and fields must have two primary operations defined on them, addition and multiplication. But a field must satisfy several properties that a ring need not satisfy. It follows immediately that every field is a ring, in fact, a commutative ring with unity.

(3) Now is a good time to answer the riddle "When is an axiom not an axiom?" For example, the commutative law of addition is a real number axiom; that means we assume (or assert) it to be true about that particular number system, without proof. In that context, it is a genuine axiom. But this law is also part of the above definitions of the terms "ring" and "field." In this context, axiom V-5 is not an axiom at all, since we are not assuming it to be true about any particular number system(s). Rather, we are simply specifying what conditions must be satisfied for a number system to qualify as a ring or a field. It would be more accurate to refer to properties rather than axioms when they occur in a definition. Whether or not it's strictly accurate, many mathematicians use the word "axiom" for properties that are part of a definition.

Example 1: Most of the familiar number systems that mathematicians work with are rings, if not fields. The real number system $\mathbb{R}$ is an ordered field; so is $\mathbb{Q}$. The complex number system $\mathbb{C}$ is a field but not an ordered one. The system $\mathbb{Z}$ is a commutative ring with unity but not a field because it does not satisfy the multiplicative inverse property. The system $\mathbb{N}$ is not even a ring. For one thing, it has no additive identity. If we consider $\mathbb{N} \cup \{0\}$, then there is an additive identity, but this structure is still not a ring because the additive inverse property is not satisfied.

Of the assertions made in this example, the one about $\mathbb{R}$ does not need to be proved because we have assumed axioms to this effect. The claims about $\mathbb{C}$ are proved in Chapter 10. The other assertions are all proved in this chapter.

Proposition 8.1: In any ring:

(a) The additive identity is unique.

(b) Each member's additive inverse is unique.

(c) If there is a multiplicative identity, it is unique.

(d) If an element has a multiplicative inverse, that inverse is unique.

Proof: The proofs of parts (a) and (b) are identical to the proofs of Theorems A-1 and A-3 of Appendix 2. (That is, those results hold for all *rings*, not just for fields, because only the ring axioms are used in their proofs.) Similarly, the proofs of parts (c) and (d) are similar to those for the uniqueness proofs for Theorems A-2 and A-4. ∎

Note how, in the context of rings, parts (c) and (d) must be stated as conditionals, as opposed to Theorems A-2 and A-4.

Lemma 8.2: If m and n are natural numbers and $m > n$, then $m - n$ is also a natural number (and so $m \geq n + 1$).

Proof: Viewing m as fixed, we use induction on n. The statement P(n) in this induction (without the part in parentheses) is, "If $m > n$, then $m - n \in \mathbb{N}$." We also use the equivalent form "$m \leq n$ or $m - n \in \mathbb{N}$."

To prove this for $n = 1$, apply Theorem 4.16(b). It says that $m = 1$ or $m - 1 \in \mathbb{N}$. Of course, this implies $m \leq 1$ or $m - 1 \in \mathbb{N}$.

For the induction step, assume $m \leq n$ or $m - n \in \mathbb{N}$. We want to prove that if $m > n + 1$, then $m - (n + 1) \in \mathbb{N}$. So assume $m > n + 1$. It follows that $m > n$, so by the induction hypothesis, $m - n \in \mathbb{N}$. Now apply Theorem 4.16(b) to the number $m - n$. So $m - n = 1$ or $(m - n) - 1 \in \mathbb{N}$. But if $m - n = 1$, then $m = n + 1$; this contradicts the assumption $m > n + 1$. So we must have $(m - n) - 1 \in \mathbb{N}$. But $(m - n) - 1 = m - (n + 1)$, so we have the desired conclusion.

The claim in parentheses follows immediately from the main claim. For if $m - n \in \mathbb{N}$, then $m - n \geq 1$, by Theorem 4.16(a). ∎

The next corollary appeared in Exercises 5.1.

Corollary 8.3: (a) $\mathbb{Z} = \mathbb{N} \cup \{0\} \cup \{-k \mid k \in \mathbb{N}\}$

(b) $\mathbb{N} \subset \mathbb{Z}$, and $\mathbb{N}$ is the set of all positive integers.

Proof: (a) Let $k \in \mathbb{Z}$. Then, by definition, $k = n - m$ for some m and $n \in \mathbb{N}$. We proceed by cases, based on trichotomy. If $k > 0$, then $n > m$, and so $k \in \mathbb{N}$ by Lemma 8.2. If $k < 0$, then $m > n$; thus $-k$, which equals $m - n$, is in $\mathbb{N}$. Finally, if $k = 0$, there's nothing to prove.

Conversely, if $k \in \mathbb{N}$, then $k = (k + 1) - 1$, which is in $\mathbb{Z}$. Also, $-k = 1 - (k + 1)$, which is in $\mathbb{Z}$. And $0 = 1 - 1$, which is in $\mathbb{Z}$.

(b) Follows immediately from (a). ∎

Lemma 8.4: The system $\mathbb{Z}$ is closed under addition and multiplication.

Proof: Let i and j be any two integers. Then, by the definition of $\mathbb{Z}$, there are natural numbers m, n, a, and b such that $i = n - m$ and $j = a - b$. Then we know that $i + j = (n - m) + (a - b) = (n + a) - (m + b)$. Since we already have that $\mathbb{N}$ is closed under addition (Theorem 4.10), this is a difference of natural numbers, so it's in $\mathbb{Z}$. Also, $ij = (m - n)(a - b) = (ma + nb) - (mb + na)$, by elementary algebra (see Exercise 2). Since $\mathbb{N}$ is closed under multiplication as well as addition (Section 4.5, Exercise 3), this is also an integer. ∎

Theorem 8.5: The integers form an ordered commutative ring with unity; that is, they satisfy all the ordered field axioms except possibly axiom V-11.

Proof: Clearly, $\mathbb{Z} \subseteq \mathbb{R}$. It follows that most of the ordered field axioms, namely the ones that don't contain an existential quantifier, are automatically established for $\mathbb{Z}$. (For example, here is a proof of the commutative law of addition for $\mathbb{Z}$: Let i and j be arbitrary integers. Then i and j are real numbers, since $\mathbb{Z} \subseteq \mathbb{R}$. Therefore $i + j = j + i$, by axiom V-5 for $\mathbb{R}$.) Thus, axioms V-3 through V-7 and V-13 through V-17 are true in $\mathbb{Z}$.

By Lemma 8.4, field axioms V-1 and V-2 are true in $\mathbb{Z}$. It remains to show that field axioms V-8, V-9 and V-10 are also true in $\mathbb{Z}$. We leave this for Exercise 6. ∎

Definitions: Let A be a ring, and $B \subseteq A$. Then B is called a **subring of A** provided B is a ring, using the same addition and multiplication operations as in A. Similarly, we may refer to a **subfield** of a particular field. As usual, if $B \subset A$, we add the word **proper** to these expressions. Some mathematicians may refer to a subfield of a ring that is not a field, but this usage is not standard.

Lemma 8.6: Suppose B is a subring of A. Then:

(a) The additive identity of B is the same as that of A.

(b) For any $x \in B$, its additive inverse in B equals its additive inverse in A.

Proof: (a) Let 0_B and 0_A denote the additive identities of B and A, respectively. Then the equation $0_B + 0_B = 0_B$ must hold in B. But since $B \subseteq A$ and B has the same addition operation as A does, this equation must also hold in A. So, working in A, we can add -0_B to both sides of this equation; this yields $0_B = 0_A$, as desired.

(b) The proof is left for Exercise 8. ∎

Lemma 8.6 may seem pointless, but it states important and subtle facts. As evidence of its subtlety, it fails completely with the word "additive" replaced by "multiplicative." In fact, A could have unity and B not or vice versa, or they could both have multiplicative identities that are not the same. Specific instances of these phenomena are given in Example 3 and in Example 6 of Section 8.3.

Lemma 8.7: Suppose A is a ring and $B \subseteq A$. Then:

(a) B is a subring of A iff field axioms V-1, V-2, V-8 and V-10 hold in B.

(b) If multiplication is commutative in A, then it is commutative in B.

Proof: (a) The forward direction is trivial. The idea of the reverse direction was already explained in the proof of Theorem 8.5: any field axiom *not* containing *existential* quantifiers that holds in A automatically holds in any subset of A.

(b) Field axiom V-6 has no existential quantifiers, so if it holds in A, it must hold in B. ∎

Example 2: The system $\mathbb{Z}$ is a subring of $\mathbb{R}$, in fact a subring with unity. This is just a restatement of Theorem 8.5. But $\mathbb{Z}$ is not a field, so it isn't a subfield of $\mathbb{R}$ (see Exercise 7).

$\mathbb{N}$ is not a ring because it does not satisfy the additive inverse property. (For instance, -3 is not in $\mathbb{N}$, by Corollary 8.3. It follows that 3 has no additive inverse in $\mathbb{N}$.) Therefore, $\mathbb{N}$ is not a subring of $\mathbb{R}$ (or of $\mathbb{Z}$).

Example 3: Let B be the set of all *even* integers. The set B is closed under addition and multiplication, contains 0, and contains the negative of each of its elements. So by Lemma 8.7, B is a commutative subring of $\mathbb{R}$ or of $\mathbb{Z}$. Theorem 8.10 generalizes this example and shows that B does not have unity. This is possibly the simplest example of a subring *without* unity of a ring *with* unity.

Mathematicians like to characterize sets and number systems as the smallest one or the largest one with a certain property. We now make this terminology precise.

Definitions: We say A is the **smallest set** with a certain property provided that A has that property and A is a subset of every set with that property. Similarly, we say A is the **largest set** with a certain property provided that A has that property and every set with that property is a subset of A.

Theorem 8.8: If the smallest (respectively, largest) set with a certain property exists, then it is unique and in fact is the intersection (respectively, union) of all sets with that property.

Proof: Let A be the smallest set with some property P. Since this means that there is at least one set satisfying P, we can form the intersection of all of them; call this intersection B. Since A is one of the sets we have intersected to form B, $B \subseteq A$. But since A is a subset of every set that is used in this intersection, we also have $A \subseteq B$. Therefore $A = B$, as desired. Uniqueness is immediate: if there were two smallest sets with a certain property, each would be a subset of the other, and this would make them equal.

The proof for the largest set is analogous. ■

Theorem 8.8 gives us a way of defining smallest and largest sets explicitly. However, there is a big if involved: the smallest or largest set with a certain property need not exist, as Example 5 shows.

Example 4: Suppose we want to form the smallest closed interval (within $\mathbb{R}$) containing the numbers 3, 7, and 15. By Theorem 8.8, we can try to do this by forming the intersection of all such intervals. It is not hard to show (see Exercise 10) that this intersection is the single interval $[3, 15]$. Since this interval is closed and contains 3, 7, and 15, it is in fact the smallest such interval.

Example 5: Now suppose we want to form the smallest *open* interval containing 3, 7, and 15. Once again, we would take the intersection of all such open intervals. But this intersection is once again the *closed* interval $[3, 15]$ (see Exercise 11). Since this is not an open interval, there simply is no smallest open interval containing these numbers.

These examples illustrate some typical behavior: operations involving intersections tend to work better with closed intervals than with open ones; for unions, it's the opposite.

Example 6: Recall the axioms for $\mathbb{N}$, group VI of Appendix 1. The first two say that the set $\mathbb{N}$ contains 1 and is closed under the operation of adding 1. Some mathematicians call such a set **progressive**. Then another way to state the set form of mathematical induction is that $\mathbb{N}$ is a subset of every progressive set. Therefore, the whole group of natural number axioms can be summarized very simply: $\mathbb{N}$ is the smallest progressive set. Equivalently, $\mathbb{N}$ is the intersection of all progressive sets.

Theorem 8.9: $\mathbb{Z}$ is the smallest subring of $\mathbb{R}$ with unity (and therefore, the intersection of all subrings of $\mathbb{R}$ with unity).

Proof: We have already shown that $\mathbb{Z}$ is a subring of $\mathbb{R}$ with unity. We must still show that if A is any subring of $\mathbb{R}$ with unity, then $\mathbb{Z} \subseteq A$. First, note that $1_A 1_A = 1_A$. Also, $1_A \neq 0_A = 0$, by axiom V-12 and Lemma 8.6(a). So we can divide both sides of the equation $1_A 1_A = 1_A$ by 1_A to obtain $1_A = 1$. In other words, $1 \in A$. It follows easily that A is obviously progressive. Thus, by the set form of mathematical induction, $\mathbb{N} \subseteq A$. Also, axiom V-8 holds in A, so 0 is in A. Finally, axiom V-10 holds in A; therefore, since every natural number is in A, so is the negative of any natural number. By cases, we conclude that every integer is in A. ■

We have been talking about subrings of $\mathbb{R}$; now let's look at some subrings of $\mathbb{Z}$ (which of course are also subrings of $\mathbb{R}$).

Notation: For any n, $n\mathbb{Z}$ denotes the set of all integer multiples of n; in symbols, $n\mathbb{Z} = \{kn \mid k \in \mathbb{Z}\}$.

Example 7: The product $1\mathbb{Z} = \mathbb{Z}$, because 1 is the multiplicative identity. Also, $(-1)\mathbb{Z} = \mathbb{Z}$. Theorem A-5 of Appendix 1 tells us that $0\mathbb{Z} = \{0\}$. And $2\mathbb{Z}$ is the set of all even integers, by the definition of what it means to be even.

Theorem 8.10: For every integer n, $n\mathbb{Z}$ is a subring of $\mathbb{Z}$. Unless $n = 1$ or -1, it is a subring without unity.

Proof: Let n be an arbitrary integer. By Lemma 8.7(a), to show that $n\mathbb{Z}$ is a subring of $\mathbb{Z}$, we need to show it is closed under addition and multiplication, contains 0, and includes the negative of each number in it. To show closure under addition, let a and $b \in n\mathbb{Z}$. Then $a = jn$ and $b = kn$, for some j and k. So $a + b = jn + kn = (j + k)n$, which is also in $n\mathbb{Z}$, since $j + k$ is also an integer. The rest of this part of the proof is left for Exercise 12(a). The proof of the second statement is left for Exercise 12(b). ■

Theorem 8.16 shows that every subring of $\mathbb{Z}$ is of the form $n\mathbb{Z}$ for some n.

The Field of Rational Numbers

The ring of integers is an extension of the natural numbers, obtained from $\mathbb{N}$ by adjoining zero and additive inverses (negatives) to obtain a number system that is closed under subtraction. It is logical to repeat this extension process with $\mathbb{Z}$, adjoining multiplicative inverses (reciprocals) to obtain the field $\mathbb{Q}$ of all rational numbers.

We have shown that $\mathbb{Z}$ is an ordered commutative ring with unity and is the smallest subring of $\mathbb{R}$ with unity. We now establish parallel results for $\mathbb{Q}$. We already know that $\mathbb{Z} \subset \mathbb{Q}$, by Exercise 11(c) of Section 5.2.

Theorem 8.11: The system $\mathbb{Q}$ is an ordered field (and so, it is a subfield of $\mathbb{R}$).

Proof: Since $\mathbb{Q}$ is a subset of $\mathbb{R}$, it is automatic (as in the proof of Theorem 8.5) that ordered field axioms V-3 through V-7 and V-13 through V-17 are true in $\mathbb{Q}$.

Closure of $\mathbb{Q}$ under addition and multiplication means simply that the sum and product of any two fractions are still fractions. This is straightforward (the verification for sums is Exercise 15 of Section 4.5). Technically, this proof also uses the fact that $\mathbb{Z}$ is a ring to show that when you add or multiply two rational numbers, you still get a quotient of *integers*.

Since $\mathbb{Z} \subseteq \mathbb{Q}$, both 0 and 1 are rational numbers, and so field axioms V-8, V-9, and V-12 hold in $\mathbb{Q}$. Also, since $x/y + (-x)/y = 0$, as is easily seen, axiom V-10 is true in $\mathbb{Q}$.

Finally, to verify axiom V-11, let x/y be any nonzero rational number. Then we must have $x \neq 0$, since otherwise $0(1/y)$ would be zero. It follows that $(x/y)(y/x) = 1$, so y/x is the multiplicative inverse of x/y. ■

Theorem 8.12: The system $\mathbb{Q}$ is the smallest subfield of $\mathbb{R}$ (and therefore, the intersection of all subfields of $\mathbb{R}$).

Proof: The previous theorem shows that $\mathbb{Q}$ is a subfield of $\mathbb{R}$. We must also show that if A is any subfield of $\mathbb{R}$, then $\mathbb{Q} \subseteq A$. To see this, first note that A is of course a subring of $\mathbb{R}$, so by Theorem 8.9, $\mathbb{Z} \subseteq A$. But then, for every nonzero integer j, $1/j$ must be in A, since A satisfies axiom V-11. Therefore, by multiplicative closure, for any two integers i and j such that $j \neq 0$, i/j is in A. So $\mathbb{Q} \subseteq A$. ■

Exercises 8.1

(1) Using only the ring axioms, prove rigorously:

(a) $(-x) + (-y) = -(x + y)$ (b) $x - y = -(y - x)$

It follows that these equations are true in every ring.

(2) Using only the ring axioms, carefully prove the equation used in the proof of Lemma 8.4: $(m - n)(k - j) = (mk + nj) - (mj + nk)$.

(3) Determine whether each of the following algebraic structures is a ring. If it is, determine whether it is commutative, has unity, or is a field.

(a) The set of all real numbers, with addition defined normally, but xy defined to equal 0, for all x and y

(b) The set of all real numbers, with addition defined normally, but xy defined to equal 1, for all x and y

(c) The set of all real numbers, with these special operations: $x \oplus y$ is defined to be $2x + 2y$, and $x \otimes y$ is defined to be $4xy$

(d) The set of all positive real numbers, with these special operations: $x \oplus y$ is defined to be xy, and $x \otimes y$ is defined to be $x^{\ln y}$

(e) $\mathbb{R} \times \mathbb{R}$, with addition and multiplication defined pointwise; that is, $(x, y) + (u, v) = (x + u, y + v)$ and $(x, y)(u, v) = (xu, yv)$

(f) The set of all functions from $\mathbb{R}$ to $\mathbb{R}$, with addition and multiplication of functions defined in the usual way (as in Exercise 13, Section 7.1)

(g) $\mathcal{P}(\mathbb{N})$, with addition and multiplication defined to be union and intersection, respectively

(h) $\mathcal{P}(\mathbb{N})$, with multiplication defined to be intersection, and addition defined to be $A \,\Delta\, B$, as in Exercise 15 of Section 5.3.

(4) Determine whether each of the following subsets of $\mathbb{R}$ is a subring of $\mathbb{R}$. (Your main tool for this should be Lemma 8.7(a).) If it's not, name at least one ring axiom that

is not satisfied. If it is, state whether it's a subring with unity; if so, state whether it's a subfield of $\mathbb{R}$.

(a) The set of integers that are multiples of 6
(b) The set of all odd integers
(c) The set of all nonnegative integers
(d) The set of all integer multiples of 1/2
(e) The set of all rational numbers
(f) The set of all numbers of the form $i \cdot 2^j$, where i and j are integers
(g) The set of all numbers of the form $i + j\sqrt{2}$, where i and j are integers
(h) The set of all numbers of the form $i + j\sqrt{2}$, where i and j are rational
(i) The set of all numbers of the form $i + j\sqrt[3]{2}$, where i and j are integers
(j) The set of all numbers of the form $i + j\sqrt[3]{2} + k\sqrt[3]{4}$, where i, j, and k are integers
(k) The set of all numbers of the form $i + j\sqrt[3]{2} + k\sqrt[3]{4}$, where i, j, and k are rational numbers

(5) (a) Determine exactly which of the ordered field axioms are satisfied by the set of positive rational numbers.

(b) Find a subset of $\mathbb{R}$, other than $\mathbb{Z}$, that satisfies all the ordered field axioms except one.

(6) Complete the proof of Theorem 8.5.

(7) (a) Prove that if n is not 0, 1, or -1, then $1/n \notin \mathbb{Z}$. You may use Corollary 8.2. But don't just assume that $1/n \notin \mathbb{N}$; prove it. This result generalizes Exercise 10(b) of Section 5.2.

(b) Prove that, in contrast to Theorem 8.5, $\mathbb{Z}$ is not a field; that is, it does not satisfy axiom V-11.

(8) Prove Lemma 8.6(b).

(9) Prove the following generalization of Lemma 8.6(a): Let B be a *subset* of a ring A. If B has an additive identity (for addition within B), then that identity element must be 0_A (and is therefore unique).

(10) Prove the claim made in Example 4.

(11) Prove the claim made in Example 5.

(12) (a) Complete the proof of the first part of Theorem 8.10.
(b) Prove the second part of Theorem 8.10. You may use Exercise 6(a).

(13) Critique the following proof. (If necessary, refer back to the instructions for this type of problem in Exercises 4.2.)

Theorem: If A and B are subrings of a ring C, then so is $A \cap B$.

Proof: By Lemma 8.7(a), we only need to verify axioms V-1, V-2, V-8, and V-10 for $A \cap B$.

To verify axiom V-1 (additive closure), assume x and $y \in A \cap B$. Then x and $y \in A$. Therefore $x + y \in A$ because A is closed under addition. Similarly, $x + y \in B$. So $x + y \in A \cap B$, as desired. The verification of axiom V-2 is almost identical.

To verify axiom V-8, note that since A and B are subrings of C, they both contain 0. Therefore, $0 \in A \cap B$. Similarly, for any $x \in A \cap B$, $-x$ must be in both A and B. So $-x$ is also in $A \cap B$. This verifies axiom V-10, completing the proof. ■

(14) Prove or find a counterexample: if A and B are subfields of a field, so is $A \cap B$.

(15) Prove or find a counterexample: if A and B are subrings of a ring, so is $A \cup B$.

(16) Prove or find a counterexample: if $\mathscr{A}$ is a collection of subrings of a ring C, then so is $\bigcap \mathscr{A}$.

(17) An element x of a ring is called **idempotent** iff $x^2 = x$. A **boolean ring** is one in which every element is idempotent.

(a) Prove that $x + x = 0$ for every x in any boolean ring.

(b) Prove that a boolean ring must be commutative.

8.2 Introduction to Number Theory

In basic mathematics, including elementary algebra, geometry, and calculus, the word "number" usually means "real number." There are several good reasons for this, one of which is that real numbers seem to be the best way of representing most of the quantities that occur in science and other applied situations. But there are also many situations that occur in which makes sense to restrict one's attention to natural numbers, integers, or rational numbers. This type of restriction can make simple-looking problems very difficult.

Example 1: Suppose you are asked to find a solution of the equation $x^2 + y^2 = 47$. Under normal circumstances, this is trivial: one solution is $x = 1$ and $y = \sqrt{46}$, and it's easy to write as many solutions as you want. But if it is specified that x and y must be rational, you suddenly have a much harder problem, which may require sophisticated methods to analyze completely.

The branch of mathematics that studies problems about natural numbers, integers and rational numbers is called **number theory**. It is one of the main branches of mathematics invented by the ancient Greeks. Among other things, number theory deals with **Diophantine equations**, which are equations whose solutions must come from $\mathbb{N}$,

$\mathbb{Z}$, or $\mathbb{Q}$, as in Example 1. Number theory is a subject in which the concepts are simple but the problems can be extremely hard. It is generally considered part of pure mathematics. Yet it has important applications, in particular, in **cryptography** and **cryptanalysis** (the theories of creating and deciphering codes), which have become highly sophisticated with the advent of powerful computers.

In this section we look at some of the classic results of number theory, some of which were first proved well over 2000 years ago. The next three definitions are given here for all integers, but most of the time they are applied only to nonnegative ones. The first one also appears in Section 6.3 but is important enough to repeat.

Definitions: For integers m and n, we say m **divides** n, denoted $m|n$, iff for some integer k, $n = km$. The notation $m|n$ may also be read "m is a **divisor** of n," "n is **divisible** by m," "m is a **factor** of n," or "n is a **multiple** of m."

An integer n is **prime** iff it is not 1 or -1 and its only factors are $1, -1, n$, and $-n$. Two integers are **relatively prime** iff the only integers that divide both of them are 1 and -1.

Theorem 8.13: (a) The number 1 divides every integer; so does -1.
(b) Every integer divides 0.
(c) If $i|j$ and $j|k$, then $i|k$.
(d) For any integers a, b, i, j, and k, if $i|j$ and $i|k$, then $i|(aj + bk)$.
(e) $j|k$ iff $(-j)|k$ iff $j|(-k)$ iff $(-j)|(-k)$.
(f) If $j|k$ and $k|j$, then $j = \pm k$.
(g) The only factors of 1 or -1 are 1 and -1.
(h) The number 1 is relatively prime to every integer; so is -1.

Proof: (a)–(e) The proofs are straightforward and are left for Exercise 5.

(f) Assume $j|k$ and $k|j$. Then $k = mj$ and $j = nk$, for some integers m and n. Substitution yields $k = mnk$. In the case that $k = 0$, we have $j = n \cdot 0 = 0$, and we are done. Otherwise we can divide both sides by k and obtain $1 = mn$, or $m = 1/n$. By Exercise 6(a) of Section 8.1, $n = \pm 1$, so $j = \pm k$.

(g) Assume $m|1$. By (a), $1|m$. So, by (f), $m = \pm 1$. Similarly, if $m|-1$, $m = \pm 1$.

(h) By part (g), since the only factors of 1 (or -1) are ± 1, the only factors that 1 (or -1) can have in common with another number are ± 1. ■

Recall that in a field, division is always possible (as long as the divisor is nonzero) and yields a quotient in the field. But $\mathbb{Z}$ is not a field and so is not closed under division. The following classic theorem and corollary state that the next-best situation holds in $\mathbb{Z}$: if you divide two integers, it's possible to find an integer quotient, but there may also be a **remainder**.

Theorem 8.14 (Division algorithm): Let n be any nonnegative integer, and m any positive one. Then there are unique integers q and r such that $n = mq + r$ and $0 \le r < m$.

Proof: This proof provides an instructive opportunity to use Theorem 5.6, the well-ordering property of $\mathbb{N}$. Given n and m as in the theorem, we first prove existence of q and r. Let $A = \{k \in \mathbb{N} \mid mk > n\}$. The set A is not empty because it contains $n + 1$ (see Exercise 6). So, by Theorem 5.6, there is a least number k in A. Define q to be $k - 1$, and $r = n - mq$.

The definition of r immediately gives $n = mq + r$. Also, by the definition of q, $mq \le n < m(q + 1) = mq + m$. By subtracting mq from all parts of this inequality, $0 \le n - mq = r < m$. We still have to show q and r are unique. In other words, we must show that if $n = mq + r$ with $0 \le r < m$, and $n = mq' + r'$ with $0 \le r' < m$, then $q = q'$ and $r = r'$. So assume the hypothesis of this implication. We then have $n = mq + r = mq' + r'$. We proceed by cases, using trichotomy on the relationship between q and q'.

Case 1: If $q = q'$, we can say $mq + r = mq + r'$. So $r = r'$, and we are done.

Case 2: Assume $q < q'$. Then $q' - q$ is a positive integer; this means that it is at least 1. We can rewrite $mq + r = mq' + r'$ as $m(q - q') = r' - r$. Since $q - q' \ge 1$ and m is positive, the left side of this equation is at least m. But since $r' < m$ and $r \ge 0$, it follows easily that $r' - r < m$. This is a contradiction, so this case is impossible.

Case 3: If $q' < q$, we similarly reach a contradiction. Thus the only possible case leads to the desired conclusion. ■

Corollary 8.15: Let n and m be any integers, with $m \neq 0$. Then there exist unique integers q and r as in Theorem 8.14, except that we must now say $0 \le r < |m|$.

Proof: We omit this proof, which consists of several cases, each of which either follows directly from Theorem 8.14 or has a proof that is very similar to the proof of that theorem (see Exercise 8). ■

Example 2: The application of Corollary 8.15 to negative dividends and divisors can be confusing. If you are asked to divide 23 by 5, obviously the quotient is 4 and the remainder is 3. But what about dividing -23 by 5? It's tempting to say that $q = -4$ and $r = -3$. But the theorem says that the remainder must be nonnegative. So it turns out that we must use $q = -5$ and $r = +2$ for this problem. See Exercise 1 for some additional problems of this sort.

In Section 8.1 we briefly discussed subrings of $\mathbb{Z}$ of the form $n\mathbb{Z}$. We can now prove an important result that tells us exactly what subrings of $\mathbb{Z}$ look like.

Theorem 8.16: Every subring of $\mathbb{Z}$ is of the form $n\mathbb{Z}$ for some unique $n \ge 0$.

Proof: Let A be any subring of $\mathbb{Z}$. Note that 0 is in A, by field axiom V-8 and Lemma 8.7(a). One possibility is that $A = \{0\}$. Then $A = 0\mathbb{Z}$, as desired. Otherwise, A contains at least one nonzero integer k. Either $k > 0$ or $k < 0$. In the latter case, $-k$ is also in A and is positive. In either case, A contains at least one positive integer. In other words, $A \cap \mathbb{N}$ is not empty. By the well-ordering property of $\mathbb{N}$ (Theorem 5.6), $A \cap \mathbb{N}$ therefore has a least element; call it n. So n is the smallest positive number in A. We show that $A = n\mathbb{Z}$.

To show $A \subseteq n\mathbb{Z}$, assume $j \in A$. We must show that j is a multiple of n. By the division algorithm, we can write $j = qn + r$. By the previous paragraph, $qn \in A$. By the additive inverse property, this means $-qn \in A$. Therefore, by closure under addition, $j - qn \in A$. That is, $r \in A$. But the remainder r must be less than the divisor n. Therefore r can't be positive, because n is the smallest positive number in A. But we also know that r is nonnegative. Thus $r = 0$. This makes j a multiple of n, as desired.

The proof that $n\mathbb{Z} \subseteq A$ is straightforward and is left for Exercise 9. Finally, we need the uniqueness of n. So suppose $A = n\mathbb{Z} = m\mathbb{Z}$. Thus $n \in m\mathbb{Z}$ and $m \in n\mathbb{Z}$. In other words, $m|n$ and $n|m$. So by Theorem 8.13(f), $m = \pm n$. If we restrict our attention to nonnegative m and n, as in this theorem, then they must be equal. ■

Definitions: For any integers m and n, we define their **greatest common divisor**, denoted GCD(m, n), to be the largest natural number that divides both m and n and their **least common multiple**, denoted LCM(m, n), to be the smallest nonnegative integer that both m and n divide.

We can also talk about the GCD and the LCM of three or more numbers.

Theorem 8.17: (a) Every pair of integers (not both zero) has a unique GCD and LCM.

(b) If GCD(m, n) = k, then there are integers a and b such that $am + bn = k$.

(c) Given any pair of integers (not both zero), every number that divides both of them also divides their GCD.

Proof: (a) Let m and n be given. To show that they have a GCD, consider the set $C = \{am + bn \mid a, b \in \mathbb{Z}\}$. (We call C the set of all **linear combinations** of m and n, with integer coefficients. It is somewhat analogous to $n\mathbb{Z}$, except that instead of consisting of all multiples of a single number n, it is **generated** by two numbers m and n.) We wish to show that C is a subring of $\mathbb{Z}$. This is a simple consequence of Lemma 8.7(a) and is left for Exercise 10. Therefore, $C = k\mathbb{Z}$ for some k, and we may choose k to be positive. (We know $k \neq 0$ since m and n are not both 0.) We claim k = GCD(m, n).

To see this, first note that both m and n are in C. So, by the definition of $k\mathbb{Z}$, k divides both m and n. And if j is any other divisor of both m and n, then j also divides any number of the form $am + bn$, by Theorem 8.13(e). In particular, $j|k$. Therefore, $j < k$. Uniqueness of the GCD is clear since there can be at most one greatest number with a certain property in $\mathbb{N}$.

The set of positive common multiples of m and n is not empty, since it includes $|mn|$. So this set has a least element, which is unique and is the LCM of m and n.

(b) Follows immediately from the definition of C and the fact that k is in C.

(c) This is proved as part of the proof of (a). ■

The proof of Theorem 8.17 is concise and elegant but gives no clue as to how to compute GCDs and LCMs. Here is the most efficient method, known as the **Euclidean algorithm.** To find the GCD of two numbers, divide the larger by the smaller and replace the larger original number by the remainder. Repeat the process. The last nonzero remainder is the GCD.

For finding LCMs, it is useful to know that GCD$(m, n) \times$ LCD$(m, n) = |mn|$. So the LCM can be found immediately once the GCD is known (see Exercises 11 and 20).

Example 3: To find the GCD of 42 and 120: $120 \div 42$ gives a remainder of 36, $42 \div 36$ gives a remainder of 6, and $36 \div 6$ gives a remainder of 0. So GCD(42, 120) = 6. LCM(42, 120) = $42 \times 120 \div 6 = 840$.

We can compute GCDs of three or more numbers by the inductive formula

$$\text{GCD}(a_1, a_2, \dots, a_{n+1}) = \text{GCD}(\text{GCD}(a_1, a_2, \dots, a_n), a_{n+1})$$

LCMs of three or more numbers can be computed similarly. This means we never need to find the GCD or LCM of more than two numbers at a time (see Exercise 15).

Example 4: Suppose we want the GCD and LCM of the numbers 120, 216, 300, and 52. By the methods just described, we find that GCD(120, 216) = 24 and LCM(120, 216) = 1080. Then

$$\begin{aligned} \text{GCD}(120, 216, 300) &= \text{GCD}(24, 300) = 12 \\ \text{LCM}(120, 216, 300) &= \text{LCM}(1080, 300) = 1080 \times 300 \div \text{GCD}(1080, 300) \\ &= 1080 \times 300 \div 60 = 5400 \end{aligned}$$

Finally,

$$\begin{aligned} \text{GCD}(120, 216, 300, 52) &= \text{GCD}(12, 52) = 4 \\ \text{LCM}(120, 216, 300, 52) &= \text{LCM}(5400, 52) = 5400 \times 52 \div \text{GCD}(5400, 52) \\ &= 5400 \times 52 \div 4 = 70{,}200 \end{aligned}$$

We have to be careful when using this method. Note that we did *not* say

$$\text{LCM}(120, 216, 300) = \text{LCM}(1080, 300) = 1080 \times 300 \div \text{GCD}(120, 216, 300)$$

because that would lead to an incorrect answer.

Lemma 8.18: (a) If k and m are relatively prime and $k|mn$, then $k|n$.
(b) If p is prime and $p|jk$, then $p|j$ or $p|k$.

Proof: (a) If k and m are relatively prime, then GCD$(k, m) = 1$, so by Theorem 8.17(b) there are integers a and b such that $ak + bm = 1$. Therefore $akn + bmn = n$. Since $k|mn$, the left side is a multiple of k. Thus so is n.

(b) Assuming that p is prime, $p|jk$ and $p \nmid j$, we need to show that $p|k$. First note that since p is prime, GCD(p, n) equals 1 or p for any n, since p has no other positive factors. But we also know that p is not a factor of j. So GCD$(p, j) = 1$. The conclusion $p|k$ then follows from part (a). ■

For the rest of this section, the word "prime" means a *positive* prime number.

Lemma 8.19: If p, q_1, q_2, ... , q_n, are all prime and $p|q_1q_2 \dots q_n$, then p must equal one of the q_is.

Proof: We prove this by induction on n. As usual, this means that for each value of n, we must view p and the q's as universally quantified.

For $n = 1$, we just need to show that if p and q are primes and $p|q$, then $p = q$. But $p|q$ means that $q = pk$ for some integer k. Then, since q is prime, k must be 1, -1, q, or $-q$. Since p and q are positive, k can't be -1 or $-q$. Also, if $k = q$, we get $p = 1$ (since $q \neq 0$), which is impossible since p is prime. Therefore $k = 1$, so $p = q$.

For the induction step, we assume the statement is true whenever there are n primes multiplied together. Then we assume that p, q_1, q_2, ... , q_{n+1} are all primes, and $p|q_1q_2 \dots q_{n+1}$. So we can write $p|(q_1q_2 \dots q_n)q_{n+1}$, and apply the previous lemma. Thus either $p|q_1q_2 \dots q_n$, or $p|q_{n+1}$. If $p|q_{n+1}$, then the argument just given for $n = 1$ shows that $p = q_{n+1}$. On the other hand, if $p|q_1q_2 \dots q_n$, then the induction hypothesis says that p must equal one of the q_is. In either case, we're done. ■

Complete Induction and the Fundamental Theorem of Arithmetic

Our next theorem is important for at least two reasons. Most significant is what it says: a natural number is completely determined by its prime factors, repeated as necessary. But the proof of the theorem is also noteworthy, because it requires a variant of mathematical induction that we have not yet seen, called **complete induction** or **course-of-values induction.**

To illustrate this principle, imagine we want to prove that every integer greater than 1 has a prime factor. (This is a weak version of Theorem 8.21.) We could attempt this by ordinary induction (starting at $n = 2$) on the statement "n has a prime factor." But then, for the induction step, we must prove that $n + 1$ has a prime factor from the assumption that n does. And this is essentially an impossible task, because knowing factors of n doesn't help us find factors of $n + 1$; in fact, consecutive numbers have no factors in common. For example, how would knowing that 2158 has a prime factor help us to know that 2159 has one?

However, we can easily show that $n + 1$ has a prime factor if we are allowed to assume not just that n does, but that *every* number (larger than 1) *up to and including* n does. And if you think about the rationale for mathematical induction, you can see that this stronger assumption ought to be allowed. That is, an induction is meant to establish some property for natural numbers one at a time, in increasing order. Therefore, at the point where you've established P(n) and want to proceed to P($n + 1$), you also know that P holds for all natural numbers less than n, and there's no harm in including this knowledge in the induction hypothesis. Compare this idea to the generalization of the method of inductive definitions given in Corollary 7.12.

The principle of induction with the usual assumption P(n) in the induction step replaced by the stronger assumption $\forall m \leq n$ P(m) is called complete induction. Let's now show that this principle follows from ordinary induction.

Lemma 8.20 (Principle of complete induction): For any proposition P(n), with n representing a natural number, we have

$$[\mathrm{P}(1) \wedge \forall n((\forall m \leq n\ \mathrm{P}(m)) \rightarrow \mathrm{P}(n+1)] \rightarrow \forall n\ \mathrm{P}(n)$$

Proof: Assume the statement in brackets, and define Q(n) to be the predicate $\forall m \leq n$ P(m). We show $\forall n$ Q(n) by ordinary induction. Since Q(n) clearly implies P(n), that suffices.

We have assumed P(1), and since 1 is the smallest natural number by Exercise 7(a) of Section 4.5, this implies Q(1). Now for the induction step, assume Q(n). By our assumption, this implies P(n + 1). Thus we have $\forall m \leq n$ P(m) and P(n + 1); and since Lemma 8.1 tells us that there are no integers between n and n + 1, we have $\forall m \leq n + 1$ P(m), which is precisely Q(n + 1). ■

Although complete induction looks more complicated than ordinary induction, it's really just a way of making life easier. In practice, it allows you to make the stronger induction hypothesis $\forall m \leq n$ P(m) whenever it suits you to do so.

Theorem 8.21 (Fundamental theorem of arithmetic): Every integer greater than 1 can be written as a product of (one or more) prime numbers. Furthermore, this factorization is unique except for the order of the factors. That is, if we specify that the prime factors must be written in nondecreasing order, then the factorization is unique.

Proof: We prove the first claim, existence of the factorization, by complete induction. We begin at $n = 2$: since 2 is prime, the claim holds for 2. Now assume that every integer between 2 and n (inclusive) can be written as a product of primes. Looking at n + 1, we have two cases. If n + 1 is prime, then we are done. If not, then we can write $n + 1 = km$, where k and m are both larger than 1 and less than n + 1 (see Exercise 13). By Lemma 8.1, we have k and $m \leq n$, and so by the induction hypothesis, k and m both can be factored into primes. Simply putting these factorizations together gives us the desired factorization of n + 1.

We still need to show the uniqueness of the factorization. So assume that $k = p_1p_2 \ldots p_m = q_1q_2 \ldots q_n$, where the p's and the q's are all primes and each factorization of k is written in nondecreasing order. We need to show that $m = n$ and that for each i between 1 and n, $p_i = q_i$. We first show that $p_1 = q_1$. If p_1 were less than q_1, then p_1 would have to equal one of the later q's, by Lemma 8.19. But this would contradict the fact that the q's are in nondecreasing order. Similar, it is impossible for q_1 to be less than p_1. By trichotomy, this means $p_1 = q_1$.

So we can cancel the first term of each of the products to obtain $p_2p_3 \ldots p_m = q_2q_3 \ldots q_n$. But then we can repeat the argument of the previous paragraph to obtain $p_2 = q_2$. Then we cancel these terms and repeat the argument to get $p_3 = q_3$, and so on.

We can continue in this manner until one or both products run out of factors. But if one runs out before the other, we would have the product of one or more primes equal to 1, which is impossible (see Exercise 14). Thus $m = n$, and we're done. ■

The uniqueness proof in Theorem 8.21 is not completely rigorous. A more rigorous proof would use induction on m or n or perhaps $m + n$ (see Exercise 16).

Corollary 8.22: Every integer greater than 1 can be written as a product of distinct prime numbers to positive integer powers. This factorization is unique except for the order of the primes.

The fundamental theorem of arithmetic provides a more elegant way of computing GCDs and LCMs. Suppose

$$m = 2^{a_1}3^{a_2} \dots p^{a_k} \quad \text{and} \quad n = 2^{b_1}3^{b_2} \dots p^{b_k}$$

where p is the largest prime factor of either m or n. (For this to make sense, some of the exponents shown may have to be zero.) Then

$$\text{GCD}(m, n) = 2^{c_1}3^{c_2} \dots p^{c_k} \quad \text{and} \quad \text{LCM}(m, n) = 2^{d_1}3^{d_2} \dots p^{d_k}$$

where each c_i is the smaller of a_i and b_i, while each d_i is the larger of a_i and b_i. This method works the same with three or more numbers. In fact, this method is often more efficient than the Euclidean algorithm for finding the GCD and LCM of several numbers that are not very large. But the Euclidean algorithm is much more efficient when the numbers are very large. ■

Example 5: Suppose we want the GCD and LCM of the numbers 120, 216, and 300. It is easy to compute that $120 = 2^3 \times 3 \times 5$, $216 = 2^3 \times 3^3$, and $300 = 2^2 \times 3 \times 5^2$. So the GCD of these numbers is $2^2 \times 3 = 12$, and their LCM is $2^3 \times 3^3 \times 5^2 = 5400$.

We haven't said anything about rational numbers for awhile. We are now in a position to prove some important facts about them.

Theorem 8.23: Every rational number can be put in lowest terms, that is, in the form j/k, with j and k relatively prime.

Proof: Let a/b be any rational number. Then set $j = a/\text{GCD}(a, b)$ and $k = b/\text{GCD}(a, b)$. Clearly $j/k = a/b$, and $\text{GCD}(j, k) = 1$. ■

Theorem 8.24: The number $\sqrt{2}$ is not rational. That is, there is no rational number r such that $r^2 = 2$.

Proof: Assume, on the contrary, that there is such a rational number. By Theorem 8.23, it can be written in lowest terms j/k, so we have $(j/k)^2 = 2$ and $\text{GCD}(j, k) = 1$. The first equation becomes $j^2 = 2k^2$. Since 2 is a factor of j^2 and 2 is prime, we can apply Lemma 8.19. Thus $2|j$; that is, j is even. So write $j = 2i$. Then $(2i)^2 = 2k^2$. Multiplying out the left side yields $4i^2 = 2k^2$, or $2i^2 = k^2$. But then $2|k^2$, so by Lemma 8.18(b), $2|k$. Thus j and k are both even; this contradicts that j and k are relatively prime. ■

Theorem 8.24 is a special case of a much more general fact: for any natural numbers m and n, if $\sqrt[m]{n}$ is not an integer, then it's not a rational number. Exercise 17 asks you to prove this.

Even though the name **Pythagoras** may be familiar to more people than any other mathematician's, very little is known of his life. This is partly due to the fact that his school was secretive and mystical and put almost nothing in writing.

Pythagoras was born around 570 B.C. and lived until some time after 500 B.C. He lived in many places and eventually founded the society now known as the Pythagoreans in Croton, a Greek colony in Italy, in around 520 B.C. This society was as much a religious cult as it was a school of scholarly inquiry. Among other things, members of the society believed in reincarnation, the mystical powers of whole numbers (indicated by their motto, "All is number"), and the "music of the spheres" ("spheres" in this case referring to the planets). They led an ascetic life, were vegetarians, and were sworn to the strictest secrecy.

At the same time, the Pythagoreans did important investigations in philosophy, science, and mathematics, notably in geometry. It is impossible to know what was done by Pythagoras himself, as opposed to his followers (both during his life and after). For instance, Pythagoras's theorem had been known empirically for at least a thousand years, but was probably first proved by Pythagoras or one of his followers. Certainly, the Pythagoreans were among the very first people to do mathematics systematically and deductively, with reasonably sound proofs. The first two books of Euclid's *Elements* were attributed to the Pythagoreans, although many scholars believe this is an exaggeration.

The Pythagoreans knew that numbers like $\sqrt{2}$ must exist in nature. (For example, consider a right-angled triangle in which each of the legs is one unit long.) They are also credited with proving Theorem 8.24. But they were quite disturbed at the discovery of irrational numbers, since it meant that there were quantities that could not be represented in terms of whole numbers. Reportedly, some members of their society attempted to suppress this information, perhaps by less-than-honorable means.

We conclude this section with a sampler of well-known theorems (with proofs not provided) and conjectures of number theory. Rather than worry now about how to prove these theorems, you should just try to appreciate the combination of simplicity and depth that characterizes many results in this subject. The list of conjectures that follows the theorems is also intended for your pleasure and fascination. Many of these conjectures look very simple, at least in the sense that they involve simple concepts and are easy to understand. And yet no one has been able to prove or disprove any of them, even though some of them were proposed several centuries ago.

Many of the theorems and conjectures in these lists involve prime numbers. Mathematicians have long been fascinated with prime numbers and the way they are arranged within $\mathbb{N}$. But it is often very difficult to prove things about prime numbers.

Some Well-Known Theorems of Number Theory

Theorem (Lagrange): Every positive integer can be written as the sum of four squares (of integers).

Theorem (claimed by Pierre de Fermat in 1640; first published proof by Leonhard Euler in 1754)**:** For any odd prime p, p is the sum of two squares iff $p \equiv 1 \pmod 4$, that is, iff $4 \mid (p - 1)$.

Fermat's little theorem: If p is prime and $p \nmid k$, then $k^{p-1} \equiv 1 \pmod p$.

Wilson's theorem (actually proved by Joseph-Louis Lagrange (!), in 1771)**:** For any positive integer n, n is prime iff $(n - 1)! \equiv -1 \pmod n$.

Dirichlet's theorem: If GCD$(m, n) = 1$, then the arithmetic sequence m, $m + n$, $m + 2n$, $m + 3n$, ... contains infinitely many primes.

Bertrand's postulate (conjectured by Joseph Bertrand but proved by Pafnuti Chebyshev in 1850; of course, it should not be called a postulate any more, but names tend to stick!)**:** If $n > 1$, then there is a prime strictly between n and $2n$.

Prime number theorem (Jacques Hadamard and C. J. de la Vallee Poussin, independently, in 1896)**:** Let $\pi(m)$ denote the number of prime numbers that are less than or equal to m. Then, for large values of m, $\pi(m)$ is approximately equal to $m/\ln m$. More precisely,

$$\lim_{m \to \infty} [(\pi(m) \ln m)/m] = 1$$

Fermat's last theorem: There are no natural numbers n, a, b, and c such that $n > 2$ and $a^n + b^n = c^n$.

The story behind this theorem is one of the most famous ones in all of mathematics. While studying Pythagorean triples (introduced in Exercise 9, Section 1.2) in about 1637, Fermat wrote a note in his copy of the works of Diophantus, saying he had a "truly wonderful proof" of the theorem. The many marginal notes in this book were discovered after Fermat's death. Every other claim he made in that book turned out to be correct. But for over 300 years after the note was discovered, despite an enormous amount of empirical evidence in favor of the conjecture, no one was able to prove it.

Pierre de Fermat (1601–1665) was born to a well-to-do and influential French family. He studied the classics, languages, and law and became a successful lawyer and judge. He never worked as a professional mathematician or as any type of academic but did mathematics strictly as a hobby. This makes his achievements even more remarkable, considering that he made major contributions to at least three branches of mathematics.

The first field of mathematics in which Fermat did important work was analytic geometry and calculus. By 1636, he had worked out a system of analytic geometry that was almost identical to Descartes's. He also appears to have been the first person to devise the general, efficient method for calculating derivatives (tangent slopes) of curves: starting with an algebraic expression for a secant slope (a "difference quotient"), simplifying that fraction, and finally letting the change in the independent variable become zero. Descartes and others had calculated tangent slopes, but their methods were primarily geometric and not applicable to a wide variety of curves.

Another field in which Fermat did important work was probability theory. In fact, he and Pascal are generally regarded as the cofounders of the subject. (To be fair, Geronimo Cardano laid some important groundwork a century earlier, and Christiaan Huygens made many important improvements to Fermat and Pascal's work in the 1650s.) Fermat and Pascal developed the theory through a very productive correspondence that continued for many years. The impetus for this development was a question about a betting game that had been posed to Pascal.

But Fermat's most important work was in the field of number theory. In addition to proving many theorems in the subject, he also made several brilliant *conjectures* that have stimulated work in number theory ever since.

Finally, in 1994, the theorem was proved by the eminent English mathematician Andrew Wiles. Wiles's proof is complex, deep, and long (about 200 pages!), using many sophisticated methods of modern mathematics that don't have any obvious connection to number theory. It took the world's best number theorists, *including Wiles*, over a year to check the proof and be sure it was correct, and the issue was in severe doubt for many months. Fermat certainly could not have understood this proof, let alone discovered a similar proof on his own. And most mathematicians doubt that Fermat saw a simple proof that thousands of later mathematicians have been unable to recreate. It is considered likely that Fermat saw the proof for the case $n = 3$ and mistakenly thought he had a way to generalize it to all larger values.

Some mathematicians now correctly call this result Wiles's theorem. But the name "Fermat's last theorem" will probably continue to predominate, in part because of the enormous length of time it has been in use and in part as a tribute to Fermat's amazing insight into number-theoretic problems.

Some Famous Conjectures of Number Theory

All the following conjectures are *believed* to be true by almost all mathematicians, and some of them have been verified by computer in many thousands or even millions of cases. Also, various interesting special cases have been proved for many of them.

The first and third conjectures in this list were discussed in Chapter 1.

Goldbach's conjecture (1742): Every even integer greater than 2 can be written as the sum of two primes.

Twin Prime conjecture: There are an infinite number of prime numbers p such that $p + 2$ is also prime.

De Polignac's conjecture: The full version of this conjecture is that every even integer is the difference of two primes *in infinitely many ways*. That is, for every even number n, there are an infinite number of primes p such that $p + n$ is also prime. Note that this also includes the twin prime conjecture as a special case ($n = 2$).

Catalan's conjecture (1842): The numbers 8 and 9 are the only consecutive natural numbers that are both powers (that is, of the form a^n with $n > 1$).

Exercises 8.2

(1) Find the correct quotient and remainder for the following divisions, in accordance with Corollary 8.15:

(a) $7 \div 4$ (b) $(-7) \div 4$
(c) $7 \div (-4)$ (d) $(-7) \div (-4)$

(2) Find the GCD and the LCM of each of the following sets of numbers, using the Euclidean algorithm.

(a) 42 and 1000 (b) 360 and 84
(c) 616 and 27 (d) 24, 60, and 88
(e) 20, 27, and 40 (f) 30, 18, 66, and 40

(3) Repeat Exercise 2 using prime factorizations instead of the Euclidean algorithm, as discussed after Corollary 8.22.

(4) For each pair of numbers given in parts (a) through (c) of Exercise 2, find a linear combination of the numbers that equals their GCD, as is guaranteed by Theorem 8.17(b). You may need to rely primarily on trial and error.

(5) Prove parts (a) through (e) of Theorem 8.13.

(6) (a) Prove the claim, made in the proof of Theorem 8.14, that $n + 1 \in A$.
(b) Prove that the quotient q obtained in this proof is nonnegative.

*(7) Reprove the existence part of Theorem 8.14, using ordinary induction on n instead of Theorem 5.6. This proof is longer than the one given in the text but has the advantage of showing how the quotient and remainder change as n is increased.

(8) Prove Corollary 8.15.

(9) Complete the proof of Theorem 8.16 by showing that $n\mathbb{Z} \subseteq A$.

(10) Prove the claim, made in the proof of Theorem 8.17, that C is a subring of $\mathbb{Z}$.

(11) Prove the formula, given before Example 3, relating LCMs to GCDs.

(12) Prove that every multiple of both m and n is also a multiple of LCM(m, n).

(13) Prove, without using Theorem 8.21 or anything after it, that if n is greater than 1 and nonprime, then $n = km$ for some k and m that are greater than 1 and less than n.

(14) Prove the claim, made in the proof of Theorem 8.21, that a product of prime numbers cannot equal 1.

(15) Prove the inductive formula given for GCDs and LCMs after Example 3.

*(16) Give a more rigorous proof of the uniqueness part of Theorem 8.21, as suggested after the theorem.

*(17) Prove the generalization mentioned after Theorem 8.24.

*(18) Review the rules of the game of Nim, introduced in Exercise 11 of Section 1.2. State and prove a theorem about which player (first or second) has a winning strategy when there are two piles of sticks, with m sticks and n sticks.

(19) Consider a rectangular billiards table (no pockets) of dimensions m units by n units, where m and n are natural numbers. A ball is shot from one of the corners at a 45° angle to the sides, and always reflects at a 45° angle whenever it hits a side.
(a) Prove that the ball will eventually hit a corner of the table.

(b) Determine how far the ball travels and how many times it hits a side of the table before it hits a corner.

(c) Prove that the first corner the ball hits cannot be the one from which it started.

(d) Determine and prove a rule for predicting which corner the ball will first hit, depending on the numbers m and n.

This problem was created by Joseph Becker and appears in the book *Mathematics: A Human Endeavor*, by Harold R. Jacobs (3rd. ed., W. H. Freeman and Co., 1994).

*(20) Consider a round-robin tournament in which n people play each other once in a game or sport like tennis, in which there are no ties. Let's denote by W_i and L_i the number of wins and losses by the ith player. It is clear that $\sum_{i=1}^{n} W_i = \sum_{i=1}^{n} L_i$. Prove the much less obvious fact that $\sum_{i=1}^{n} W_i^2 = \sum_{i=1}^{n} L_i^2$.

Hint: One way to prove this is by **double induction**, an induction proof within an induction proof. The main induction can be on n, the number of players. For the induction step, think of a tournament with $n + 1$ players as a tournament with n players, after which one more player arrives and plays against the original n players. Within the induction step, you can use induction on the number of games the new player wins.

Critique the proofs in the remaining exercises. (If necessary, refer to the instructions for this type of problem in Exercises 4.2.)

(21) **Corollary to Theorem 8.17:** Given any integers m and n (not both zero) and j, j is a linear combination of m and n iff j is a multiple of GCD(m, n).

Proof: The proof of Theorem 8.17(a) shows that C, the set of all linear combinations of m and n, consists of all multiples of GCD(m, n).

(22) **Theorem:** Assuming the results of exercises 11 and 15: for any nonzero integers m, n, and k, $\text{GCD}(m, n, k) \times \text{LCM}(m, n, k) = |mnk|$.

Proof: $\text{LCM}(m, n, k) = \text{LCM}(\text{LCM}(m, n), k)$ By the result of Exercise 15
$= |k|\, \text{LCM}(m, n) \div \text{GCD}(\text{LCM}(m, n), k)$ By the result of Exercise 11
$= |k|\, (|mn| \div \text{GCD}(m, n)) \div \text{GCD}(\text{LCM}(m, n), k)$ Again by Exercise 11
$= |mnk| \div [\text{GCD}(m, n)\, \text{GCD}(\text{LCM}(m, n), k)]$
$= |mnk| \div \text{GCD}(m, n, k)$

*8.3 More Examples of Rings and Fields

Abstract algebra is one of the most important branches of mathematics, and it would be impossible to give more than a very brief introduction to it in this text. Appendix 2 and Section 8.1 attempt to convey some of the flavor of abstract algebra, but they stress familiar number systems such as $\mathbb{Z}$, $\mathbb{Q}$, and $\mathbb{R}$. In this section we give examples of rings and fields that are quite different from the standard number systems, in an attempt to show you how much variety there can be within one type of algebraic structure.

Remember that a field must satisfy several properties beyond what a ring must satisfy. It follows that there is a greater variety of rings than of fields. Yet there is quite a bit of variety even among fields.

Example 1 (A noncommutative ring): Remember that addition in a ring must be commutative by definition but multiplication need not be. There are many ways to define a noncommutative ring. The most important examples (though perhaps not the simplest) are rings of **matrices**.

For any natural number n, let M_n be the set of all n by n matrices with real-valued entries. Then, with the usual addition and multiplication operations on matrices, M_n is a ring with unity; and this ring is noncommutative if $n > 1$.

Showing that M_n is a ring is not difficult, but it is rather long and tedious. For the most part, the properties pertaining to addition are pretty trivial to show, whereas the ones pertaining to multiplication are not. In particular, a somewhat messy computation is required to show that multiplication is associative. Exercise 1 asks you to carry out some of this.

Clearly, the additive identity of M_n is the matrix in which all the entries are 0. This is called the n by n **zero matrix**. Also, the additive inverse of any matrix B is the matrix usually denoted by $-B$, obtained from B by multiplying each entry by -1. It's less obvious what the multiplicative identity is; but if you have ever studied matrices, you probably know that it's the so-called n by n **identity matrix**, which has 1s down the main diagonal and 0s everywhere else. For example, the 3 by 3 identity matrix is

$$\begin{pmatrix} 1 & 0 & 0 \\ 0 & 1 & 0 \\ 0 & 0 & 1 \end{pmatrix}$$

Perhaps you also recall that not every nonzero matrix is **invertible**. In other words, the multiplicative inverse property V-11 does not hold in M_n. For example, the matrix

$$\begin{pmatrix} 1 & 0 \\ 0 & 0 \end{pmatrix}$$

does not have a multiplicative inverse in M_2 (see Exercise 2).

To show that multiplication in M_n is not commutative when $n > 1$, a bit of trial and error should suffice. For example, in M_2 we have

$$\begin{pmatrix} 1 & 0 \\ 0 & 0 \end{pmatrix} \begin{pmatrix} 0 & 1 \\ 0 & 0 \end{pmatrix} \neq \begin{pmatrix} 0 & 1 \\ 0 & 0 \end{pmatrix} \begin{pmatrix} 1 & 0 \\ 0 & 0 \end{pmatrix}$$

Exercise 3 asks you to find examples for larger values of n.

Technically, the rings we have been discussing should be denoted $M_n(\mathbb{R})$, to make it clear that we are allowing real coefficients in these matrices. More generally, we can take any ring A and define $M_n(A)$, the set of all n by n matrices with entries in A. This generalization changes nothing that we've said in the context of real entries: $M_n(A)$ is a ring; if A is a ring with unity, so is $M_n(A)$; and $M_n(A)$ is noncommutative for $n > 1$, regardless of whether A is commutative (provided A has more than one member). We can think of M_n as a method, or **operator**, for defining new rings from old ones.

Example 2 (Rings without unity, commutative and otherwise): A simple example of a ring with no multiplicative identity is the set of all even integers. More generally, for any $n > 1$, Theorem 8.10 shows that $n\mathbb{Z}$ is a ring without unity. The rings $n\mathbb{Z}$ are commutative. Are there also noncommutative rings without unity? The answer is yes, and one simple way to construct one is to combine the ideas in this example and Example 1. For instance, $M_2(2\mathbb{Z})$ is such a ring (see Exercise 4).

Example 3 (Polynomial rings): Examples 3, 4, and 5, like Example 1, are not just examples of specific rings; they also provide important methods for constructing new rings from old ones. For any ring A, let $A[x]$ denote the set of all polynomials in the one variable x, with coefficients in A. This notation looks a lot like function notation, and it may be read "A of x" (or "A bracket x"). But note that $A[x]$ is certainly *not* a function, though it may be viewed as a *set* of functions.

You presumably know what polynomials are and how to add, subtract, and multiply them. Actually, there are two reasonable interpretations of what the word "polynomial" means here. We can view a polynomial, like $x^2 + 3x - 5$, as a function from A to A; or we can view it as just a formal expression, a sequence of symbols. Most of the results about polynomial rings work equally well under either interpretation, but the usual viewpoint in abstract algebra is the second one.

We leave the following results to Exercise 5: for any ring A, the set $A[x]$ is also a ring. Furthermore, if A is commutative, so is $A[x]$; and if A has unity, then so does $A[x]$. These facts are all straightforward to show. One small but important point to remember is that constants count as polynomials. So the additive identity of A works as the additive identity of $A[x]$, and similarly for multiplicative identities. Also, the usual convention is that the variable x automatically commutes with all elements in A, even if A is noncommutative. For example, if a and $b \in A$, then $(ax)(bx) = abx^2$ by definition.

Obviously, it's not very important what letter is used for the variable of our polynomials. In other words, there's no significant difference between $A[x]$ and $A[y]$. On the other hand, we can define structures like $A[x, y]$ and $A[x, y, z]$—sets of polynomials with two or three variables, or more. These are also rings and have many similarities to $A[x]$, but also have some important differences.

Definitions: Two nonzero elements x and y of a ring such that $xy = 0$ are called **zero-divisors**. A commutative ring with unity that has *no* zero-divisors is called an **integral domain**.

Every field is an integral domain, by Exercise 12 of Section 4.5. If A is a field, then there are many similarities between the rings $A[x]$ and $\mathbb{Z}$. Both of them are integral domains but not fields. Both satisfy a unique factorization property similar to Theorem

8.21. And both of them satisfy a property involving division with remainder, similar to Theorem 8.14 (see Theorem 10.10). An algebraist would sum this up by saying that both $A[x]$ and $\mathbb{Z}$ are **Euclidean domains**.

Example 4 (Fields of rational functions): We just said that $A[x]$ is not a field, even if A is. To see this, consider the polynomial x. This is not the zero polynomial, so if $A[x]$ were a field, x would have to have a reciprocal. That is, there would have to be a polynomial $p(x)$ such that $x\,p(x) = 1$. This is impossible (see Exercise 6).

If A is an integral domain, it is not difficult to define a structure similar to $A[x]$ that is a field: let $A(x)$ be the set of all **rational functions** in the variable x with coefficients in A. A rational function is a quotient of two polynomials, with nonzero denominator. As with polynomials, there are two permissible interpretations of this: we can think of a rational function as a function or as a symbolic expression.

However, there is a subtle problem with the latter approach. For example, $1/x$ and $5x/5x^2$ are technically different expressions, but they should not be considered two different rational functions. They should be viewed as two different ways of writing the same rational function. The preferred and most rigorous way of handling this point involves equivalence relations, based on the idea that two numerical fractions like 1/2 and 3/6 may be called equivalent to formalize the fact that they represent the same real number. This idea was introduced in Exercise 8 of Section 6.2.

The relationship between $A[x]$ and $A(x)$ is very similar to that between $\mathbb{Z}$ and $\mathbb{Q}$. In both cases, we take an integral domain, which is a rather well-behaved ring, and obtain a **field of quotients** (with nonzero denominators) from the integral domain. The construction of $\mathbb{Q}$ from $\mathbb{Z}$ in this way is carried out in Section 9.5.

Like the notation $A[x]$, $A(x)$ is also read "A of x," though it does not stand for a function either. This is an example of some genuinely ambiguous mathematical notation; but it's universal, so we're probably stuck with it.

Example 5 (Product rings): Here is another way of constructing new rings. Given any two rings A and B, we can form the cartesian product $A \times B$. Then, we can turn $A \times B$ into an algebraic structure by performing addition and multiplication coordinate by coordinate or pointwise. You are probably familiar with this type of addition, since it is the usual algebraic way of adding vectors; and we are going to do multiplication in the same way. For example, in $\mathbb{R} \times \mathbb{R}$ (or $\mathbb{Z} \times \mathbb{Z}$, $\mathbb{Q} \times \mathbb{R}$, and so on) we have $(3, 5) + (-2, 3) = (1, 8)$, $(3, 5)(-2, 3) = (-6, 15)$, and so on.

The following results are all fairly straightforward to verify (see Exercise 8): for any rings A and B, $A \times B$ (with addition and multiplication defined as above) is also a ring. If A and B are both commutative, then so is $A \times B$; and if A and B both have unity, then so does $A \times B$.

The ring $A \times B$ defined in this manner is called the **direct product** of the rings A and B. Strangely enough, it can also be called the **direct sum** of A and B. It is possible to extend this notion to the product and sum of three or more rings, or even an infinite number of rings. When an infinite number of rings are involved the terms "direct product" and "direct sum" no longer mean the same thing.

The algebra of product rings is pretty straightforward. For example, the additive identity of $A \times B$ is (0, 0), as you might guess. Technically, this notation is a bit sloppy. If A and B are different rings, then the two 0s in this (0, 0) are not the same. The first is the additive identity of A, and the second is the additive identity of B. The same goes for multiplicative identities. Naturally, in $A \times B$ the additive inverse of (x, y) is $(-x, -y)$.

Even if A and B are fields, their product is not. To see this, note that (1, 0) and (0, 1) are both nonzero but their product is (0, 0). In other words, a product ring $A \times B$ always has zero-divisors, as long as A and B each has at least one nonzero element. Thus $A \times B$ cannot be an integral domain, let alone a field. Generally, product rings like $\mathbb{R} \times \mathbb{R}$ or $\mathbb{Z} \times \mathbb{Z}$ are straightforward to work with; but their having zero-divisors makes them quite different from rings like $\mathbb{Z}$, $\mathbb{R}$, and $\mathbb{Z}[x]$.

Example 6: Now we can give examples of two phenomena mentioned after Lemma 8.6. First, consider the ring $A = \mathbb{Z} \times 2\mathbb{Z}$. Since $2\mathbb{Z}$ does not have unity, neither does A. But $\mathbb{Z} \times \{0\}$ is a subring of A that does have unity, because (1, 0) is its multiplicative identity.

Now let B be the ring $\mathbb{Z} \times \mathbb{Z}$. It has unity; (1, 1) is its multiplicative identity. But its subring $\mathbb{Z} \times \{0\}$ has a different multiplicative identity, namely (1, 0).

Modular Arithmetic

We spend the rest of this section studying an extremely important type of ring, for which we give two equivalent definitions. The main concept was discussed a bit in Section 6.2, but we did not go into the algebra involved. In this discussion, n always denotes a positive integer.

Definition: The **ring of integers modulo *n***, denoted $\mathbb{Z}_n$ (not to be confused with $n\mathbb{Z}$!) is defined as follows: the elements of the ring are the integers $0, 1, \dots, n - 1$. (So there are exactly n members.) To add or multiply two numbers in $\mathbb{Z}_n$, first add or multiply them in the usual way. Then divide this result by n. The remainder obtained is the sum or product in $\mathbb{Z}_n$.

Example 7: In $\mathbb{Z}_7$ we have $2 + 3 = 5$, $5 + 4 = 2$, $2 + 5 = 0$, $4 \times 4 = 2$, and $6 \times 2 = 5$. In $\mathbb{Z}_6$, we have $5 + 4 = 3$, $2 \times 2 = 4$, $2 \times 3 = 0$, and $4 \times 5 = 2$.

The algebra of $\mathbb{Z}_n$ is called **modulo *n* arithmetic**. Recall that the equivalence relation $a \equiv b \pmod{n}$ means that $a - b$ is an integer multiple of n. So the calculations in the previous example can be rewritten $2 + 3 \equiv 5 \pmod{7}$, $5 + 4 \equiv 2 \pmod{7}$, $2 + 5 \equiv 0 \pmod{7}$, $2 \times 3 \equiv 0 \pmod{6}$, and $4 \times 5 \equiv 2 \pmod{6}$.

One easy way to think of modular arithmetic is as clock arithmetic. Imagine an ordinary clock with hands, except with the numeral 0 on the top instead of 12. Then everyone would know what you meant if you said that, on the clock, $9 + 3 = 0$ or $10 + 5 = 3$. This is just modulo 12 arithmetic.

The ring $\mathbb{Z}_n$ is commutative, with unity provided $n > 1$. Exercise 11 asks you to prove this and some other basic facts about $\mathbb{Z}_n$.

Alternative Definition: Consider the equivalence relation $a \equiv b \pmod{n}$ on $\mathbb{Z}$. The elements of $\mathbb{Z}_n$ are defined to be the equivalence classes under this relation. Let $[k]$ denote the equivalence class of k. Addition and multiplication in $\mathbb{Z}_n$ are defined by the apparently simple rules $[a] + [b] = [a + b]$ and $[a][b] = [ab]$, for all a and $b \in \mathbb{Z}$. But we soon see that there are pitfalls inherent in such definitions.

Remarks: (1) The set $[k]$ is the set of all integers that differ from k by an integer multiple of n. For example, if $n = 8$, then $[3] = \{\dots, -21, -13, -5, 3, 11, 19, 27, \dots\}$.

(2) Just as before, $\mathbb{Z}_n$ contains exactly n elements. For the n equivalence classes $[0], [1], \dots, [n - 1]$ are all different, and these are the only ones.

(3) Under the alternative definition, the "numbers" of $\mathbb{Z}_n$ are *sets* of numbers (equivalence classes). This takes some getting used to, but it's a very important way of defining new algebraic structures. An algebraic structure whose members are equivalence classes of members of another algebraic structure is called a **quotient structure**. Our alternative definition of $\mathbb{Z}_n$ provides an example of a **quotient ring**.

"Well-Definedness" of Operations on Quotient Structures

What are the pitfalls involved in the second definition of $\mathbb{Z}_n$? The tricky point is that we must define how to add and multiply equivalence classes, not single numbers. To add or multiply two classes, it would seem logical to choose one number from each class, add or multiply them, and then take the equivalence class of the result. For example, suppose we want to add [6] and [8] in $\mathbb{Z}_9$. Since $6 + 8 = 14$, the answer is [14]. But we could also note that $24 \in [6]$, $-55 \in [8]$, and $24 + (-55) = -31$, so the answer is $[-31]$. Does this mean we get two different solutions to the same addition problem? The answer is an emphatic *no*! For $14 \equiv -31 \pmod{9}$, so $[14] = [-31]$. So the two solutions are the same; the simplest way to express the sum might be [5].

☞ The preceding calculation illustrates an extremely important point: when you define algebraic operations in a quotient structure, you have to make sure these operations are **well defined**. The simple rules for addition and multiplication in $\mathbb{Z}_n$ (in the alternative definition) are valid only if the right side of these equations does not change when you pick different members of the equivalence classes on the left side (see Exercise 12).

Can you see how the two definitions of $\mathbb{Z}_n$ are essentially the same? In the first definition, the elements of $\mathbb{Z}_n$ are the numbers $0, 1, \dots, n - 1$. In the second definition, the elements are the equivalence classes $[0], [1], \dots, [n - 1]$. The definitions of addition and multiplication are set up so that, except for the brackets, the first definition and the second definition produce exactly the same algebra. For instance, in $\mathbb{Z}_{10}$, with the first definition we'd have $7 + 5 = 2$ and $7 \cdot 4 = 8$. With the second definition we'd have $[7] + [5] = [2]$ and $[7][4] = [8]$. So the two versions of $\mathbb{Z}_n$ are the same in a *structural* sense. If there's any difference between them, it's just a matter of notation, or the "names" given to the elements. The technical term for this is to say the two versions of $\mathbb{Z}_n$ are **isomorphic**. This important notion is defined and discussed in Section 8.4.

Why have we defined $\mathbb{Z}_n$ in two different ways? Why didn't we just stick with the first, simpler definition? One answer is that defining $\mathbb{Z}_n$ as a quotient structure is of more general and theoretical interest than the first definition. We use quotient structures extensively in Section 9.4. Also, using both definitions allowed us to introduce the concept of isomorphism.

Finite Fields

Now we come to perhaps the most interesting feature of the rings $\mathbb{Z}_n$. If n is composite, $\mathbb{Z}$ has zero-divisors and so is not an integral domain. For example, in $\mathbb{Z}_6$, we have $3 \times 2 = 0$, but neither 3 nor 2 is zero. But now suppose that n is prime. Then there do not exist two natural numbers less than n whose product is a multiple of p, so $\mathbb{Z}_n$ is an integral domain. But we can say more: if p is a prime and m is any natural number less than p, then there's a k such that $mk \equiv 1 \pmod{p}$ (see Exercise 13). Thus, if n is prime, every nonzero number in $\mathbb{Z}_n$ has a reciprocal, so $\mathbb{Z}_n$ is actually a field.

Example 8: In $\mathbb{Z}_7$, we have $1 \times 1 = 1$, $2 \times 4 = 1$, $3 \times 5 = 1$, and $6 \times 6 = 1$. So every nonzero number in $\mathbb{Z}_7$ has a reciprocal. It's easy to carry out similar verifications for other small prime values of n (see Exercise 14).

The fields $\mathbb{Z}_p$ (p being prime) are quite important in abstract algebra. For our purposes, they are being presented primarily as examples of small fields. The only fields we have discussed before these are $\mathbb{Q}$ and $\mathbb{R}$, which are infinite sets. It is interesting to note that there is field with only two elements, one with three elements, and so on. Remember that a field must obey all the usual laws of addition, subtraction, multiplication, and division. A natural question is: if n is not prime, is there a field with n elements? Although we do not have the means to prove the answer to this question, we include it in the following theorem, which also summarizes the basic facts about the rings $\mathbb{Z}_n$.

Theorem 8.25: (a) For any n, $\mathbb{Z}_n$ is a commutative ring. If $n > 1$, it also has unity.

(b) If n is not prime, then $\mathbb{Z}_n$ has zero-divisors and so is not a field, or even an integral domain.

(c) If n is prime, then $\mathbb{Z}_n$ is a field.

(d) For any natural number n, a field with n elements exists iff n equals a prime number to a positive integer power.

(e) Suppose $n = p^m$. Then there is only one field with n elements, structurally speaking. (More technically, any two fields with n elements are *isomorphic*, a term defined in the next section.) This field has **characteristic p**, which means that any number added to itself p times equals zero. This means that if $m > 1$, the field with n elements does *not* look like $\mathbb{Z}_n$, for if $m > 1$, then $p < n$, and so 1 added to itself p times in $\mathbb{Z}_n$ is nonzero.

Proof: (a), (b), and (c) These parts were discussed previously and some of them are assigned in the exercises.

(d) and (e) These proofs are beyond the scope of this book. ■

It is worth taking some time to understand what parts (d) and (e) of Theorem 8.25 say. If n is a prime number, then there is essentially just one field with n elements, which is $\mathbb{Z}_n$. If n is not prime but is a power of a prime, like 4 or 32 or 81, then there is a field with n elements but it does not look like $\mathbb{Z}_n$. For example, in the field with 81 elements, $1 + 1 + 1 = 0$. Exercise 15 asks you to investigate finite fields in more detail. On the other hand, if n has more than one prime factor, like 6 or 10 or 50, then there is no field with n elements.

Exercises 8.3

Several exercises ask you to prove something is a ring. In these proofs, it is important to use your judgment to avoid getting bogged down in petty details. If a certain ring property is particularly obvious, it is fine to say so rather than prove it rigorously. Of course, there are potential pitfalls to this practice, but every mathematics student needs to learn to make this sort of judgment.

(1) (a) Explain briefly why all the field axioms pertaining only to addition hold in the system M_n.

(b) By direct computation, show that field axioms V-4 and V-7 hold in M_2.

(2) Show that the 2 by 2 matrix $\begin{pmatrix} 1 & 0 \\ 0 & 0 \end{pmatrix}$ does not have a multiplicative inverse.

(3) (a) Find 3 by 3 matrices B and C such that $BC \neq CB$.

(b) Generalize part (a) to n by n matrices.

(4) As claimed in Example 3, show that the ring $M_2(2\mathbb{Z})$ is not commutative. Also show that this ring has no multiplicative identity.

(5) Without getting bogged down in too many details, prove the claims about the ring $A[x]$ made in the third paragraph of Example 3.

(6) Every nonzero polynomial has a **degree**—the highest power of x occurring in it, with a nonzero coefficient. (The degree of the zero polynomial is usually left undefined.) What is the degree of the product of two nonzero polynomials? Prove your assertion.

(7) Prove that the polynomial x has no multiplicative inverse in the ring $A[x]$.

(8) Prove the claims made about $A \times B$ in the second paragraph of Example 5.

(9) Explain how the definition of product rings in Example 5 can be generalized to define the ring $(A_1 \times A_2 \times \ldots \times A_n)$, where each A_i is a ring.

(10) (a) Let A be a ring, and B be any *set*. Recall that A^B is the set of all functions from B to A. Since A is a ring, we can add and multiply any two such functions in the usual way (as in Exercise 13 of Section 7.1). Prove that A^B is a ring with these operations. Rings of this type are called **rings of functions** and have many applications.

(b) Under what circumstances, if any, is this ring an integral domain or a field?

(11) Prove that $\mathbb{Z}_n$, as it is first defined in Example 7, is a commutative ring and has unity provided $n > 1$.

(12) (a) Prove that whenever $a \equiv a'$ (mod n) and $b \equiv b'$ (mod n), then $a + b \equiv a' + b'$ (mod n) and $ab \equiv a'b'$ (mod n). This shows that the definitions of addition and multiplication given in the second definition of $\mathbb{Z}_n$ are in fact well defined.

(b) We can also try to define an absolute value function on $\mathbb{Z}_n$ by the rule $|[a]| = [a]$ if $a \geq 0$, and $|[a]| = [-a]$ if $a < 0$. Show that this function is *not* well defined.

(13) Prove the claim made in the discussion of $\mathbb{Z}_n$ that if p is prime and $0 < m < p$, then there is a k such that $km \equiv 1$ (mod p).

(14) Find the multiplicative inverse of every nonzero number in $\mathbb{Z}_5$, $\mathbb{Z}_{11}$, and $\mathbb{Z}_{13}$.

*(15) (a) Work out complete addition and multiplication tables for the field with four elements.

(b) Repeat part (a) for the field with nine elements.

Critique the proofs in the remaining exercises. (If necessary, refer to the instructions for this type of problem in Exercises 4.2.)

(16) **Theorem:** For any ring A and any nonzero f and g in $A[x]$, the degree of $f + g$ is the larger of the degree of f and the degree of g.

Proof: Assume f and g are nonzero polynomials in $A[x]$. Let ax^m and bx^n be the highest power terms in f and g, respectively. If $m > n$, then ax^m is the highest power term in $f + g$. Thus $f + g$, like f, has degree m, which is the larger of the degree of f and the degree of g. Similarly, if $n > m$, then $f + g$, like g, has degree n, which is the larger of the degree of f and the degree of g. Finally, if $m = n$, then $f + g$ has the same degree as both f and g.

(17) **Theorem:** If A is a ring with unity, $x \in A$, and $x^2 = x$, then x must be 0 or 1.

Proof: If $x^2 = x$, then $x^2 - x = 0$; this factors to $x(x - 1) = 0$. Therefore $x = 0$ or $x - 1 = 0$. The latter equation becomes $x = 1$.

*8.4 Isomorphisms

In Section 8.3 we mentioned the notion of isomorphism, which is supposed to express that two algebraic structures are essentially the same. It is now time to make this precise.

Definitions: An **isomorphism** between two rings A and B is a bijection f from A to B such that, for all x and $y \in A$,

$$f(x + y) = f(x) + f(y) \quad \text{and} \quad f(xy) = f(x)f(y)$$

For these equations to make sense, the addition and multiplication operations mentioned on the left sides of the equations must take place in A, and those on the right must take place in B. From now on we usually do not mention this point.

Two rings are said to be **isomorphic** iff there is an isomorphism between them.

Example 1: Let $A = \mathbb{R} \times \mathbb{Z}$, and $B = \mathbb{Z} \times \mathbb{R}$, product rings as defined in the previous section. Then A and B are isomorphic, an isomorphism (the only one) being given by $f((x, y)) = (y, x)$. More generally, for any rings C and D, the product rings $C \times D$ and $D \times C$ are isomorphic (see Exercise 1).

Example 2: The ring $\mathbb{Z}_6$ is isomorphic to $\mathbb{Z}_2 \times \mathbb{Z}_3$. An isomorphism is given by $f(x) = (a, b)$, where a and b are the remainders when x is divided by 2 and 3, respectively. For example, $f(3) = (1, 0)$ and $f(5) = (1, 2)$. Exercise 6 asks you to verify that f is an isomorphism, and Exercise 7 gives a generalization of this example.

Remarks: (1) It takes a while to get a feel for what it means for two rings to be isomorphic. First of all, an isomorphism is a bijection, so the rings are in one-to-one correspondence. But we have additional conditions that say that any algebraic relationship among elements in A is preserved when these elements are mapped to B. For example, if A is commutative, B must also be commutative (and conversely). Mathematicians like to think of isomorphic rings as structurally identical. They also like to view an isomorphism as a renaming; except for this renaming of elements, isomorphic rings have exactly the same addition and multiplication tables.

The example of an isomorphism mentioned in Section 8.3 illustrates these ideas. There we had two versions of $\mathbb{Z}_n$; one had members $\{0, 1, \ldots, n - 1\}$, the other $\{[0], [1], \ldots, [n - 1]\}$. The function f from the first $\mathbb{Z}_n$ to the second, defined by $f(x) = [x]$, is clearly a bijection. But it is easy to show (see Exercise 5) that it is also an isomorphism. Therefore, except for the way the elements are named—one with brackets, the other without—there is no difference between these two versions of $\mathbb{Z}_n$.

(2) If the word "onto" in the above definition is relaxed to "into," then f is called an **isomorphic embedding**. In this situation, it is straightforward to show (see Exercise 2) that the range of f must be closed under addition and multiplication; this makes it a subring of B. In other words, an isomorphic embedding from A to B is the same as an isomorphism between A and a subring of B.

If we weaken the definition further by dropping the words "one-to-one," then f is called a **homomorphism.**

(3) The concept of isomorphism is useful for every type of algebraic structure but must be modified accordingly. The type of isomorphism we have defined is **ring isomorphism**, since it specifies conditions for addition and multiplication. In contrast, a group has only one algebraic operation defined on it, so a **group isomorphism** would need to satisfy only one algebraic condition. On the other hand, if we have two *ordered* rings, we could have an **ordered ring isomorphism** between them. This would have to be a ring isomorphism and also be **order preserving**: whenever $x < y, f(x) < f(y)$.

Theorem 8.26: Assume f is an isomorphism between rings A and B. Then:

(a) $f(0) = 0$

(b) For every $x \in A, f(-x) = -f(x)$.

(c) A is commutative iff B is.

(d) A has unity iff B does; and, in this case, $f(1) = 1$.

(e) For every $x \in A$, x has a multiplicative inverse iff $f(x)$ does; and, in this case, $f(x^{-1}) = [f(x)]^{-1}$.

(f) Thus, A is a field iff B is.

Proof: We prove just parts (a) and (b) and leave the rest for Exercise 3.

(a) More precisely, we must show that $f(0_A) = 0_B$. But $f(0_A) = f(0_A + 0_A) = f(0_A) + f(0_A)$. Subtracting $f(0_A)$ from both sides of this equation yields $0_B = f(0_A)$.

(c) Assume A is commutative, and x and y are any members of B. Since f is onto B, we have $f(u) = x$ and $f(v) = y$ for some u and $v \in A$. So $xy = f(u)\, f(v) = f(uv) = f(vu) = f(v) f(u) = yx$, as desired. ■

Theorem 8.27: (a) For any ring A, id_A is an isomorphism between A and A.

(b) The inverse of any isomorphism is an isomorphism.

(c) The composition of any two isomorphisms is an isomorphism.

(d) Therefore, being isomorphic is an equivalence relation on the class of all rings.

Proof: See Exercise 4. ■

Example 3: Using Theorem 8.26 *and* careful thinking about algebraic properties, one can often quickly see that particular rings are *not* isomorphic. For instance, $\mathbb{Q}$ and $\mathbb{Z}$ are not, because only $\mathbb{Q}$ is a field. Also, $\mathbb{Z}$ and $2\mathbb{Z}$ are not, because only $\mathbb{Z}$ has unity (but see Exercise 8).

We can also see that $\mathbb{R}$ and $\mathbb{Q}$ are not isomorphic, simply because $\mathbb{R}$ has greater cardinality than $\mathbb{Q}$. A more algebraic argument might use the fact (proved in Chapter 9) that only $\mathbb{R}$ has a square root of 2. Let's show this in more detail: assume f is an isomorphism between $\mathbb{R}$ and $\mathbb{Q}$. Then $f(1) = 1$, by Theorem 8.26(d). So $f(2) = f(1+1) = f(1) + f(1) = 1 + 1 = 2$. Now, since $\sqrt{2}\,\sqrt{2} = 2$, $f(\sqrt{2})f(\sqrt{2}) = f(2) = 2$. This would mean that $f(\sqrt{2})$ is a square root of 2 in $\mathbb{Q}$, and this violates Theorem 8.24.

The ability to see reasons why rings are not isomorphic comes with experience (see Exercises 11 and 12).

Theorems 8.9 and 8.12 are very similar results, characterizing $\mathbb{Z}$ and $\mathbb{Q}$ as special subsystems of $\mathbb{R}$. Using the terminology of isomorphisms, we can prove important generalizations of these theorems, after proving two preliminary lemmas.

Lemma 8.28: Let A be any ring with x and $y \in A$. Then:
(a) $(-x)y = x(-y) = -(xy)$
(b) $(-x)(-y) = xy$
Proof: See Exercise 9. ■

Lemma 8.29: Let A be an ordered ring with unity. Then:
(a) A has no zero-divisors.
(b) If $x \in A$ and $x^2 = x$, then $x = 0$ or $x = 1$.
(c) If B is a subring of A with unity, then B has the same multiplicative identity as A.

Proof: (a) Assume x and $y \in A$, both nonzero. We must prove that $xy \neq 0$. We do this by cases, using axiom V-17: if x and $y > 0$, we have $xy > 0y = 0$, so $xy \neq 0$. If $x < 0$ and $y > 0$, then $xy < 0y = 0$, so $xy \neq 0$. The case $x > 0$ and $y < 0$ is similar. Finally, if x and $y < 0$, then $-x$ and $-y$ are positive (why?). We have $xy = (-x)(-y)$ by Lemma 8.28(b), and $(-x)(-y) > 0$ by the first case of this proof. So $xy > 0$, and thus is nonzero.

(b) If $x^2 = x$, then $0 = x^2 - x = x(x-1)$. By (a), $x = 0$ or $x - 1 = 0$. So $x = 0$ or $x = 1$.

(c) Let B be any subring of A with unity. By Lemma 8.6(a), B has the same additive identity as A, which we continue to denote 0. If c is the multiplicative identity of B, then $c^2 = c$. By part (b), $c = 0$ or $c = 1$. The former case is ruled out by axiom V-12, so $c = 1$, as desired. ■

Theorem 8.30: For any ordered ring with unity A, there is a unique isomorphism (in fact, an ordered ring isomorphism) between $\mathbb{Z}$ and a subring of A. This subring is also the smallest subring of A with unity.

Proof: We define $f: \mathbb{Z} \to A$. First of all, we define f on $\mathbb{N}$, inductively: $f(1) = 1_A$; and for the induction step, $f(n+1) = f(n) + 1_A$. We also let $f(0) = 0_A$, and $f(-n) = -f(n)$, $\forall n \in \mathbb{N}$. This completes the definition of f. Since A is ordered, $0_A < 1_A$, and therefore $f(n) < f(n+1)$, $\forall n \in \mathbb{N}$. An easy induction then shows that if $0 < m < n$, then $0_A < f(m) < f(n)$. Using several cases, we can extend this to yield that whenever m and $n \in \mathbb{Z}$ and $m < n$, then $f(m) < f(n)$ (see Exercise 13). In other words, f is order preserving (and therefore it is also one-to-one).

We must still prove that f preserves addition and multiplication. To show that $f(m+n) = f(m) + f(n)$, the easiest way is to first assume m and $n > 0$, and prove it by induction on m or n. The extension to zero and negative values of m and/or n is again easily handled by cases. The details of this argument, and the one for multiplication, are also left for Exercise 13. Thus f is an isomorphic embedding, as desired.

To prove f is unique, assume g is any isomorphic embedding from $\mathbb{Z}$ to A. We show $g = f$. Let B be the range of g. Then B is a subring of A. By Theorem 8.26(d), B has unity and $g(1) = 1_B$. But Lemma 8.29(c) says that $1_B = 1_A$, so $g(1) = 1_A = f(1)$. A simple induction then yields that $g(n) = f(n)$, $\forall n \in \mathbb{N}$. By Theorem 8.26(a) and Lemma 8.6(a), $g(0) = 0_A = f(0)$. Finally, for any $n \in \mathbb{N}$, Theorem 8.26(b) and Lemma 8.6(b) yield $g(-n) = -g(n) = -f(n) = f(-n)$. Thus $g(n) = f(n)$ for all integers n; that is, $g = f$.

Finally, let C be any subring of A with unity. Since $1_C = 1_A$, C contains $f(1)$. The same reasoning as in the previous paragraph shows that C contains $f(n)$, $\forall n \in \mathbb{N}$; $f(0)$; and $f(-n)$, $\forall n \in \mathbb{N}$ (see Exercise 13). Thus $\text{Rng}(f) \subseteq C$. This shows that $\text{Rng}(f)$ is the smallest subring of A with unity. ■

There are various ways to describe the meaning of this theorem. We can say that every ordered ring with unity contains a *copy* of $\mathbb{Z}$. Mathematicians might also say, "Every ordered ring with unity contains $\mathbb{Z}$, up to isomorphism." The words "up to isomorphism" may even be omitted, but technically they should not be.

The simplest way to paraphrase the last part of the theorem is that $\mathbb{Z}$ is the smallest ordered ring with unity. Similarly, the next theorem tells us that every ordered field contains a copy of $\mathbb{Q}$ and that $\mathbb{Q}$ is the smallest ordered field.

Theorem 8.31: For any ordered field A, there is a unique isomorphism (in fact, an ordered ring isomorphism) between $\mathbb{Q}$ and a subfield of A. This subfield is also the smallest subfield of A.

Proof: This proof is a minor modification or extension of the proof of Theorem 8.30. We want to define an isomorphic embedding $f\colon \mathbb{Q} \to A$. The definition of f on $\mathbb{Z}$ is precisely as before. We then want to extend f to the rest of $\mathbb{Q}$ by the obvious rule $f(m/n) = f(m)/f(n)$. However, we have to be careful about this, because one rational number can be represented by many different fractions. We have to show that this rule can't give more than one output for a single rational number: note that if m/n and a/b are equal fractions, then $mb = na$. Therefore, since f is an isomorphism on $\mathbb{Z}$, $f(m)f(b) = f(n)f(a)$ in A. We thus obtain $f(m)/f(n) = f(a)/f(b)$, so $f(m/n) = f(a/b)$. So f is indeed a *function* on $\mathbb{Q}$. *[This is really a well-definedness argument of the type discussed in the previous section.]* The rest of the modifications needed to complete this proof are left for Exercise 14. ■

Corollary 8.32: Every ordered field with no proper subfields is isomorphic to $\mathbb{Q}$. (Therefore, any two such ordered fields are isomorphic to each other.)

Proof: Let A be an ordered field with no proper subfields. Let $f\colon \mathbb{Q} \to A$ be the isomorphic embedding of Theorem 8.31. Then $\text{Rng}(f)$ must be a subfield of A (see Exercise 2). Thus it is all of A, so $\mathbb{Q}$ and A are isomorphic. The claim in parentheses follows from Theorem 8.27(d). ■

The fact that $\mathbb{Q}$ has no proper subfields was already established by Theorem 8.12. What Corollary 8.32 adds to this is that $\mathbb{Q}$ is, up to isomorphism, the *unique* ordered field with no subfields. Similarly, we could use Theorem 8.30 to conclude that $\mathbb{Z}$ is, up to isomorphism, the unique ordered ring with unity with no subrings with unity.

Exercises 8.4

(1) Prove the general assertion made about product rings in Example 1.

(2) Show that the range of an isomorphic embedding must be a ring, as asserted in the second remark of this section.

(3) Prove the remaining parts of Theorem 8.26.

(4) Prove Theorem 8.27.

(5) Prove the claim, made in the first remark of this section, that f is an isomorphism between the two versions of $\mathbb{Z}_n$.

(6) Show that the function f defined in Example 2 is an isomorphism between $\mathbb{Z}_6$ and $\mathbb{Z}_2 \times \mathbb{Z}_3$.

*(7) Generalizing Example 2, prove that if m and n are positive and relatively prime, then $\mathbb{Z}_{mn}$ is isomorphic to $\mathbb{Z}_m \times \mathbb{Z}_n$.

(8) Show that the function $f: \mathbb{Z} \to 2\mathbb{Z}$ defined by $f(n) = 2n$ satisfies all the conditions of isomorphism except the one involving multiplication. So we say that $\mathbb{Z}$ and $2\mathbb{Z}$ are isomorphic *as groups under addition* but not isomorphic as rings (as pointed out in Example 3).

(9) Prove Lemma 8.28. Be careful not to assume any more about A than the fact that it is a ring.

(10) Show that Lemma 8.29 fails if the word "ordered" is deleted. In fact, find a single ring with unity in which all three parts are false.

(11) Explain why each of the following pairs of rings is *not* isomorphic. ***Hint:*** You might want to refer to Example 3 for some ideas.

(a) $2\mathbb{Z}$ and $\mathbb{Z}_2$
(b) $\mathbb{Z}$ and $\mathbb{Z} \times \mathbb{Z}$
(c) $\mathbb{Z}_2 \times \mathbb{Z}_2$ and $\mathbb{Z}_4$
(d) $m\mathbb{Z}$ and $n\mathbb{Z}$, where $0 < m < n$
(e) $\mathbb{Z}$ and $\mathbb{Z}[x]$

*(12) Determine whether each of the following pairs of rings is isomorphic. If so, specify an isomorphism between the rings. If not, explain why not.

(a) $\mathbb{Z} \times \mathbb{Z}$ and the ring of **diagonal** 2 by 2 matrices with integer coefficients (that is, matrices whose top right and bottom left entries are 0)

(b) $\mathbb{Z} \times \mathbb{Z}$ and $\mathbb{Z} \times \mathbb{Q}$

(c) $\mathbb{Z} \times 4\mathbb{Z}$ and $2\mathbb{Z} \times 2\mathbb{Z}$

(d) The function rings $\mathbb{R}^{\mathbb{N}}$ and $\mathbb{R}^{\mathbb{Z}}$, as defined in Exercise 10 of Section 8.3

*(13) Prove the three omitted steps mentioned in the proof of Theorem 8.30.

*(14) Complete the proof of Theorem 8.31, making appropriate modifications to the proof of Theorem 8.30.

(15) Prove that if f is a bijection from a *ring* A to a *set* B, then there is a unique way to define addition and multiplication on B to make f a ring isomorphism.

(16) The function f defined by $f(n) = 2n$ is a bijection between $\mathbb{Z}$ and E, the set of even integers. Use f and the previous exercise to define a ring structure on E that makes the resulting ring isomorphic to $\mathbb{Z}$.

(17) The function f defined by $f(x) = e^x$ is a bijection between $\mathbb{R}$ and $\mathbb{R}^+$, the set of positive reals. Use f and Exercise 15 to define a ring structure on $\mathbb{R}^+$ that makes the resulting ring isomorphic to $\mathbb{R}$.

*(18) Let C be a fixed set. Given any A and $B \subseteq C$, define AB to be $A \cap B$, and $A + B$ to be $A \,\Delta\, B$, as defined in Exercise 15 of Section 5.3.

(a) Show that these operations make $\wp(C)$ into a boolean ring (see Exercise 17, Section 8.1).

(b) Prove that this ring is isomorphic to the function ring $(\mathbb{Z}_2)^A$.

Suggestions for Further Reading: Two excellent introductory texts on abstract algebra are Fraleigh (1994) and Herstein (1996). For more thorough coverage of number theory, see Burn (1982), Burton (1997), Niven and Zuckerman (1980), or the classic Hardy and Wright (1979). A very clear treatment of the integers as an algebraic structure is given in Feferman (1989). For a short, well-written outline of Wiles's proof of Fermat's last theorem, see the November 1997 *Scientific American.*

Chapter 9

The Real Number System

9.1 The Completeness Axiom

The development of calculus in the seventeenth century was certainly one of the most important events in the history of mathematics. Not only does calculus provide methods and solutions for a wide variety of problems in mathematics, science, and engineering; many other branches of mathematics invented since then also owe their existence, at least partially, to problems that arose in calculus. In particular, the need to clarify the concept of limits led, in the eighteenth and nineteenth centuries, to an intensive study of the real number system, functions, sequences, and finally sets. The rigorous study of the foundations of calculus and the real number system is called **real analysis**. In this chapter we examine some basic concepts of this important subject.

Completeness of the Real Number System

We have already discussed some of the similarities and differences between the system of rational numbers $\mathbb{Q}$ and the system of real numbers $\mathbb{R}$. The major similarity is algebraic: they are both ordered fields. One of the main differences was mentioned in Section 7.5: $\mathbb{R}$ has greater cardinality than $\mathbb{Q}$. Theorem 8.24 indicates a more concrete difference: numbers like $\sqrt{2}$ cannot be in $\mathbb{Q}$. In this chapter we prove that such numbers are in $\mathbb{R}$.

In this section we pinpoint the special property of $\mathbb{R}$ that sets it apart from $\mathbb{Q}$. We begin with some pictures, since this special property is more geometric than algebraic. If you were asked to draw a picture of the set of real numbers, you would know exactly what to do. We all learn that the real number system can be represented as a straight line, extending without limit in both directions (see Figure 9.1). This representation makes sense because our intuition and experience tell us that any possible length or distance corresponds to a positive real number. So if we draw a line and then choose an origin, a unit of length, and a direction on the line to be considered positive, it's reasonable to believe that the points on the line are in one-to-one correspondence with the set of all real numbers (with the bijection given by the directed distance from the origin).

If you were then asked to draw a picture of the set of integers as a subset of $\mathbb{R}$, this would also be no problem. The integers are spaced one unit apart from each other on the

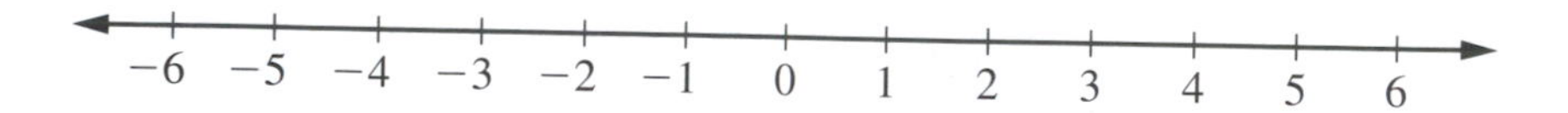

Figure 9.1 The usual real number line

number line, so it's easy to single them out. (Technically, $\mathbb{Z}$ is called a **discrete** subset of $\mathbb{R}$; that means every number in it is **isolated**—separated from all other numbers in the set by a positive distance, as shown in Figure 9.2.) But now suppose you were asked to show the set of all rationals as a subset of a number line. How would you do that? There's no way to do it. Within every interval in $\mathbb{R}$, there are an infinite number of rational numbers but also an infinite number of irrationals (see Exercises 13 and 14). So there are rational numbers "all over" the real line, but there are also "holes" in $\mathbb{Q}$, all over. The purpose of this discussion is to pinpoint the difference between $\mathbb{Q}$ and $\mathbb{R}$. Intuitively, it's simply that while $\mathbb{Q}$ is full of holes, $\mathbb{R}$ has no holes in it; it's a complete line. The axiom that states this property of $\mathbb{R}$ rigorously is number V-18 in our axiom system. It is the one real number axiom that is not included in the ordered field axioms.

In the second half of the nineteenth century, when many of the world's best mathematicians were attempting to develop a rigorous theory of real numbers and functions, several equivalent forms of the completeness axiom were postulated. We need to define a few concepts before we can actually discuss the axiom.

Definitions: Let $A \subseteq \mathbb{R}$, and $x \in \mathbb{R}$. We say that x is an **upper bound** of A iff x is equal to or greater than every number in A; that is, $\forall y \in A\ (x \geq y)$. Also, a set of real numbers is called **bounded above** iff it has an upper bound. Similarly, x is a **lower bound** of A iff $\forall y \in A\ (x \leq y)$, and a set of real numbers is called **bounded below** iff it has a lower bound. A set that is both bounded above and bounded below is simply called **bounded**.

Recall that the word "bounded" has already been defined for subsets of $\mathbb{N}$, in Section 7.5. Every subset of $\mathbb{N}$ is bounded below (by the number 1, for instance), so for these sets "bounded" means "bounded above."

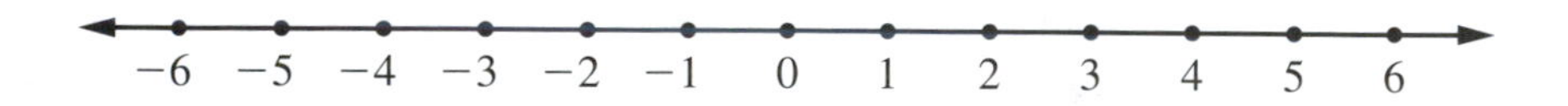

Figure 9.2 $\mathbb{Z}$, a discrete, infinite subset of $\mathbb{R}$

Example 1: The interval $[-5, 3)$ is bounded. The number 3 is an upper bound of this set, and so is any greater number. Similarly, -5, -73, and -5.2 are some lower bounds of this set. More generally, every interval that does not involve the symbols ∞ or $-\infty$ is bounded. Recall that intervals that do involve those symbols are *called* unbounded, because they are in fact unbounded.

Example 2: The set $\mathbb{R}$ is *not* bounded above or below. No number x is an upper bound of $\mathbb{R}$, because $x < x + 1$. Similarly, $\mathbb{R}$ has no lower bounds either.

Definitions: Let $A \subseteq \mathbb{R}$, and $x \in \mathbb{R}$. We say that x is **the least upper bound** of A or **the supremum** of A (written $x = \text{LUB}(A)$ or $x = \text{Sup}(A)$) iff x is an upper bound of A and x is less than every other upper bound of A. Similarly, x is **the greatest lower bound** of A or **the infimum** of A (written $x = \text{GLB}(A)$ or $x = \text{Inf}(A)$) iff x is a lower bound of A and x is greater than every other lower bound of A.

Clearly, a set can't have more than one supremum or infimum (recall Exercise 16(a) of Section 6.3). This justifies the use of the word "the" in these definitions.

Example 3: Let $A = \{-2, 0, 3, 5.2, 8\}$. Then 8 is an upper bound of A, and so is any number greater than 8; but no number less than 8 is an upper bound of A. Therefore $\text{Sup}(A) = 8$. Similarly, $\text{Inf}(A) = -2$.

Let's generalize this simple example.

Theorem 9.1: Assume $A \subseteq \mathbb{R}$.

(a) If A has a largest (respectively, smallest) member, then that member of A is $\text{Sup}(A)$ (respectively, $\text{Inf}(A)$).

(b) If A is a nonempty finite set, then A has a largest member and a smallest one. Therefore, A has a least upper bound and a greatest lower bound.

Proof: (a) Obvious.

(b) Best proved by induction on the cardinality of A (see Exercise 2). ■

Example 4: Let A be the empty set. Then every number is both an upper bound and a lower bound of A. It follows that A does not have a supremum or an infimum.

Example 5: The set $\mathbb{N}$ has a least member, namely 1. Therefore, $1 = \text{Inf}(\mathbb{N})$. But we soon see (Theorem 9.4) that $\mathbb{N}$ is not bounded above, and so has no supremum.

Example 6: Let A be the set of all negative real numbers. Then A is unbounded below; that is, it has no lower bounds. But 0 is an upper bound of A, and so is any positive number. Since there is no largest negative number (why?), no negative number is an upper bound of A. It follows that $\text{Sup}(A) = 0$. Exercise 3 asks you to show the details of this. Note that $\text{Sup}(A)$ is not in A. This is a common phenomenon.

Completeness axiom: If a nonempty set of real numbers has an upper bound, then it has a least upper bound.

An ordered field satisfying the completeness axiom is called **complete.** So the entire group V of axioms can be summarized as saying that $\mathbb{R}$ is a complete ordered field. It can be proved that *every* complete ordered field is isomorphic to $\mathbb{R}$. Essentially, that means $\mathbb{R}$ is the *only* complete ordered field.

The completeness axiom is supposed to express rigorously the idea that there are no holes in the real number system. How does it say that? To see this, imagine some set of real numbers A that is bounded above (see Figure 9.3). Now think of an upper bound of A as a sort of wall, to the right of the entire set A. Since A has many upper bounds, we can think of the wall as *movable.* We can move the wall as much as we please to the right and still have an upper bound of A. But we can only move the wall a certain amount to the left and still have an upper bound of A. When the wall is moved as far as it can be to the left without going to the left of any member of A, the wall is at Sup(A), which may or may not be in A. Geometrically, the completeness axiom says that wherever this wall ends up in this process, there's a real number at that point. In other words, the real numbers are everywhere on the line.

After the next simple technical fact, we demonstrate the power of the completeness axiom to prove the existence of irrational numbers in $\mathbb{R}$.

Theorem 9.2: The completeness axiom is equivalent (using the rest of the axiom system) to this statement: every nonempty set of real numbers with a lower bound has a greatest lower bound.

Proof: For the forward direction, assume the completeness axiom. Let A be any set of reals with a lower bound. Define $-A$ to be $\{-x \mid x \in A\}$. If u is any lower bound of A, it is easy to show that $-u$ is an upper bound of $-A$. So, by completeness, $-A$ has a supremum y. It is then easy to show that $-y$ is the greatest lower bound of A. The details of this proof are left for Exercise 4.

The proof of the reverse direction is similar. ■

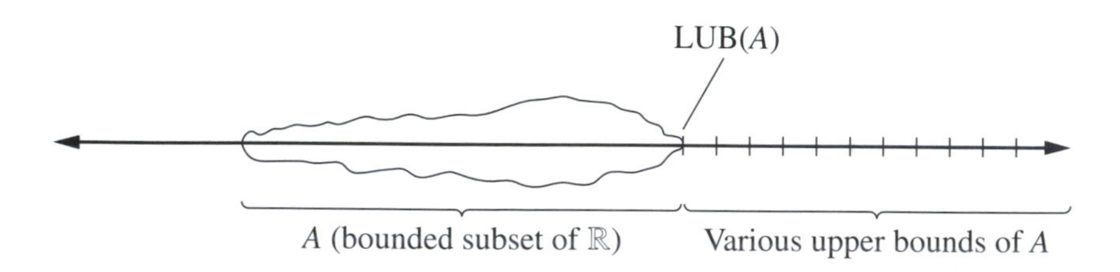

Figure 9.3 Illustration of the completeness axiom

Theorem 9.3: The number $\sqrt{2}$ is in $\mathbb{R}$. That is, there's a real $z > 0$ such that $z^2 = 2$.

Proof: Let A be the set of all positive real numbers whose square is less than 2. Then A is bounded above, for instance by 2 (see Exercise 6). So by the completeness axiom, A has a supremum; call it z. Our goal is to show that $z = \sqrt{2}$.

Since A contains positive numbers like 1, z is clearly positive. We must also show that $z^2 = 2$. We do this by trichotomy and indirect proof; that is, we show that each of the inequalities $z^2 < 2$ and $z^2 > 2$ is impossible. In order not to get bogged down in the details of the proof to follow, let's outline the main idea behind it: what makes the proof work is that the function $f(x) = x^2$ is **continuous**. Essentially that says that when x is changed by a small amount, the quantity x^2 also does not change by very much.

First assume that $z^2 < 2$. We show there is a number larger than z that is in A. Let $c = 2 - z^2$. We know $z \geq 1$, so $c \leq 1$. Let $d = c/3z$. Then d is positive, and we have

$$\begin{aligned}(z + d)^2 &= (z + c/3z)^2 = z^2 + 2c/3 + (c/3z)^2 \\ &\leq z^2 + 2c/3 + (c/3)^2 \leq z^2 + 7c/9 < z^2 + c = 2\end{aligned}$$

So $(z + d)^2 < 2$; this means that $z + d \in A$. But this would mean that z is not an upper bound of A, contradicting the assumption.

Now assume that $z^2 > 2$. We show there is a number less than z that is an upper bound of A. This time, let $c = z^2 - 2$, and then let $d = c/2z$. Both c and d are positive. We have $(z - d)^2 = (z - c/2z)^2 = z^2 - c + d^2 > z^2 - c = 2$. We claim that $z - d$ is an upper bound of A. Let x be any member of A. Then $x^2 \leq 2$, by definition of A, and we just showed that $(z - d)^2 > 2$. It follows easily (see Exercise 6) that $x < z - d$. Since x was an arbitrary member of A, $z - d$ is an upper bound of A. But this contradicts the fact that z is the *least* upper bound of A.

Since both $z^2 < 2$ and $z^2 > 2$ are impossible, we have $z^2 = 2$. ∎

The proof of Theorem 9.3 is involved, but the ideas used are fairly simple. It would be a good idea to study it until you are certain that you understand it. The theorem can be generalized to show that for any positive real number x and any natural number n, $\sqrt[n]{x}$ is a real number. The next result is another important consequence of the completeness axiom.

Theorem 9.4: The set $\mathbb{Z}$ is unbounded in both directions.

Proof: We first show, indirectly, that $\mathbb{Z}$ is unbounded above. So assume $\mathbb{Z}$ has an upper bound. Then it has a supremum; call it c. Since $c - 1 < c$, $c - 1$ cannot be an upper bound of $\mathbb{Z}$. In other words, for some integer k, $k > c - 1$. But then $k + 1 > c$, and $k + 1$ is also an integer. This contradicts the assumption that c is an upper bound of $\mathbb{Z}$. So $\mathbb{Z}$ is unbounded above. Similar reasoning shows it's unbounded below (see Exercise 7). ∎

The statement of Theorem 9.4 is one version of the **Archimedean property**, and an ordered field in which $\mathbb{Z}$ is unbounded is called **Archimedean**. So the theorem says that $\mathbb{R}$ is Archimedean. So is $\mathbb{Q}$, since it is a subfield of $\mathbb{R}$. The original form of the Archimedean property, an axiom for the ancient Greeks, is the subject of Exercise 8.

Archimedes (287?–212 B.C.) was the most productive mathematician of ancient Greece and one of the greatest of all time. He introduced a variety of brilliant and innovative methods and advanced many branches of mathematics far beyond the work of his contemporaries. Among other things, he calculated (with rigorous proofs) areas and volumes of numerous planar regions, surfaces, and solids with curved boundaries. This work became the foundation for the theory of integration some nineteen hundred years later.

Archimedes was also a brilliant engineer who devoted much energy to designing war machines for King Hieron of Syracuse to use against the siege by the Romans under Marcellus. It was reported that he used a system of mirrors reflecting sunlight to set Roman ships on fire, but most historians doubt this. Another anecdote has it that Archimedes, while at a public bathhouse, suddenly saw how he could use principles of water displacement to determine whether Hieron's crown was pure gold. He then ran home, naked, shouting, "Eureka" ("I have found it").

Even Archimedes's death was extraordinary. When the Romans finally overtook Syracuse, he was apparently drawing geometric figures in the sand, oblivious to the war going on around him. When a Roman soldier walked toward him, he reportedly said, "Don't disturb my circles," upon which the enraged soldier killed him. Marcellus, who had greatly admired Archimedes, built him a tomb displaying the figure of a sphere inscribed inside a cylinder.

Decimal Representation of Real Numbers

A computer or calculator will tell you that $\sqrt{2} = 1.4142\ldots$. What exactly does this mean? You may know that the decimal expansion of $\sqrt{2}$ or of any irrational number never terminates and never goes into a permanent repeating pattern. So there is no way to evaluate $\sqrt{2}$ numerically, except to give its decimal expansion to a finite number of places, followed by "...". Let's now use the completeness axiom to analyze the meaning of such decimals more precisely.

Consider again the set used in the proof of Theorem 9.3: $A = \{x \in \mathbb{R} \mid x > 0$ and $x^2 \leq 2\}$. Since $1^2 = 1$ and $2^2 = 4$, $1 \in A$ but $2 \notin A$. So $\sqrt{2}$ is between 1 and 2. Going now to the first decimal place, we find that $(1.4)^2 = 1.96$, and $(1.5)^2 = 2.25$. So $1.4 \in A$ but $1.5 \notin A$. So $\sqrt{2}$ is between 1.4 and 1.5; that is, $\sqrt{2} = 1.4\ldots$. Increasing the accuracy to the next decimal place, it turns out that $1.41 \in A$ but $1.42 \notin A$; so we can say that $\sqrt{2} = 1.41\ldots$. Now let $B = \{1, 1.4, 1.41, 1.414, \ldots\}$, the set of all the decimal approximations to $\sqrt{2}$. What we've just said is that $B \subseteq A$, but none of the numbers obtained by adding 1 to the last decimal place of any member of B is in A. More important, it can be shown that Sup(B) = Sup(A) (see Exercise 9); and we just proved that Sup(A) = 2.

In summary, when we say that a real number equals some positive nonterminating decimal, this technically means that the number is the supremum of all the truncated (cut-off) approximations to that decimal. (For a negative decimal, change "supremum" to "infimum.") We continue our discussion of decimals in the next section.

Exercises 9.1

(1) Determine, with brief explanation (as opposed to proof), the supremum and infimum of the following sets of real numbers if they exist.

(a) The set of all negative numbers
(b) $\mathbb{Q} \cap [3, 8]$
(c) $\{x \mid x^2 \leq 7\}$
(d) $\{x \mid x^2 < 7\}$
(e) $\{x \mid x^3 < 7\}$
(f) $\{1/n \mid n \in \mathbb{N}\}$

(2) Prove Theorem 9.1.

(3) Fill in the details of the claim, in Example 6, that Sup(A) = 0.

(4) (a) Complete the proof of the forward direction of Theorem 9.2.
(b) Prove the reverse direction of Theorem 9.2.

(5) Show that if the symbol $\geq$ is changed to $>$ in the definition of an upper bound, then the completeness axiom becomes false.

(6) (a) Prove the claim in the proof of Theorem 9.3 that 2 is an upper bound of A.
(b) Prove the later claim in the proof of Theorem 9.3 that $x < z - d$.

(7) Complete the proof of Theorem 9.4 by showing that $\mathbb{Z}$ is unbounded below.

(8) The classical statement of the Archimedean property is: Given any two positive real numbers a and b, there is a natural number n such that $na > b$. Without using completeness, prove the equivalence of this statement with Theorem 9.4.

(9) Prove the claim that Sup(A) = Sup(B) in the last paragraph of this section.

(10) Prove that for any $x \in \mathbb{R}$ and $A \subseteq \mathbb{R}$, x = Sup(A) iff x is an upper bound of A and, for every $u > 0$, there is a member of A that is greater than $x - u$.

(11) Prove or disprove: if Sup(A) = x and Sup(B) = y, then Sup($A - B$) = $x - y$.

(12) For any A and $B \in \mathbb{R}$, let $A + B = \{x + y \mid x \in A \text{ and } y \in B\}$. Prove or disprove: If Sup(A) = a and Sup(B) = b, then Sup(A + B) = $a + b$. (Note that, in contrast to the definition of $A - B$, the definition of $A + B$ is based on addition of numbers.)

*(13) (a) Prove that every interval in $\mathbb{R}$ contains a rational number.
(b) Using the result of part (a), show that every interval in $\mathbb{R}$ in fact contains an infinite set of rational numbers.

*(14) Prove that every interval in $\mathbb{R}$ contains an infinite set of *irrational* numbers. You may use the result of Exercise 13, whether or not you did it. Also, you may use the fact that if r is rational and nonzero, then $r\sqrt{2}$ is irrational.

9.2 Limits of Sequences and Sums of Series

Recall the discussion of sequences in Section 7.4. In this chapter, the word "sequence" always refers to a sequence of real numbers, unless stated otherwise.

Intuitively, when we say that a sequence a has a limit L, we mean that the terms a_n eventually get "arbitrary close" to L. But this is not at all rigorous. It took mathematicians a long time to find a correct definition of this concept, but eventually this was achieved: to say that the limit of a is L should mean that, given any target interval around L, no matter how small, there is a term in the sequence beyond which *all* terms are inside that target interval. Here is the rigorous statement.

Definitions: For any sequence a and real number L, we say that $\lim_{n \to \infty} (a_n) = L$, or simply $\lim (a_n) = L$ or $a_n \to L$, iff $\forall \varepsilon > 0\ \exists m\ \forall n > m\ (|a_n - L| < \varepsilon)$.

(The use of the Greek letter ε (epsilon) in this definition is standard; it has nothing to do with the modified epsilon (∈) we are using to denote set membership.)

The notation $\lim_{n \to \infty} (a_n) = L$ is read "the **limit** as n approaches infinity of a_n is L." The notation $a_n \to L$ is read "a_n **approaches** L" or "a_n **converges** to L."

A sequence that has a limit is called **convergent**; otherwise, it's called **divergent**.

Every mathematician needs to understand this definition, so you should probably spend some energy on it now. The variable ε represents a distance extending left and right from the number L, forming a target interval. The variable m represents a term in the sequence beyond which all terms are within a distance ε of L. Note that the quantifier structure of this definition specifies that m is to be found in terms of ε, or, if you wish, as a *function* of ε. The smaller ε is, the farther you may need to go in the sequence before the terms stay within that distance of L. But if the terms do approach L, it should be possible to find such an m for any positive ε, no matter how small.

Example 1: Consider first a *constant* sequence, like $a_n = 5$. We claim that $\lim (a_n) = 5$. To show this, given any positive number ε, just let $m = 1$. Since all the terms of this sequence equal 5, it is trivial that all the terms after the first are within a distance ε of 5. Note that we can say that a_n approaches 5, even though it seems strange to say that numbers that *are* 5 are *approaching* 5.

Example 2: Consider the sequence (1, 1/2, 1/3, 1/4, ...) given by the formula $a_n = 1/n$. It should be easy to guess that this sequence converges to 0. To prove this, let $\varepsilon > 0$ be given. By Theorem 9.4, there is an m such that $m > 1/\varepsilon$. Pick such an m. Then if $n > m$, we have $n > 1/\varepsilon$, which becomes $1/n < \varepsilon$. We also know that $1/n > 0$ (by Theorem A-13), so $|a_n - 0| = |a_n| = |1/n| = 1/n < \varepsilon$. This shows that $a_n \to 0$.

Example 3: Recalling the discussion of $\sqrt{2}$ in the previous section, consider the sequence (1.4, 1.41, 1.414, 1.4142, ...), in which the nth term is $\sqrt{2}$ to n decimal places, rounded down or truncated. (To state this more rigorously, a_n is the largest fraction whose numerator is an integer, whose denominator is 10^n, and whose square is less than 2.) It should be obvious that this sequence converges to $\sqrt{2}$. In fact, by its definition, the difference between a_n and $\sqrt{2}$ is no more than 10^{-n}. Exercise 11 asks you to prove the limit of this sequence is $\sqrt{2}$.

We have now seen three ways to think of an irrational number like $\sqrt{2}$ numerically: as a decimal; as the supremum (or infimum) of a set, as in Theorem 9.3; or as the limit of a sequence. It can also be thought of as an infinite sum, as we see shortly.

Example 4: Consider the sequence (1, 2, 1, 2, ...). This is a typical example of an **oscillating sequence**. It can be defined by the formula $a_n = [3 + (-1)^n]/2$. You might think this sequence should have two limits, 1 and 2. But it has no limit. To prove this, we must find ε in terms of any L such that the terms never stay within a distance ε of L. Since successive terms are always 1 apart, $\varepsilon = 1/2$ works for any L (see Exercise 5). This example illustrates the following result.

Theorem 9.5 (Uniqueness of limits): A sequence cannot have more than one limit.

Proof: Assume, on the contrary, that some sequence (a_n) has two limits L_1 and L_2, with $L_1 > L_2$. Let $\varepsilon = (L_1 - L_2)/2$. By the assumption, there exist m_1 and m_2 representing how far you have to go in the sequence before the terms stay within a distance ε of L_1 and L_2, respectively. But now let n be any integer greater than both m_1 and m_2. Then a_n is within a distance ε of both L_1 and L_2. Figure 9.4 shows the impossibility of this, but an algebraic argument is more rigorous: note that $2\varepsilon = L_1 - L_2 = (L_1 - a_n) + (a_n - L_2)$. But both of the expressions in parentheses have absolute value less than ε; this implies they are less than ε. This violates the triangle inequality, Theorem A-15(b). ■

It would be easy to fill a book discussing the theory of limits of sequences and series. We restrict ourselves to a few of the major results.

Definition: A sequence (a_n) is called **bounded above** iff its range, that is the set $\{a_1, a_2, a_3, ...\}$, is bounded above as a subset of $\mathbb{R}$. Similarly, a sequence may be called **unbounded above, bounded below, bounded** (in both directions), and so on.

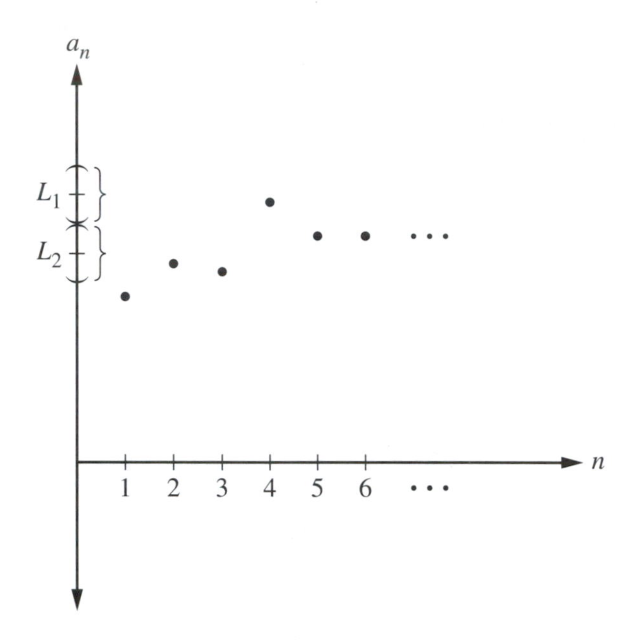

Figure 9.4 Illustration of the proof of Theorem 9.5

Theorem 9.6: If a sequence converges, it must be bounded.

Proof: Suppose that (a_n) converges to L. Then, using $\varepsilon = 1$ in the definition of what this means, choose an m such that all the terms after the mth term are between $L - 1$ and $L + 1$. Now consider the finite set $\{a_1, a_2, \dots, a_m, L - 1, L + 1\}$. By Theorem 9.1(b), this set has a smallest member and a largest member; let's call them b and c, respectively. It is now easy to show (see Exercise 4) that b is a lower bound for the range of (a_n) and c is an upper bound for it. ■

The converse of Theorem 9.6 is false, as the oscillating sequence of Example 4 demonstrates. But the next theorem shows that the converse holds under the additional assumption that the sequence does not oscillate at all.

Definitions: A sequence is called **increasing** iff each term (after the first) is equal to or greater than the one before it; symbolically, $\forall n\ (a_{n+1} \geq a_n)$. A sequence is **decreasing** iff $\forall n\ (a_{n+1} \leq a_n)$. A sequence is **monotone** iff it's increasing or decreasing.

These definitions allow two or more terms of a sequence to be equal. For example, a constant sequence is both increasing and decreasing. The terms **strictly increasing**, **strictly decreasing**, and **strictly monotone** may be used to exclude equal terms. It is clear (and easily proved by induction on n) that if a is increasing, then $a_n \geq a_m$ whenever $n > m$; and the analogous condition holds for decreasing sequences.

Theorem 9.7 (Monotone convergence theorem): If a sequence is monotone and bounded, it must be convergent.

Proof: Let (a_n) be a bounded, monotone sequence. So (a_n) could be monotone increasing or monotone decreasing. We do the case of a monotone increasing sequence and leave the other case for Exercise 6.

Since (a_n) is bounded, the set $\{a_n | n \in \mathbb{N}\}$ has an upper bound. Therefore, by completeness, it has a supremum; call it L. We claim that L is also the limit of the sequence (a_n). To prove this, let $\varepsilon > 0$ be given. There must be a term a_m of the sequence that is larger than $L - \varepsilon$; otherwise $L - \varepsilon$ would be an upper bound of the set of terms, and this would contradict the fact that L is its supremum. But if $a_m > L - \varepsilon$, then for every $n > m$, we also have $a_n > L - \varepsilon$, since the terms never get smaller. Also, for every n we have $a_n \leq L$, which implies that $a_n < L + \varepsilon$. So we have that for every $n > m$, $|a_n - L| < \varepsilon$. Since ε was arbitrary, we have $\lim (a_n) = L$. ■

Definition: The sum, difference, and product of any two sequences are formed in the obvious term-by-term way. We can also define the quotient of two sequences provided all terms in the denominator are nonzero. (Since a sequence is by definition a function from $\mathbb{N}$ to $\mathbb{R}$, these combinations of sequences are technically combinations of functions, as defined in Exercise 13 of Section 7.1.)

The next important technical result says that limits of sequences are *preserved under* these ways of forming new sequences. Part (a) says that the limit of a sum of convergent sequences equals the sum of their individual limits, and so on.

Theorem 9.8: Suppose (a_n) and (b_n) are sequences, $\lim a_n = L_1$, and $\lim b_n = L_2$.

(a) $\lim (a_n + b_n) = L_1 + L_2$

(b) $\lim (a_n b_n) = L_1 L_2$

(c) $\lim (k a_n) = kL_1$, for any constant k

(d) $\lim (a_n - b_n) = L_1 - L_2$

(e) $\lim (a_n / b_n) = L_1/L_2$, provided that L_2 and all of the b_ns are nonzero

Proof: (a) Given $\varepsilon > 0$, $\varepsilon/2$ is also a positive number. Therefore, since $a_n \to L_1$, there is an m_1 such that for any $n > m_1$, $|a_n - L_1| < \varepsilon/2$. Similarly, there is an m_2 such that for any $n > m_2$, $|a_n - L_2| < \varepsilon/2$. Let m be the larger of m_1 and m_2. Then, for any $n > m$,

$$\begin{aligned} |(a_n + b_n) - (L_1 + L_2)| &= |(a_n - L_1) + (b_n - L_2)| \\ &\leq |a_n - L_1| + |b_n - L_2| \qquad \text{By the triangle inequality} \\ &< \varepsilon/2 + \varepsilon/2 = \varepsilon \end{aligned}$$

(b) Let $\varepsilon > 0$ be given. We must find m such that for any $n > m$, $|a_n b_n - L_1 L_2| < \varepsilon$. Note that $|a_n b_n - L_1 L_2| = |a_n b_n - L_1 b_n + L_1 b_n - L_1 L_2| \leq |a_n b_n - L_1 b_n| + |L_1 b_n - L_1 L_2|$, by the triangle inequality. So, as in (a), it suffices to find m such that for any $n > m$, each of the last two expressions between the absolute value symbols must be less than $\varepsilon/2$.

We deal with the second expression first. We have $|L_1 b_n - L_1 L_2| = |L_1(b_n - L_2)| = |L_1||b_n - L_2|$, by Theorem A-15(a). This is a product in which the first factor is constant and the second factor gets small; this means that the product must also get small. More rigorously, $\varepsilon/[2(|L_1| + 1)]$ is positive, so we can find m_1 such that, for any $n > m_1$, $|b_n - L_2| < \varepsilon/[2(|L_1| + 1)]$. But then $|L_1||b_n - L_2| \leq (|L_1| + 1)|b_n - L_2| < \varepsilon/2$, as desired.

Now consider the first expression. As in the previous paragraph, we can write $|a_nb_n - L_1b_n| = |b_n(a_n - L_1)| = |b_n||a_n - L_1|$, by Theorem A-15(a). Now the first factor of our product is no longer constant, but the fact that it converges is good enough: by Theorem 9.6, the sequence b_n is bounded. That means there's a number M such that $|b_n| \leq M$, for all n. Then we can find a number m_2 such that for any $n > m_2$, $|a_n - L_1| < \varepsilon/2M$. It follows that for any $n > m_2$, $|b_n||a_n - L_1| < \varepsilon/2$.

Finally, let m be the maximum of m_1 and m_2. For any $n > m$, both $|a_nb_n - L_1b_n|$ and $|L_1b_n - L_1L_2|$ are less than $\varepsilon/2$, as desired.

(c) This follows immediately from part (b), letting (a_n) be the constant sequence all of whose values are k. For then L_1 also equals k.

(d) Since $a_n - b_n = a_n + (-1)b_n$, this follows from parts (a) and (c).

(e) First consider the special case in which all the a_ns equal 1. The statement becomes $\lim (1/b_n) = 1/L_2$, with the same restrictions as in the general case. We leave the proof of this for Exercise 9.

Now note that $a_n/b_n = a_n(1/b_n)$. Therefore, the general case of (e) follows from (b) and the special case just considered. ∎

Remarks: (1) It is instructive to think about why the proof of part (b) is so much harder than that of part (a). In part (a) we are adding two quantities, each of which can be off from its limit by some amount. Think of the absolute value of the difference between each quantity and its limit as the error in that quantity. When we add two quantities, the error in the sum can be at most the sum of the two errors, by the triangle inequality. Therefore, if we want a total error less than ε, it suffices to make sure that each separate error is less than $\varepsilon/2$.

In part (b) we are dealing with a product, and it is simply not the case that the error in a product can be at most the product (or the sum) of the two errors. To get a feel for this situation, see Figure 9.5. It is natural to illustrate a product as the area of a rectangle. Figure 9.5 shows the case where both limits are positive and so are both errors. Then L_1L_2 is the area of the smaller rectangle, while a_nb_n is the area of the larger one. The difference is the shaded L-shaped region, and in the proof of part (b), we divided it up into two parts as shown. The proof shows that, if we are allowed to make the linear errors $|a_n - L_1|$ and $|b_n - L_2|$ as small as we please, we can also make the L-shaped area smaller than any given positive number.

One similarity in the proofs of (a) and (b) is that they are both "$\varepsilon/2$ proofs." Instead, we could have divided the L-shaped region into three rectangles and given an "$\varepsilon/3$ proof" for (b). This sort of terminology is frequently used in the study of limits.

(2) In part (e) of the theorem, we don't really need to specify that all the b_ns are nonzero. Instead we can simply agree to form the quotient sequence only with terms a_n/b_n for which b_n is nonzero. Since L_2 is nonzero, there can be only a finite number of zero terms in the sequence b (why?). So we still get an *infinite* sequence of quotients even if these terms are omitted.

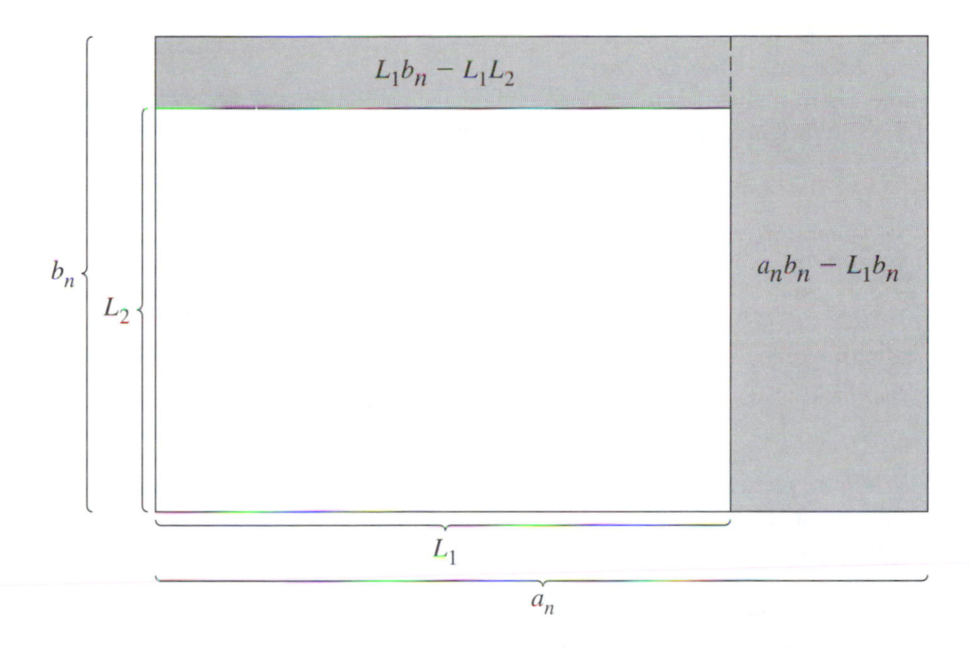

Figure 9.5 The error in a product shown as an area (Theorem 9.8(b))

We conclude this subsection by listing several equivalents of the completeness property, all of which have already been mentioned. In fact, all parts of this theorem are proved earlier in this section or in Section 9.1, except for the converse of the monotone convergence theorem. Another interesting equivalent of completeness is discussed at the end of this section.

Theorem 9.9: In any ordered field (that is, using the ordered field axioms), the following statements are equivalent.

(a) The completeness property.
(b) Every nonempty set of numbers with a lower bound has an infimum.
(c) Every bounded, increasing sequence converges.
(d) Every bounded, decreasing sequence converges.

Proof: Theorem 9.2 states that (a) and (b) are equivalent, and Theorem 9.7 states that (a) implies (c) and (d). It remains only to prove that (c) and (d), separately, imply (a) or (b). We outline the proof that (c) implies (a); the proof that (d) implies (b) is similar.

Assume (c), and let A be a nonempty set with an upper bound. Let b be any member of A, and c any upper bound of A. Using these numbers, construct a bounded, increasing sequence whose limit must be Sup(A). Specifically, let a_n be the largest number of the form $b + (c - b)k/10^n$ such that $k/10^n \leq 1$ and a_n is *not* greater than *every* member of A. *[Think of $k/10^n$ as an n-place decimal between 0 and 1. For example, a_2 is the largest percent of the way we can go from b to c without reaching a number that is greater than every number in A.]*

We claim that (a_n) is a bounded, increasing sequence and that its limit must be Sup(A). The proofs of these claims are left for Exercise 10. ■

Infinite Series and Decimals (Optional Material)

Definition: Let (a_n) be any sequence. When we refer to the **series** $a_1 + a_2 + a_3 + \ldots$ or $\sum_{n=1}^{\infty} a_n$ or simply $\sum a_n$, we mean the new sequence (s_n), called the **sequence of partial sums** of the sequence (a_n), defined by $s_n = \sum_{i=1}^{n} a_i$; that is,

$$s_1 = a_1\,,\quad s_2 = a_1 + a_2\,,\quad s_3 = a_1 + a_2 + a_3\,,\quad \text{and so on.}$$

When we refer to the **sum of the series** $\sum a_n$ or simply the value of $\sum a_n$, we mean the limit of the sequence (s_n). Similarly, when we say that the series $\sum a_n$ converges or diverges, we mean that the sequence (s_n) does so.

☞ There is some genuine ambiguity in the terminology and notation introduced in these definitions. If (a_n) is a sequence, does the notation $\sum a_n$ denote a new sequence (the sequence of partial sums) or a single number (the *limit* of the sequence of partial sums)? It can mean either, but in practice it probably means a single number most of the time. For example, if (b_n) were another sequence and we wrote $\sum a_n = \sum b_n$ we certainly would *not* mean that the two sequences of partial sums are equal, term by term. Rather, we would be saying that their limits were equal.

Example 5: Suppose we are asked to evaluate the series $\sum 2^{-n}$, that is, $1/2 + 1/4 + 1/8 + 1/16 + \ldots$. By the definition of the sum of a series, this is the same as finding the limit of the sequence (1/2, 3/4, 7/8, 15/16, ...). It appears obvious that the limit is 1, and we shortly prove that this is so.

If you have taken a couple of years of calculus, you probably studied infinite sequences and series. You may remember that it can be quite difficult to determine whether a series converges and even harder to determine its sum if it does converge. One of the few types for which it's easy to evaluate the sum is the geometric series.

Theorem 9.10: (a) If $|r| < 1$, then $\lim r^n = 0$.

(b) If $|r| < 1$, the geometric series $\sum ar^{n-1}$ converges to $a/(1 - r)$.

Proof: (a) We proceed by cases, using trichotomy. If $r = 0$, all the terms are 0; this makes the limit 0, as in Example 1.

Next, assume $r > 0$. An easy induction on n shows that the sequence (r^n) is then positive and strictly decreasing. So by the monotone convergence theorem, the sequence has a limit L, which is also the infimum of the terms. Since the terms are all positive, 0 is a lower bound for them, and therefore $L \geq 0$. Let's now show L can't be positive: if

$L > 0$, consider the number L/r. Since $0 < r < 1$, $L/r > L$. Therefore, since L is the greatest lower bound of the terms of the sequence, L/r cannot be a lower bound. So, for some n, $r^n < L/r$. Now multiply both sides by r to obtain $r^{n+1} < L$. This contradicts the assumption about L. Since this shows L is not positive, the only remaining possibility is $L = 0$.

Finally, assume $r < 0$. Now we have to be a bit more careful because the sequence (r^n) is not monotone; the terms alternate sign. So let p be the positive number $|r|$, or $-r$. We still have $|p| < 1$, so the argument of the previous paragraph applies: p^n converges to 0. But for every n, $r^n = p^n$ or $r^n = -p^n$ (easily proved by induction). So $|r^n - 0| = |p^n - 0|$. Therefore, r^n also converges to 0.

(b) We must show that the sequence (s_n) converges to $a/(1 - r)$, where s_n is the sum of the first n terms of the series. But Theorem 4.14 says that $s_n = a(1 - r^n)/(1 - r)$. By part (a), as $n \to \infty$, $r^n \to 0$; so it seems obvious that $s_n \to a/(1 - r)$. To prove it rigorously we use Theorem 9.8: we have $r^n \to 0$. Therefore, $1 - r^n \to 1 - 0 = 1$; $a(1 - r^n) \to a$; and $a(1 - r^n)/(1 - r) \to a/(1 - r)$. ∎

Example 6: Returning to Example 9, we see that $1/2 + 1/4 + 1/8 + 1/16 + \ldots$ is geometric, with $a = r = 1/2$. Therefore, the infinite sum is $(1/2)/(1 - 1/2) = 1$. Similarly, $1 - 1/2 + 1/4 - 1/8 + \ldots$ is an **alternating** geometric series, with $a = 1$ and $r = -1/2$. Therefore, by the formula, the infinite sum is $2/3$.

We now resume our discussion of decimal numbers. Earlier we said that a decimal like 1.4142... can be viewed as the supremum of the set of truncated decimals 1, 1.4, 1.41, ... Equivalently, it can be viewed as the limit of the sequence of these truncated decimals. Finally, it's clear from our definition of series that it can also be viewed as the infinite sum $1 + 0.04 + 0.01 + 0.004 + \ldots$. This last approach is the way we normally think of decimals. It can also be used to convert *repeating* decimals to fractions.

Example 7: Evaluate the following decimals as fractions:

(a) 0.55555 ...
(b) 0.99999 ...
(c) 0.37373737 ...
(d) 2.3628282828 ...

Solution: (a) $0.55555 \ldots = 0.5 + 0.05 + 0.005 + \ldots$, a geometric series with $a = 0.5$ and $r = 0.1$. By Theorem 9.9, this equals $0.5/(1 - 0.1) = 5/9$.

(b) Similarly, we find that $0.99999 \ldots = 9/9 = 1$.

(c) $0.37373737 \ldots = 0.37 + 0.0037 + .000037 + \ldots$, a geometric series with $a = 0.37$ and $r = .01$. So its value is $0.37/(1 - 0.01) = 37/99$.

(d) To turn this into a fraction, we must split it up as $2.36 + 0.0028282828 \ldots$. The second part of this is a geometric series with $a = 0.0028$ and $r = 0.01$. So its value is $28/9900$ or $7/2475$. If desired, we can then turn the expression $2.36 + 7/2475$ into a single fraction.

The technique used in Example 7 can be generalized to show that *every* repeating decimal (that is, a decimal in which some finite sequence of digits repeats itself forever) is a rational number. The converse is also true: every fraction has a decimal expansion that is repeating. (A terminating decimal like 3.74 is considered to be repeating.) More specifically, a fraction with denominator m has a decimal form whose repeating part is less than m digits long (see Exercise 15).

Example 7(b) is interesting because it shows that a single number can have two different decimal forms: $1 = 0.99999 \ldots$. There are plenty of numbers like this, for instance, $2.58 = 2.5799999 \ldots$. But it turns out that the only way this occurs is that any number with a terminating decimal also has a decimal form ending in an infinite string of 9s (see Exercise 16).

Let's now summarize, without proofs, the main points we've made about decimals.

Theorem 9.11: (a) Every real number has a decimal form.

(b) Every decimal expression denotes a unique real number.

(c) If we exclude decimal expressions ending in an infinite string of 9s, we can add the words "in a unique way" at the end of part (a); this means there is a one-to-one correspondence between real numbers and the included decimal expressions.

(d) A number is rational iff its decimal form is repeating.

Parts (a), (b), and (c) of Theorem 9.11 require the completeness axiom for their proof. Loosely, (b) says that there are no holes in $\mathbb{R}$, and (a) says $\mathbb{R}$ is archimedean. Conversely, taken together, (a) and (b) imply the completeness axiom. Thus, we could add the conjunction of (a) and (b) or (a) and (c) to the list of equivalents of completeness given in Theorem 9.9.

We have also mentioned that every complete ordered field is isomorphic to $\mathbb{R}$. Theorem 9.11 provides a very powerful way of viewing this: up to isomorphism, $\mathbb{R}$ is *the only* ordered field whose members correspond perfectly to decimal expressions.

Exercises 9.2

(1) Modify the argument used in Example 2 to show that the sequence defined by $a_n = 1/\sqrt{n}$ converges to 0.

(2) For each of the following sequences, decide whether it converges or diverges. If it converges, determine its limit and write a formula for m in terms of ε, as in the definition of limits. If it diverges, write a formula for ε in terms of L that shows there is no limit. You need not prove your assertions.

(a) $a_n = (-2/3)^n$ (b) $a_n = (1/2)^n + (3/4)^n$

(c) $a_n = \sin(\pi n/2)$ (d) $a_n = (\cos n)/n$

(3) Give an example of each of the following, or explain briefly why none exists.

(a) A sequence that is unbounded above and unbounded below

(b) A decreasing sequence of positive numbers that diverges

(c) A convergent sequence whose terms alternate sign

(d) A sequence of irrational numbers that converges to a rational number

(e) A sequence of rational numbers that converges to an irrational number

(4) Prove the claims made in the last sentence of the proof of Theorem 9.6.

(5) Give a more rigorous proof that the sequence defined in Example 4 has no limit, by carefully *negating* the symbolic statement that the sequence *does* have a limit.

(6) Prove Theorem 9.7 for decreasing sequences. You may use the result for increasing sequences.

(7) Using Theorem 9.8, prove that if (a_n) diverges and (b_n) converges, then:
(a) $(a_n + b_n)$ diverges.
(b) If $\lim (b_n)$ is nonzero, then $(a_n b_n)$ diverges.
(c) If $k \neq 0$, then (ka_n) diverges.

(8) This exercise is also related to Theorem 9.8. Assume that (a_n) and (b_n) both diverge. Prove or disprove (by a counterexample) each of the following.
(a) $(a_n + b_n)$ must diverge. (b) $(a_n b_n)$ must diverge.

(9) Prove the special case of Theorem 9.8(e) that was not proved in the text.

(10) Prove the claims in the last sentence of the proof of Theorem 9.9.

*(11) Prove that the sequence defined in Example 3 converges to $\sqrt{2}$.

(12) For each of the following, determine whether it's a convergent geometric series. If it is, evaluate its sum.
(a) $1/2 - 1/6 + 1/18 - 1/54 + \dots$ (b) $-5 + 5 - 5 + 5 - \dots$
(c) $\sum \frac{1}{n}$ (d) $\sum \frac{1}{2^n + 1}$

(13) Determine the infinite sum of the telescoping series $\sum \frac{1}{n^2 + n}$. You will first need to do Exercise 13 of Section 4.5.

(14) Rewrite each of the following decimals as a fraction or a sum of two fractions:
(a) 2.10101010 ... (b) 327.27272727 ...

(15) Prove that a fraction with denominator m has a decimal form whose repeating part is less than m digits long.

(16) (a) Prove that every terminating decimal equals a decimal that ends in an infinite string of 9s.
*(b) Show that the situation described in part (a) is the only way that one number can have two different decimal representations.

*(17) Prove the result, mentioned in Section 7.5, that $\mathbb{R}$ has the same cardinality as $\wp(\mathbb{N})$ and therefore is uncountable.

9.3 Limits of Functions and Continuity

Section 9.2 discusses limits of *sequences*. Although the theory of sequences and series is an important part of real analysis, it is not as central as the study of the limits of *functions*. In particular, calculus is totally based on this notion. Throughout this section, the letters f, g, and h denote functions whose domains and ranges are subsets of $\mathbb{R}$.

Definition: If f is a function and a and L are real numbers, we say $\lim_{x \to a} f(x) = L$ (read, "the **limit** as x approaches a of $f(x)$ is L") iff

$$\forall \varepsilon > 0\ \exists \delta > 0\ \forall x\ (0 < |x - a| < \delta \rightarrow |f(x) - L| < \varepsilon)$$

Remarks: (1) If you understand the definition of the limit of a *sequence*, you will be able to understand this definition, too. In both definitions the letter ε defines a target interval around the number L. You can also think of ε as the error or tolerance by which you're allowed to miss L. The major difference between the two definitions is the domains involved. Remember that a sequence is technically a function whose domain is $\mathbb{N}$; the letter n denotes an arbitrary input and a_n denotes the corresponding output. When we write $\lim a_n = L$, it means that no matter how small a target we pick around L, the terms of the sequence are eventually all in the target interval if n is allowed to get large enough.

Now we are talking about functions whose domains consist of real numbers. For these, the most important type of limit is the type in which the domain value gets closer and closer to some fixed value, rather than becoming very large (but see Exercise 4). So when we write $\lim_{x \to a} f(x) = L$, it means that no matter how small a target interval we pick around L, the values of the function are eventually all in the target interval, if the domain variable x is allowed to get close enough to a.

(2) Note the symbols "0 <" in the definition of the limit of a function. They play an important role. Without these symbols, the hypothesis of the implication would always be true if $x = a$, no matter what value is chosen for δ. The effect of omitting those symbols would be that the condition $\lim_{x \to a} f(x) = L$ would imply $f(a) = L$ (see Exercise 2). But as the definition is written, the hypothesis of the implication is false if $x = a$; this means that the implication is automatically true. The result is that the statement $\lim_{x \to a} f(x) = L$ says *nothing* about $f(a)$. That is, $f(a)$ does not have to equal L, be close to L, or even be defined. But we soon see that for "reasonable" functions, $\lim_{x \to a} f(x)$ and $f(a)$ never have different values.

(3) The condition $|f(x) - L| < \varepsilon$ in the limit definition is meant to imply, automatically, that $f(x)$ is defined. (This is sensible; for example, if I say "My brother is tall," this statement cannot be true unless my brother exists.) The net result of this is that if $\lim_{x \to a} f(x)$ exists, there must be some interval centered at a in which $f(x)$ is defined at every value of x, except possibly at a.

Example 1: Define $f: \mathbb{R} \to \mathbb{R}$ by $f(x) = 2x + 5$. We claim that $\lim_{x \to 3} f(x) = 11$. (Note that $f(3) = 11$, which is no coincidence, since this f is a "reasonable" function.) To show that a function has a particular limit, it's necessary to find δ in terms of ε. So let $\varepsilon > 0$ be given. We claim $\delta = \varepsilon/2$ works. To prove this, assume $0 < |x - 3| < \delta$. Then

$$|f(x) - 11| = |(2x + 5) - 11| = |2x - 6| = 2\,|x - 3| < 2(\varepsilon/2) = \varepsilon$$

Why did we choose $\delta = \varepsilon/2$ in this argument? Figure 9.6 shows why. The graph of $y = 2x + 5$ is a straight line through the point (3, 11), with a slope of 2. We are given a target interval extending from $11 - \varepsilon$ to $11 + \varepsilon$, which may be drawn on the y axis. How close does x need to be to 3 to keep the function values in this interval? Well, since the

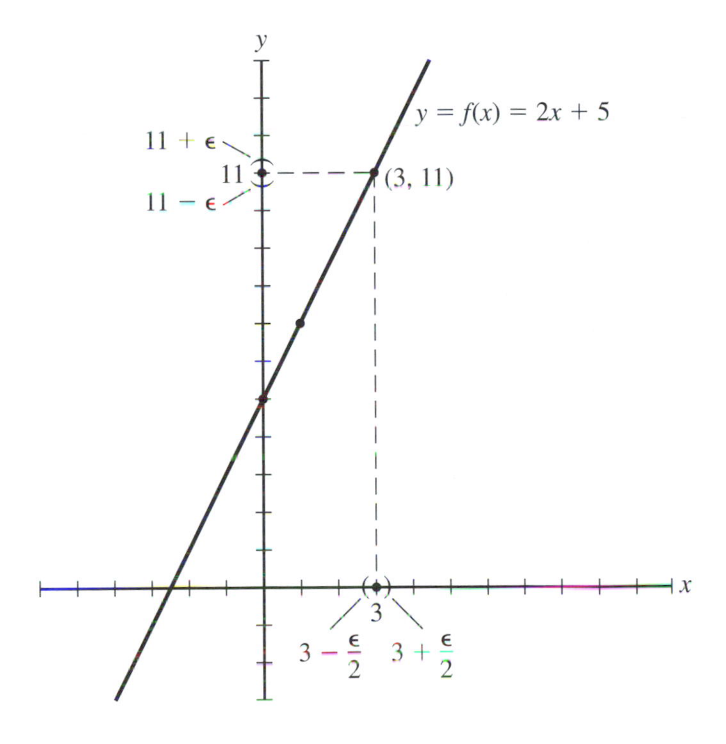

Figure 9.6 Illustration of Example 1

slope of the line is 2, y changes twice as fast as x. Therefore, if we want to keep y within a certain distance of 11, we should keep x within half that distance of 3.

Example 2: Now define a function $g: \mathbb{R} \to \mathbb{R}$ by cases:

$$g(x) = \begin{cases} 2x+5 & \text{if } x \neq 3 \\ 8 & \text{if } x = 3 \end{cases}$$

The graph of g (Figure 9.7) is the same as the graph of f, except that the point (3, 11) has been moved to (3, 8). But although $g(3) \neq 11$, we still have $\lim_{x \to 3} g(x) = 11$. The proof of this is just like the argument in the previous example, except now we must also use the fact that if $0 < |x - 3| < \delta$, then $x \neq 3$ (see Exercise 3).

Example 3: Now define a function h that is the same as the functions f and g of Examples 1 and 2, except that $h(3)$ is *undefined*. So its graph is a straight line with a single point missing. Then we still have $\lim_{x \to 3} h(x) = 11$. The proof is as in Example 2.

Actually, the three functions f, g, and h don't just have the same limit as x approaches 3; they have the same limit as x approaches any number. Theorem 9.12 generalizes this.

Theorem 9.12: Suppose $f(x)$ and $g(x)$ are functions that differ at only a finite number of values of x. (These may be values of x at which the two functions have

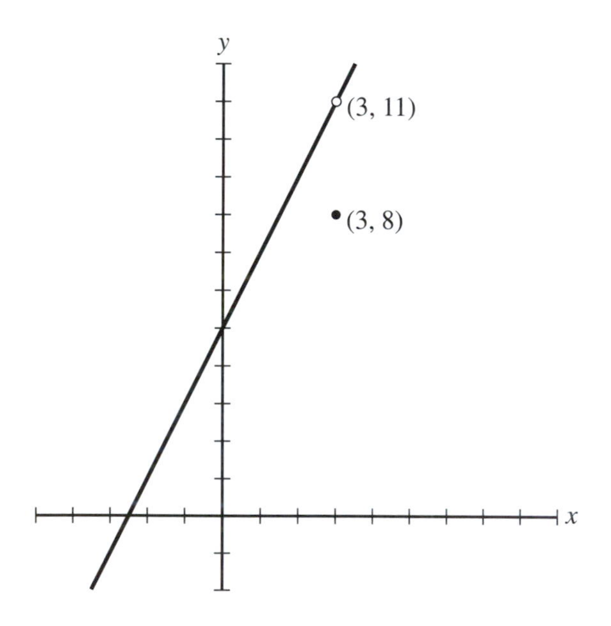

Figure 9.7 Graph of the function g of Example 2

different values or at which one function is defined and the other isn't.) Then, for any numbers a and L,

$$\lim_{x \to a} f(x) = L \quad \text{if and only if} \quad \lim_{x \to a} g(x) = L$$

Proof: Let $x_1, x_2, \ldots, x_n$ be the x values where f and g differ, and assume $\lim_{x \to a} f(x) = L$. We must prove that $\lim_{x \to a} g(x) = L$. Consider the set of *positive* numbers in the list $|a - x_1|, |a - x_2|, \ldots, |a - x_n|$. In other words, if a equals one of the x_is, exclude that $|a - x_i|$, which is 0, from the set. By Theorem 9.1(b), this set of positive numbers has a smallest member, which must be positive. Let's call it b. Then note that if x is any number such that $0 < |x - a| < b$, then $f(x) = g(x)$ or both $f(x)$ and $g(x)$ are undefined. Now let $\varepsilon > 0$ be given. Since $\lim_{x \to a} f(x) = L$, there is a $\delta > 0$ such that $|f(x) - L| < \varepsilon$ whenever $0 < |x - a| < \delta$. Now let δ' be the smaller of the two positive numbers δ and b. *[Note how existential specification is used here but not even mentioned.]* Since $\delta' \le \delta$, $|f(x) - L| < \varepsilon$ holds whenever $0 < |x - a| < \delta'$. But we also have $\delta' \le b$, so by what we just said about b, $|g(x) - L| < \varepsilon$ whenever $0 < |x - a| < \delta$. Since ε was arbitrary and we have found a positive number δ' in terms of it, making this condition true, $\lim_{x \to a} g(x) = L$.

The proof that $\lim_{x \to a} g(x) = L$ implies $\lim_{x \to a} f(x) = L$ is similar. ■

This result may seem strange at first. It says you can change the graph of any function at one point, a hundred points, or a billion points without changing any of its limits!

The next two theorems (and their proofs) are analogous to Theorems 9.5 and 9.8.

Theorem 9.13 (Uniqueness of limits): Limits of functions are unique. That is, if $\lim_{x \to a} f(x) = L_1$ and $\lim_{x \to a} f(x) = L_2$, then $L_1 = L_2$.

Proof: See Exercise 8. ■

Example 4: Consider the function f defined by

$$f(x) = \begin{cases} x - 1 & \text{if } x > 4 \\ 5 & \text{if } x < 4 \end{cases}$$

(see Figure 9.8). You might think this function should have two limits as x approaches 4, but Theorem 9.13 says that can't occur. In fact, f has no limit as $x \to 4$. It is possible to define **one-sided limits**; then we can say $\lim_{x \to 4^+} f(x) = 3$, and $\lim_{x \to 4^-} f(x) = 5$. But a function does not have a limit if it has unequal one-sided limits, as in this case (see Exercise 5). Compare this example to Example 4 of Section 9.2.

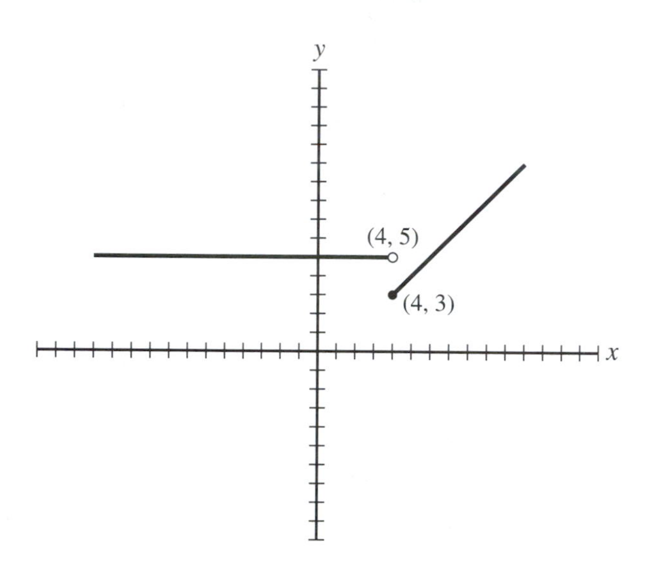

Figure 9.8 Graph of the function f of Example 4

Example 5: Let

$$f(x) = \begin{cases} 1 & \text{if } x \text{ is rational} \\ 0 & \text{if } x \text{ is irrational} \end{cases}$$

Then f is a perfectly valid function from $\mathbb{R}$ to $\mathbb{R}$, but it is what mathematicians call pathological. There is no way even to graph it. It is not hard to show that this function has no limit or even a one-sided limit as x approaches any number (see Exercise 6).

When we introduced the sum, product, and so on, of functions (Exercise 13, Section 7.1), we required the two functions to have the same domain. This is not necessary. If $f: A \to \mathbb{R}$ and $g: B \to \mathbb{R}$, then $f+g$, $f-g$, and fg are defined as functions with domain $A \cap B$. The domain of f/g must be restricted further to the set $A \cap \{x \in B \mid g(x) \neq 0\}$. Also, we can always define $g \circ f$; its domain is a certain subset of A (see Exercise 16(b) of Section 7.2).

Theorem 9.14: Suppose $\lim_{x \to a} f(x) = L_1$ and $\lim_{x \to a} g(x) = L_2$. Then:

(a) $\lim_{x \to a} (f+g)(x) = L_1 + L_2$, $\lim_{x \to a} (f-g)(x) = L_1 - L_2$, and $\lim_{x \to a} (fg)(x) = L_1 L_2$.

(b) If in addition $L_2 \neq 0$, then $\lim_{x \to a} (f/g)(x) = L_1/L_2$.

Proof: This proof is nearly identical to that of Theorem 9.8 (see Exercise 9(a)). ∎

Continuity

We have made a couple of references to reasonable functions. It is now time to make this precise. The technical term for what we have meant by "reasonable" is "continuous."

Definitions: Suppose that f: $D \to \mathbb{R}$ (where $D \subseteq \mathbb{R}$), and $a \in D$. We say that f is **continuous at a** iff

$$\forall \varepsilon > 0 \ \exists \delta > 0 \ \forall x \ (|x - a| < \delta \land x \in D \to |f(x) - f(a)| < \varepsilon)$$

We say that f is **continuous on B** (where B is an *interval* contained in D) iff it's continuous at every number in B. And we simply say that f is **continuous** iff it's continuous on its entire domain.

According to this definition, we do not call a function continuous unless its domain is an interval (bounded or unbounded). This restriction is not standard, but many awkward situations are avoided by including it. For instance, without this restriction, every function whose domain is a finite set would be continuous (see Exercise 11).

Continuity is probably the most important concept in the subjects of real analysis. Intuitively, a function is continuous iff a small change in the input creates only a small change in the output. Graphically, to say that f is continuous at a means that the graph of f is *unbroken* at $x = a$. Even simpler, a function is continuous iff its graph is unbroken.

You can see from the similarity of their definitions that continuity must be related to limits. This relationship forms the basis for Theorem 9.15.

Definitions: Suppose $A \subseteq \mathbb{R}$ and $x \in A$. We say x is an **interior point** of A iff there is an *open* interval I such that $x \in I$ and $I \subseteq A$. We say A is **open** iff every member of A is an interior point of A.

Intuitively, x is an interior point of A iff A contains all points that are sufficiently close to x; and A is open iff it has no points that are on its edge, or boundary. Section 9.4 makes these ideas more precise.

Example 6: Let A be any open interval, such as $(1, 4)$. Then every x in A is an interior point of A, because A itself is an open interval that contains x and is a subset of A. Therefore, A is an open set.

Example 7: The empty set is open, trivially. Also, $\mathbb{R}$ is open, for given any $x \in \mathbb{R}$, the open interval $(x - 1, x + 1)$ contains x and is a subset of $\mathbb{R}$.

Example 8: A closed interval like $[1, 4]$ is not open, because its endpoints are in the set but are not interior points of the set (see Exercise 10).

Theorem 9.15: (a) If $\lim_{x \to a} f(x) = f(a)$, then f is continuous at a.

(b) If a is an *interior* point of Dom(f), then the converse of (a) also holds.

(c) If Dom(f) is an open interval, then f is continuous iff $\lim_{x \to a} f(x) = f(a)$, for every x in Dom(f).

Proof: For conciseness, let $D = \text{Dom}(f)$.

(a) Assume that $\lim_{x \to a} f(x) = f(a)$. Then, given any $\varepsilon > 0$, there is a $\delta > 0$ such that $\forall x\, (0 < |x - a| < \delta \rightarrow |f(x) - f(a)| < \varepsilon)$. But since $f(a) - f(a) = 0$, we can drop the symbols "0 <" in this condition. Also, $[(P \rightarrow R) \rightarrow (P \land Q \rightarrow R)]$ is a tautology, so we have $\forall x\, (|x - a| < \delta \land x \in D \rightarrow |f(x) - f(a)| < \varepsilon)$. Thus f is continuous at a.

(b) Assume a is an interior point of D and f is continuous at a. There is some open interval containing a, which we can write as $(a - c, a + d)$, which is a subset of D. Let b be the smaller of c and d. To show that $\lim_{x \to a} f(x) = f(a)$, let $\varepsilon > 0$ be given. Since f is continuous at a, there is a $\delta > 0$ such that $\forall x\, (|x - a| < \delta \land x \in D \rightarrow |f(x) - f(a)| < \varepsilon)$. Now let δ' be the smaller of δ and b. Then the condition $\forall x$ (...) still holds with δ replaced by δ'. But also, by the definition of b, $|x - a| < \delta'$ automatically implies that $x \in D$. In other words, $\forall x\, (|x - a| < \delta' \rightarrow |f(x) - f(a)| < \varepsilon)$. Since ε was arbitrary and we found an appropriate δ' in terms of it, $\lim_{x \to a} f(x) = f(a)$.

(c) Follows immediately from parts (a) and (b). ■

Note that part (c) of this theorem *does* apply to functions whose domain is $\mathbb{R}$.

Example 9: Let's determine whether the functions discussed earlier in this section are continuous. Example 1 considers the function $f: \mathbb{R} \rightarrow \mathbb{R}$ defined by $f(x) = 2x + 5$ and shows that $\lim_{x \to 3} f(x) = f(3)$. A similar argument shows that $\lim_{x \to a} f(x) = f(a)$ for every real number a. Thus f is continuous, by Theorem 9.15(c).

Example 10: The function g of Example 2 isn't continuous since $\lim_{x \to 3} g(x) \neq g(3)$. However, g is continuous at every number but 3. Also, since $\lim_{x \to 3} g(x)$ is defined, we say that g has a **removable discontinuity** at $x = 3$. In other words, the break in the graph of g can be fixed by moving just one point to fill in the hole.

Example 11: The function h of Example 3 *is* continuous at every number in its domain. But since its domain is not an interval, we do not call it a continuous function. We could say that its graph is broken only because its *domain* is broken. This function, like the function g of Example 10, has a removable discontinuity at $x = 3$, because the break in its graph can be fixed by inserting one point.

Example 12: The function f of Example 4 is also continuous at every number in its domain but is not continuous under our definition because its domain is not an interval. Since $\lim_{x \to 4} f(x)$ does not exist, f has a **nonremovable discontinuity** at $x = 4$, which is considered more severe than the discontinuities described in Examples 10 and 11. That is, the break in this graph cannot be fixed by moving or inserting one point.

Example 13: The function f of Example 5 is not continuous at any number; we say it is **nowhere continuous**. This explains why it's impossible to graph (although there are also continuous functions that are essentially impossible to graph).

☞ With the experience of Theorem 9.15 and a variety of examples at our disposal, this is a good time to carefully compare the definitions of the concepts "f has a limit as $x \to a$" and "f is continuous at a." The definitions are very similar, but each one has one short but important condition that is missing from the other.

The symbols "$0 <$" are the ones that are special to the notion of f having a limit as $x \to a$. As mentioned earlier, the presence of these symbols in the definition is what allows $\lim_{x \to a} f(x)$ to exist without being equal to $f(a)$. Without these symbols, having a limit would always imply continuity.

The symbols "$\in D$" are found in the definition of continuity only. Without these symbols, continuity could only occur at interior points of the domain of a function. One could adopt that viewpoint and omit these symbols, but most mathematicians prefer to allow continuity to occur at endpoints of a domain as well as interior points.

In other words, Theorem 9.15(b) provides a nice biconditional, but only at interior points of a function's domain. If the domain of a function is not open, we must be careful about determining continuity.

Example 14: Let $f(x) = \sqrt{x}$. The natural domain of f is $[0, \infty)$, which is not open because 0 is not an interior point. Or consider $g(x) = \sqrt{1 - x^2}$. The natural domain of g is the closed interval $[-1, 1]$, which is not open. Both f and g have unbroken graphs (see Figure 9.9), and it is sensible to call them continuous. Therefore, our definition is set up to make these functions continuous. Yet $\lim_{x \to 0} f(x)$ does not exist; neither do $\lim_{x \to 1} g(x)$ and $\lim_{x \to -1} g(x)$. (To be sure, *one-sided* limits do exist at these endpoints.)

There is no perfect or completely standard set of definitions for limits and continuity in the case of noninterior points. To be honest, all definitions of limits and continuity lead to some awkward situations. The definitions in this book are chosen specifically to make "continuous" equivalent to the graphical concept "unbroken."

Our next goal is to show that *discontinuity* is an uncommon phenomenon.

Theorem 9.16: (a) Every constant function $f(x) = c$ is continuous.
(b) The identity function $f(x) = x$ is continuous.
(c) The function $f(x) = |x|$ is continuous.

Proof: (a) and (b) These simple proofs are left for Exercise 12.

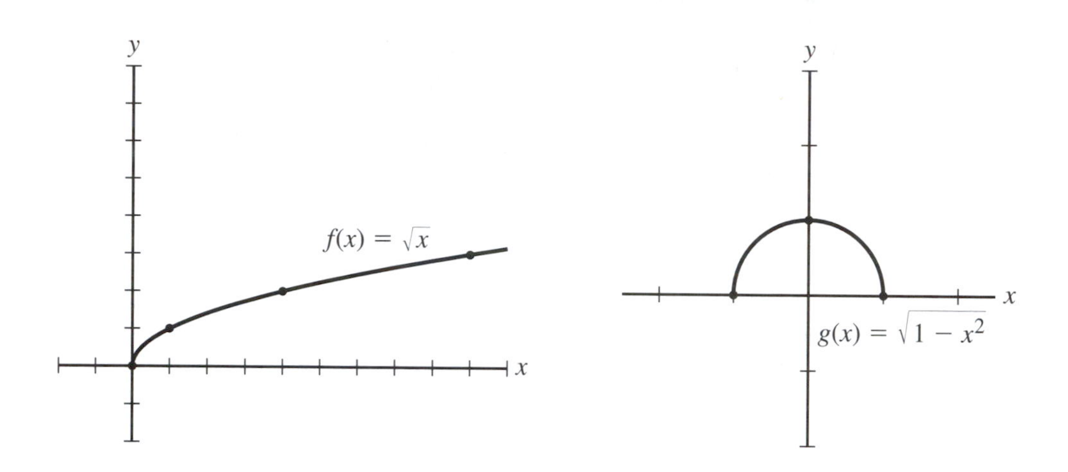

Figure 9.9 Two continuous functions whose domains are not open (Example 14)

(c) It suffices to show (by Theorem 9.15(c)) that $\lim_{x \to a} |x| = |a|$, for every real a. Given any a and any $\varepsilon > 0$, let $\delta = \varepsilon$. Now assume $0 < |x - a| < \delta$. A simple proof by cases, also left for Exercise 12, shows that $|x - a| \geq ||x| - |a||$. Thus $||x| - |a|| < \varepsilon$, so we are done. ■

Theorem 9.17: (a) If f and g are continuous at a number a, then so are $f + g$, $f - g$, and fg.

(b) If, in addition, $g(a) \neq 0$, then f/g is also continuous at a.

(c) If g is continuous at a and f is continuous at $g(a)$, then $f \circ g$ is continuous at a.

Proof: As with Theorem 9.14, the proofs of (a) and (b) are almost identical to the proof of Theorem 9.8 (see Exercise 9(b)).

(c) Assume the givens. Let $b = g(a)$ and $c = f(b) = (f \circ g)(a)$. Now let any $\varepsilon > 0$ be given. Since f is continuous at b, there is an $\varepsilon' > 0$ such that, whenever $u \in \text{Dom}(f)$ and $|u - b| < \varepsilon'$, $|f(u) - c| < \varepsilon$. But we also know that g is continuous at a. So there is a $\delta > 0$ such that, whenever $x \in \text{Dom}(g)$ and $|x - a| < \delta$, $|g(x) - b| < \varepsilon'$. But if $|g(x) - b| < \varepsilon'$, then $|f(g(x)) - c| < \varepsilon$. Also, $\text{Dom}(f \circ g) \subseteq \text{Dom}(g)$. Therefore, whenever $x \in \text{Dom}(f \circ g)$ and $|x - a| < \delta$, then $|(f \circ g)(x) - (f \circ g)(a)| < \varepsilon$. ■

Part (c) of this theorem is not similar to any part of Theorems 9.8 or 9.14, but we could have included an analogous statement for compositions in Theorem 9.14. The proof of (c) seems involved, but the idea behind it is simple. Using the variables $y = g(x)$ and $z = f(y)$, we want to keep z within a given tolerance ε of c. We know we can achieve this if we keep y within some corresponding amount ε' of b. But then we can view ε' as a tolerance for y. We know we can keep y within ε' of b (which in turn keeps z where we want it) if we keep x within some corresponding amount δ of a.

Corollary 9.18: If f and g are continuous at every number in their domains, then so are the functions $f+g$, $f-g$, fg, f/g, and $f \circ g$.

Proof: This follows immediately from Theorem 9.17. ■

Theorem 9.19: (a) Every polynomial function is continuous.

(b) Every rational function is continuous at every number in its domain.

Proof: (a) Using Theorem 9.16(b) and the part of Corollary 9.18 concerning products, it follows (by a simple induction on n) that every function of the form $f(x) = x^n$, where n is a natural number, is continuous. By the same part of Corollary 9.18, we then get that every function of the form $f(x) = cx^n$ is continuous. Finally, since a polynomial is by definition a sum of such functions and constants, we can apply Theorem 9.16(a) and the part of Corollary 9.18 involving sums to get what we want.

(b) Since a rational function is a quotient of polynomials, this is immediate from part (a) and Corollary 9.18. ■

Keep in mind why part (b) of Theorem 9.19 has a restriction that (a) does not. Every polynomial has domain $\mathbb{R}$ and is continuous on $\mathbb{R}$. But a rational function like $1/x$ may have a broken domain, preventing it from being a continuous function under our definition.

The last four results say that many important functions are continuous at every number in their domains, but they don't come close to telling the whole story. It also turns out that functions of the form $f(x) = x^n$ are continuous at every number in their domains, even for noninteger values of n. (So, for example, the function $\sqrt{x}$ is continuous; see Exercise 16.) Furthermore, all the standard trigonometric, exponential, and logarithm functions are continuous at every number in their domains. And Corollary 9.18 guarantees that all reasonable combinations of such functions are also continuous at every number in their domains.

We have been discussing functions whose domains and ranges are subsets of $\mathbb{R}$. But limits and continuity can also be defined for functions from a subset of $\mathbb{R}^m$ to $\mathbb{R}^n$, where m and n are any natural numbers. (The only change needed is to change the expressions involving absolute value in the definitions to expressions for distance between points.) It turns out that results 9.16 through 9.19 and the comments in the previous paragraph still hold for functions with these more general domains and ranges.

In short, *just about every function you have ever worked with or will ever work with is continuous at every member of its domain.* The only somewhat common exceptions are functions defined by cases and functions whose definitions involves integers in some way. Most of these exceptions are **piecewise continuous**; that is, their graphs consist of a finite number or at worst a countable number of unbroken pieces.

Example 15: One of the few standard examples of a function with discontinuities in its domain is the greatest integer function $\lfloor x \rfloor$, defined in Section 6.2: $\lfloor x \rfloor$ equals the greatest integer that is less than or equal to x. The graph of this function is shown in Figure 9.10. The graph jumps at every integer value of x, and since these numbers are in the domain of the function, the function is not continuous (see Exercise 13).

Example 16: A function from real life that is very similar to the greatest integer function is the function for first-class postage. As of this writing, the formula for first-class postage is 32 cents for the first ounce or part thereof, and 23 cents for each additional ounce or part thereof. This may be considered a function from the positive reals to $\mathbb{R}$. Its graph is shown in Figure 9.11. Like the greatest integer function, this function is discontinuous at every integer in its domain, so it is not a continuous function. Both of these functions are examples of **step functions**.

We conclude this section with one of the most useful and important properties of continuous functions. Even though this property is visually obvious, the proof is nontrivial and requires the completeness axiom.

Theorem 9.20 (Intermediate value theorem): If f is continuous on $[a, b]$ and y is strictly between $f(a)$ and $f(b)$, then there is a number c in (a, b) such that $f(c) = y$.

Proof: We do the proof for the case $f(a) < f(b)$. The case $f(a) > f(b)$ is analogous. (If $f(a) = f(b)$, there's nothing to prove.) Let $A = \{x \in [a, b] \mid f(x) < y\}$. Then A is nonempty because a is in it, and A is bounded above by b. So by completeness, A has a supremum; call it c. We claim that $f(c) = y$. To see this, first note that $c \in [a, b]$, so $f(c)$ is defined. We then want to show that neither $f(c) < y$ nor $f(c) >$ is possible.

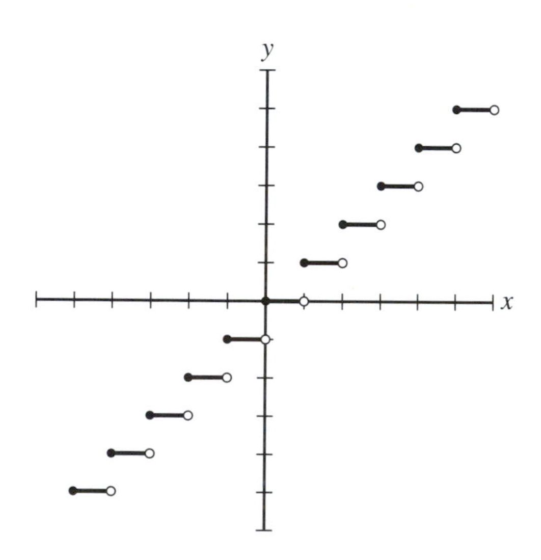

Figure 9.10 Graph of the function $f(x) = \lfloor x \rfloor$

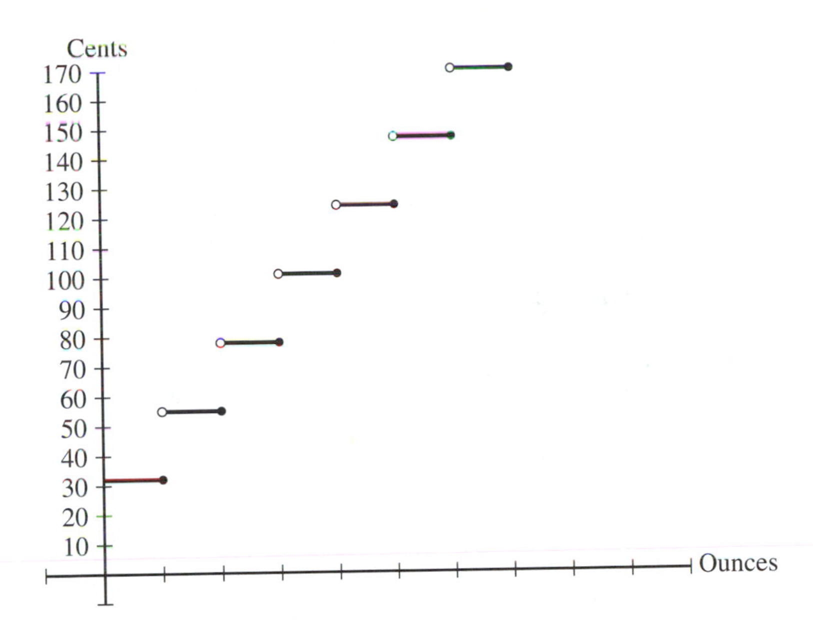

Figure 9.11 Graph of the first-class postage function

So first assume $f(c) < y$. Since $y < f(b)$, this means $c \neq b$; therefore, $c < b$. Let $\varepsilon = y - f(c)$. Since f is continuous at c, there's an interval around c on the X axis in which all the values of f are within ε of $f(c)$. In particular, this means there's a number $x > c$ such that $f(x) < y$. But this contradicts the fact that c is an upper bound for A.

The proof that $f(c) > y$ is impossible is left for Exercise 14. ■

Perhaps this proof reminds you of the proof of Theorem 9.3. That makes sense, because Theorem 9.3 is a special case of this theorem. Can you see what f, a, b, and y could be to yield Theorem 9.3 from Theorem 9.20? (See Exercise 15.)

Corollary 9.21: If f is continuous on $[a, b]$ and $f(a)$ and $f(b)$ are of opposite sign, then f has a root between a and b.

Exercises 9.3

(1) Give an example (which could be an algebraic definition and/or a graph) of each of the following situations, or explain briefly why it cannot exist.

(a) A function that has a limit as x approaches 2 but is not continuous at 2

(b) A function that is continuous at 2 but has no limit as x approaches 2

(c) A function f that is continuous at all numbers in its domain while $f \circ f$ is not

(2) Show that if the symbols "0 <" were omitted from the definition of the statement $\lim_{x \to a} f(x) = L$, then this statement would imply that $f(a) = L$.

(3) Carry out the proof that $\lim_{x \to 3} g(x) = 11$, in Example 2.

(4) Suppose f is a function, and a and $L \in \mathbb{R}$. In the text, we define what is meant by $\lim_{x \to a} f(x) = L$. Give rigorous definitions of the following related statements:

$$\lim_{x \to +\infty} f(x) = L \quad \text{and} \quad \lim_{x \to a} f(x) = +\infty$$

Hint: Both definitions, especially the first, can be closely modeled after the definition of limits of sequences in Section 9.2. Your definitions may not use the symbol ∞, since this symbol does not represent a real number.

Similarly, one can define statements such as $\lim_{x \to -\infty} f(x) = L$ and $\lim_{x \to +\infty} f(x) = -\infty$.

(5) (a) Prove that the function f of Example 4 has no limit as x approaches 4.

(b) Define the statements $\lim_{x \to a^-} f(x) = L$ and $\lim_{x \to a^+} f(x) = L$.

The notation $x \to a^-$ (respectively, $x \to a^+$) may be read "x approaches a from the left" (respectively, "from the right").

(c) Prove that, in Example 4, $\lim_{x \to 4^+} f(x) = 3$, and $\lim_{x \to 4^-} f(x) = 5$.

(d) Prove that, in general, $\lim_{x \to a} f(x) = L$ iff $\lim_{x \to a^-} f(x) = L$ and $\lim_{x \to a^+} f(x) = L$.

(6) Prove the claim made in Example 5. You may use the results of Exercises 13 and 14 of Section 9.1.

(7) Prove the claim made in Example 6 about the function h.

(8) Prove Theorem 9.13.

(9) (a) Prove the first two assertions in Theorem 9.14(a).
(b) Prove the first two assertions in Theorem 9.17(a).

(10) (a) Show that no closed interval $[a, b]$, where $a < b$, is open.
(b) Show that every interval of the form (a, ∞) or $(-\infty, a)$ is open.
(c) Show that no interval of the form $[a, \infty)$ or $(-\infty, a]$ is open.

*(11) Prove that a function whose domain is a finite set or $\mathbb{Z}$ is continuous at every number in its domain.

(12) (a) Prove Theorem 9.16(a) and (b).

(b) Complete the proof of Theorem 9.16(c), by showing $|x - a| \geq ||x| - |a||$.

(13) Prove the claim made in Example 15 that the function $\lfloor x \rfloor$ is not continuous.

(14) Complete the proof of Theorem 9.20 as indicated in the text.

(15) Answer the questions in the paragraph before Corollary 9.21.

(16) Suppose $f: A \to \mathbb{R}$, where $A \subseteq \mathbb{R}$. We say f is **increasing** iff, for any a and b in A, $a < b$ implies $f(a) \leq f(b)$. If the symbol $\leq$ in this definition is changed to $\geq$, $<$, or $>$, we get the definition of f's being **decreasing**, **strictly increasing**, or **strictly decreasing**. Also, f is called **monotone** iff it is increasing or decreasing.

(a) Prove that for *sequences*, these definitions are equivalent to the ones in Section 9.2.

(b) Prove that if f is strictly monotone, then f is one-to-one. Does it follow that f^{-1} is also strictly monotone?

(c) Prove or find a counterexample to the converse of (b).

*(d) By common sense and/or some examples, decide whether you think the converse of (b) holds for *continuous* functions. Then prove or disprove your claim.

*(17) Prove that if f is continuous and strictly monotone, then f^{-1} is continuous.

(18) Prove that the function $\sqrt{x}$ is continuous. You may use the result of the previous exercise, whether or not you did it.

*(19) (a) Prove the following, which is an analog of the monotone convergence theorem for functions: if f is monotone and a is an interior point of $\text{Dom}(f)$, then $\lim_{x \to a^-} f(x) = L$ and $\lim_{x \to a^+} f(x)$ exist.

(b) Must the one-sided limits in part (a) be equal? Prove they are, or find a counterexample.

(c) Prove that if these one-sided limits are equal, then f is continuous at a.

Critique the proofs in the next two exercises. (If necessary, refer to Exercises 4.2 for the instructions for this type of problem.)

(20) **Theorem:** If $\lim_{x \to a} f(x) = L$ and $\lim_{x \to L} g(x) = L'$, then $\lim_{x \to a} g \circ f(x) = L'$.

Proof: As x gets closer and closer to a, the corresponding values of $f(x)$ get closer and closer to L. In turn, the corresponding values of $g(f(x))$, which by definition is $g \circ f(x)$, get closer and closer to L'.

(21) **Theorem:** If f is continuous at a then, for every sequence (b_n) of elements of $\text{Dom}(f)$ such that $b_n \to a$, the sequence $(f(b_n))$ converges to $f(a)$.

Proof: Assume f, a, and (b_n) are as above. Let $\varepsilon > 0$ be given. Then there's a $\delta > 0$ such that whenever $x \in \text{Dom}(f)$ and x is within δ of a, $f(x)$ is within ε of L. But because $b_n \to a$, there's also an m such that whenever $n > m$, b_n is within δ of a. So, whenever $n > m$, $f(b_n)$ is within ε of L. This shows that $(f(b_n)) \to L$.

*(22) Prove the converse of Exercise 21, assuming also that $a \in \text{Dom}(f)$.

*9.4 Topology of the Real Line

Like set theory, topology is a relatively new branch of mathematics. To be sure, problems of a topological nature were studied and solved as early as the 1700s. But topology did not really blossom into a full-fledged subject until the twentieth century.

It is not easy to characterize topology precisely. Certainly, it is more geometric than algebraic. Like geometry, topology deals with things that can be seen: real figures in real space. But topology is concerned with more general properties of figures than geometry is. Topology can be done even in very abstract settings where there is no measure of distance or angle. For example, in geometry, a ball is very different from a cube, because a cube has edges and corners. But to a topologist, they are equivalent because a rubber ball can in theory be molded to form a cube without breaking the rubber, and vice versa. On the other hand, a ball is topologically very different from a torus (doughnut shape), because a torus has a hole through the middle, and you can't mold a rubber ball into a torus without tearing the rubber (see Figure 9.12).

In spite of its closeness to geometry, topology also has connections to many other branches of mathematics. It relies heavily on techniques from set theory and algebra. In turn, it has important applications to subjects as diverse as combinatorics, differential equations, and, as we soon see, real analysis.

Exercise 16 takes you through a famous problem that helped create this subject.

Definition: Recall the definitions of the terms *interior point* and *open set* in Section 9.3. We also say that A is a **neighborhood of x** iff A is open and $x \in A$.

☞ Mathematicians often say that something occurs in a neighborhood of x. Do *not* use the phrase "in *the* neighborhood of x." These words would imply that there is only one neighborhood of a given point, an impression that can lead to mistakes.

From the examples in Section 9.3, we know that every open interval is an open set. Now let's make the connection between open intervals and open sets precise.

Theorem 9.22: A subset of $\mathbb{R}$ is open iff it is the union of a collection of open intervals. (The collection involved may be finite or infinite; also, it may be empty.)

Proof: For the forward direction, assume A is an open subset of $\mathbb{R}$. Let $\mathscr{C}$ be the collection of all open intervals that are subsets of A. Then $\bigcup\mathscr{C}$ is also a subset of A, clearly. But since A is open, any member of A is an interior point of A; this means it is in at least one of the sets in $\mathscr{C}$, which in turn puts it in $\bigcup\mathscr{C}$. Therefore, $A = \bigcup\mathscr{C}$.

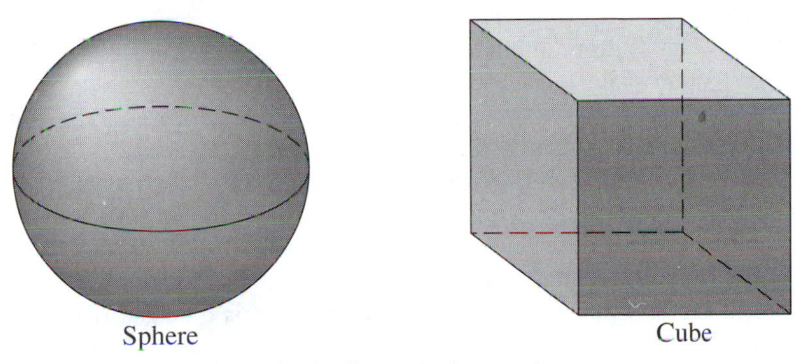

Topologically equivalent surfaces

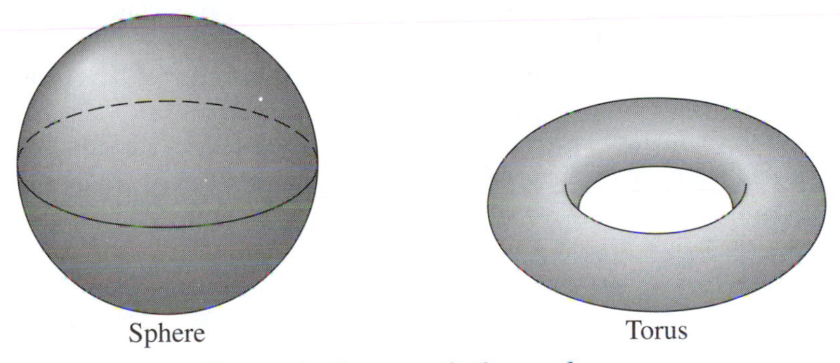

Topologically nonequivalent surfaces

Figure 9.12 Illustration of some basic concepts of topology

Conversely, assume $A = \bigcup \mathscr{C}$, where $\mathscr{C}$ is a collection of open intervals. Then for any x in A, we must have some B in $\mathscr{C}$ for which x is in B. So x is an interior point of A. Since x was arbitrary, A is open. ■

Example 1: Let A be the set of all negative real numbers. Then A is open. This can be proved by showing that every point of A is an interior point or by applying Theorem 9.12. For instance, A is the union of the collection of open intervals $\{(-n, 0) \mid n \in \mathbb{N}\}$. Even simpler, $A = (-\infty, 0)$, which is a single unbounded open interval.

Theorem 9.23: (a) The union of any collection of open sets is open.

(b) The intersection of any *finite* collection of open sets is open.

(c) The sets $\varnothing$ and $\mathbb{R}$ are open.

Proof: (a) This follows immediately from Theorem 9.22, since if each open set is a union of open intervals, then so is any union of open sets.

(b) We prove this for the intersection of two open sets. The generalization to any finite collection of open sets is then a simple induction (see Exercise 2).

Let $A = B_1 \cap B_2$, with B_1 and B_2 open. Assume $x \in A$. Then x is in each B_i, and since B_i is open, x is an interior point of it. Thus, for each i, we can select an open interval (c_i, d_i) that contains x and is a subset of B_i. So $c_i < x < d_i$, for each i. Now let c be the larger of the c_is, and let d be the smaller of the d_is. We leave it for you (Exercise 2) to prove that (c, d) is an open interval that contains x and is a subset of A. So x is an interior point of A. Since x was arbitrary, A is open.

(c) These simple facts were noted in Example 7 of Section 9.3. ■

The restriction in part (b) to finite collections is necessary (see Exercise 3).

You may wonder why we have grouped these three facts into a theorem. In the introduction to this section we mentioned that topology is a very general subject. The set $\mathbb{R}$ is not the only one for which one talks about open subsets. For example, to develop topology for the plane $\mathbb{R} \times \mathbb{R}$, or $\mathbb{R}^2$, simply replace "interval" with "disk," and "$\mathbb{R}$" with "$\mathbb{R}^2$," in everything we've said so far. (An open disk is the inside of a circle, not including the boundary; see Figure 9.13.) Then Theorems 9.22 and 9.23 still hold (see Exercise 4). For three-dimensional space, replace "interval" with "sphere," and "$\mathbb{R}$" with "$\mathbb{R}^3$," and once again the theorems still hold. Also, the remark after the definition of open sets still applies to these two other examples.

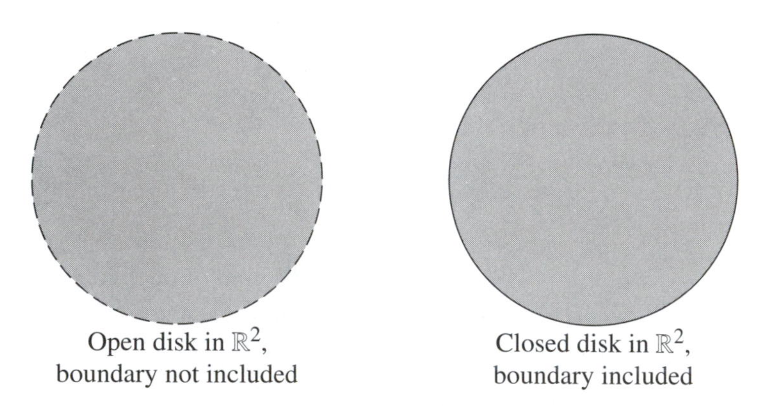

Figure 9.13 Open and closed disks

What we have just described are three examples of **topological spaces**. By definition, a topological space consists of a set X (for example, $\mathbb{R}^2$) and a specification of which subsets of that set are to be called open (for example, unions of open disks; in general, some subset of $\wp(X)$), such that Theorem 9.23 holds. So the relevance of that theorem is that it embodies the definition of this important concept.

There is an amazing variety of topological spaces; see Exercise 5 for a few. In order not to get too far off the track, we concentrate on the topology of $\mathbb{R}$, but most of our definitions and results are applicable to all topological spaces.

Definition: A subset A of $\mathbb{R}$ is called **closed** iff its complement A' is open.

Theorem 9.24: (a) The intersection of any collection of closed sets is closed.

(b) The union of any *finite* collection of closed sets is closed.

(c) The sets $\varnothing$ and $\mathbb{R}$ are closed.

(d) Every closed interval is closed; so is every one-point set.

Proof: These facts follow directly from the definition of closed and Theorems 9.22 and 9.23. The proofs are left for Exercise 6. ■

☞ Note that "closed" does *not* mean "not open." On the one hand, we have seen that some sets, $\varnothing$ and $\mathbb{R}$ to be precise, are "clopen" (simultaneously closed and open). On the other hand, many sets are neither open nor closed.

Also, be aware that in a certain sense closed sets can be much more complicated than open sets. In particular, we cannot say (in analogy to Theorem 9.22) that every closed set is the intersection or union of closed intervals. The Cantor set, defined in Section 5.3, is a closed set that is quite complicated and virtually impossible to visualize. It is much easier to visualize its complement.

Definitions: For $x \in \mathbb{R}$ and $A \subseteq \mathbb{R}$, we say that x is an **accumulation point** of A iff every neighborhood of x contains at least one number *other than* x that is in A. A member of A that is not an accumulation point of A is called an **isolated point** of A.

The words *other than* x in this definition are vital. Without them, every member of A would automatically be an accumulation point of A. This is not the intended meaning; in fact, $x \in A$ is neither a necessary nor a sufficient condition for x to be an accumulation point of A. (This is analogous to the situation with LUBs, GLBs and limits.) For x to be an accumulation point of A means there are lots of numbers close to x in A. We now state this more precisely.

Theorem 9.25: The number x is an accumulation point of A iff every neighborhood of x contains an infinite number of members of A.

Proof: See Exercise 7. ■

Theorem 9.26: A subset of $\mathbb{R}$ is closed iff it contains all its accumulation points.

Proof: Let $A \subseteq \mathbb{R}$. Our proof is a chain of iff's: A is closed iff A' is open, iff every

x in A' is interior to A', iff for every x in A' there's an open interval containing it that is a subset of A', iff *no* x in A' is an accumulation point of A, iff A contains all of its accumulation points. ∎

Theorem 9.26 provides a way of defining "closed" without reference to openness.

Example 2: Let $A = (3, 6]$. Then A is not closed because it fails to contain one of its own accumulation points, namely 3. This is another way of saying that 3 is in A' but is not an interior point of A', so A' is not open. Is A' closed?

Let $B = (-2, 1) \cup \{3\}$. All numbers in the closed interval $[-2, 1]$ are accumulation points of B. The number 3 is the only isolated point of B. Is B open? Is B closed? (See Figure 9.14.)

Example 3: Let $A = \{1/n \mid n \in \mathbb{N}\}$. Is A open? Is A closed? What are the accumulation points of A? If we take the union of A and the set of all its accumulation points, do we get a closed set? (See Exercise 8.)

Example 4: What are the accumulation points of $\mathbb{Q}$? What are the accumulation points of $\mathbb{R} - \mathbb{Q}$? Are these sets closed? Are they open? (See Exercise 9.)

We next prove two versions of a famous theorem that is one of the cornerstones of analysis. It is topological in nature but its proof requires completeness; this makes it specific to $\mathbb{R}$. However, it also holds for the plane and other spaces of the form $\mathbb{R}^n$.

Theorem 9.27 (Bolzano-Weierstrass theorem for sets): Every bounded, infinite subset of $\mathbb{R}$ has an accumulation point.

Proof: Let A be a bounded, infinite subset of $\mathbb{R}$. Since A is bounded, it is contained in some interval $[c, d]$. Let $B = \{x \mid A \cap [c, x] \text{ is finite}\}$. Then c is in B, but d is not. By completeness, B has a least upper bound y. It follows that every neighborhood of y contains an infinite number of members of A (see Exercise 10). Hence, by Theorem 9.25, y is an accumulation point of A. ∎

For the other version of the Bolzano-Weierstrass theorem, we need the notion of a **subsequence** of a sequence. This can be defined rigorously (see Exercise 11), but an informal definition should suffice: given an (infinite) sequence, a subsequence is formed by deleting an arbitrary set of terms, provided that an infinite number of terms are left.

Figure 9.14 The set *B* of Example 2

Karl Weierstrass (1815–1897) was a prime example of a talent that could not be suppressed. He was a gifted student in many areas, but he concentrated his studies in public finance and law, under the influence of his father. He showed little interest in these subjects, and left school with poor grades and no degree in 1838. He had spent much of his university career hanging out at taverns and becoming a skilled fencer, but he had also begun to read and think seriously about mathematics.

In 1841 Weierstrass began a fourteen-year career as a secondary school teacher. He taught a wide variety of subjects, including botany, gymnastics, and calligraphy. During this period he had little contact with other mathematicians, but he did a great deal of pioneering mathematical work, especially in analysis. Finally, at the age of forty-one, his accomplishments became recognized to the extent that he obtained a faculty position in Berlin. He went on to become one of the most important mathematicians of the nineteenth century and one of the most brilliant and popular mathematics teachers ever.

Weierstrass's most important work was in the foundations of analysis (that is, the real number system, calculus, and the theory of functions). Bernhard Bolzano and Augustin-Louis Cauchy had made important progress in developing a fully rigorous and satisfactory definition of limits; Weierstrass completed the task. He pioneered the study of complex-valued functions as power series and made important contributions to the theory of convergence of real and complex functions. He showed how to define, via a power series, a function that is continuous everywhere on $\mathbb{R}$ but differentiable nowhere; however, he graciously allowed one of his students to publish this result. Weierstrass always strove for clarity and logical coherence, and his contemporaries coined the phrase "Weierstrassian rigor" to describe his style. In addition to being a superb lecturer, he was also very fair and democratic toward students. For example, he ended the requirement that Ph.D. theses be in Latin.

Theorem 9.28 (Bolzano-Weierstrass theorem for sequences): Every bounded infinite sequence has a convergent subsequence.

Proof: Let (a_n) be a bounded infinite sequence. This means that the *set* $\{a_n \mid n \in \mathbb{N}\}$, or simply $\{a_n\}$ for short, is bounded. We have two cases to consider.

Case 1: The set $\{a_n\}$ is finite. Say $\{a_n\} = \{b_1, b_2, \dots, b_k\}$. For each of these b_is, we can consider how many times it occurs in the sequence (a_n). If each of the b_is occurred only a finite number of times, then Theorem 7.22(d) would be violated, since (a_n) is an infinite sequence. So at least one of the b_is occurs an infinite number of times in (a_n). Pick any one such b_i. Then all we need to do is cross out all terms of the sequence (a_n) that don't equal this particular b_i. This gives us a subsequence of (a_n) that is constant and so converges.

Case 2: The set $\{a_n\}$ is infinite. Then, by Theorem 9.27, it has an accumulation point L. By Theorem 9.25, this tells us that every neighborhood of L contains infinitely many members of $\{a_n\}$. We define a subsequence (b_n) of (a_n) inductively, as follows: let b_1 be the first term in (a_n) that differs from L by less than 1. Let b_2 be the first term in (a_n), occurring *after* the chosen occurrence of b_1, that differs from L by less than 1/2. In general, having defined b_1 through b_n, let b_{n+1} be the first term in (a_n), occurring after the chosen occurrence of b_n, that differs from L by less than $1/(n+1)$. Theorem 9.25 guarantees that we can continue this process indefinitely and get an infinite subsequence of (a_n); and it is clear that (b_n) converges to L. ■

Whereas Theorem 9.28 links topology to the material in Section 9.2, Theorem 9.29 gives a topological characterization of continuous functions. This important result, like many results in this chapter, still holds if the two "$\mathbb{R}$"s mentioned in it are changed to "$\mathbb{R}^m$" and "$\mathbb{R}^n$," for any natural numbers m and n. Furthermore, in the study of functions whose domains and ranges are arbitrary topological spaces, where the definition of continuity given earlier doesn't work because there might be no such thing as the distance or difference between two points, the condition given in Theorem 9.29 is the standard *definition* of continuity.

Theorem 9.29: Suppose $f: A \to \mathbb{R}$, where A is an open subset of $\mathbb{R}$. Then f is continuous at every number in A iff for every open $B \subseteq \mathbb{R}$, $f^{-1}(B)$ is open.

Proof: Assume f and A are as stated. For the forward direction, assume f is continuous and $B \subseteq \mathbb{R}$ is open. Let a be any member of $f^{-1}(B)$; we want to show a is an interior point of this set. By the definition of this set, $f(a) \in B$. Since B is open, there is an open interval (c, d) that contains $f(a)$ and is a subset of B. Let ε be the smaller of $f(a) - c$ and $d - f(a)$. This means that $(f(a) - \varepsilon, f(a) + \varepsilon) \subseteq (c, d) \subseteq B$. Using this ε and the definition of continuity, there is a $\delta > 0$ such that $|f(x) - f(a)| < \varepsilon$ whenever $|x - a| < \delta$. That is, whenever $a - \delta < x < a + \delta$, $f(a) - \varepsilon < f(x) < f(a) + \varepsilon$. In other words, every number in $(a - \delta, a + \delta)$ gets mapped into $(f(a) - \varepsilon, f(a) + \varepsilon)$, which is a subset of B. So we have $(a - \delta, a + \delta) \subseteq f^{-1}(B)$; so a is an interior point as desired.

The proof of the reverse direction is left for Exercise 12. ■

Example 5: Let $f(x) = x^2$. From the previous section we know that f is continuous, since it is a polynomial. Find f^{-1} of the open sets $(2, 5)$, $(0, \infty)$, $(-4, 2)$, and $(-1, 3) \cup (5, 7)$. Are all of these open? Now find f^{-1} of at least three *closed* sets. What do you notice?

Now choose several open sets and determine whether the (*forward*) image of these sets under f is open. See Exercises 13 through 15.

Connected Sets

In topology, the most important types of sets are open sets and closed sets. Another important topological concept is the notion of **connected sets**. Intuitively, a set of points is connected if and only if it's in one piece. But how can we make this rigorous? For subsets of $\mathbb{R}$, connectedness is easy to define in terms of intervals. However, the concept of intervals does not exist even in $\mathbb{R}^2$, whereas the following works in all situations.

Definitions: A subset A of $\mathbb{R}$ (or of any topological space) is called **disconnected** iff there are open sets V and W such that:

(i) V and W are disjoint
(ii) $A \subseteq V \cup W$
(iii) Neither V nor W is disjoint from A.

Otherwise, A is called **connected.**

Example 6: If A has just one member, then it must be connected, because if V and W are disjoint, at least one of them must be disjoint from A. Similarly, $\varnothing$ is connected.

Example 7: A two-point set like $\{3, 8\}$ is disconnected: we can let $V = (2, 4)$ and $W = (6, 19)$, for instance. Also, a set that is the union of two separated intervals, like $[-2, 1] \cup [2, 3)$, is disconnected. So is the complement of any one-point set, and so is $\mathbb{Q}$ (see Exercise 19).

Now we present a more difficult result that depends on the completeness property of $\mathbb{R}$, as must any result that exhibits a contrast between $\mathbb{R}$ and $\mathbb{Q}$.

Lemma 9.30: A subset A of $\mathbb{R}$ is connected iff for all c and d in A with $c < d$, $[c, d] \subseteq A$.

Proof: For the forward direction, we prove its contrapositive: assume the right side of the biconditional is false. That means there are numbers c and d such that $c < d$, $c \in A$, and $d \in A$, but $[c, d] \not\subseteq A$. The last condition implies there's a number b such that $c < b < d$ and $b \notin A$. Let $V = (-\infty, b)$ and $W = (b, \infty)$. Using these sets, A is disconnected.

For the reverse direction, we use indirect proof. Assume the right side of the biconditional is true but A is disconnected. So let V and W be sets that are as in the definition of disconnectedness for A; let $u \in A \cap V$ and $v \in A \cap W$. Since V and W are disjoint, $u \neq v$; and it does no harm to assume $u < v$, since a separate proof for $u > v$ would be identical. Now, let $B = \{x < v \mid x \in V\}$. The set B is bounded above by v and nonempty since it contains u. Thus, B must have a least upper bound, which we'll call w. Obviously, $w \in [u, v]$. By our assumption, the whole interval $[u, v]$ is contained in A. Thus w is in A, and so it must be in either V or W. We now proceed by cases.

Case 1: Assume $w \in V$. Then of course $w < v$, since $v \in W$. But since V is open, there must be some open interval (b, c) that contains w and is a subset of V. Then any number that is larger than w but smaller than both v and c is in B. This violates the supposition that w is an upper bound for B.

Case 2: Assume $w \in W$. So $[w, v] \subseteq W$. But since W is open, there must be some open interval (b, c) that contains w and is a subset of W. Thus $(b, v] \subseteq W$, which makes b an upper bound for B. This violates the supposition that w is the supremum of B.

Since both cases lead to contradictions, we are done. ∎

Theorem 9.31: (a) For any $A \subseteq \mathbb{R}$, A is connected iff $A = \varnothing$, A has exactly one member, or A is an interval. In particular, $\mathbb{R}$ is connected.

(b) The only subsets of $\mathbb{R}$ that are both open and closed are $\varnothing$ and $\mathbb{R}$.

Proof: (a) We prove both directions of this result using Lemma 9.30. For the forward direction, assume that A is a connected subset of $\mathbb{R}$. First of all, if $A = \varnothing$, we are done. If $A \neq \varnothing$, we proceed by several cases.

Case 1: Assume A is bounded above and below. Then A has a supremum and an infimum; call them u and v, respectively. So A contains no numbers less than u or greater than v. That is, $A \subseteq [u, v]$. We now show that $(u, v) \subseteq A$: assume $x \in (u, v)$, so $u < x < v$. Since $u = \text{Inf}(A)$ and $u < x$, x is not a lower bound of A. So there is a number $c \in A$ such that $c < x$. Similarly, there is a number $d \in A$ such that $x < d$. By Lemma 9.30, $x \in A$. So we have $(u, v) \subseteq A \subseteq [u, v]$. By considering the various subcases based on whether u and/or v is in A, it is clear that A must be (u, v), $[u, v)$, $(u, v]$, or $[u, v]$.

Case 2: Assume A is bounded above but not below. Let v be $\text{Sup}(A)$. Then an argument similar to that of case 1 shows that A is $(-\infty, v)$ or $(-\infty, v]$ (see Exercise 20).

Case 3: Assume A is bounded below but not above. Then A is either (u, ∞) or $[u, \infty)$, where $u = \text{Inf}(A)$ (see Exercise 20).

Case 4: Assume A is unbounded above and below. Then it follows that $A = \mathbb{R}$ (again, this is left for Exercise 20).

The reverse direction can be established by verifying that the empty set and any type of interval satisfy the condition of Lemma 9.30 (see Exercise 21).

(b) We have already seen that $\varnothing$ and $\mathbb{R}$ are both open and closed. If V were another "clopen" subset of $\mathbb{R}$, we could let $W = V'$; then V and W would establish the disconnectedness of $\mathbb{R}$, and this would contradict part (a). ∎

Theorem 9.31(b) can be generalized as follows: a topological space X is connected (as a subset of itself) iff the only "clopen" subsets of X are $\varnothing$ and X (see Exercise 22).

Exercises 9.4

(1) Determine whether each of the following sets is (i) open, (ii) closed, (iii) connected. Explain briefly.

(a) $(-\infty, 3]$ (b) $[-2, 7)$ (c) $\mathbb{R} - \{5\}$

(d) $(1, 2) \cup (2, 3)$ (e) $[1, 2) \cup [2, 3]$ (f) $\mathbb{Z}$

(2) Complete the proof of Theorem 9.23(b) as indicated in the text.

(3) Give an example of an infinite collection of open sets whose intersection is not open.

(4) (a) Prove that the plane is a topological space, as claimed after Theorem 9.23. In other words, show that Theorem 9.23 holds if $\mathbb{R}$ is replaced by $\mathbb{R}^2$ and open means "a union of open disks."

(b) Prove an analog of Theorem 9.22 for $\mathbb{R}^2$.

(5) Prove that each of the following defines a topological space. That is, show that Theorem 9.23 holds with $\mathbb{R}$ replaced by X and "open" interpreted as described:

(a) X is any set, and the only open sets are $\varnothing$ and X. This is called the **trivial** topology on X.

(b) X is any set, and *all* subsets of X are open. This is called the **discrete** topology on X.

(c) $X = \mathbb{R}$. The open sets are $\varnothing$, $\mathbb{R}$, and for each real b, the interval (b, ∞).

(6) Prove Theorem 9.24.

*(7) Prove Theorem 9.25.

(8) Answer the questions asked in Example 3.

(9) Answer the questions asked in Example 4.

(10) Prove the indicated missing step in the proof of Theorem 9.27.

*(11) Give a rigorous definition of the notion of a subsequence of a sequence, as discussed before Theorem 9.28.

(12) Prove the reverse direction of Theorem 9.29.

(13) Answer the questions asked in the first paragraph of Example 5.

(14) Prove or find a counterexample: if f is a continuous function as in Theorem 9.29, then the image of every open set under f is open.

*(15) Prove Theorem 9.29 with "open" changed to "closed" in all three places.

(16) Here is a problem that played a role in the creation of the subject of topology: the city of Königsberg, Germany, had seven bridges, as shown in Figure 9.15. People had wondered whether it was possible to go for a walk that would cross every bridge exactly once. In 1736, using a simple and original argument, Leonhard Euler proved that this could *not* be done.

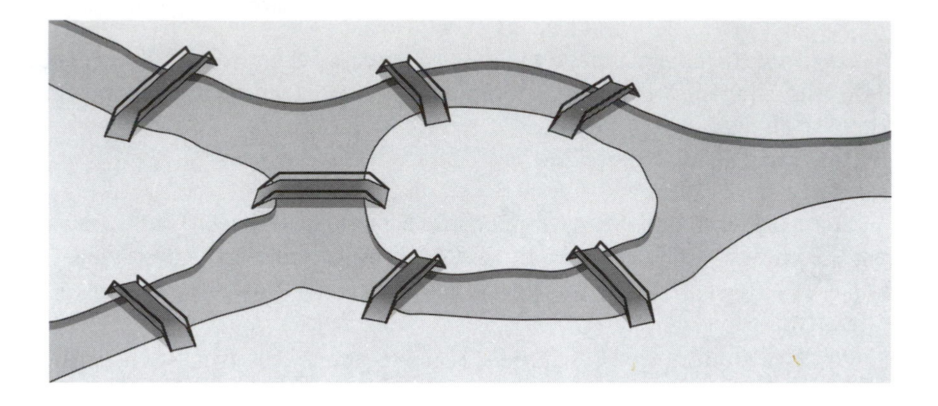

Figure 9.15 The bridges of Königsberg

Prove Euler's result. First, assume there is a path that crosses every bridge exactly once. Next, count the bridges from the upper shore in the figure. Imagine that the walk neither starts nor finishes on this shore. What can you conclude? Now, count the bridges from each of the bodies of land in the figure, and derive a contradiction.

Critique the proofs in the next two exercises. (If necessary, refer to Exercises 4.2 for the instructions for this type of problem).

(17) **Theorem:** A point x is an accumulation point of $A \cup B$ iff x is an accumulation point of A or x is an accumulation point of B.

Proof: x is an accumulation point of $A \cup B$ iff every neighborhood of x contains an infinite set of members of $A \cup B$, iff every neighborhood of x contains an infinite set of members of A or every neighborhood of x contains an infinite set of members of B, iff x is an accumulation point of A or x is an accumulation point of B.

(18) **Theorem:** A point x is an accumulation point of $A \cap B$ iff x is an accumulation point of A and x is an accumulation point of B.

Proof: Analogous to the proof in the previous exercise.

(19) Prove that the examples given in Example 7 are disconnected.

(20) Prove cases 2, 3 and 4 of the forward direction of the proof of Theorem 9.31(a).

(21) Prove the reverse direction of Theorem 9.31(a).

(22) Prove the generalization mentioned in the remark following Theorem 9.31. You may use Theorems 9.23 and 9.24, since these hold in every topological space.

(23) A subset of $\mathbb{R}^2$ or $\mathbb{R}^3$ is called **convex** iff for every pair of points in the set, the entire line segment between those points is also contained in the set. Note that in $\mathbb{R}$, a line segment is just an interval; in other words, Lemma 9.30 shows that connectedness and convexity are the same in $\mathbb{R}$. (By the way, convexity is not strictly a topological notion, since the idea of a straight line is geometric rather than topological.)

(a) Draw several figures in the plane that appear to be connected but are definitely not convex. (Don't try to prove these results.)

*(b) Outline a proof of the important result that a convex set must be connected. ***Hint:*** Try to adapt the proof of the reverse direction of Lemma 9.30.

(c) Using part (b), deduce that $\mathbb{R}^2$ and $\mathbb{R}^3$ are connected.

*9.5 The Construction of the Real Numbers

Although it doesn't matter for most practical purposes, mathematicians like to be very clear about which objects are *primitive* (assumed to exist) and which objects are *defined* from primitive ones. In many situations, there is more than one reasonable approach and none of them is *the* correct one, or even superior to the others. Naturally, the less you assume to exist, the more work remains to define new objects and prove things about them. On the other hand, if you assume the existence of lots of abstract, unintuitive objects, your results are considered less trustworthy as a result.

Our approach to number systems in this book has been to view the real numbers and the natural numbers as primitive. This approach has made it fairly simple to define $\mathbb{Z}$ and $\mathbb{Q}$, the other important subsystems of $\mathbb{R}$. But the concept of the real numbers is somewhat sophisticated, and many mathematicians prefer not to assume their existence and all their important properties. This objection was more serious in the last century, when the theory of functions, sequences, and sets was not as developed as it is now, so real numbers were not understood as clearly as they are now.

One alternative, perhaps the severest one, to assuming the existence of $\mathbb{R}$ is the pure set-theoretic approach. We have already mentioned this approach to the foundations of mathematics, in which *no* number systems are assumed to exist. Instead, the only primitive objects are sets and, from the axioms of set theory, all the familiar number systems are defined and their properties are proved. We do not describe this approach, because it is technically complicated and also it is unnatural to think of numbers as special types of sets.

Instead, we devote this section to outlining an intermediate approach to number systems. In this approach, the natural numbers are the only numbers taken to be primitive. From the axioms for $\mathbb{N}$ (together with set theory), the number systems $\mathbb{Z}$, $\mathbb{Q}$, and $\mathbb{R}$ can successively be defined, or "constructed." This approach reflects a point of view that many famous mathematicians, starting with Pythagoras, have espoused: that the natural numbers are the most fundamental objects in mathematics and mathematics should be based on them, as much as possible. The most famous quote supporting this viewpoint came from Kronecker: "God created the natural numbers; all the rest is the work of man." Kronecker and others of like mind would have preferred to construct $\mathbb{R}$ from $\mathbb{N}$ without using more abstract notions like infinite sets or sequences. Unfortunately, this cannot be done.

A bit of clarification: in this approach we assume not only the natural number axioms in our axiom system, but also that addition, multiplication, and $<$ are defined on $\mathbb{N}$, satisfying all the ordered field axioms except V-8, V-10, V-11 and V-12. It is possible to define these operations and relations instead of assuming their existence, but we won't.

The Construction of the Integers and the Rationals

In the approach to number systems that we just outlined, the first step should be the construction of $\mathbb{Z}$ from $\mathbb{N}$. However, this is definitely the least interesting of the three constructions involved. It is also very similar to (but less instructive than) the construction of $\mathbb{Q}$ from $\mathbb{Z}$. So we omit this first construction. The interested reader, after studying the construction of $\mathbb{Q}$ from $\mathbb{Z}$, can see Exercise 1 for an outline of the omitted construction. We assume that $\mathbb{Z}$ satisfies all the ordered field axioms except V-11 and that $\mathbb{N}$ is the set of all positive elements of $\mathbb{Z}$. In a full treatment of the construction of $\mathbb{Z}$, all these properties would have to be proved.

So let's move on to the construction of the rational numbers from the integers. Recall that the main difference between $\mathbb{Z}$ and $\mathbb{Q}$ is that $\mathbb{Z}$ is not closed under division. If we simply consider all possible fractions of integers, just as formal expressions, we can define the number system $\mathbb{Q}$ from these expressions.

Definition: Let $\mathbb{Q}^*$ be the set of all expressions j/k, where j and $k \in \mathbb{Z}$ and $k \neq 0$.

Remarks: (1) One may object that the word "expression" is too vague to be part of a mathematical definition. The usual way to counter this is to be more set theoretic: define $\mathbb{Q}^*$ to be $\mathbb{Z} \times (\mathbb{Z} - \{0\})$, the set of all ordered pairs of integers in which the second member $\neq 0$. We can then write j/k as an abbreviation for the ordered pair (j, k).

(2) We call the elements of $\mathbb{Q}^*$ **fractions**. Perhaps you are wondering why we are not calling the set of all fractions $\mathbb{Q}$. The answer is that the representation of rational numbers as fractions contains unwanted duplication. For example, 1/2 and 3/6 are different *expressions,* so they are distinct elements of $\mathbb{Q}^*$. But we consider these fractions to be numerically equal; that is, we don't want to think of these fractions as representing or being different numbers. There are several possible solutions to this sticky point:

(a) One remedy, which is fine for informal purposes, is to *define* equality on the set $\mathbb{Q}^*$ in a special way that allows different fractions to be equal. But if we are thinking of fractions as technically being ordered pairs, it's not technically correct to say we consider (1, 2) and (3, 6) to be equal.

(b) A more rigorous approach is to let Q be the set of all fractions in *lowest terms* (with positive denominator, say). Then the duplication is eliminated.

(c) Another rigorous approach is to define an *equivalence relation* on Q*, by which any fractions that should be numerically equal are lumped together. Even though this method entails more work than (b), we use it because it is more instructive and more generally useful:

Definition: We define the relation $\sim$ on Q* by: $j/k \sim m/n$ iff $jn = km$.

Note that $\sim$ is defined strictly in terms of multiplication of integers, not division or fractions. This is important because we only have $\mathbb{Z}$ to work with.

Theorem 9.32: The relation $\sim$ is an equivalence relation on Q*.
Proof: This is Exercise 8 of Section 6.2. ■

As usual, the equivalence class (under $\sim$) of a fraction j/k is denoted $[j/k]$.

Definition: We define Q to be the set of all equivalence classes under $\sim$. Elements of Q are called **rational numbers**.

Note the subtleties of our terminology here. We are saying that 1/2 and 3/6 are two different fractions. Instead of calling them equal, we say they are equivalent under $\sim$. The equivalence class [1/2], which contains both of them as well as 4/8, $(-7)/(-14)$, and so on, is called a single rational number. Of course, *after* Q is constructed, the equation $1/2 = 3/6$ becomes a correct one in this number system.

Since the viewpoint of this section is that Q is not a subsystem of an already defined system, we must define all relevant operations and relations on the new system *and* show that they behave as they should. It's simple to define algebraic operations and relations on Q*, since this set just consists of fractions. But Q consists of equivalence classes of fractions. So we are attempting to define a *quotient structure,* as discussed in Section 8.3; and we must make sure that the operations and relations we define on it are *well defined.*

Definitions: Addition, multiplication, and the relation < on Q are defined as follows.

(a) To add two rational numbers, choose any fraction from each one, add them in the usual way, and take the equivalence class of the fractional sum. In other words, $[j/k] + [m/n] = [(jn + km)/kn]$.

(b) To multiply two rational numbers, choose any fraction from each one, multiply them in the usual way, and take the equivalence class of the fractional product. In other words, $[j/k][m/n] = [jm/kn]$.

(c) To determine whether one rational number is less than another, choose any fraction *with positive denominator* from each one (of course, this is always possible). Then, assuming k and $n > 0$, $[j/k] < [m/n]$ iff $jn < mk$.

Theorem 9.33: The definitions we have given for Q are well defined. That is, for any two rational numbers x and y, the rational numbers $x + y$ and xy are uniquely defined. Also, the truth or falsity of the condition $x < y$ does not depend on which fractions are chosen for x and y.

Proof: We prove only the part about multiplication and leave the others for Exercise 3. Let x and y be rational numbers. To evaluate xy, choose fractions $j/k \in x$ and $m/n \in y$. Then, by definition, $xy = [jm/kn]$. Now consider another pair of choices, $j'/k' \in x$ and $m'/n' \in y$. So we also have $xy = [j'm'/k'n']$. To show the desired uniqueness, we must prove that $[jm/kn] = [j'm'/k'n']$, that is, $jm/kn \sim j'm'/k'n'$. But since j/k and j'/k' are both in x, we have $j/k \sim j'/k'$, which means that $jk' = j'k$. Similarly, $mn' = m'n$. But combining these equations gives $jk'mn' = j'km'n$, which becomes $(jm)(k'n') = (j'm')(kn)$. By definition, this says $jm/kn \sim j'm'/k'n'$. ■

Let's summarize what we've done so far in this section. We defined the set of all fractions Q*, the equivalence relation ~ on Q* (which expresses the familiar notion of two fractions being numerically equal), and the set of rational numbers Q (consisting of the equivalence classes of ~). We then defined +, ×, and < on Q, on the basis of the usual algebra of fractions. And finally we showed the vital fact that these definitions are well defined.

This completes the *definition* of the system of rational numbers. But we can't say our construction is complete until we prove that the number system we've created behaves as it should. The next two theorems serve this purpose.

Theorem 9.34: The algebraic structure Q, with +, ×, and < as we have defined them, is an ordered field.

Proof: There is nothing hard about this proof, but it is tedious since there are so many ordered field axioms. We prove a few and leave several others for Exercise 5. By the way, we don't have to worry about the closure axioms, since Theorem 9.32 showed that for any rational numbers x and y, $x + y$, and xy are unique rational numbers.

Commutativity of Addition: We must show $x + y = y + x$, for any rationals x and y. If $x = [j/k]$ and $y = [m/n]$, then

$$\begin{aligned} x + y &= [j/k] + [m/n] = [(jn + km)/kn] \\ &= [(mk + nj)/nk] \qquad \text{By the commutativity of + and} \times \text{in } \mathbb{Z} \\ &= [m/n] + [j/k] = y + x \end{aligned}$$

Additive Identity: We claim that [0/1] is the additive identity in Q, since $[j/k] + [0/1] = [(j \cdot 1 + k \cdot 0)/k \cdot 1] = [j/k]$.

Transitivity of <: Assume $x < y$ and $y < z$. We must show that $x < z$. Choose fractions for x, y, and z with positive denominators: say $x = [j/k]$, $y = [m/n]$, and $z = [a/b]$. Then by the definition of < on Q, we have $jn < mk$ and $mb < an$. Therefore, $jnb < mkb$ and $mbk < ank$. By the commutativity of multiplication and transitivity of < in N, this gives us $jnb < ank$, or $jbn < akn$. From this it is easy to show (see Exercise 4) that $jb < ak$. By the definition of < on Q, this says $x < z$. ■

The final thing we should show about this new $\mathbb{Q}$ is that it looks just like the old $\mathbb{Q}$, the one we defined as a subset of $\mathbb{R}$. However, since we are not assuming the existence of $\mathbb{R}$ in this section, we should try to accomplish this without directly referring to the old $\mathbb{Q}$, as much as possible. Theorem 9.35 attempts to do this.

Theorem 9.35: (a) Theorem 8.31 still holds for our new $\mathbb{Q}$.
(b) Our new $\mathbb{Q}$ has no proper subfields.
(c) Every ordered field with no proper subfields is isomorphic to our new $\mathbb{Q}$.

Proof: (a) Like the original version of Theorem 8.31, the proof of (a) is a simple extension of the proof of Theorem 8.30. Given an isomorphic embedding $f: \mathbb{Z} \to A$, where A is an ordered field, we can extend f to the set of all fractions $\mathbb{Q}^*$ by the rule $f(m/n) = f(m)/f(n)$. The proof of Theorem 8.31 verified that this rule gives equal outputs for numerically equal fractions. In the terminology of this section, this says precisely that this f with domain $\mathbb{Q}^*$ yields a well-defined function on the quotient structure $\mathbb{Q}$. The rest of the proof is identical to that of Theorem 8.31.

(b) One way to prove (b) is to modify the proof of Theorem 8.9 (see Exercise 6). Or we can use part (a): since the identity map is an isomorphism between $\mathbb{Q}$ and $\mathbb{Q}$, part (a) implies that $\mathbb{Q}$ is the smallest subfield of $\mathbb{Q}$. So $\mathbb{Q}$ has no proper subfields.

(c) Since Theorem 8.31 holds for our new $\mathbb{Q}$, so does Corollary 8.32. ■

Remarks: (1) If we assume the existence of $\mathbb{R}$ for the purpose of directly comparing the new $\mathbb{Q}$ with the old one, then Corollary 8.32 and Theorem 9.35 immediately tell us that the two $\mathbb{Q}$s are isomorphic. Of course, this is to be expected.

(2) The concept of isomorphism allows us to settle another point about our new $\mathbb{Q}$. The system of integers is supposed to be a subset of the rationals. But, literally, $\mathbb{Z}$ is not a subset of the $\mathbb{Q}$ we have just defined, since this $\mathbb{Q}$ consists of sets of ordered pairs of integers. However, by Theorems 8.30 and 9.33, this $\mathbb{Q}$ contains an isomorphic copy of $\mathbb{Z}$, and that's essentially the same thing as containing $\mathbb{Z}$.

(3) Section 8.3 mentions that $\mathbb{Z}$ is an integral domain and that one can construct what's called the **field of quotients** of any integral domain. Our definition of $\mathbb{Q}$ from $\mathbb{Z}$ in this section is precisely that construction. So if you understand how we have just defined $\mathbb{Q}$, you understand one of the most important techniques of abstract algebra.

Construction of the Real Numbers

There are at least two standard ways to construct $\mathbb{R}$ from $\mathbb{Q}$. The most common one, based on what are called **Cauchy sequences** of rational numbers, is similar to our construction of $\mathbb{Q}$ in that a real number is defined to be an equivalence class of these sequences. But in order not to overload you with equivalence relations, we use a different approach due to Richard Dedekind. Dedekind's method is more awkward algebraically than the method of Cauchy sequences, but it is more geometric and so perhaps more intuitive. The letters r, s, and t continue to stand for rational numbers.

Definitions: A **Dedekind cut** (or simply a **cut**) is a nonempty, proper subset A of $\mathbb{Q}$ with the following properties:

(i) A has *no* largest element.

(ii) Whenever $r \in A$ and $s < r$, then $s \in A$.

A **real number** is defined to be a Dedekind cut, and the set of all reals is denoted $\mathbb{R}$.

When we talk about the complement of a cut A, we mean relative to $\mathbb{Q}$. That is, A' means $\mathbb{Q} - A$. Note that A' is also a nonempty, proper subset of $\mathbb{Q}$.

Here is the idea behind Dedekind cuts: if A is a cut, then A and A' partition $\mathbb{Q}$ into two nonempty subsets. Furthermore, by property (ii) in the definition of a cut, every number in A is less than every number in A'. So A and A' can be thought of as the *left* and *right* sets of the partition, respectively. In other words, a cut is essentially a way of cutting the set of rational numbers into two pieces, as if it were done with a pair of scissors at one point.

What does this have to do with real numbers? To see this, think of $\mathbb{R}$, as usual, as a line, with $\mathbb{Q}$ forming a sort of skeleton of points that occupies the whole line in a sense but also has holes all over the place. Remember that we already have $\mathbb{Q}$ constructed and we are trying to define $\mathbb{R}$. We might like to say that a real number is defined to be a point on the line, but this would be a circular definition because we haven't defined what a line is. But a point on the line can be thought of as a place where you can put your (perfectly thin) scissors and cut the line in two, and thereby cut $\mathbb{Q}$ in two pieces. Therefore, Dedekind's idea of defining real numbers as cuts was a brilliant way of being both geometric and rigorous (see Figure 9.16).

Note that if we cut the real line at a point, the cut can be made at a rational number or an irrational number. If it's made at an irrational number, then $\mathbb{Q}$ is unambiguously split into two pieces. But if the cut is made at a rational number, does that number fall to the left or to the right? The answer to this question doesn't matter as long as we are consistent about it. Property (i) in the definition of a cut says we always assume that such a number falls to the right. This should be borne in mind in what follows.

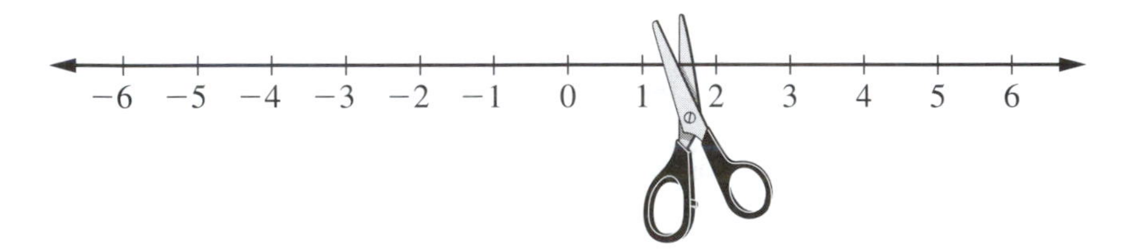

Figure 9.16 A Dedekind cut

Notation: For each rational r, let $r^* = \{s \in Q \mid s < r\}$.

Theorem 9.36: (a) For any rational r, r^* is a cut.

(b) r^* is the unique cut such that the least member of its complement is r.

(c) If $r < s$, then $r^* \subset s^*$ (hence, $r^* \neq s^*$).

Proof: See Exercise 7. ■

We identify each rational number r with the cut r^*. More rigorously, Theorem 9.36(a) tells us we have a *function* * from Q to R, where we write r^* instead of $*(r)$. And Theorem 9.36(c) implies that this function is one-to-one. So, if we denote the range of * by Q′, we have a bijection between Q and Q′, which we soon show is an isomorphism. (Of course, we can't show it's an isomorphism until we define addition, multiplication, and < on our constructed R.)

Thus we have a situation similar to the one that occurred in the previous subsection: when we constructed Q from Z, we didn't literally make Z a subset of Q, but we defined a subset of Q that was isomorphic to Z and so could be identified with it. In the same way, Q is not literally a subset of the R we are constructing, but the crucial thing is that we can identify Q with Q′. Theorem 9.43 will make this rigorous.

We can say there are two types of cuts: if A is a cut, then A cannot have a largest member, but A' may or may not have a smallest one. If A' does have a smallest member, say r, then Theorem 9.36(b) says that $A = r^*$, which means that A is a *rational* real number. But if A' has no least member, then A is not in Q*, and A is an *irrational* real.

Example 1: Consider $5^* = \{r \in Q \mid r < 5\}$. We know that this is a cut, and we think of it as the number 5 in the real number system we are constructing.

Example 2: The set $\{r \in Q \mid r \leq 5\}$ is not a cut, because it has a largest element and so violates property (i) in the definition of a cut.

Example 3: The set $\{r \in Q \mid r^2 < 2\}$ is also not a cut, because it violates property (ii) in the definition of a cut; -1 is in the set, but -2 is not.

Example 4: We can modify Example 3 by letting $A = \{r \in Q \mid r^2 < 2 \text{ or } r < 0\}$. It is easy to show (see Exercise 8) that A is a cut. To figure out the numerical value of A, think about where you would need to cut the real number line so that the set to the left of the cut would be precisely A. Clearly, the cut would have to be made at the number $\sqrt{2}$. In fact, we soon show that A is $\sqrt{2}$ in the R we are constructing. The proof is similar to that of Theorem 9.3 and is also left for Exercise 8.

Example 4 relates the use of cuts, as a way of defining real numbers, to the material in Section 9.1 that showed how irrational numbers like $\sqrt{2}$ can be defined as least upper bounds of sets. Indeed, for any cut, the real number it defines is just its least upper bound. So instead of using cuts, we could have used arbitrary bounded nonempty sets of rationals to define real numbers. But different sets of rational numbers can have the same least upper bound, so this alternative approach would have required equivalence classes again. The nice thing about using cuts is that equivalence relations and classes are avoided.

It is now time to define the algebraic structure on ℝ. We resume using the letters x, y, and z for real numbers; but remember that they are now cuts.

Definitions: For any x and y, $x < y$ means $x \subset y$. Also, $x + y$ denotes the set $\{r + s \mid r \in x \text{ and } s \in y\}$.

The intuition behind these definitions should be clear to you. Since each real number is being represented as the set of rationals to the left of it, "less than" corresponds to "proper subset." Addition is also logical: for example, if $x = \{r \mid r < 3\}$ and $y = \{s \mid s < 6\}$, it's clear that $x + y$ turns out to be $\{r \mid r < 9\}$, as it should be.

The definition of multiplication is much more complex than that of addition; in fact, this is the only unappealing feature of the cut approach to reals. We therefore prove some things about addition and the ordering before defining multiplication.

Lemma 9.37: For any reals x and y, $x + y$ is also a real number.

Proof: The set $x + y$ is the set of all sums of a number in x and a number in y. Since x and y are both nonempty, so is $x + y$. Next, note that both x and y are proper subsets of ℚ. So say $r \notin x$ and $s \notin y$. By property (ii) in the definition of a cut, it follows that every number in x is less than r, and every number in y is less than s. Therefore, $r + s \notin x + y$, so $x + y \subset$ ℚ.

Now we show that $x + y$ does not have a greatest element: assume $t \in x + y$. Then $t = r + s$, for some $r \in x$ and $s \in y$. But since x and y are cuts, neither of them has a largest element. So there are $r' \in x$ and $s' \in y$ with $r' > r$ and $s' > s$. Then $r' + s' \in x + y$, and $r' + s' > t$. So t is not the greatest element in $x + y$. Since t was an arbitrary member of $x + y$, $x + y$ has no greatest element.

The proof that $x + y$ satisfies property (ii) is left for Exercise 9. ■

Lemma 9.38: (a) The system ℚ is Archimedean; that is, ℕ is unbounded in ℚ.

(b) For any cut A and any positive rational r, there is an $s \in A$ with $s + r \notin A$.

Proof: (a) This is simple: if r is a rational number, then $r = m/n$ for some integers m and n. Then $|m| + 1$ is a natural number that is larger than r.

(b) Let any cut A and positive rational r be given. Let $t \in A$ and $t' \notin A$. Since A is a cut, $t' > t$. By part (a), there is a natural number n such that $n \geq (t' - t)/r$. Since $r \geq 0$, $nr \geq t' - t$, and $t + nr \geq t'$. Therefore, $t + nr \notin A$. By the well-ordering property of ℕ, there is a least such n. Using this n, let $s = t + nr - r$. So $s + r \notin A$, and by the choice of n, it follows that $s \in A$. ■

Theorem 9.39: With the addition operation and ordering relation we have defined, ℝ satisfies all the real number axioms (group V of Appendix 1) that do not mention multiplication. The additive identity is the cut 0^*. The additive inverse of any cut x is defined as follows: let $B = \{-r \mid r \notin x\}$. If B has no largest element, then $-x = B$. Otherwise, $-x$ consists of all of B except its largest element.

Proof: We go through the various axioms, in order:

Additive Closure: Lemma 9.37 proved this.

Additive Associativity: Given cuts x, y, and z, we must show that $(x + y) + z = x + (y + z)$. Assume $r \in (x + y) + z$. By definition, this means $r = s + t$, for some $s \in x + y$ and $t \in z$. Similarly, $s = u + v$ for some $u \in x$ and $v \in y$. So $r = (u + v) + t = u + (v + t)$. But note that $v + t \in y + z$, and so $r \in x + (y + z)$. So we have shown that $(x + y) + z \subseteq x + (y + z)$. The proof of the other direction is the same.

Additive Commutativity and Additive Identity: These are left for Exercise 10.

Additive Inverses: We first need to show that whenever x is a cut, so is $-x$. We leave this for Exercise 10.

Given any x, we must also show that $x + -x = 0^*$, the set of all negative rationals. So first assume $t \in x + -x$. Then $t = r + s$, with $r \in x$ and $s \in -x$. By our description of $-x$, we must have $s = -s_1$, for some $s_1 \in x'$. But from the facts that $r \in x$ and $s_1 \in x'$, we know that $r < s_1$. Therefore $t = r + s = r + -s_1 < 0$, so $t \in 0^*$. So we have $x + -x \subseteq 0^*$.

Conversely, assume $t \in 0^*$. Thus $t < 0$. We apply Lemma 9.38(b) to the cut x and the positive number $-t/2$. Thus we have $s \in x$ and $s - t/2 \notin x$. So $s - t > s - t/2$, which implies that $s - t \notin x$ and $s - t$ is not the smallest number in x'. Therefore, $t - s$ is in $\{-r \mid r \notin x\}$, and $t - s$ is not the largest number in this set. So, by definition of $-x$, $t - s$ is in $-x$. Since $s \in x$, we obtain that $t \in x + -x$, as desired.

(Axiom V-12 mentions multiplication indirectly, since it's about the multiplicative identity, so we leave it until later.)

Irreflexivity: By definition, $x < x$ would mean $x \subset x$, which is impossible.

Transitivity: The relation $<$ is transitive because $\subset$ is transitive.

Trichotomy: This is easiest to prove indirectly; we leave it for Exercise 10.

Axiom V-16: Let x, y, and z be cuts such that $x < y$. We want to show that $x + z < y + z$. If $r \in x + z$, then $r = s + t$ for some $s \in x$ and $t \in z$. Since $x \subset y$, $s \in y$. So $r \in y + z$. Thus we have $x + z \subseteq y + z$.

It remains to find a rational in $y + z$ that is not in $x + z$. Let $r \in y$ and $r \notin x$. Since y has no largest element, we can find $s \in y$ with $s > r$. Let $c = s - r$. Since $c > 0$, use Lemma 9.38(b) to get $t \in z$ such that $t + c \notin z$. Since $s \in y$ and $t \in z$, $s + t \in y + z$. We claim that $s + t \notin x + z$, for any number in x is less than $s - c$ and any number in z is less than $t + c$.

Completeness: Surprisingly, this is one of the easier axioms to verify. Let A be a set of reals with an upper bound y. Note that A is a set of sets. Exercise 10 asks you to show that $\bigcup A$ is a cut and is the least upper bound of A. ■

Corollary 9.40: With $-x$ as described in the previous theorem:

(a) $-x$ is the unique additive inverse of x.

(b) $-(-x) = x$

(c) $-(0^*) = 0^*$

(d) $x > 0^*$ iff $-x < 0^*$

(e) $x < 0^*$ iff $-x > 0^*$

(f) Trichotomy (with *exclusive* or) holds for cuts.

Proof: (a) and (b) follow directly from Theorem 9.39, as in Theorems A-3 and A-6 of Appendix 2. The remaining parts are also easy and are left for Exercise 11. ■

The main purpose of Corollary 9.40 is to justify the following definition by cases.

Definition: The product xy of any two reals x and y is defined as follows.
(a) If either x or y is 0^*, then so is xy.
(b) If x and $y > 0^*$, then $xy = \{r \mid r \le 0\} \cup \{st \mid s \in x, t \in y$, and s and $t > 0\}$.
(c) If x and $y < 0^*$, then $xy = (-x)(-y)$, as defined in part (b).
(d) If $x < 0^*$ and $y > 0^*$, then $xy = -[(-x)y]$, where $(-x)y$ is defined as in (b).
(e) If $x > 0^*$ and $y < 0^*$, then $xy = -[x(-y)]$, where $x(-y)$ is defined as in (b).

Example 5: Let $x = 3^*$ and $y = 7^*$. If xy were defined to be $\{st \mid s \in x$ and $t \in y\}$, it would be $\mathbb{Q}$, which is not a cut. But the restrictions in (b) of the definition of xy prevent s and t from being negative, with the result that $xy = 21^*$.

Theorems 9.41 and 9.42 show that our constructed $\mathbb{R}$ looks exactly as it should; they represent the culmination of this subsection, analogous to Theorems 9.34 and 9.35.

Theorem 9.41: With the addition, multiplication, and ordering relation we have defined, $\mathbb{R}$ is a complete ordered field. The multiplicative identity is 1^*. We defer the definition of multiplicative inverses until it comes up in the proof.

Proof: In view of Theorem 9.39, we need to verify only those ordered field axioms that pertain to multiplication. Since the definition of multiplication was by cases, all these proofs must be by cases, and some of them have quite a few cases to consider. We gloss over or omit some of the proofs because they are very long and tedious.

Multiplicative Closure: We must show that whenever x and y are cuts, then so is xy. We first deal with the case x and $y > 0^*$. So 0^* is a proper subset of both x and y. Therefore x and y both contain some positive numbers. By definition, xy contains 0 and all negative numbers, so it's nonempty. Also, if a and b are positive rationals with $a \notin x$ and $b \notin y$, then clearly every positive number in xy is less than ab. This shows $xy \subset \mathbb{Q}$.

To show xy has no greatest element, if a and $b > 0$, $a \in x$, and $b \in y$, then $ab \in xy$. So xy contains positive numbers. But if $r > 0$ and $r \in xy$, then by definition $r = ab$ for some positive a and b with $a \in x$ and $b \in y$. But then there are rationals a' and b' such that $a' > a$, $a' \in x$, $b' > b$, and $b' \in y$. It follows that $a'b' \in xy$ and $a'b' > r$.

Finally, we need that $r \in xy$ and $s < r$ imply $s \in xy$. If $s \le 0$, then $s \in xy$ by definition. If $s > 0$, then $r > 0$ and we can write $r = ab$ with a and b as in the previous paragraph. Then $s = a(bs/r)$. But bs/r is a positive rational less than b, so it's in y. Therefore, $s \in xy$. This completes the proof when x and $y > 0^*$.

In every other case, xy is defined to be 0^*, which we know is a cut, another product of two positive cuts, or minus the product of two positive cuts. By Theorem 9.39, minus a cut is always a cut. Thus, in all cases, xy is a cut.

Multiplicative Associativity: This is a simple but tedious argument with several cases. We omit it.

Multiplicative Commutativity: We leave this for Exercise 12.

Distributivity: This involves many detailed cases. We omit it.

Multiplicative Identity: We have to show $x \cdot 1^* = x$. We know that $1^* > 0$, so we just need to consider cases on x. We leave the case $x > 0^*$ for Exercise 12. We have $0^* \cdot 1^* = 0^*$ by definition. Finally, assume $x < 0^*$. Then,

$0^* = 0^* \cdot 1^*$	By the previous case
$= (x + -x) \cdot 1^*$	By Theorem 9.39
$= x \cdot 1^* + (-x) \cdot 1^*$	By the distributive property
$= x \cdot 1^* + -x$	By Corollary 9.40(e) and the first case

We can now add x to both sides of the equation $0^* = x \cdot 1^* + -x$ to obtain $x = x \cdot 1^*$.

Multiplicative Inverses: We need to show that every nonzero cut x has a reciprocal. We do the case $x > 0^*$, and leave the case $x < 0^*$ for Exercise 12. Given $x > 0^*$, let $y = \{r \in \mathbb{Q} \mid \exists s \in x' \ (r < 1/s)\}$. Then y is a cut; we leave the proof of this for Exercise 12. We now show that $xy = 1^*$, so $y = x^{-1}$. *[Before continuing, you might want to verify this definition with a simple example: for instance, if $x = 2^*$, then $y = (1/2)^*$.]* Assume $r \in xy$. If $r \leq 0$, then $r \in 1^*$ by the definition of 1^*. If $r > 0$, then $r = ab$, with $a \in x$, $b \in y$, and a and $b > 0$. By the definition of y, $b < 1/c$ for some $c \notin x$. So $a < c$; thus $1/c < 1/a$. So $b < 1/a$, and therefore $ab < 1$; that is, $r \in 1^*$, and this shows $xy \subseteq 1^*$.

Conversely, assume $r \in 1^*$. So $r < 1$ and $1/r > 1$. By a modification of Lemma 9.38(b) (see Exercise 13), there is an $s \in x$ such that $s/r \notin x$. Choose such an s and also $t \in x$ with $t > s$. This implies $r/t < r/s$, and since r/s is the reciprocal of s/r, $r/t \in y$. Therefore $r = t(r/t) \in xy$. So $1^* \subseteq xy$.

Axiom V-12: Since $0 \in 1^*$ but $0 \notin 0^*$, $0^* \neq 1^*$.

Axiom V-17: One way to prove this axiom is to use the obvious fact that the product of two positive cuts is positive. We leave the details for Exercise 12. (Be careful not to assume that $(-x)y = -(xy)$. If you need this fact, prove it.) ■

So we now know that the $\mathbb{R}$ we have constructed satisfies all the properties that the reals are supposed to satisfy. To strengthen this, we conclude this section with a result that is similar to Corollary 8.32. It says that there is, up to isomorphism, only one field satisfying all the real number axioms.

Theorem 9.42: Every complete, ordered field is isomorphic to $\mathbb{R}$. Therefore, any two such fields are isomorphic.

Proof: We just outline the proof. Let A be any complete ordered field. By Theorem 8.31, there is an isomorphic embedding $g: \mathbb{Q} \to A$. We use g to define an isomorphism $f: \mathbb{R} \to A$. If $x \in \mathbb{R}$, then $x \subset \mathbb{Q}$, and x has an upper bound. If r is an upper bound of x, then $g(r)$ is an upper bound of the set $g(x)$ (that is, $\{g(s) \mid s \in x\}$), because g is order preserving. Since A is complete, $g(x)$ has a least upper bound. Let $f(x)$ be the least upper bound of $g(x)$.

It is obvious that f is order preserving and so one-to-one. It is also straightforward to show that f preserves addition and multiplication, so we omit these proofs (see

Exercise 14). The final step is to show that f is onto A, which we do indirectly. Assume $y \in A$ but $y \notin \mathrm{Rng}(f)$. If y is greater than every member of Rng(g), then A would not be archimedean, and this would violate Theorem 9.4. On the other hand, if y is equal to or less than some members of Rng(g), then $\{r \in \mathbb{Q} \mid g(r) < y\}$ is easily seen to be a cut, and f must map this cut to y. ■

This proof actually shows that f is the *unique* order-preserving isomorphism between $\mathbb{R}$ and A. Also, it should be noted that Theorem 9.42 can be proved in much the same way without any reference to Dedekind cuts and our new $\mathbb{R}$.

Finally, we prove that the set $\mathbb{Q}'$ defined earlier is indeed an isomorphic copy of the rational numbers.

Theorem 9.43: The function $f: \mathbb{Q} \to \mathbb{R}$ defined by $f(r) = r^*$ is an isomorphic embedding (of *ordered* fields).

Proof: By Theorem 9.36(c), this f is order preserving (and thus one-to-one). It remains to show $(r + s)^* = r^* + s^*$ and $(rs)^* = r^*s^*$ (see Exercise 15) ■.

Exercises 9.5

*(1) (a) Using the construction of $\mathbb{Q}$ from $\mathbb{Z}$ as a guide, show how to define or construct $\mathbb{Z}$ from $\mathbb{N}$. ***Hint:*** Instead of using fractions of integers, use *differences* of *natural numbers* as the basis of $\mathbb{Z}$. Then define the sensible equivalence relation that expresses numerical equality of these expressions.

(b) Define +, ×, and < on $\mathbb{Z}$.

(c) Prove that the operations + and × and the relation < are well defined.

(2) When working with rational numbers, it is useful to know that every fraction can be put in lowest terms. A more rigorous statement of this fact is that for every r in $\mathbb{Q}$ there are unique $n \in \mathbb{N}$ and $m \in \mathbb{Z}$ such that $r = \{km/kn \mid k \in \mathbb{Z}\}$. Prove this. You will need some results from Section 8.2.

(3) Prove the parts of Theorem 9.33 involving addition and <.

(4) (a) Complete the proof that < is transitive (in Theorem 9.34) by proving this lemma: whenever c, d, and n are integers such that $cn < dn$ and $n > 0$, then $c < d$.

(b) Explain why part (a) cannot be proved simply by dividing both sides of the inequality $cn < dn$ by n.

(5) Prove the following parts omitted from the proof of Theorem 9.34: $\mathbb{Q}$ satisfies ordered field axioms V-6, V-7, V-11, V-13, and V-16.

(6) Prove Theorem 9.35(b) by modifying the proof of Theorem 8.9.

(7) Prove Theorem 9.36.

(8) Show that the set $A = \{r \in \mathbb{Q} \mid r < 0 \text{ or } r^2 < 2\}$, defined in Example 3, is a cut, and also that $A^2 = 2^*$.

(9) Prove the final claim in the proof of Lemma 9.37.

*(10) (a) Prove the additive commutativity part of Theorem 9.39.
(b) Prove the additive identity part of Theorem 9.39.
(c) Prove the additive inverses part of Theorem 9.39.
(d) Prove the trichotomy part of Theorem 9.39.
(e) Finish the completeness part of Theorem 9.39.

(11) Prove parts (c) through (e) of Corollary 9.40.

*(12) (a) Prove the multiplicative commutativity part of Theorem 9.41.
(b) Prove the $x > 0^*$ case of the multiplicative identity part of Theorem 9.41.
(c) Prove that y is a cut, in the multiplicative inverse part of Theorem 9.41.
(d) Prove the $x < 0^*$ case of the multiplicative inverse part of Theorem 9.41.
(e) Prove the part of Theorem 9.41 for axiom V-17.

*(13) Prove this modification of Lemma 9.38(b) that is used in the proof of Theorem 9.41: if x is a cut, $x > 0^*$, and $r > 1$, then there is an $s \in x$ such that $rs \notin x$.

(14) Show that f preserves addition and multiplication in the proof of Theorem 9.42.

(15) Show that $(r + s)^* = r^* + s^*$ and $(rs)^* = r^*s^*$ and thus complete the proof of Theorem 9.43.

Suggestions for Further Reading: For a more complete introduction to real analysis and topology, see Rudin (1976), Feferman (1989), or Royden (1988). Rudin's book carries out the construction of the reals via Dedekind cuts in more detail than this text does.

Chapter 10

The Complex Number System

10.1 Complex Numbers

Most of the number systems we have discussed so far are subsets of the real number system. Now we investigate the most important number system that is not a subset of $\mathbb{R}$.

Imaginary and complex numbers have an interesting history. As you might guess, imaginary numbers were first encountered in the solving of quadratic equations, since the quadratic formula often includes an expression denoting the square root of a negative number. In the sixteenth century, the Italian mathematicians Geronimo Cardano and Rafael Bombelli first discussed the appearance of imaginary numbers in the solution to various quadratic and cubic equations. But these expressions were dismissed as meaningless curiosities by them and for over 200 years thereafter. Finally, after Euler, Gauss, and others began to work with these expressions seriously, the complex number system gained full acceptance in the nineteenth century.

It turns out to be very fruitful to expand the real number system by adjoining a quantity whose square is a negative number. It's not considered important to worry about whether such a quantity exists. One of the most important developments in mathematics in the last two hundred years has been the realization that mathematical systems don't have to be directly based on the real world to be worth studying. In fact, new branches of mathematics that seem completely useless and unrelated to reality often turn out to have important applications. Complex numbers are a perfect example of this phenomenon.

Definitions: A **complex number** is an expression of the form $a + bi$, where a and b are in $\mathbb{R}$. We define addition and multiplication of complex numbers in the obvious way:

$$(a + bi) + (c + di) = (a + c) + (b + d)i$$

$$(a + bi)\cdot(c + di) = (ac - bd) + (ad + bc)i$$

The set of all complex numbers is denoted by $\mathbb{C}$. In this chapter, the variable z always denotes a complex number.

Leonhard Euler (1707–1783) was born and raised in Basel, Switzerland. By the age of twenty he had displayed enough mathematical talent and ability to receive an appointment at the new St. Petersburg Academy in Russia. During the course of his career, Euler also had academic positions in Basel and Berlin, but he spent the bulk of his career—over thirty years, including the last seventeen years of his life—in St. Petersburg. Euler's personality was not to everyone's liking, and he felt much more accepted in St. Petersburg than elsewhere.

Euler was a true genius, the most important mathematician of the eighteenth century, probably the most versatile mathematician of all time, and definitely the most prolific. He published over 500 books and papers in his lifetime, and another 360 or so were published posthumously. He worked in every branch of mathematics and had extensive knowledge of many sciences, history, theology, languages, and more. Euler had an amazing memory, on which he relied greatly during the last fifteen years of his life. His productivity was barely diminished during that period, even though he was totally blind.

It is difficult to synopsize Euler's mathematical achievements. He made enormous advances in calculus, the theory of sequences and series, and many areas of applied mathematics. He made important contributions in algebra, number theory, and geometry. He crystallized the modern concept of a function, and his solution to the Königsberg bridge problem (Section 9.4, Exercise 16) helped create the subject of topology. Euler also invented a great deal of useful mathematical notation, including e for the main constant of exponential functions, i for the imaginary unit, $f(x)$ for functions, and Σ for summation.

Euler probably used nonrigorous methods more than any other important mathematician. Such methods include the formal manipulation of infinite series without careful attention to convergence, and the use of infinite and infinitesimal quantities. Euler's exceptional insight—and maybe a bit of luck—enabled him to do this without making any serious errors. In contrast, Archimedes used similar methods to great advantage but always backed up his discoveries with rigorous proofs. On the other hand, Johannes Kepler is a perfect example of a mathematician who also used infinitesimals to advantage but occasionally made mistakes in doing so.

Remarks: (1) The definition of a complex number is similar to the construction of the rationals in Chapter 9, in that we define something to be a type of expression.

Again, it is more rigorous to define a complex number to be an ordered pair of real numbers and then to agree to write $a + bi$ for the complex number (a, b). One benefit of the rigorous definition is that it makes it clear when two complex numbers are *equal*:

$$a + bi = c + di \quad \text{iff} \quad a = c \text{ and } b = d.$$

(2) Note that we have defined algebraic operations on C but no ordering relation. Exercise 2 makes it clear why not.

(3) We *identify* any real number a with the complex number $a + 0i$. As in Chapter 9, the technical meaning of such a statement is that if we define a function f from R to C by mapping any real a to the complex number $a + 0i$, then f is an isomorphism between R and the range of f. (Since C does not have an ordering defined on it, f is just a ring isomorphism, not an ordered ring isomorphism.) This allows us to view R as a subset of C. The proof that this f is an isomorphism is simple (see Exercise 3).

Definitions: For any complex number $a + bi$, a is called its **real part** and b (not bi!) is called its **imaginary part**. Just as a complex number whose imaginary part is 0 is considered to be a real number, a complex number whose real part is 0 is called a **pure imaginary number.**

For any complex number $z = a + bi$, its **conjugate**, denoted $\overline{z}$, is $a - bi$.

Theorem 10.1: For any complex numbers, we have:

(a) $z = \overline{\overline{z}}$

(b) If $z = a + bi$, then $z\overline{z}$ is the real number $a^2 + b^2$.

(c) $\overline{(z_1 + z_2)} = \overline{z_1} + \overline{z_2}$

(d) $\overline{(z_1 z_2)} = \overline{z_1}\,\overline{z_2}$

(e) $\overline{(z^n)} = (\overline{z})^n$, for every natural number n

Proof: We prove (c) and leave the other parts for Exercise 4. Say $z_1 = a + bi$ and $z_2 = c + di$. So $z_1 + z_2 = (a + c) + (b + d)i$, and

$$\overline{(z_1 + z_2)} = (a + c) - (b + d)i = (a - bi) + (c - di) = \overline{z_1} + \overline{z_2} \quad \blacksquare$$

Parts (c) and (d) of this theorem, together with the simple fact that the function $f(z) = \overline{z}$ is a bijection on C, show that this function is in fact an *isomorphism* on C. This tells us that any algebraic property of the number i is also true of $-i$. In other words, there is no *structural* way of distinguishing between i and $-i$.

One important use of conjugates is in computing reciprocals and quotients of complex numbers, as Theorem 10.2 demonstrates.

Theorem 10.2: The system C is a field. Naturally, the additive identity is $0 + 0i$, the multiplicative identity is $1 + 0i$, and the additive inverse of any number $a + bi$ is

$-a + (-b)i$. Less obviously, if $z = a + bi \neq 0$, then its multiplicative inverse is $(1/z\bar{z})\bar{z}$, that is, $(a/z\bar{z}) + (-b/z\bar{z})i$.

Proof: We must show that C, with the algebraic operations defined earlier and the identity and inverse elements specified above, satisfies all twelve field axioms. Most of these are quite simple and we leave them to Exercise 5. Naturally, we must use the fact that $\mathbb{R}$ is a field to prove that C is also one. Let's verify the multiplicative inverse property for C: if $z \neq 0$, then $z = a + bi$, where either a or b is nonzero. Therefore, $z\bar{z} = a^2 + b^2$, a positive real number. But then $(1/z\bar{z})z\bar{z} = 1$, which means $[(1/z\bar{z})\bar{z}]z = 1$, by the associative and commutative laws of multiplication. So $(1/z\bar{z})\bar{z}$ is the multiplicative inverse of z. ■

Example 1: The proof of Theorem 10.2 indicates how to compute z^{-1} for any nonzero z: just write the fraction $1/z$ and then multiply its numerator and denominator by $\bar{z}$. This creates a denominator that is a positive real number, and then it's easy to express the answer in the form of a complex number. For example,

$$(3 + 2i)^{-1} = 1/(3 + 2i) = (3 - 2i)/[(3 + 2i)(3 - 2i)] = (3 - 2i)/13 = 3/13 + (-2/13)i$$

$$(1 - i)^{-1} = 1/(1 - i) = (1 + i)/[(1 - i)(1 + i)] = (1 + i)/2 = 0.5 + 0.5i$$

$$i^{-1} = 1/i = (-i)/[i(-i)] = -i/1 = -i$$

$$(2 + i)/(2 - i) = (2 + i)^2/[(2 - i)(2 + i)] = (2 + i)^2/5 = (3 + 4i)/5 = 0.6 + 0.8i$$

Definition: The **absolute value** or **modulus** of any complex number $z = a + bi$, denoted $|z|$, is the real number $\sqrt{a^2 + b^2}$. (For real numbers, this is consistent with the usual definition of absolute value; see Exercise 13.)

Trigonometric Representation of Complex Numbers

Much of the beauty and the usefulness of complex numbers stems from the geometric and trigonometric ways of viewing them. The geometric approach is simply this: since a complex number is technically an ordered pair of real numbers, it can be plotted as a point in a coordinate plane. More technically, there is a one-to-one correspondence between C and the set of all points in a coordinate plane. The **complex plane** is drawn just like an ordinary rectangular coordinate system, except that the axes are usually labeled Re and Im, instead of x and y (see Figure 10.1). Note the following simple properties of this representation:

(1) Pure real numbers are plotted on the real (horizontal) axis.

(2) Pure imaginary numbers are plotted on the imaginary (vertical) axis.

(3) For any z, its conjugate $\bar{z}$ is its mirror image across the real axis.

(4) For any z, $|z|$ is the distance from the origin to z (or the *magnitude* of z if it's viewed as a vector).

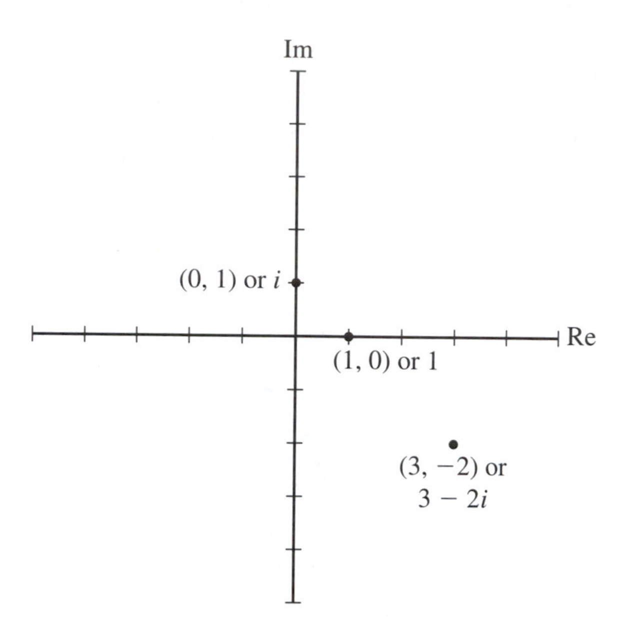

Figure 10.1 The complex plane

(5) For any z_1 and z_2, their sum as complex numbers is the same point as their sum as vectors. So addition of complex numbers can be done geometrically by the usual parallelogram method, if desired.

(6) For any complex number z and *real* number c, the complex number cz is the same point as the scalar c times the vector z.

Thus the geometric representation of complex numbers is helpful for visualizing operations on $\mathbb{C}$. But the main benefits come from the trigonometric representation. Let $z = a + bi$ be any complex number. Then the point (a, b) in the complex plane can also be represented in polar coordinates (r, θ) (see Figure 10.2). In polar coordinates, r is normally allowed to be negative; but when used in conjunction with complex numbers, we insist that $r \geq 0$. With this restriction, r is uniquely determined by z, but θ is not.

Recall these rules for converting polar coordinates to rectangular and vice versa:

$$a = r \cos \theta \qquad\qquad b = r \sin \theta$$

$$r = \sqrt{a^2 + b^2} = |z| \qquad \theta = \arctan (b/a)$$

Thus we have $z = a + bi = r \cos \theta + (r \sin \theta)i = r(\cos \theta + i \sin \theta)$.

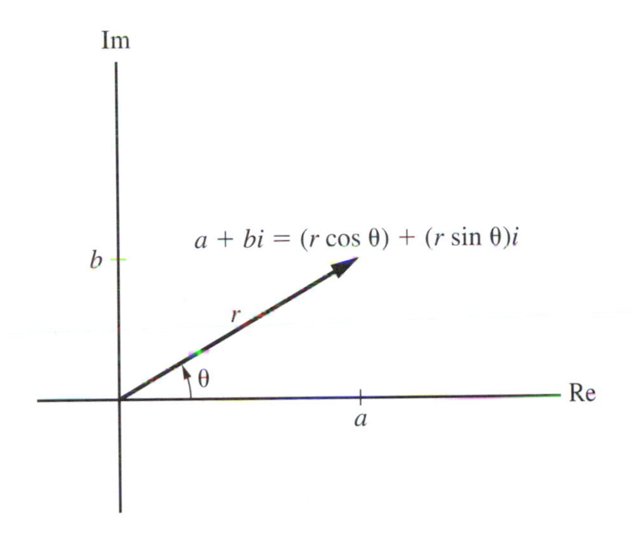

Figure 10.2 Trigonometric (or polar) form of a complex number

Definitions: The last expression, $r(\cos \theta + i \sin \theta)$, is called the **trigonometric form** of z, and the angle θ is called the **argument** of z, abbreviated $\arg(z)$. (Since θ is not unique, the uses of the word "the" in this definition are technically inaccurate. This is fine, provided we are always careful to remember that a complex number has many possible arguments.)

Notation: We write cis θ as an abbreviation for the expression "$\cos \theta + i \sin \theta$." This is a harmless and convenient piece of mathematical slang. Note that the point specified by cis θ is the unit vector in the direction θ.

The following theorem and its corollaries are the main point of trigonometric form. since they give simple formulas for products, quotients, powers, and roots of complex numbers. For the rest of this section, we always assume $z = r \operatorname{cis} \theta$, $z_1 = r_1 \operatorname{cis} \theta_1$, and $z_2 = r_2 \operatorname{cis} \theta_2$.

Theorem 10.3: For any z_1 and z_2, $z_1 z_2 = (r_1 r_2) \operatorname{cis} (\theta_1 + \theta_2)$.

Proof: This is a straightforward computation, but it requires use of the sum-of-angles formulas:

$$\sin (\theta_1 + \theta_2) = \sin \theta_1 \cos \theta_2 + \cos \theta_1 \sin \theta_2$$
$$\cos (\theta_1 + \theta_2) = \cos \theta_1 \cos \theta_2 - \sin \theta_1 \sin \theta_2$$

See Exercise 6. ■

Corollary 10.4: If $z \neq 0$, then $z^{-1} = (1/r)\text{ cis }(-\theta)$.

Proof: By Theorem 10.3, we have $(r\text{ cis }\theta)[(1/r)\text{ cis }(-\theta)] = 1\text{ cis }0 = 1$. So the second factor is the multiplicative inverse of the first. ■

Corollary 10.5: If $z_2 \neq 0$, then $z_1/z_2 = (r_1/r_2)\text{ cis }(\theta_1 - \theta_2)$.

Proof: This follows immediately. ■

Corollary 10.6 (De Moivre's formula): For any z and any $n \in \mathbb{N}$, $z^n = r^n\text{ cis }(n\theta)$. And if $z \neq 0$, n can be any integer.

Proof: Both parts of this proof are easy inductions on n and are left for Exercise 7. (Remember that if z is nonzero, we have $z^0 = 1$ and $z^{-n} = 1/(z^n)$, by definition.) ■

Example 2: Theorem 10.3 and its corollaries often shorten computations with complex numbers. For instance,

$$(1+i)^8 = [\sqrt{2}\text{ cis }(\pi/4)]^8 = (\sqrt{2})^8\text{ cis }(2\pi) = 16\text{ cis }0 = 16$$

$$(1+i)/(1-i) = [\sqrt{2}\text{ cis }(\pi/4)]\,/\,[\sqrt{2}\text{ cis }(-\pi/4)] = 1\text{ cis }(\pi/2) = i$$

$$(1+\sqrt{3}\,i)^9 = [2\text{ cis }(\pi/3)]^9 = 2^9\text{ cis }(9\pi/3) = 512\text{ cis }\pi = -512$$

But whereas the above formulas for products, quotients, and powers are useful time-savers, Theorem 10.7 makes it possible to do problems that are virtually impossible without using the trigonometric form. It is the most important application of De Moivre's formula.

Theorem 10.7: If $z \neq 0$ and n is any natural number, then z has exactly n nth roots in $\mathbb{C}$. Specifically, these are the numbers

$$\sqrt[n]{r}\text{ cis }[(\theta + 2\pi k)/n] \quad \text{for } k = 0, 1, \ldots, n-1$$

These numbers form a regular n-sided polygon, centered at the origin.

Proof: Assume the complex number x is an nth root of z; this means that $x^n = z$. If we write $x = s\text{ cis }\phi$, then this becomes $s^n\text{ cis }(n\phi) = z = r\text{ cis }\theta$. Therefore $s^n = r$ (since the modulus of a complex number is unique), and since both r and s are nonnegative real numbers, this becomes $s = \sqrt[n]{r}$.

We also must have $\text{cis }(n\phi) = \text{cis }\theta$, but this does not imply that $n\phi = \theta$. Two angles have the same trigonometric functions (and therefore the same "cis") iff they are **coterminal**; that means they differ by a multiple of 2π. So $n\phi = \theta + 2\pi j$, or $\phi = (\theta + 2\pi j)/n$, for some integer j.

We have shown that any nth root of z must be of the form $\sqrt[n]{r}\text{ cis }[(\theta + 2\pi j)/n]$, for some $j \in \mathbb{Z}$. Conversely, an easy computation shows that every number of this form is an nth root of z. So the set of nth roots of z consist precisely of these numbers.

Since j can be any integer, it appears that we have found an infinite set of nth roots of z. But there is duplication, for if two integers j and k differ by a multiple of n, then

$(\theta + 2\pi j)/n$ and $(\theta + 2\pi k)/n$ differ by a multiple of 2π, so the two nth roots of z obtained by using j and k are equal. Recall that $\mathbb{Z}_n = \{0, 1, \ldots, n-1\}$. From the division algorithm, it follows that for every integer j, there's a unique k in $\mathbb{Z}_n$ such that j and k differ by a multiple of n (see Exercise 9). So if we restrict j to $\mathbb{Z}_n$ when listing the nth roots of z, we get every nth root, without duplication.

The claim about the polygon is easily seen: there are exactly n nth roots of z, and they are all the same distance $\sqrt[n]{r}$ from the origin. Furthermore, the angle between any one of the roots and the next one counterclockwise, measured from the origin, is obviously $2\pi/n$. *[This paragraph is not very rigorous, but it is possible to make the proof of the claim about the polygon as rigorous as desired.]* ∎

Example 3: Let's find all six 6th roots of -1. Since $-1 = 1 \text{ cis } \pi$, the complex 6th roots of -1 are the numbers cis $[(\pi + 2\pi k)/6]$, where $k = 0, 1, 2, 3, 4$, and 5. These numbers are plotted in Figure 10.3; note that they form a regular hexagon centered at the origin, as Theorem 10.7 predicts. It is also easy to express these numbers algebraically: they are the numbers $\pm i$ and $\pm \sqrt{3}/2 \pm (1/2)\, i$.

Example 4: Let's find the eight complex 8th roots of 1 (or, as they are called, the **8th roots of unity**). As Figure 10.4 shows, these form a regular octagon centered at the origin, with one of its vertices at (1, 0). It is not hard to express these numbers algebraically (see Exercise 10).

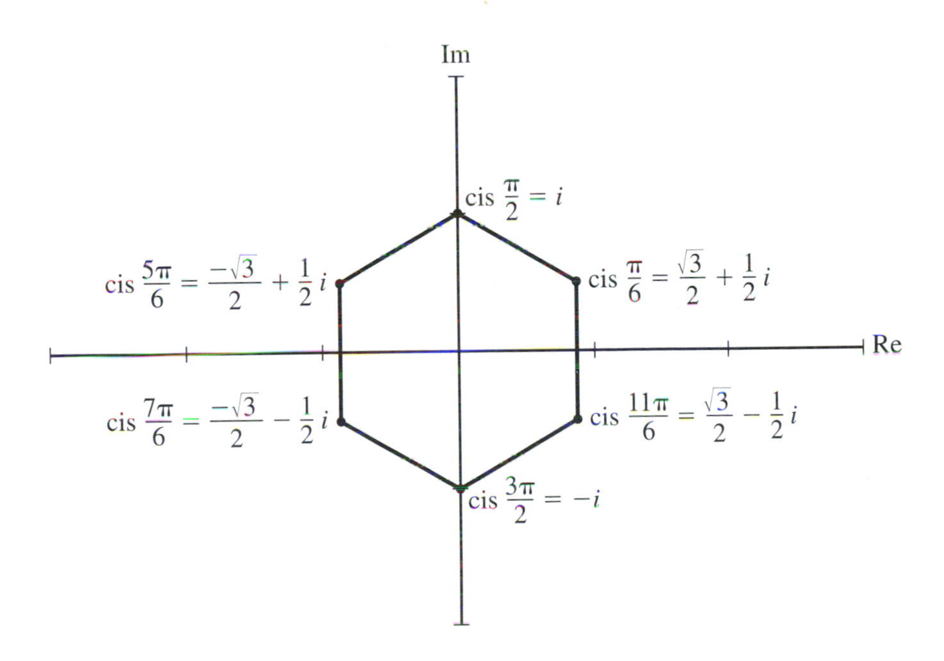

Figure 10.3 The 6th roots of -1

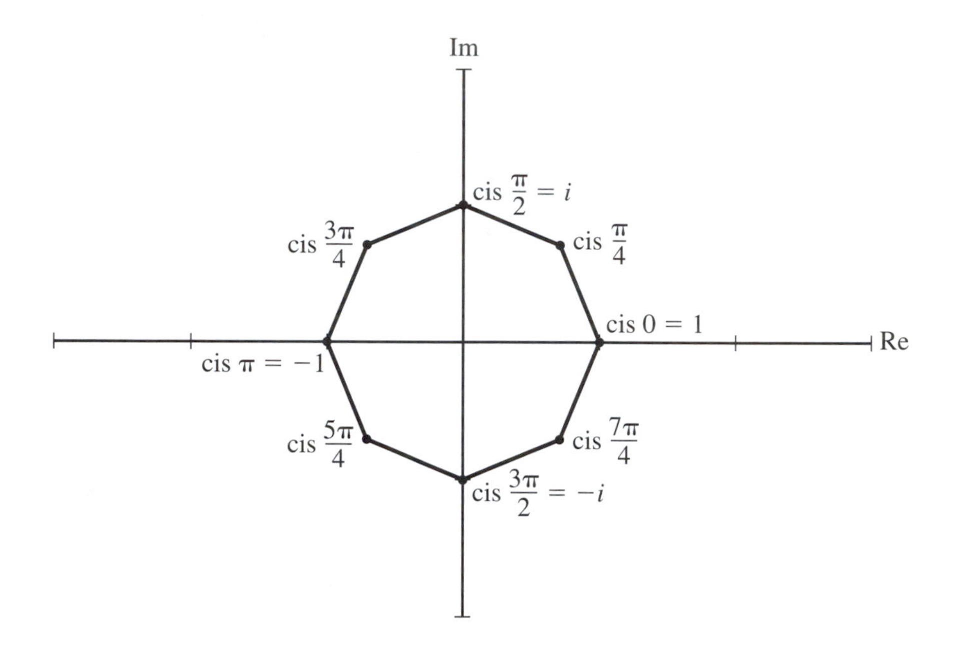

Figure 10.4 The 8th roots of unity

Similarly, the seven complex 7th roots of unity are pictured in Figure 10.5. But since $2\pi/7$ is not an angle whose sine and cosine are expressible in a simple form, there is no way to express the 7th roots of unity without trigonometric functions.

Exercises 10.1

(1) Evaluate the following expressions, obtaining an answer in the form $a + bi$. Do not use Theorem 10.3 and Corollaries 10.4 through 10.6.

(a) $(1 + i)(1 + 2i)(1 + 3i)$ (b) $(3 + 2i)(5 - i)(3 - 2i)$
(c) $(1 - i)/(1 + i)$ (d) $(3 - 2i)/(3 + 4i)$

(2) Show that the field C cannot be made into an ordered field; that is, there is no way to define a relation $<$ on C such that all the ordered field axioms hold.

(3) Show that the function $f: \mathbb{R} \to \mathbb{C}$ defined by $f(x) = x + 0i$ is an isomorphic embedding, as asserted in the first remark of this section.

(4) Prove parts (a), (b), (d), and (e) of Theorem 10.1.

(5) Show that field axioms V-2, V-7, V-9, and V-10 hold in C.

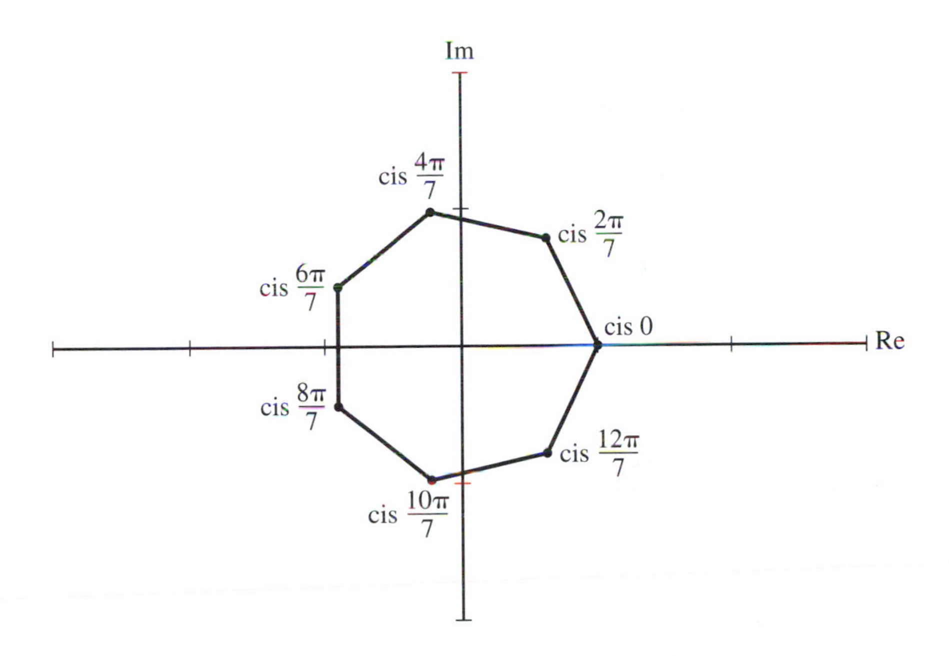

Figure 10.5 The 7th roots of unity

(6) Prove Theorem 10.3.

(7) Prove Corollary 10.6.

(8) Evaluate the following expressions, using Theorem10.3 and Corollaries 10.4 through 10.6. Obtain answers that do not mention any trigonometric functions.

(a) $(\sqrt{3} - i)^5$ (b) $[(1 - i)/(1 + i)]^{10}$

(c) $(2 \text{ cis } 20°)^7/(8 \text{ cis } 50°)$ (d) $(\text{cis } (-9°))^{1000}$

(9) Prove the claim made in the third paragraph of the proof of Theorem 10.7, involving the division algorithm.

(10) Express the eight complex 8th roots of unity in algebraic form (with no mention of sines or cosines).

*(11) Let z be an nth root of unity (that is, $z^n = 1$). Then z is called a **primitive** nth root of unity iff *every* nth root of unity is an integer power of z.

(a) Find all the primitive nth roots of unity for $n = 6$, $n = 7$, and $n = 8$.

(b) Let z be an nth root of unity. Prove that z is primitive iff the numbers $z, z^2, z^3, \ldots, z^n$ are all distinct.

(c) Show that for every $n > 1$, there is a primitive nth root of unity.

(d) Let z be an nth root of unity. By Theorem 10.7, this means $z = \text{cis}\,(2\pi k/n)$, for some integer k. Find a simple number-theoretic condition on the numbers k and n that is equivalent to z being primitive.

(12) Find all the complex cube roots of i, using each of the following methods:

(a) Use Theorem 10.7; this is by far the easiest way.

(b) Write $(a + bi)^3 = i$, expand the left side, and set the real and imaginary parts of both sides of the resulting equation equal to each other. You should then be able to solve for a and b.

(c) Write $z^3 = i$, or $z^3 - i = 0$. View this as a polynomial equation in the variable z. Notice that $(-i)^3 = i$. In other words, $-i$ is a solution to this equation; this means that $z - (-i)$ or $z + i$ is a *factor* of $z^3 - i$. (This conclusion is based on the **factor theorem**, which you have probably encountered; it is proved in Section 10.2.) Divide $z^3 - i$ by $z + i$ to obtain a second-degree polynomial, whose roots can be found by the quadratic formula.

Not only is (a) the easiest of these methods; notice also that (b) and (c) would be quite impossible if we were dealing with, say, 6th roots instead of cube roots.

(13) Show that if a, or $a + 0i$, is any real complex number, then the value of $|a|$ according to the definition given in this section matches its value based on the definition for real numbers given in Appendix 2.

*(14) Although not every real number has a real square root, Theorem 10.7 tells us that every complex number has at least one complex square root. Yet we have not defined a *function* $\sqrt{}$ on C. Try to define such a function, with domain C. Explain the difficulties you encounter if any.

(15) Recall the Taylor series formulas

$$e^x = \sum_{n=0}^{\infty} x^n/n!$$

$$\cos x = \sum_{n=0}^{\infty} (-1)^n x^{2n}/(2n)!$$

$$\sin x = \sum_{n=0}^{\infty} (-1)^n x^{2n+1}/(2n+1)!$$

In elementary mathematics, the variable x is restricted to real numbers in all these functions. Euler and his contemporaries were the first to realize that the domain of such functions could be extended to all of C by using their Taylor series.

Using these Taylor series, prove that $e^{i\theta} = \text{cis}\,\theta$. Therefore, the trigonometric form of a complex number can be rewritten in the elegant form $re^{i\theta}$. This formula is one of Euler's most brilliant and useful discoveries. The idea that e to an imaginary exponent equals a combination of trigonometric functions is anything but obvious! One particularly pretty consequence, obtained by letting $\theta = \pi$, is the fact that $e^{i\pi} = -1$.

*10.2 Additional Algebraic Properties of C

In this section we take a more detailed look at what makes the complex number system special, and in a way superior even to the reals. Let's begin this discussion by recalling that R is *complete*. A question that then comes to mind is: if it's complete, why are we bothering to enlarge it by defining C? One answer to this is that the completeness property is primarily a *geometric* property, not an algebraic one. (Remember that the intuitive content of the completeness property is that the real number system has no holes, which is a statement about the representation of R on a number line.)

On the other hand, the completeness property does have important algebraic consequences. Section 9.1 shows that, in R, every nonnegative number has an nth root, for every natural number n, and every negative number has a cube root, a fifth root, a seventh root, and so on. Furthermore, we will see that every polynomial (with real coefficients) of *odd* degree has a root in R. All these important algebraic results about R depend essentially on the completeness property, as evidenced by the fact that none of these results is true in Q.

But note that R is nice only with respect to roots and polynomials of odd degree. The complex number system is clearly an attempt to remedy the lack of even roots of negative numbers. This might not be worth doing if it just shifted the difficulty to another level. In other words, when we introduce the imaginary unit i and essentially define it to be a square root of -1, we immediately get a number system in which every negative number has a square root, but is there any reason to think the number i will have a square root? And if not, will it then be necessary to add a new symbol denoting a square root of i, and then another symbol denoting a square root of that number, and so on ad infinitum? However, Theorem 10.7 showed a surprising fact about C: by merely adding in square roots of negative numbers, we get a number system in which all numbers, even the ones just included, have square roots. We will now proceed to make this idea more precise.

Definition: A field F is called **algebraically closed** iff every nonconstant polynomial in $F[x]$ (that is, every polynomial with coefficients in F) has a root in F. (Recall that a **root** or **zero** of a polynomial g is simply a number r such that $g(r) = 0$.)

Example 1: The system Q is not algebraically closed, because the polynomial $x^2 - 2$ has no root in Q. This polynomial has roots in R, namely $\pm\sqrt{2}$; however, the polynomial $x^2 + 1$ has no root in R, so R is not algebraically closed either.

Theorem 10.8 states the single most important feature of the complex number system. Its proof, which is due to Gauss, is quite profound and will not be included here.

Theorem 10.8 (Fundamental theorem of algebra): The system C is algebraically closed.

Theorem 10.9: If a field F is algebraically closed, then for any natural number n, every number in F has an nth root in F.

Proof: For any $a \in F$ and $n \in \mathbb{N}$, consider the polynomial $x^n - a$, which is in $F[x]$. A root of this polynomial is clearly the same thing as an nth root of a. ■

Of course, Theorem 10.9 tells us nothing new about $\mathbb{C}$, since Theorem 10.7 told the whole story about nth roots in $\mathbb{C}$.

Theorem 10.10 and Corollaries 10.11 through 10.13 are basic, indispensable tools in the study of polynomials. In these, it is assumed that the number c and all the coefficients of the polynomial mentioned are members of some field F.

Notation: For any polynomial f, $\deg(f)$ denotes its **degree**, the highest power of x occurring in it. The degree of a constant polynomial is 0, except that the degree of the polynomial 0 is usually left undefined.

In Section 8.3, it is mentioned that the division algorithm for integers (Theorem 8.14 and Corollary 8.15) can be adapted to polynomials. Here is the polynomial version. Note that the reference to degree is the only change in the statement. Theorem 10.10 and Corollaries 10.11 through 10.13 hold for polynomials with coefficients in any field, not just ones with complex coefficients.

Theorem 10.10: If f and g are polynomials and $g \neq 0$, then there are unique polynomials Q and R such that

$$f = Qg + R \quad \text{and} \quad \text{either } \deg(R) < \deg(g) \text{ or } R = 0$$

Proof: This proof is similar to that of Theorem 8.14; here is an outline of the modified proof. We first prove the existence of a correct quotient and remainder. Given polynomials f and $g \in \mathbb{C}[x]$ with $g \neq 0$, let $P = \{f - pg \mid p \in \mathbb{C}[x]\}$. Then P is a nonempty set of polynomials. If $0 \in P$, there's a polynomial p such that $f - pg = 0$. Let Q be any such p, and let $R = 0$. Clearly, this Q and R satisfy the conditions of the theorem.

If $0 \notin P$, then every polynomial in P has a degree, which is in $\mathbb{N} \cup \{0\}$. It is a simple corollary of Theorem 5.6 that $\mathbb{N} \cup \{0\}$, like $\mathbb{N}$, is well ordered (see Exercise 2). Therefore, some number m is the smallest degree of any member of P. Let R be any member of P whose degree is m, and let Q be a polynomial such that $f - Qg = R$. Then we are done (with existence) if we can show $m < \deg(g)$.

Assume, on the contrary, that $m \geq n$, where $n = \deg(g)$. Let the leading terms of R and g be ax^m and bx^n, respectively. Let h be the polynomial $(a/b)x^{m-n}$. Note that hg has leading term ax^m, like R. So let $p = Q + h$. Then $f - pg = f - (Q + h)g = f - Qg - hg = R - hg$, which is a polynomial of degree less than m, a contradiction.

The proof of uniqueness is left for Exercise 3. ■

Corollary 10.11 (Remainder theorem): When a polynomial f is divided by a polynomial of the form $x - c$, where c is any number, the remainder is the number $f(c)$.

Proof: Let f and c be given. By Theorem 10.10, there is a quotient polynomial Q and a remainder R (which is just a *number* because $x - c$ has degree 1) such that $f(x) = Q(x)(x - c) + R$. This equation says the two sides are equal as formal expressions, but this means they must also be equal as *functions*. That is, they stay equal when any number is substituted for x. Substituting $x = c$ yields that $f(c) = R$, as desired. ■

Corollary 10.12 (Factor theorem): For any polynomial f and any number c, $x - c$ is a factor of f iff c is a root of f.

Proof: See Exercise 4. ■

Corollary 10.13: A polynomial of degree n has at most n roots.

Proof: If a polynomial had more than n roots, then the factor theorem would assert that it must have more than n different first-degree factors, which would make its degree more than n (see Exercise 5). ■

Theorem 10.14: A field F is algebraically closed iff every nonconstant polynomial in $F[x]$ can be written as a product of first-degree factors in $F[x]$.

Proof: The reverse direction is immediate, because if a polynomial has even one first-degree factor, say $ax + b$, then the polynomial must have a root, specifically $-b/a$. So if every polynomial can be factored, then every polynomial has a root. For the forward direction, the conclusion can be proved by induction on the degree of the polynomial (see Exercise 6). ■

Corollary 10.15: If F is algebraically closed, then every nonconstant polynomial in $F[x]$ can be written as a product of the form $c(x - r_1)(x - r_2) \dots (x - r_n)$. Here, n is the degree of the polynomial, the r_ks are its roots, and c is its leading coefficient (the coefficient of its highest-power term).

Proof: Assume F is algebraically closed, and f is a polynomial of degree n in $F[x]$. Then, by the previous theorem, f can be written as a product of first-degree factors in $F[x]$, and there must be exactly n such factors in any such factorization of f. But a first-degree factor $ax + b$ can always be rewritten in the form $a(x - r)$, where $r = -b/a$. By doing this for all the factors and combining all the numerical terms, we can get the factorization of f into the desired form. The fact that the r_ks are the roots of f follows directly from the factor theorem. ■

In Corollary 10.15, each r_k is a root of f, and the number of factors of the form $x - r_k$ occurring is equal to the degree of f. So it is tempting to conclude that, in an algebraically closed field like C, a polynomial of degree n must have *exactly* n roots. Some textbooks make this assertion, but it is misleading: in Corollary 10.15, nothing guarantees that the r_ks are all distinct; and if they are not, the polynomial clearly has fewer than n roots.

A more accurate count of the roots of a polynomial requires some additional terminology: if r is a root of a polynomial f, its **multiplicity** is the highest power of $(x - r)$ that is a factor of f. (A root of multiplicity 1 is called a **single** or **ordinary** root.

A root of multiplicity 2 or more is called a **multiple** or **repeated** root.) Then the following is a simple consequence of Corollary 10.15: in an algebraically closed field, a polynomial of degree n must have exactly n roots if each root is counted as many times as its multiplicity. If there are no repeated roots, then there are n distinct roots.

Example 2: Most people would say that the polynomial $x^2 - 2$ cannot be factored. What they mean is that it cannot be factored into first-degree factors with *rational* coefficients, since it has no roots in $\mathbb{Q}$. But it has roots in $\mathbb{R}$, so it can be factored if real coefficients are allowed:

$$x^2 - 2 = (x - \sqrt{2})(x + \sqrt{2})$$

Example 3: The polynomial $x^2 + 1$ cannot be factored using *real* coefficients, since it has no real roots. But since it has roots i and $-i$ in $\mathbb{C}$, it can be factored as $(x - i)(x + i)$.

Example 4: Similarly, there are no fourth roots of -1 in $\mathbb{R}$, so the polynomial $x^4 + 1$ has no first-degree factors with real coefficients. But by Theorem 10.7, we can compute the complex fourth roots of -1: they are $(1 + i)/\sqrt{2}$, $(1 - i)/\sqrt{2}$, $(-1 + i)/\sqrt{2}$, and $(-1 - i)/\sqrt{2}$. Therefore, we can write the following factorization:

$$x^4 + 1 = [x - (1 + \mathrm{i})/\sqrt{2}\,][x - (1 - \mathrm{i})/\sqrt{2}\,][x - (-1 + \mathrm{i})/\sqrt{2}\,][x - (-1 - \mathrm{i})/\sqrt{2}\,]$$

This factorization is not pretty, but it's good to know that it's possible.

Note that we have only asserted that $x^4 + 1$ has no first-degree factors with real coefficients, so it can't be written as a first-degree factor times a third-degree factor, with real coefficients. But we haven't said anything about whether it can be written as a product of two second-degree factors with real coefficients. Perhaps surprisingly, it can be factored in this way, as we soon see.

Example 5: Here is an example involving repeated roots. Consider the polynomial $x^6 - x^4 - x^2 + 1$. We can first factor this as $(x^2 - 1)(x^4 - 1)$, and then further as $(x^2 - 1)^2(x^2 + 1)$, and finally as $(x - 1)^2(x + 1)^2(x + i)(x - i)$. Thus this sixth-degree polynomial has only four distinct roots; but two of them have multiplicity 2, so we have the correct count: $2 + 2 + 1 + 1 = 6$.

Theorem 10.16: Suppose f is a polynomial in $\mathbb{R}[x]$ and $z \in \mathbb{C}$. Then $f(\bar{z}) = \overline{f(z)}$. In particular, if z is a root of f, then so is $\bar{z}$.

Proof: The fact that $f(\bar{z}) = \overline{f(z)}$ follows directly from Theorem 10.1, but the proof requires several stages. If f is of the simple form x^n, it follows immediately from Theorem 10.1(e). Then, if f is of the form cx^n, where c is *real*, we must apply Theorem 10.1(d). (Note that $\bar{c} = c$.) Finally, if f is any polynomial, it is a sum of terms of the form cx^n, so apply Theorem 10.1(b). The second claim follows from the first, since $\bar{0} = 0$. ■

Corollary 10.17: (a) For any polynomial in $\mathbb{R}[x]$, the total number of its nonreal roots must be even.

(b) Any polynomial in $\mathbb{R}[x]$ can be written as a product of first- and second-degree factors in $\mathbb{R}[x]$.

Proof: (a) Follows immediately from Theorem 10.16.

(b) We use complete induction on the degree of the polynomial. Note that if $\deg(f) \le 2$, there's nothing to prove. For the induction step, assume the result holds for all polynomials of degree $\le n$. Let $f \in \mathbb{R}[x]$, with $\deg(f) = n + 1$. By Corollary 10.15, f can be written in the form $c(x - r_1)(x - r_2) \ldots (x - r_{n+1})$, where the r_ks are complex and c is real. If all the r_ks are real, we're done. Otherwise *[note that this is a proof by cases]*, some r_k is a nonreal number $a + bi$. By Theorem 10.16, there's a j such that $r_j = a - bi$. By direct calculation, $[x - (a + bi)][x - (a - bi)] = x^2 - 2ax + a^2 + b^2$, which has real coefficients, and this quadratic is a factor of f. Thus, $f(x) = [x^2 - 2ax + a^2 + b^2]g(x)$, where the quotient g is also in $\mathbb{R}[x]$ and has degree $n - 1$. By the induction hypothesis, g can be written as a product of first- and second-degree factors. Therefore, so can f. ■

Corollary 10.17, as stated, does not deal with multiplicity (see Exercise 12). Contrast this result with Theorem 10.14: if we allow complex coefficients in the factors, every nonconstant polynomial "splits completely" into first-degree factors. If we restrict ourselves to real coefficients in the factors, we get the next-best thing—every nonconstant polynomial splits into factors of degree at most 2.

However, it should be mentioned that these results are not "constructive." That is, just because we have a result guaranteeing that every polynomial has a root or can be factored in a certain way does not mean that there is any feasible method to find its roots or factors in an exact form. (It's always possible to find decimal approximations for roots, but that's not as desirable as an answer in some exact algebraic form.) The only polynomials whose roots and factors are simple to find are those of degree 2 or lower. For polynomials of degree 3 and 4, there are formulas for finding roots, but they are quite messy. And for polynomials of degree 5 or higher, a famous theorem of Abel shows that there is no general method for factoring them or finding their roots in terms of radicals.

Corollary 10.18: Every polynomial of *odd* degree in $\mathbb{R}[x]$ has a real root.

Proof: If a polynomial in $\mathbb{R}[x]$ has odd degree, the factorization mentioned in Corollary 10.17 must contain at least one first-degree factor. Setting this factor to zero yields a real root. ■

Example 6: Let's reexamine the polynomial $x^4 + 1$. Earlier we factored it as $[x - (1 + i)/\sqrt{2}\,][x - (1 - i)/\sqrt{2}\,][x - (-1 + i)/\sqrt{2}\,][x - (-1 - i)/\sqrt{2}\,]$.

Since these factors are arranged in pairs as required by the proof of Corollary 10.17, we can multiply each pair of factors to obtain factors in $\mathbb{R}[x]$:

$$x^4 + 1 = (x^2 - \sqrt{2}\,x + 1)(x^2 + \sqrt{2}\,x + 1)$$

This example shows the importance of specifying which field must be used for the coefficients in factoring problems. With complex coefficients allowed in the factors, $x^4 + 1$ can be factored completely into linear factors. With only real coefficients, we can still get quadratic factors. But with the coefficients restricted to rational numbers, it can't be factored at all.

We now examine $\mathbb{C}$ from a more theoretical viewpoint. Intuitively, we want to show that $\mathbb{C}$ is not just *any* algebraically closed field that contains all of $\mathbb{R}$; it is the *smallest* algebraically closed extension of $\mathbb{R}$.

Definitions: Suppose E is a subfield of a field F and $a \in F$. We say a is **algebraic over** E iff a is a root of some nonzero polynomial in $E[x]$. If every member of F is algebraic over E, then F is called an **algebraic extension of E.** The word **transcendental** means "not algebraic," applied to either elements of a field or to extensions. When a complex number is called simply algebraic or transcendental, the words "over $\mathbb{Q}$" are understood.

We denote the set of all algebraic complex numbers $\mathbb{A}$. Then $\mathbb{A} \cap \mathbb{R}$ is the set of all algebraic real numbers.

Example 7: Is $\mathbb{R}$ an algebraic extension of $\mathbb{Q}$? Every rational number is obviously algebraic: if $a \in \mathbb{Q}$, then a is a root of the polynomial $x - a$, which is certainly in $\mathbb{Q}[x]$. But many irrational numbers are also algebraic. For instance, $\sqrt[3]{5}$ is algebraic, since it is a root of the polynomial $x^3 - 5$.

But not every real number is algebraic; this means that $\mathbb{R}$ is a transcendental extension of $\mathbb{Q}$. The easiest way to prove this is by a cardinality argument: $\mathbb{Q}$ is countable (as we proved in Section 7.5), and from this it follows that $\mathbb{Q}[x]$ is countable. Since each polynomial can have only a finite number of roots, $\mathbb{A}$ is also countable (see Exercise 8). But we know that $\mathbb{R}$ is uncountable. Therefore, $\mathbb{R}$ and $\mathbb{C}$ are not algebraic extensions of $\mathbb{Q}$. In fact, the set of transcendental real numbers must be uncountable, by Theorem 7.24(b). So most numbers are transcendental.

Aside from this cardinality argument, it can also be proved (though not easily) that various specific numbers such as π and e are transcendental.

Definition: We say F is an **algebraic closure of E** iff F is algebraically closed and is also an algebraic extension of E.

Intuitively, if F is an algebraic closure of E, then F contains all the algebraic numbers it possibly can (over E or even over itself) and no transcendental ones (over E). This makes algebraic closures very special and helps to explain Theorems 10.21(b) and Theorem 10.22.

Theorem 10.19: The system $\mathbb{C}$ is an algebraic closure of $\mathbb{R}$.

Proof: By Theorem 10.8, $\mathbb{C}$ is algebraically closed. And, in the proof of Corollary 10.17, we noted that any complex number $a + bi$ is a root of the polynomial $x^2 - 2ax + a^2 + b^2$, which is in $\mathbb{R}[x]$. So $\mathbb{C}$ is an algebraic extension of $\mathbb{R}$. ■

Of course, since we haven't proved Theorem 10.8, we have skipped the lion's share of the proof of Theorem 10.19. Another result that is related to Theorem 10.19 is that $\mathbb{A}$ is an algebraic closure of $\mathbb{Q}$. It is not even obvious that $\mathbb{A}$ is a field (see Exercise 9); proving that $\mathbb{A}$ is algebraically closed is significantly harder.

Theorem 10.20: The system $\mathbb{C}$ is the *unique* algebraic closure of $\mathbb{R}$, in the following sense: if F is any other algebraic closure of $\mathbb{R}$, then there is an isomorphism g between $\mathbb{C}$ and F that leaves $\mathbb{R}$ fixed (meaning that $g(x) = x$, $\forall x \in \mathbb{R}$).

Proof: Assume F is any algebraic closure of $\mathbb{R}$. Since F is algebraically closed, it contains an element j that is a root of the polynomial $x^2 + 1$; that is, $j^2 + 1 = 0$. Since $\mathbb{R} \subseteq F$ and F is a field, F contains all numbers of the form $a + bj$, with a and $b \in \mathbb{R}$. Define g by $g(a + bi) = a + bj$. The simple proofs that $g: \mathbb{C} \to F$, g preserves addition and multiplication, g is one-to-one, and g leaves $\mathbb{R}$ fixed are left for Exercise 10.

It remains to prove that g is onto F. Assume $y \in F$. Since F is algebraic over $\mathbb{R}$, there is a polynomial h in $\mathbb{R}[x]$ such that $h(y) = 0$. By Corollary 10.17, h can be factored into first- and second-degree factors in $\mathbb{R}[x]$, and so y must be a root of one of these factors (since a field has no zero-divisors). But if y is a root of a first-degree polynomial in $\mathbb{R}[x]$, then y is real, and thus $y \in \text{Rng}(g)$. And if y is a root of a second-degree polynomial in $\mathbb{R}[x]$, then the quadratic formula can be adapted to F to yield that y has the form $a + bj$, for some a and $b \in \mathbb{R}$. In this case also, $y \in \text{Rng}(g)$. ■

Theorem 10.21 states important generalizations of Theorems 10.19 and 10.20, but their proofs are beyond the scope of this book.

Theorem 10.21: (a) Every field has an algebraic closure.

(b) If F and F' are both algebraic closures of some field E, then there is an isomorphism g between F and F' that leaves E fixed.

The intuitive meaning of part (b) is that F and F' are the same (isomorphic), not just as fields but also as extensions of E. On the basis of this result, one normally refers to *the* algebraic closure of a given field.

We conclude by stating, without proof, a theorem that rephrases the remark before Theorem 10.19 in a more rigorous way (Exercise 11 includes part of the proof).

Theorem 10.22: The following are equivalent:

(a) The field F is an algebraic closure of E.

(b) The field F is a *maximal* algebraic extension of E; that is, F is an algebraic extension of E, and no proper extension of F is an algebraic extension of E.

(c) The field F is a *minimal* algebraically closed extension of E; that is, E is a subfield of F, F is algebraically closed, and there is no algebraically closed field F' such that $E \subseteq F' \subseteq F$.

Exercises 10.2

(1) (a) Prove that if $f(x)$ and $g(x)$ are polynomials and $\deg(f) > \deg(g)$, then $\deg(f+g) = \deg(f)$.

(b) If $\deg(f) = \deg(g)$, what can be concluded about $\deg(f+g)$? Justify your claim.

(2) Prove the claim, in the proof of Theorem 10.10, that $\mathbb{N} \cup \{0\}$ is well ordered.

(3) Complete the proof of Theorem 10.10 by showing the uniqueness of Q and R.

(4) Prove Corollary 10.12.

(5) Prove Corollary 10.13.

(6) Prove the forward direction of Theorem 10.14.

(7) Factor the following polynomials as completely as possible, using factors (i) in $\mathbb{Q}[x]$, (ii) in $\mathbb{R}[x]$, and (iii) in $\mathbb{C}[x]$:

(a) $x^4 - 2$ (b) $x^6 - 1$

(c) $(x^2 - 2)(x^2 - x + 2)$ (d) $x^6 + x^5 + x^4 - x^2 - x - 1$

*(8) Prove the claims in Example 7 that $\mathbb{Q}[x]$ and the set of all algebraic real numbers are both countable. ***Hint:*** Use Theorem 7.24 and Corollary 7.25.

*(9) It was mentioned in the text that $\mathbb{A}$, the set of algebraic complex numbers, is a field. The main steps in proving this are verifying the closure and inverse axioms.

(a) Prove that $\mathbb{A}$ satisfies the additive inverse axiom; that is, if z is algebraic, so is $-z$.

(b) Show that the following combinations of algebraic numbers are algebraic, by actually finding an appropriate polynomial of which each one is a root:

(i) $\sqrt{2}\,\sqrt[3]{5}$ (ii) $\sqrt{2} + \sqrt{3}$

(iii) $\sqrt[3]{2} + \sqrt{3}$ (iv) $\sqrt{2} + i$

(10) Prove the steps omitted in the proof of Theorem 10.20.

(11) (a) Prove that an algebraically closed field has no proper algebraic extensions.

(b) Using part (a), prove that (a) implies (b) in Theorem 10.22.

(12) Prove that if $f \in \mathbb{R}[x]$, $z \in \mathbb{C}$ and z is a root of f, then $\bar{z}$ is a root of f *with the same multiplicity as* z.

(13) Prove that r is a multiple root of a polynomial f iff r is a root of both f and f', the derivative of f.

Suggestions for Further Reading: Most of the references for Chapters 8 and 9 have some material on complex numbers. For a thorough treatment of the complex number system and the algebraic notions introduced in Section 10.2, see Feferman (1989). For an introduction to the subject of complex analysis, see Boas (1987).

Appendix 1

A General-Purpose Axiom System for Mathematics

Like most reference sections, Appendix 1 is not meant to be read all at once. Instead, you should try to become familiar with it, section by section, as you study the related chapters in the text. Appendix 1 is also meant to be a valuable tool for your use whenever you are trying to write a proof or even read one, not just during this course but for your entire mathematical career.

An asterisk (*) next to an axiom or rule of inference indicates that it is technically superfluous. In other words, everything you can prove with these axioms and rules included can be proved without them.

There are a number of footnotes to this axiom system, which are intended to provide important clarification.

Rules of Inference[1]

(1) Propositional consequence (PC): You may assert any statement that is a propositional consequence of previous steps in the proof.

*(2) Conditional proof: If you can prove the statement Q from the assumption P (and perhaps other assumptions), you may conclude the single statement $P \rightarrow Q$, without the assumption P.

*(3) Substitution of equivalent statements: From statements $P \leftrightarrow Q$ and S[P], you may assert S[P/Q]. (Here, S[P] is any statement containing P as a substatement, S[P/Q] results from replacing *some or all* of the occurrences of P in S[P] by Q, and no free variable of P or Q becomes bound in S[P] or S[P/Q].)

(4) Universal generalization (UG): From a proof of $P(x)$ which uses no assumptions involving the *free* variable x, you may conclude $\forall x\, P(x)$.

*(5) Existential specification (ES): From a statement $\exists x\, P(x)$, you may assert $P(c)$, where c is a *new constant symbol* that does *not* appear in the final step of the proof.

Axioms

I. Propositional axioms

* All tautologies [2]

II. Quantifier axioms [3]

(1) Universal specification (US): $\forall x\, P(x) \rightarrow P(t)$, where t is any term or expression of the same sort as the variable x and no free variable of t becomes bound when P(t) is formed. [4]

(2) De Morgan's law for quantifiers: $\sim \forall x\, P(x) \leftrightarrow \exists x \sim P(x)$

*(3) De Morgan's law for quantifiers: $\sim \exists x\, P(x) \leftrightarrow \forall x \sim P(x)$

*(4) Existential Generalization (EG): $P(t) \rightarrow \exists x\, P(x)$, with the same restrictions on t as in universal specification.

III. Equality axioms

(1) Reflexivity: $x = x$

*(2) Symmetry: $x = y \rightarrow y = x$

*(3) Transitivity: $(x = y \wedge y = z) \rightarrow x = z$

(4) Substitution of equals: $x = y \rightarrow [P(x) \leftrightarrow P(y)]$, where $P(y)$ results from replacing *some or all* of the occurrences of x in $P(x)$ with y and neither x nor y is quantified in $P(x)$. [5]

IV. Set axioms

In these axioms, the variables A, B, C, and D denote sets but w, x, y, and z can be any sort of objects, not necessarily real numbers. This group of axioms is included here primarily for completeness; most of them are not discussed in the text.

(1) Extensionality: $A = B \leftrightarrow \forall x\, (x \in A \leftrightarrow x \in B)$

(2) Pairing: $\forall x, y\, \exists A\, \forall z\, [z \in A \leftrightarrow (z = x \vee z = y)]$. (Less formally, this says: for every x and y, the set $\{x, y\}$ exists.)

(3) There is a set $\mathbb{R}$ of all real numbers. [6]

*(4) For every x and y, the ordered pair (x, y) exists.

*(5) $(w, x) = (y, z)$ iff $w = y$ and $x = z$.

(6) Power set axiom: $\forall A\, \exists B\, \forall C\, (C \in B \leftrightarrow C \subseteq A)$. (Less formally, this says: for every set A, $\wp(A)$ exists.)

(7) Union axiom: $\forall \mathscr{A}\, \exists B\, \forall x\, [x \in B \leftrightarrow \exists C\, (C \in \mathscr{A} \land x \in C)]$. (Less formally, this says: for every set of sets $\mathscr{A}$, the union of all the sets in $\mathscr{A}$ ($\bigcup \mathscr{A}$) exists.)

*(8) Separation axiom: For every proposition $\mathrm{P}(x)$ and every set A, the set $\{x \in A \mid \mathrm{P}(x)\}$ exists.

(9) Replacement axiom: For every proposition $\mathrm{P}(x, y)$ and every set A,

$$[\forall x \in A\, \exists! y\, \mathrm{P}(x, y)] \rightarrow \exists B\, \forall y\, [y \in B \leftrightarrow \exists x \in A\, \mathrm{P}(x, y)]$$

(Less formally, this says: if $\mathrm{P}(x, y)$ defines a function whose domain is the set A, then its range is also a set.)

(10) Foundation axiom: $\forall A\, [A \neq \varnothing \rightarrow \exists B \in A\, (B \cap A = \varnothing)]$

(11) Axiom of choice (AC): For every collection $\mathscr{A}$ of nonempty sets, there is a function f such that, for every B in $\mathscr{A}$, $f(B) \in B$.

V. Real number axioms

In these axioms, x, y, and z denote real numbers. Axioms 1–12 of this group are called the **field axioms**, while axioms 1–17 are called the **ordered field axioms**.

(1) Additive closure: $\forall x, y\, \exists z\, (x + y = z)$

(2) Multiplicative closure: $\forall x, y\, \exists z\, (x \cdot y = z)$

(3) Additive associativity: $x + (y + z) = (x + y) + z$

(4) Multiplicative associativity: $x \cdot (y \cdot z) = (x \cdot y) \cdot z$

(5) Additive commutativity: $x + y = y + x$

(6) Multiplicative commutativity: $x \cdot y = y \cdot x$

(7) Distributivity: $x \cdot (y + z) = (x \cdot y) + (x \cdot z)$ and $(y + z) \cdot x = (y \cdot x) + (z \cdot x)$[7]

(8) Additive identity: There is a number, denoted 0, such that for all x, $x + 0 = x$.[8]

(9) Multiplicative identity: There is a number, denoted 1, such that for all x, $x \cdot 1 = 1 \cdot x = x$.[7,8]

(10) Additive inverses: For every x there is a number, denoted $-x$, such that $x + (-x) = 0$.[8]

(11) Multiplicative inverses: For every nonzero x there is a number, denoted x^{-1}, such that $x \cdot x^{-1} = x^{-1} \cdot x = 1$.[7,8]

(12) $0 \neq 1$

(13) Irreflexivity of $<$: $\sim (x < x)$

(14) Transitivity of $<$: If $x < y$ and $y < z$, then $x < z$

(15) Trichotomy: Either $x < y$, $y < x$, or $x = y$

(16) If $x < y$, then $x + z < y + z$

(17) If $x < y$ and $0 < z$, then $x \cdot z < y \cdot z$ and $z \cdot x < z \cdot y$[7]

(18) Completeness: If a nonempty set of real numbers has an upper bound, then it has a *least* upper bound.

VI. Natural number axioms

In these axioms, m and n denote natural numbers and A denotes a set.

(1) $1 \in \mathbb{N}$[9]

(2) If $m \in \mathbb{N}$, then $m + 1 \in \mathbb{N}$.

(3) Mathematical induction (set form): $[1 \in A \land \forall n\, (n \in A \rightarrow n + 1 \in A)] \rightarrow \mathbb{N} \subseteq A$

*(3′) Mathematical induction (statement form):

$$[\mathrm{P}(1) \land \forall n\, (\mathrm{P}(n) \rightarrow \mathrm{P}(n+1))] \rightarrow \forall n\, \mathrm{P}(n)$$ [10]

Footnotes

(1) Various other derived rules of inference, which may be used as if they were part of the axiom system, are given in Section 4.2.

(2) For a list of many useful tautologies, see Appendix 3.

(3) Many useful laws of quantifier logic, which may be used as if they were axioms, are given in Table 4.2.

(4) We are calling US an axiom, but there are an infinite number of possibilities for the proposition P. So it is technically not a single axiom, but rather an infinite list of axioms. Such a list is called an **axiom schema**. Since the propositional variable P may *not* be quantified, there is no way to turn US into a single axiom. As you can easily verify, our axiom system contains several axiom schemas, including the statement form of mathematical induction.

(5) For a slightly less general but more convenient version of this axiom, see Theorem 4.7.

(6) As is mentioned in Section 9.5, all of mathematics can be developed within pure set theory, in which case there would not be any axioms specifically involving $\mathbb{N}$ or $\mathbb{R}$. However, for this to work there must still be an axiom that guarantees that there is an infinite set. The standard form of this axiom is called the axiom of infinity. If that axiom were included in our axiom system, then axiom IV-3 and all of axiom groups V and VI would become superfluous.

(7) You may have noticed an apparent lack of uniformity in our axiom system. Axioms V-7, V-9, V-11, and V-17 allow for the possibility that multiplication might not be commutative, in spite of axiom V-6. But no such allowances are made for the possible noncommutativity of *addition* in axioms V-8, V-10, and V-16. Why is that? The answer is that there are important algebraic structures called **rings** (see Chapter 8) in which multiplication is not necesarily commutative, making the extra provisions for multiplication necessary. But addition is virtually always taken to be commutative.

(8) To make them more readable, field axioms V-8 through V-11 are given here in a somewhat nonrigorous form. Each of them is really a combination of a genuine axiom and a *definition*. For instance, axiom V-8 states that there is an additive identity (a number such that adding it to any given number has no effect), and also defines 0 to be an additive identity. But, as discussed in Section 3.4, the definition of 0 should be made only *after* proving that there is a *unique* additive identity. More correctly, axiom V-8 should read $\exists y\ \forall x\ (x + y = x)$. Appendix 2 provides a rigorous treatment of identities and inverses.

(9) In this group of axioms, it is understood that 1 refers to the same number and + refers to the same operation as in the real number axioms (see Theorem 4.9).

(10) The two forms of mathematical induction are equivalent to each other. See Exercise 11 of Section 5.1.

Appendix 2

Elementary Results about Fields and Ordered Fields

This appendix is intended to serve three purposes:

(1) We use it as one main source of examples to illustrate the methods of proof that are discussed in Chapter 4, because it contains many proofs using principles that are part of high school algebra and so should be quite familiar to you. Also, many of the exercises in Section 4.4 involve the material in this appendix.

(2) The results of this appendix are used to help study number systems in Unit 3, since all the theorems in this appendix can be viewed as results about familiar number systems like the real numbers and the rational numbers.

(3) It is educational in its own right because it illustrates the typical way in which mathematicians develop a subject rigorously from a set of axioms. The important thing about such a development is that it should assume nothing except the initial axioms and definitions. In this appendix, we prove many things that are completely "obvious." But, as we have been saying all along, this type of caution is essential in mathematics.

The theorems in this appendix are divided into two groups. In the first group, we prove various simple consequences of the *field axioms*, axioms V-1 through V-12 in Appendix 1. In the second group, we enlarge our list of axioms to include all the *ordered field axioms*, V-1 through V-17 in Appendix 1.

Even if you have never heard of field axioms or ordered field axioms, you are acquainted with them. The field axioms are the most basic laws of addition, subtraction, multiplication, and division, and the ordered field axioms include these plus the basic laws of inequalities. Not surprisingly, a number system that satisfies all the field (respectively, ordered field) axioms is called a **field** (respectively, an **ordered field**).

The most important and familiar ordered field is the real number system; the next is the rational number system.

The Field Axioms

Before we do any proofs, let's take a more thorough look at the field axioms. According to the convention introduced in Section 4.3, a mathematical variable that appears unquantified in an axiom is assumed to be universally quantified. This convention has been used to make field axioms V-3 through V-7 more concise, but keep in mind that the universal quantifiers are assumed.

Notice that the first six axioms in the list are in pairs, one for addition and one for multiplication. If you have never seen closure properties before, you might wonder what the first two field axioms mean. All they say is that the sum and product of any two numbers are defined.

Axioms V-3 through V-7 state very simple, well-known properties of addition and multiplication. Be sure to keep the difference between commutativity and associativity straight. The former says that the order of the *numbers* in a certain type of expression doesn't affect the value of the expression. The latter says that the order in which the *operations are performed* doesn't affect the value. It is the associative laws that allow us to omit parentheses in expressions such as xyz and $4 + 3 + 8$.

Axiom V-8 states that there is an **additive identity**, a number that has no effect when it's added to any number. Axiom V-9 makes the corresponding statement for multiplication. The axioms don't say it, but it's easy to prove *uniqueness* of these special numbers; we do this in the first two theorems of this appendix.

Axioms V-10 and V-11 are the so-called **inverse axioms**, which state the familiar properties of the negative of any number and the reciprocal of any nonzero number. As with the identities, we show uniqueness of inverses (Theorems A-3 and A-4).

The last field axiom probably looks silly to you, but it's necessary. Strangely enough, you can't prove that $1 \neq 0$ from the other axioms. And although there are many unusual fields that are considered interesting, a number system in which $1 = 0$ would not be very interesting (see Exercise 18 of Section 4.4).

We now prove several theorems from the axioms. Proving theorems can be very difficult, in part because there are no cut-and-dried rules for what to do next. One of the most common frustrations is not knowing how to begin a proof. Well, proofs from the field axioms tend to be pretty straightforward. Usually, they should begin with an axiom that relates to what you're trying to prove. For example, if the statement to be proved mentions 0, try starting with the axiom that most directly describes the special nature of 0: axiom V-8. If the statement to be proved mentions subtraction, try starting with the definition of subtraction and the only axiom that mentions negatives.

In proofs from the field axioms, if you start with the right axiom, it gives you an equation to work with. You usually find that the result you want can then be obtained by performing a few simple algebraic steps on both sides of that equation. The various proof methods that involve logic more directly, like indirect proof, proof by cases, existential specification, and so on, are not used much in the first part of this appendix.

Let's also say a bit about the *meaning* of these theorems. Suppose we prove some statement from the field axioms. Then the proven statement must be a logical consequence of those axioms (by the soundness of our axiom system, discussed in Section 4.1). That is, the proven statement must be true whenever the field axioms are true. In other words, if we prove a statement from the field axioms, the statement must be true in every field, including the real number system, the rational number system, the complex number system, and so on. Similarly, when we prove something from the ordered field axioms, it must be true in every ordered field. So it must be true in the real and rational number systems, but not necessarily in the complex number system (see Exercise 2 of Section 10.1).

Recall the discussion in Section 4.1 involving formal, informal, and good proofs. Most of the proofs in this appendix are written in good, relatively informal style, but a few formal proofs are also included. You can immediately see how unwieldy the formal proofs are. You should also try to convince yourself that our informal proofs really are outlines of formal proofs. Exercises 7 and 8 of Section 4.4 ask you to turn formal proofs in this appendix into good proofs.

Proofs from the Field Axioms

Our first four results show that the identities and inverses described in axioms V-8 through V-11 are actually unique.

Theorem A-1: There is a unique additive identity; in symbols, $\exists! y\, \forall x\, (x + y = x)$.

Proof: Recall, from Section 3.4, that this type of proof normally has two parts—proving *existence* and proving *uniqueness*. In this case, the existence part is an axiom (V-8), so nothing further needs to be said.

To prove uniqueness, we use Theorem 3.4(c). That is, we assume that there are two additive identities and prove that they must be equal. So assume y and z are both additive identities. That is, $\forall x\, (x + y = x)$ and $\forall x\, (x + z = x)$. Then we can specify the first quantified x to be z, and the second one to be y. This yields $z + y = z$ and $y + z = y$. Thus $y = z + y = y + z = z$ (by the equality axioms and the commutative property of addition), as desired. ■

Theorem A-2: There is a unique multiplicative identity.

Proof: This result is completely analogous to the previous one. We therefore leave its proof for Exercise 5 of Section 4.4. ■

Recalling the discussion in Section 3.4 about the relationship between uniqueness and definitions, the following definitions are now justified:

Definitions: Henceforth, we use the standard symbols 0 and 1 to denote the additive and multiplicative identities, respectively.

Theorem A-3: Each number has a unique additive inverse.

Proof: This proof is also very similar to that of Theorem A-1. Let x be any number. By axiom V-10, x has an additive inverse. To show uniqueness, assume y and z are both additive inverses of x. So $x + y = 0$ and $x + z = 0$. Therefore, $x + y = x + z$. By adding y to both sides of this equation and applying axioms V-3 and V-5, we get

$0 + y = 0 + z$. Finally, applying axiom V-8 to both sides of this equation yields $y = z$, as desired. ■

Theorem A-4: Each nonzero number has a unique multiplicative inverse.
Proof: See Exercise 6 of Section 4.4. ■

Theorems A-3 and A-4 justify the next pair of definitions. It would have been possible to include uniqueness of the identities and inverses in the axioms, except that it's usually best to have the axioms say as little as possible. So our approach, proving uniqueness, is generally preferred.

Definitions

For any number x, its additive inverse is denoted $-x$.
For any nonzero number x, its multiplicative inverse is denoted x^{-1}.

Now is as good a time as any to define several well-known, simple abbreviations:

Definitions

Subtraction: $x - y$ means $x + (-y)$.
Division: x/y means $x \cdot (y^{-1})$.
We write xy to mean $x \cdot y$.
The numeral 2 means $1 + 1$, 3 means $2 + 1$, 4 means $3 + 1$, and so on.
The notation x^2 means $x \cdot x$, x^3 means $x \cdot x \cdot x$, and so on.

A more rigorous and complete definition of exponents is given in Section 7.1.

For the next theorem we give both a formal proof and an informal proof, for contrast.

Theorem A-5: Any number times 0 equals 0.
Formal Proof

(1)	$x + 0 = x$	Axiom V-8, US, modus ponens
(2)	$x(x + 0) = x \cdot x$	Axiom III-4 on step 1
(3)	$x(x + 0) = x \cdot x + x \cdot 0$	Axiom V-7, US, modus ponens
(4)	$x \cdot x + x \cdot 0 = x \cdot x$	Steps 2 and 3, axiom III-3, modus ponens
(5)	$(x \cdot x + x \cdot 0) + -(x \cdot x) = x \cdot x + -(x \cdot x)$	Step 4, axiom III-3, modus ponens
(6)	$x \cdot x + -(x \cdot x) = 0$	Axiom V-8, US, modus ponens
(7)	$(x \cdot x + x \cdot 0) + -(x \cdot x) = 0$	Steps 5 and 6, axiom III-3, modus ponens
(8)	$(x \cdot x + x \cdot 0) + -(x \cdot x) = x \cdot 0 + [x \cdot x + -(x \cdot x)]$	Axiom V-3 and axiom V-5)
(9)	$x \cdot 0 + [x \cdot x + -(x \cdot x)] = x \cdot 0 + 0$	Step 6, axiom III-4, modus ponens
(10)	$x \cdot 0 + 0 = x \cdot 0$	Axiom V-8, US, modus ponens
(11)	$x \cdot 0 = 0$	Steps 7, 8, 9, and 10, axiom III-3
(12)	$\forall x\, (x \cdot 0 = 0)$	UG on step 11

Informal Proof: Let x be any real number. By axiom V-8, we know that $x + 0 = x$. Multiplying both sides by x and applying the distributive law, we get $x \cdot x + x \cdot 0 = x \cdot x$.

Now add $-(x{\cdot}x)$ to both sides. Using the associative, commutative, identity, and inverse axioms of addition, this simplifies to $x{\cdot}0 = 0$. Since x was arbitrary, this holds for every value of x. ■

The informal proof is all you would expect to find for a proof of this theorem in a book, and for most purposes it's better than the formal proof. Various small steps are omitted, but any reader who understands the field axioms can easily fill in the gaps. It is not only more readable than the formal proof, but it also makes the logic of the proof clearer. We start by stating the only axiom that mentions 0 and then multiply both sides by x; this is sensible because it gives us an equation involving $x{\cdot}0$. (Multiplying both sides by 0 would also do that, but if you try that, you find it just doesn't lead anywhere.) The rest of the proof is straightforward simplification.

Theorem A-6: For any number x, $-(-x) = x$.
Proof: Given any x, we know that $x + (-x) = 0$. By commutativity, $(-x) + x = 0$. In other words, x is an additive inverse of $-x$. But by Theorem A-3, there is only one additive inverse of $-x$, which is denoted $-(-x)$. So $-(-x) = x$. ■

We conclude the first part of this appendix with something a bit different—an incorrect proof. It's not very incorrect. The theorem is provable, and the proof given is basically correct, except that it assumes one or two facts that are neither axioms, definitions, nor previous theorems. The incorrectly assumed steps certainly look "obviously true" and in fact are pretty easy to prove from the field axioms. But assuming "obviously true" statements without bothering to prove them is the surest road to mathematical damnation! (See Exercise 9 of Section 4.4.)

Theorem A-7: $\forall x, y, z\, [(y \neq 0 \wedge z \neq 0) \rightarrow (xz)/(yz) = x/y]$
Proof: Let x, y, and z be any numbers, with y and z nonzero. By definition, $(xz)/(yz)$ means $(xz)(yz)^{-1}$. By the associative and commutative laws of multiplication, this becomes $xy^{-1}(z{\cdot}z^{-1})$. By the inverse and identity axioms, this equals xy^{-1}, which by definition is the same as x/y. ■

Proofs Using the Ordered Field Axioms

The field axioms encompass the main properties of addition, subtraction, multiplication, and division, but this is not all there is to the basic algebra of $\mathbb{R}$. Another important part involves **inequalities**—statements involving the symbols $<$, $>$, $\leq$ and $\geq$. The additional axioms that turn a field into an ordered field are axioms V-13 through V-17 of our axiom system. Like the field axioms, all these axioms express simple, familiar rules. Note that even though there are four inequality symbols, the only one that the axioms mention is $<$. The other three symbols can be defined in terms of this one and equality, as follows.

Notation: We write $x > y$ to mean $y < x$, $x \leq y$ to mean $x < y$ or $x = y$, and $x \geq y$ to mean $y \leq x$.

The first axiom of the group says that no number is less than itself, and the second states the familiar transitive property. A structure satisfying these two axioms is called an **irreflexive partial ordering**. The next axiom is the **trichotomy** axiom, which essentially says that given any two unequal numbers, one of them must be less than the other. Remember that the word "or" in mathematics means the inclusive or. So, as stated, the trichotomy axiom allows the possibility that two or all of the three disjuncts could be true at once. But Theorem A-9 disproves this. A structure satisfying axioms V-13 through V-15 is called an **irreflexive total ordering**. Total orderings are also known as **linear** or **simple** orderings. Note that none of these axioms mentions any of the four algebraic operations, and a structure need not have any kind of algebra defined on it to be an ordering of some kind. For example, if a bunch of people are waiting on line, you could say they form a totally ordered set of people; yet there's no way to add or multiply two people.

Orderings are discussed in greater detail in Section 6.3.

The last two ordered field axioms express algebraic properties of inequalities. Axiom V-16 states the familiar fact that you can add (and therefore subtract) anything to both sides of an inequality. Axiom V-17 says that an inequality is preserved if both sides are multiplied by a positive number. Notice that the usual law about multiplying both sides of an inequality by a negative number is missing; Exercise 27 of Section 4.4 asks you to prove this law.

We now prove a number of theorems from the ordered field axioms, which then must be true in every ordered field. Their proofs are logically more complex than those in the previous group. There are several reasons for this. For one thing, the trichotomy axiom and the definitions of $\leq$ and $\geq$ are disjunctions; and when one has disjunctions to work with, the method of proof by cases usually becomes very important. Proving a statement about all real numbers using cases based on trichotomy is a particularly powerful technique. Also, there are more implications and negations involved in the ordered field axioms than in the field axioms, and so conditional proof and indirect proof are more often used in these proofs.

Theorem A-8: The conditions $x < y$ and $y < x$ cannot both be true.

Formal Proof

(1) Assume $x < y$ and $y < x$	Assumption for indirect proof
(2) $x < x$	Axiom V-14, US, modus ponens on step 1
(3) $\sim (x < x)$	Axiom V-13, US, modus ponens
(4) Contradiction	Steps 3 and 4, conjunction rule
(5) $\sim (x < y \land y < x)$	Indirect proof
(6) $\forall x, y \sim (x < y \land y < x)$	UG on step 5 ■

The property stated in Theorem A-8 is called the **strong antisymmetry** property. Some books include this as an axiom, but as this proof shows, it follows from our irreflexive partial ordering axioms. (That is, the only axioms required for this proof are V-13 and V-14). Strong antisymmetry is sometimes expressed in the equivalent form $\forall x, y\ (x < y \rightarrow y \nless x)$.

Theorem A-9: Trichotomy is actually an *exclusive* or; that is, no two of its three disjuncts can be true at once.

Proof: We use indirect proof for each pair of disjuncts. First assume $x = y$ and $x < y$. But then by axiom III-4, $x < x$; this contradicts axiom V-13. By similar reasoning, the assumption that $x = y$ and $y < x$ is impossible. Finally, the previous theorem showed that $x < y$ and $y < x$ cannot both occur. ■

Theorem A-10: (a) If $x \leq y$ and $y < z$, then $x < z$.

(b) If $x < y$ and $y \leq z$, then $x < z$.

(c) If $x \leq y$ and $y \leq z$, then $x \leq z$.

Proof: (a) Assume $x \leq y$ and $y < z$. The inequality $x \leq y$ is a disjunction, so we proceed by cases. Case 1: Assume $x < y$. So we have $x < y$ and $y < z$. Therefore, $x < z$, by axiom V-14. Case 2: Assume $x = y$. So we have $x = y$ and $y < z$. Therefore, $x < z$, by axiom III-4. Since both cases lead to the desired conclusion, we are done.

(b) This proof is almost identical to that of part (a), so we omit it (see Exercise 19 of Section 4.4).

(c) See Exercise 20 of Section 4.4. ■

The next theorem is proved in a manner that was mentioned earlier: the trichotomy axiom is used to set up a proof by cases. There is no other simple way to prove it.

Theorem A-11: $\forall x\,(x^2 \geq 0)$

Proof: Let x be any number. We proceed by cases, using trichotomy. Case 1: If $x = 0$, then $x^2 = 0 \cdot 0 = 0$, by Theorem A-5. And of course $0 \geq 0$. Case 2: If $x > 0$, then by axiom V-17, $x^2 > 0x = 0$, again by Theorem A-5. In other words, $x^2 > 0$, so of course $x^2 \geq 0$. Case 3, in which, $x < 0$, is left for Exercise 21 of Section 4.4. ■

Corollary A-12: $1 > 0$

Proof: We leave this for Exercise 22 of Section 4.4. ■

Theorem A-13: $x > 0$ iff $x^{-1} > 0$.

Formal Proof: We prove only the forward direction, leaving the other direction for Exercise 23 of Section 4.4.

(1) Assume $x > 0$.	Assumption for conditional proof
(2) Assume $\sim(x^{-1} > 0)$.	Assumption for indirect proof
(3) $x^{-1} > 0 \lor x^{-1} = 0 \lor x^{-1} < 0$	Axiom. V-15
(4) $x^{-1} = 0 \lor x^{-1} < 0$	Propositional consequence of steps 2 and 3

We now do a proof by cases on the previous disjunction.

(5) Case 1: Assume $x^{-1} = 0$.	
(6) $x \neq 0$	Theorem. A-9 applied to step 1
(7) $x \cdot x^{-1} = 1$	Axiom V-11 and step 6
(8) $x \cdot x^{-1} = 0$	Theorem. A-5 and step 5
(9) $0 = 1$	Axiom. III-3, steps 7 and 8
(10) $0 \neq 1$	Axiom. V-12
(11) Contradiction	Conjunction rule on steps 9 and 10

(12) Case 2: Assume $x^{-1} < 0$.	
(13) $x \cdot x^{-1} < x \cdot 0$	Axiom V-17, steps 1 and 12
(14) $x \cdot 0 = 0$	Theorem. A-5
(15) $x \cdot x^{-1} = 1$	Axiom V-11
(16) $1 < 0$	Axiom III-4, steps 13, 14, and 15
(17) $1 > 0$	Theorem A-12
(18) Contradiction	Theorem A-8, steps 16 and 17

Since both cases lead to a contradiction, the indirect proof is complete. ■

Theorem A-14: (a) There is no largest number.
(b) There is no smallest number.

Proof: (a) We want to prove $\sim \exists x \forall y (x \geq y)$. By De Morgan's laws for quantifiers it suffices to prove the equivalent statement $\forall x \exists y (y > x)$, and that's what we do. (Note that we're starting this proof backward. But we do the rest of it forward.)

So let x be given. By Theorem A-12, we know that $1 > 0$. By axiom V-16, this yields $x + 1 > x$. So by EG, we have $\exists y (y > x)$. *[That is, we are using $x + 1$ as the term t.]* Since x was arbitrary, we can assert $\forall x \exists y (y > x)$, as desired.

The proof of part (b) is nearly identical, so we omit it. ■

Let's now define **absolute value**, a concept that makes sense in any ordered field. From a logical standpoint, the most interesting aspect of this concept is that it's a **definition by cases**. Like a proof by cases, a definition by cases must be based on a known disjunction. In this case, the disjunction is $x \geq 0$ or $x < 0$. Definitions by cases are discussed in Section 7.1.

Definition: For any number x in any ordered field,

$$|x| = \begin{cases} x & \text{if } x \geq 0 \\ -x & \text{if } x < 0 \end{cases}$$

Not surprisingly, most proofs involving absolute value require proof by cases. Here are several simple but useful results of this type.

Theorem A-15: (a) $|xy| = |x|\,|y|$
(b) **Triangle inequality:** $|x + y| \leq |x| + |y|$
(c) For any x and any $b > 0$,
(i) $|x| = b$ iff $x = \pm b$
(ii) $|x| < b$ iff $-b < x < b$
(iii) $|x| > b$ iff $x > b$ or $x < -b$

Proof: (a) In the case that either x or y is zero, then so is xy, by Theorem A-5. Thus both sides of the equation equal zero.

In the case that x and y are both positive, then so is xy, by axiom V-17 and Theorem A-5. So both sides of the equation equal xy.

If $x > 0$ and $y < 0$, the same axiom and theorem yield $xy < 0$. So the left side is $-(xy)$, while the right side is $x(-y)$. Using Exercise 11 of Section 4.4, it is simple to show these are equal.

The case $x < 0$ and $y > 0$ is analogous to the previous case.

Finally, consider the case that both x and y are negative. Then, from Exercise 28 of Section 4.4 and Theorem A-5, we get $xy > 0$. So the right side is xy. The left side is $(-x)(-y)$, which must equal xy by Exercise 11 of Section 4.4 and Theorem A-6.

Exercise 24 of Section 4.4 asks you to fill in the details of this proof of part (a).

(b) and(c): These are also left for Exercise 24 of Section 4.4. ■

The triangle inequality is a particularly useful fact, as Chapter 9 shows. By the way, this result still holds if x and y represent vectors (of any dimension) and the vertical lines represent magnitude. This explains its name (see Figure A-1). In other words, it says that a straight line is the shortest distance between two points.

We conclude this appendix with a somewhat more difficult theorem. As a result, our proof of it omits more simple steps than any of our previous proofs. But it can still be considered to be an outline of a formal proof. Exercise 25 of Section 4.4 asks you to fill in the details.

Theorem A-16: Every cubic (third-degree) polynomial achieves both positive and negative values.

Proof: A cubic polynomial is an expression of the form $ax^3 + bx^2 + cx + d$, where $a \neq 0$. So let such an expression be given. *[We are doing a UG proof on a, b, c, and d. Technically, these coefficients are mathematical variables, even though we don't think of them as varying.]* To show the polynomial achieves positive values means to show there is a value of x for which the polynomial is positive. *[So we use EG.]* The idea is to pick x sufficiently large (and with the same sign as a) so that the term ax^3 dominates the other terms of the polynomial.

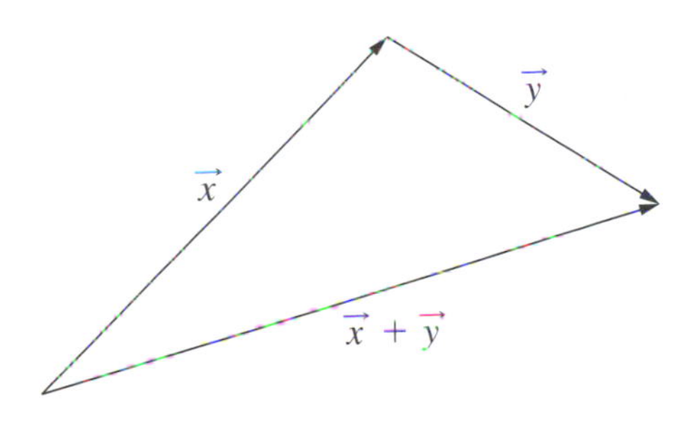

Figure A-1 The triangle inequality in two dimensions

So let x be the number $3(|a| + |b| + |c| + |d| + 1)/a$. To show this works, we proceed by cases on the sign of a. If $a > 0$, note that $x > 3|b|/|a| = 3|b|/a$. Therefore, $x^3 > 3|b|x^2/a$. So $ax^3 > 3|b|x^2$, and thus $ax^3/3 > |b|x^2$. Similarly, we can show that $ax^3/3 > |cx|$, and $ax^3/3 > |d|$. (For these last two inequalities, it's necessary to use the fact that $x > 1$). Putting these together and applying Theorem A-13 yields that $ax^3 > |bx^2 + cx + d|$, which guarantees that $ax^3 + bx^2 + cx + d > 0$. The proof if $a < 0$ is almost identical.

To show the polynomial achieves negative values, simply let x be the negative of what it was before. The proof is then almost identical to the one for positive values. ■

By the way, Theorem A-16 generalizes to any polynomial of odd degree. But the proof of this more general theorem is significantly more difficult.

Appendix 3

Some of the More Useful Tautologies

This list of tautologies is identical to Table 2.5.

(1) $P \vee \sim P$	Law of the excluded middle
(2) $\sim(P \wedge \sim P)$	Law of noncontradiction

Some implications

(3) $(P \wedge Q) \rightarrow P$	Basis for simplification
(4) $(P \wedge Q) \rightarrow Q$	Basis for simplification
(5) $P \rightarrow (P \vee Q)$	Basis for addition
(6) $Q \rightarrow (P \vee Q)$	Basis for addition
(7) $Q \rightarrow (P \rightarrow Q)$	
(8) $\sim P \rightarrow (P \rightarrow Q)$	
(9) $[P \wedge (P \rightarrow Q)] \rightarrow Q$	Basis for modus ponens
(10) $[\sim Q \wedge (P \rightarrow Q)] \rightarrow \sim P$	Basis for modus tollens
(11) $[\sim P \wedge (P \vee Q)] \rightarrow Q$	
(12) $P \rightarrow [Q \rightarrow (P \wedge Q)]$	
(13) $[(P \rightarrow Q) \wedge (Q \rightarrow R)] \rightarrow (P \rightarrow R)$	Transitivity of implication
(14) $(P \rightarrow Q) \rightarrow [(P \vee R) \rightarrow (Q \vee R)]$	
(15) $(P \rightarrow Q) \rightarrow [(P \wedge R) \rightarrow (Q \wedge R)]$	
(16) $[(P \leftrightarrow Q) \wedge (Q \leftrightarrow R)] \rightarrow (P \leftrightarrow R)$	Transitivity of equivalence

Equivalences for rewriting negations

(17) $\sim(P \wedge Q) \leftrightarrow \sim P \vee \sim Q$	De Morgan's law
(18) $\sim(P \vee Q) \leftrightarrow \sim P \wedge \sim Q$	De Morgan's law
(19) $\sim(P \rightarrow Q) \leftrightarrow P \wedge \sim Q$	

Equivalences for replacing connectives

(20) $(P \rightarrow Q) \leftrightarrow (\sim P \vee Q)$
(21) $(P \leftrightarrow Q) \leftrightarrow [(P \rightarrow Q) \wedge (Q \rightarrow P)]$
(22) $(P \leftrightarrow Q) \leftrightarrow [(P \wedge Q) \vee (\sim P \wedge \sim Q)]$

Other equivalences

(23)	$\sim\sim P \leftrightarrow P$	Law of double negation
(24)	$(P \rightarrow Q) \leftrightarrow (\sim Q \rightarrow \sim P)$	Law of contraposition
(25)	$[(P \rightarrow Q) \wedge (P \rightarrow R)] \leftrightarrow [P \rightarrow (Q \wedge R)]$	
(26)	$[(P \rightarrow R) \wedge (Q \rightarrow R)] \leftrightarrow [(P \vee Q) \rightarrow R]$	Basis for proof by cases
(27)	$[P \rightarrow (Q \rightarrow R)] \leftrightarrow [(P \wedge Q) \rightarrow R]$	
(28)	$[P \rightarrow (Q \wedge \sim Q)] \leftrightarrow \sim P$	Basis for indirect proof
(29)	$[P \wedge (Q \vee R)] \leftrightarrow [(P \wedge Q) \vee (P \wedge R)]$	Distributive law
(30)	$[P \vee (Q \wedge R)] \leftrightarrow [(P \vee Q) \wedge (P \vee R)]$	Distributive law

Solutions and Hints to Selected Exercises

Exercises 1.2

(1) (a) For instance, if $n = 3$, then $n^2 - n + 41 = 47$, a prime number.

(b) There is a number that, when substituted for n in this polynomial, makes it obvious that the polynomial's value is a multiple of this number.

(2) For instance, $4 = 2 + 2$ and $100 = 47 + 53$.

(3) (a) For instance, $10 = 17 - 7$.

(b) If either conjecture were false, there would have to be a particular number (a so-called **counterexample**) for which it fails. Suppose you were asked to test whether a certain number, say 1234567890, is a counterexample to Goldbach's conjecture. What would you have to do? In contrast, what would you have to do to test whether this number is a counterexample to de Polignac's conjecture?

(4) (b) This can easily be proved using calculus. A more elementary, algebraic proof can be obtained by starting with the desired inequality and simplifying it until a known fact is obtained. This constitutes a proof, provided that you make sure that all the steps of your simplification are reversible.

(d) One approach to this is algebraic, using the standard equations of a line and a circle. It is also possible to prove this with careful geometric reasoning.

(8) (c) If a number n is the product of two primes, what is the *largest* possible value for the *smaller* of the two primes? Then, what can be similarly concluded if n is the product of three or more primes?

(9) (a) Try some other odd number for the smallest member of the triple, and then use algebra to solve for the other members.

(b) In Pythagoras's Theorem, let $c = b + 1$ and $a = 2m$ (an arbitrary even number).

(c) Proceed as in part (b), but with $a = 2m + 1$ (an arbitrary odd number).

(11) (a) First determine who has the winning strategy if the two piles have the same number of sticks. Start with one stick each, then two, and so on.

(b) It is surprising how little this rule change affects the answer to the problem, both in terms of who wins from which starting configurations and also what the winning strategies are.

Exercises 2.1

(1) (b)

P	Q	P ∨ Q	P ↔ (P ∨ Q)
T	T	T	T
T	F	T	T
F	T	T	F
F	F	F	T

(c) You might think this must be a contradiction, but it isn't.

(e) This is a tautology. A witty example of this tautology is the old saying "If I had butter, then I'd have bread and butter—if I had bread."

(2) (d) This is one of several that is borderline, since it involves a value judgment.

(h) It would seem sensible to say that this statement is not a proposition because it is not true or false *now*. Yet, the predominant view is that such statements about the future are propositions.

(j) Most (but not all!) mathematicians believe that mathematical statements about integers are true or false, even ones that have not yet been proved or disproved.

(3) (a) For many problems of this sort, it may be easiest to interpret a symbolic statement using the words "true" and "false." For example, part (a) can be read "If it's not the case that P and Q are both true, then P and Q are both false." Is this implication necessarily correct?

(c) Tautology.

(d) Neither.

(4) (d) Equivalent.

(e) Not equivalent

(j) Not equivalent.

(6) (a) →

(8) (b) Start by considering the verbal meaning of this connective.

(9) (b) True.

(d) True.

(10) (f) The atomic components of this statement are, "The customer is wearing shoes," "The customer is wearing a shirt," and "The customer gets served."

(g) This statement is *not* a disjunction!

(11) (b) Four.

Exercises 2.2

(2) (d) If a number is nonnegative, then it has a square root.
(f) If a function is one-to-one, then it has an inverse function.

(3) (d) There is no single correct answer to this. One reasonable answer is "John's lack of happiness is necessary for Mary's happiness."

(4) (a) Form 7: It is difficult to make this sound sensible. One try is "My being happy for a day is a necessary condition for me to be reading a good book." Perhaps you can do better than this!
(b) Form 1: "Your apologizing implies that I will pay you." Converse: "My paying you implies that you have apologized."

(5) (b) Form 4: "My going for a hike today and my finishing my paper this morning are equivalent."

(6) (c) "If you eat your cake, then you can't have it."
(f) Even though it is awkward, the only logically correct answer is the negation of an implication.

(8) (c) Sometimes true and sometimes false.
(d) Also sometimes true and sometimes false! Can you give an example of a person for whom this statement is true? (However, some people might prefer to think of this conditional as necessarily false.)

(9) Give some thought to whether a conditional or a biconditional yields a more reasonable answer.

Exercises 2.3

(1) (c) If we know Q and we also know that Q implies P, then what exactly do we know, *in total*?

(2) (a) What is the simplest form of the negation of $P \wedge \sim Q$?

(4) (c) Not valid. (e) Valid.

(5) (b) Derive both $P \rightarrow R$ and $P \rightarrow \sim R$. Then use the method described in Proof Preview 3.
(d) Tautology 13 is useful for this derivation.

(12) (d) $(\sim P \wedge Q \wedge R) \vee (\sim P \wedge \sim Q \wedge R) \vee (\sim P \wedge \sim Q \wedge \sim R)$

(13) Use the result of Exercise 11 and one of the tautologies in Table 2.5.

(16) Show that no statement built up from these connectives can be equivalent to $\sim P$ by considering the row of a truth table in which all propositional variables are true.

(17) By Exercise 13, it suffices to show that the truth functions of $\sim P$ and $P \wedge Q$ can be represented by statements built up using only $|$.

Exercises 3.2

(1) (c) Not grammatically correct.
(d) Grammatically correct.

(3) (b) A simple answer which more or less has the correct meaning is $\forall m\ \exists n\ (n > m)$. But a literal symbolic translation should begin with a negation.
(c) An odd number is a number *of the form* $2k + 1$. Make sure your answer has the right free variable(s).
(h) $\forall x, y\ [(x < y \rightarrow \exists z\ (x < z < y)) \wedge (x > y \rightarrow \exists z\ (x > z > y))]$. It would be possible to omit either of the conjuncts.

(5) (b) The only free variable is z. The statement is true if $z = -1$, false if $z = 1$.
(d) The only free variable is m. To answer the other part, consider the graph of the equation $y = x^2 - x$.

(6) (a) Your answer could begin "Every nonnegative number has ... ."

(7) (b) $P(x, y) \wedge \sim W(x)$. Having written this, we can abbreviate it to $F(x, y)$.
(c) This statement involves a third person! Should the variable representing this person be free or bound?

(8) Note that these English statements contain no free variables. Therefore, your answers should not contain any free variables either!
(d) First, let $L(x, v)$ mean "x likes v," where x stands for a person and v stands for a vegetable.
(j) Note that this statement contains no connectives, but it has quantifiers for three different things. Therefore, you must begin by defining an atomic statement with three free variables.
(k) In this statement, "all cases" must be interpreted as "for all pairs of people." Your answer might begin $\sim \forall x, y$, followed by a statement without quantifiers.

Exercises 3.3

(1) Yes, this assumption has been made.

(5) Part (a) says that every number can be squared; part (b) says something quite different.
(c) False. (e) True.

(f) In words, is there a number x such that the equation $yx^2 = y$ is always true but the equation $yx = y$ is not always true?

(6) Four of the answers from Exercise 5 stay the same. The others change from true to false.

(7) (b) True. You could use Theorem 3.1 to justify this.

(c) False. The clearest way to justify this is to define a specific P, as well as domains for the variables, to make the whole statement false.

(d) True. You could justify this with an explanation that is similar to the proof of Theorem 3.2

(e) The change in the connective makes this very different from part (d)!

(8) (b) $\forall x \sim \mathrm{P} \wedge \exists y\ \mathrm{Q}$.

(9) (b) $\forall x{\in}\mathbb{R}\ \exists m, n{\in}\mathbb{Z}\ (m < x < n)$.

(10) (a) This problem is very similar to Proof Preview 4. We want to prove $\forall x\ \exists z\ (z > x \text{ and } z > x^2)$. As in the text, we could define z using absolute values. Is it possible to use a simpler definition such as $z = x^2 + 1$?

Exercises 3.4

(1) (a) This proof requires nothing more than moving the ~ inward, as in Example 8 of Section 3.3.

(b) Outline of proof: If, for some particular x, we have $\forall y\ [\mathrm{P}(y) \leftrightarrow y = x]$, we can then let $y = x$. From this, $\mathrm{P}(x)$ follows.

(2) You can't do this problem using the quantifier $\exists!$. You have to figure out how to say these things from scratch. For part (c), you need to use at least one ellipsis (...) in your answer.

(3) Again, you can't use $\exists!$ here, since that quantifier specifically means "there is exactly one." Instead, express these statements as combinations of the statements from Exercise 2.

(5) (a) True. (e) False.

(7) (a) $\forall x\ [\exists! y\ (\mathrm{P}(y, x) \wedge \mathrm{W}(x)) \wedge \exists! z\ (\mathrm{P}(z, x) \wedge \sim \mathrm{W}(z))]$

(c) Since this means "Nobody has three or more grandmothers," your answer could begin $\sim \exists x, u, v, w$.

(8) (c) $\forall A, B\ [A \neq B \rightarrow \exists! L\ (\mathrm{On}(A, L) \wedge \mathrm{On}(B, L))]$

(g) The fact that B is between A and C can be expressed using the distances $\overline{AB}$, $\overline{BC}$, and $\overline{AC}$.

(9) (b) Only one of the four equality axioms holds for the symbol $<$.

(11) (b) The absolute value signs don't make it harder to define the correct z or to show that the equation is satisfied using this value of z. But they do make it harder to prove this z is unique. You might need to break that proof up into two or more cases.

(12) (a) To say that there's a unique minimum means that there's a unique value of x with a certain property.

Exercises 4.2

(1) For example, if we know $P \leftrightarrow Q$ and $P \vee R$, then $Q \vee R$ can easily be proved by cases.

(3) (b) One good method is to use cases based on the disjunction $P \vee \sim S$.

(4) (a) I can't make you happy today.

(5) (a) There's no reason to make this complicated. You could define propositional variables H and T such that the premises are $H \rightarrow T$ and $\sim T$.

(6) $P \vee Q \vee R$ can be viewed as an abbreviation for $(P \vee Q) \vee R$. Therefore, the derivation of ordinary proof by cases given in the text can be used repeatedly to show the validity of proof by cases with three or more cases.

(8) Yes, this problem is as easy as it seems.

(9) (a) One of the tautologies in Appendix 3 is helpful for this formal proof.

(10) You may use the fact that n must be even or odd. How does one normally proceed in a proof when one knows a disjunction?

(11) The statement $y \leq z$ is not atomic but is an abbreviation for a longer statement.

(13) (iv)

(16) The theorem is true, but what about the logic used in the proof?

Exercises 4.3

(1) Your proof could begin as follows: Assume $A \subseteq B$ and $B \subseteq C$. By definition, this means $\forall x\ (x \in A \rightarrow x \in B)$ and $\forall x\ (x \in B \rightarrow x \in C)$. Now let x be given and assume $x \in A$. By US on the first two assumptions, we also have $x \in A \rightarrow x \in B$ and $x \in B \rightarrow x \in C$ ... .

(3) If you want to do this symbolically, introduce predicates S(t), I(t), and M(t), where the variable t stands for a day of this week. Common sense should tell you that it's not possible to prove a conclusion involving M(t) if none of the assumptions involves M(t). Based on the sense of the words, what conjunction should, for any t, imply M(t)? This implication needs to be one of your assumptions.

(6) There isn't much to this problem, but note that the quoted theorems do contain hidden quantifiers.

(9) Forward direction of number 19: Assume $\exists x$ (P $\rightarrow$ Q), that is $\exists x$ (P(x) $\rightarrow$ Q(x)). By ES, let c satisfy P(c) $\rightarrow$ Q(c). To prove the right side of the biconditional, assume $\forall x$ P, that is, $\forall x$ P(x). By US, P(c). Modus ponens then gives us Q(c). By EG, we have $\exists x$ Q(x). By conditional proof, applied twice, we are done.

(13) Both directions use ES and proof by cases, but in the opposite order.

Exercises 4.4

(4) For this problem to have any content, your proof must be pretty formal. Use Axiom III-4, with P(x) being the statement t(x) = t(x).

(18) More specifically, you could prove that any given x equals one specific number. It is also helpful to note that Theorem A-5 does not require axiom V-12 in its proof.

(20) This should be done as a proof by cases with four cases or, if you prefer, two cases with two subcases each.

(26) See the hint for Exercise 20. Also, note that if any case of a proof by cases leads to a contradiction, that's as good as if it leads to the desired conclusion.

(28) Try to figure out a way to apply axiom V-17 to this situation.

(29) This cannot be proved with the field axioms alone! If you prove this without using any of axioms V-13 through V-17, sit down and look for the error in your proof.

(31) The proof is more long and tedious than hard. It's a proof by cases, with quite a few cases to consider.

Exercises 4.5

(1) (b) This is an arithmetic series with $a = 2$ and $d = 3$. But what is n, the number of terms? Finding n is a necessary step in the solutions of all parts of this problem. Note that theorems 4.13 and 4.14 implicitly include formulas for the n-th term of arithmetic and geometric sequences.

(9) (b) The next to last step is $S(1 - r) = a(1 - r^n)$.

(11) You might want to use Theorem 4.16(a) to prove that $1 \notin \mathbb{N}$.

(13) (b) This series is called a **telescoping series**. Can you see why?

(15) (c) The correct conjecture should remind you of the formula for $\int x^k\,dx$. (Perhaps you recall that integrals and sums are closely related.)

(17) In the induction step, you have to use the inequality $n < 2^n$. What should you do with it? Adding one to both sides is not very helpful. Try something different. At some point in your proof, it could be helpful to use Theorem 4.16(a).

(19) (b) A meeting with $n + 1$ people may be viewed as a meeting at which there are n people, and then one more shows up.

(20) For the induction step, you could use a variant of the hint for Exercise 19(b).

Exercises 5.1

(1) (c) $\{0\}$; this can be verified using calculus.
(e) $\{-2, 2\}$.

(2) (e) There is no particular pattern to these numbers. Just write a disjunction after the vertical line.

(6) (c) Read the second disjunct carefully!
(d) You could do this by commonsense reasoning or by considering the parabola $y = x^2$.

(8) (c) False, because [5, 2] is the empty set under our definition. Some authors would prefer to say that [5, 2] is undefined. It might seem sensible to let [5, 2] = [2, 5], but this viewpoint is rarely adopted.

(9) (c) False in $\mathbb{N}$. To determine whether the statement is true in $\mathbb{Q}$, factor the left side of the equation and try some simple fractions for x and y.

(11) Note that the set form of induction is a single axiom, whereas the statement form is a list of axioms. For one direction, show that, for a certain choice of the statement P(n), the set form of induction is an instance of the statement form. For the other direction, show that each instance of the statement form follows from the set form. You may interpret $\mathbb{N} \subseteq A$ to mean "$\forall n\ (n \in A)$," where n is a natural number variable.

(13) (a) By extensionality, it suffices to show that any member of either of these sets must be in the other. It might help to use Theorem 4.16.

Exercises 5.2

(1) (c) Grammatically incorrect. (i) Grammatically correct statement.

(3) For the forward direction of Theorem 5.3(a), show that the assumption that $x \in A \cap A'$ leads to a contradiction. Technically, this implies $x \in \varnothing$!

(6) (c) **Proof:** $A \subseteq B$ iff $\forall x\,(x \in A \rightarrow x \in B)$
iff $\sim \exists x \sim (x \in A \rightarrow x \in B)$
iff $\sim \exists x\,(x \in A \wedge x \notin B)$
iff $\sim \exists x\,(x \in A - B)$
iff $A - B = \varnothing$.

(7) (c) When dealing with intervals, pay attention to endpoints!
(f) What is $\mathbb{R} \cup \mathbb{N}$?

(10) (a) You may use the result of Exercise 13 of Section 5.1.

(12) (b) A Venn diagram could be very helpful here too. Note that if you add the number of members in A, B, and C separately, you count three parts of the Venn diagram twice and one part three times.

(14) The stated theorem is correct. What about the proof?

(15) (b) This requires careful definition unraveling. For the left side, we would say $x \in A \,\Delta\, (B \,\Delta\, C)$ iff $x \in [A - (B \,\Delta\, C)] \cup [(B \,\Delta\, C) - A]$ iff $x \in A - (B \,\Delta\, C)$ or $x \in (B \,\Delta\, C) - A$. From there, you might want to proceed by cases. If you can prove that the left side is a subset of the right, you should be able to justify the other direction without much additional work.

(17) (b) For instance, parts (a) and (b) of Theorem 5.3 are dual.

Exercises 5.3

(1) (c) Your answer should have four elements.
(d) The easiest way to do this is to use a part of Theorem 5.9.

(2) (a) False; 3 is not a set.

(3) (b) Usually not true, but it could be!
(f) Use a part of Theorem 5.9.

(4) (c) We have $x \in \bigcup(\wp(A))$ iff $\exists B \in \wp(A)\,(x \in B)$
iff $\exists B\,(B \subseteq A$ and $x \in B)$.

Is this equivalent to $x \in A$? If it's not clear to you, draw a picture or try some simple examples.

(9) We have $x \in (\bigcap_{i \in I} A_i)'$ iff $x \notin \bigcap_{i \in I} A_i$ iff $\sim \forall i \in I\,(x \in A_i)$. Now you need to similarly unravel membership in the set on the right side.

(13) (a) From the hypothesis of this implication, derive both P(0) and $\forall n \in \mathbb{N}\,(\mathrm{P}(n) \rightarrow \mathrm{P}(n+1))$. Then you can use ordinary induction. Feel free to use Theorem 4.16 and/or the result of Exercise 13 of Section 5.1.

(16) For which value of n is the *induction step* (as opposed to the statement being proved!) suspicious?

(17) First show that, in our definition of the Cantor set, A_k consists of all numbers in [0, 1] that must have a 1 in the kth place of their base 3 expansion.

Exercises 6.1

(1) (e) $R \cap S = \{8, 4\}$, so $\mathrm{Dom}(R \cap S) = \{8\}$.
(f) $\{2, 8\}$

(2) (h) It is probably easiest to graph this one quadrant at a time.
(i) You should probably solve the equation for y. But don't lose sight of the restrictions inherent in the original equation.

(3) (c) Simple reasoning shows the domain is $\{x \mid |x| \leq 5\}$ or $[-5, 5]$.
(e) The domain of this relation is $\mathbb{R}$; you need not prove this. Use calculus to find the range.

(4) (c) The answer is 6. Note that there is a simpler way of writing this set.
(d) There is also a simpler way of writing this set.

(6) Part (a) is a direct application of Theorem 6.1, but part (b) is not. It requires a more subtle use of the product rule for counting.

(8) (a) That is, by definition, $(x, y, z) = (u, v, w)$ iff $((x, y), z) = ((u, v), w)$.

(11) (a) If we know something is a relation, it makes sense to use the notation (x, y), rather than x, to denote an arbitrary member of it, since we know it consists of ordered pairs.
(b) $x \in \mathrm{Dom}(R^{-1})$ iff $\exists y\,[(x, y) \in R^{-1}]$ iff $\exists y\,[(y, x) \in R]$ iff $y \in \mathrm{Rng}(R)$.

(14) Try to adapt the proofs for Exercise 13. Be careful!

(15) Consider the analogous equation for numbers. The similarity is interesting.

Exercises 6.2

(1) (b) Not an equivalence relation.

(c) An equivalence relation. One equivalence class is the set of all of Pablo Picasso's daughters' children, provided that this set is not empty.

(2) (e) Recall the well-ordering property of $\mathbb{N}$.

(3) (a) This is an equivalence relation. $[3] = \{-3, 3\}$.

(d) Not an equivalence relation; for one thing, it's not reflexive.

(5) Technically, R is an equivalence relation and a relation on $\mathbb{N}$, but not an equivalence relation on $\mathbb{N}$.

(6) Technically, parts (a) and (b) are both false.

(7) When verifying symmetry and transitivity, all ordered pairs of the form (k, k) can be ignored. Do you see why?

(9) (a) To show reflexivity, let $k = 0$. To show symmetry, assume A is a translate of B. Then there is a number k with a certain property. What number can then be used to show that B is a translate of A?

(d) For example, does it work for open intervals?

(12) (a) Here is the part of the proof for symmetry: assume R and S are symmetric, and let $(x, y) \in R \cap S$. That means $(x, y) \in R$ and $(x, y) \in S$. Thus $(y, x) \in R$ and $(y, x) \in S$, because R and S are symmetric. Therefore, $(y, x) \in R \cap S$.

Note that to prove symmetry, we only needed to assume symmetry. The same holds for reflexivity and transitivity.

(19) For part (b) especially, you might want to analyze this in terms of partitions instead of equivalence relations.

Exercises 6.3

(1) (a) Six ordered pairs.

(5) For example, it's not total when $A = \mathbb{N}$, because $\{1\}$ and $\{2\}$ are not subsets of each other.

(11) (b) Assume R and S are total, and denote the lexicographic ordering by L. Let (a, b) and (a', b') be any members of $A \times B$. Since R is total, we know aRa' or $a'Ra$. By

propositional logic, this becomes (aRa' and $a \neq a'$) or ($a'Ra$ and $a \neq a'$) or $a = a'$. We proceed by cases on this disjunction: If the first disjunct holds, then $(a, b)L(a', b')$. If the second disjunct holds, then $(a', b')L(a, b)$. Finally, if $a = a'$, we must then use cases on the fact that bSb' or $b'Sb$.

(16) (a) Assume x and z are both least elements under a particular ordering on A. Then $\forall y \in A\ (x \leq y)$ and $\forall y \in A\ (u \leq y)$. Thus $x \leq u$ and $u \leq x$, so $x = u$.

(17) (a) Assume R is a well-ordering. For any x and y in its domain, consider the set $\{x, y\}$.

Exercises 7.1

(1) (g) To find the range, ask yourself what values the denominator can have.

(i) Unlike part (h), where the equation can easily be solved for y, that is not feasible in this problem or in part (j). You'll need some calculus to determine whether these two are functions.

(2) (d) $(3 \sin x)/2$.

(7) (g) How many outputs does f have for the input $x = 0$?

(l) Recall (or look up) the extreme value theorem of calculus. Also recall that every polynomial defines a continuous function on $\mathbb{R}$.

(9) (a) This is essentially a restatement of Lemma 6.5(b).

(11) The forward direction follows immediately from the substitution of equals axiom.

(13) (d) For example, note that if $A = \mathbb{R}$ and $f: A \to \mathbb{R}$ is defined by $f(x) = x$, then f has no multiplicative inverse, because the function $1/x$ does not have domain A.

(14) This problem is easy to understand if you think in terms of graphs.

Exercises 7.2

(5) Since $\mathbb{R}$ and $\mathbb{N}$ are infinite sets, the result given in Exercise 2 does not help for this problem and the next one. For this problem, you might find appropriate functions in a calculus book or a high school algebra book. For Exercise 6, you will have to be more creative.

(9) (b) The "grandparenthood" relation.

(e) The "half-siblinghood" relation (with equality included).

(12) (a) Does $b^{\log_b xy} = b^{\log_b x + \log_b y}$? And, if so, does this yield the desired result?

(13) (d) This function is not one-to-one. This can be established using calculus or simply by finding two numbers with the same image under f.

(14) (a) What kind of values does the derivative of g take on?
Parts (b) and (c) might require some trial and error, but parts (d) and (e) do not.

(16) (a) By definition, $x \in \text{Dom}(R \circ S)$ iff $\exists y\, [(x, y) \in R \circ S]$ iff $\exists y\, \exists z\, [(x, z) \in S$ and $(z, y) \in R]$. Clearly this implies that $x \in \text{Dom}(R)$.

Exercises 7.3

(2) (a) $(f \circ g)(x) = f(g(x)) = f(x^2 - 5) = (x^2 - 5) + 3 = x^2 - 2$. Therefore, $((f \circ g) \circ h)(x) = (f \circ g)(1 - 2x) = (1 - 2x)^2 - 2 = 4x^2 - 4x - 1$.

(3) (b) It is simple to show that $\text{Rng}(g \circ f) \subseteq \text{Rng}(g)$.

(5) Assume that (x, y) and (x, z) are both in $f \cup g$. By definition of union, this creates four cases. Each of them easily leads to the desired conclusion that $y = z$.

(12) (a) (i) The number of permutations on this set is not zero!

(17) For instance, consider Theorem 7.9(e). If the left side is empty, does it follow that the right side is also empty?

(19) (b) $\mathbb{N} \cup \{0\}$ (d) $\mathbb{Z} - \{0\}$

(21) If you find the images and preimages of the *vertices* of A, B, and C and connect them in the logical order, you get the right answers.

Exercises 7.4

(2) (c) It is true that $a_n = 1 + 2 + 3 + \ldots + n$, but this is not a closed formula.

(6) (b) $b_2 = 2 \sec^2 x \tan x$.

(7) (a) The neatest way to express the answer is with four cases, depending on a certain remainder.
(c) If you take several derivatives of f, you should see a simple pattern. The same is true of part (d) except that the pattern is more complicated.

(11) (d) By part (b), you'd be done if you could prove that $2^n > n^2$ whenever $n \geq 4$. However, this statement is false. But a slight adjustment will turn it into a true statement that you can prove by induction and then use in conjunction with part (b).

(12) (a) It is allowable to use induction on n, starting at m.

(b) Assume (x, y) and (x, z) are both in $\bigcup_{n \in \mathbb{N}} f_n$. Then, for some m and k, $(x, y) \in f_m$ and $(x, y) \in f_k$. By part (a), this implies that both (x, y) and (x, z) are in f_{m+k}.

Exercises 7.5

(3) (b) Let $f(m, n) = mk + n$.

(4) (f) This is easy: how many functions are there from $\mathbb{R}$ to $\{5\}$?

(5) (b) First use Proposition 7.14(b). Then show that $\mathbb{N}_k \sim \mathbb{N}$ contradicts part (a).

(9) (c) Like Theorem 7.22(d), this is best proved by induction. In this case, the key step involving two sets has already been proved in Section 6.1.

(10) (b) The fact that A is bounded iff $A \prec \mathbb{N}$ follows from negating both sides of part (a). Also, to say that A is bounded just says that $A \subseteq \mathbb{N}_k$ for some k. This impales that A is finite.

(15) For this problem and the next two, you might first determine whether the claimed result holds for all finite sets.

(18) Given $f: A \to B$ and $g: C \to D$, there's only way to use these two functions to define a function from $A \times C$ to $B \times D$. What is it?

(19) Recall that a function from A to B automatically induces one from $\wp(A)$ to $\wp(B)$.

(20) (c) The most familiar bijection is a trigonometric function, but there are many other possibilities.

(23) What is a sensible way to take any given k-tuple of elements from a set B and use it to define a function from $\{1, 2, 3, \ldots, k\}$ to B?

(24) (b) Use the proof of Theorem 7.24(c).
(c) Assign a pair of numbers to each customer and use Theorem 7.24(a).

Exercises 7.6

(1) (a) 584. Because of the restrictions, this problem requires the sum rule as well as the product rule. For example, you could separately count the slates with Joyce as president and the slates with someone else as president.

Another approach (which still uses the product and sum rules) is to count all possible slates of three different members, and then subtract the number of prohibited ones.

(2) (b) A committee is understood to be a set of people, not a sequence. But is the club choosing a set or a sequence of *committees*?

(5) (b) This type of problem requires careful reasoning. Imagine that the letters of "goose" are on tiles or cards. If the 2 o's are distinguishable (perhaps different colors), we know that there are 5! permutations. Now, call two permutations equivalent if they differ at most in the placement of the o's. How many permutations are in each equivalence class?

(12) (b) One function, $\varnothing$.

(13) (c) For example, a member of $(C \times D)^C$ is a function from C to $C \times D$. What is the "natural" way to use such a function to define an ordered pair of functions, one from C to C and the other from C to D?

Exercises 7.7

(1) (a) If you already know how to do this for natural numbers, then it's only necessary to define a priority between positive numbers and negative numbers.

(3) Use induction on the number of sets in the collection. The essence of the initial step and the induction step is just existential specification.

(5) Given a collection $\mathscr{A}$ of nonempty sets, it is easy to define a choice function on $\mathscr{A}$ if we first specify a well-ordering on $\bigcup\mathscr{A}$.

(8) (b) We already know that $\mathbb{R} \sim (0, 1)$. To show that $(0, 1) \preceq \mathbb{R} - \mathbb{Q}$, it suffices to map every rational number between 0 and 1 to a distinct irrational number > 1.

(10) The theorem is correct, but its proof requires the axiom of choice and is quite involved.

Exercises 8.1

(3) (b) Check the distributive laws.
(d) Among other things, note that this multiplication operation is commutative.

(4) (d) Not a subring.
(f) Subring with unity. Is it a subfield?
(i) There is an important difference between this problem and part (g).

(10) Let A be the intersection of all closed intervals that contain 3, 7, and 15. Each such interval must be a superset of $[3, 15]$. It follows that $[3, 15] \subseteq A$. The converse is even simpler since $[3, 15]$ itself is a closed interval containing 3, 7, and 15.

(14) If the theorem in Exercise 13 is correct, then very little more is required to prove this result.

(17) (a) Consider $(x + x)^2$.

Exercises 8.2

(2) (d) The GCD is 4.

(5) (b) Use Theorem A-5.

(9) This result is basically a generalization of Theorem 8.9.

(12) Show that the set of integers that are multiples of both m and n is a subring of $\mathbb{Z}$. Then apply Theorem 8.16.

(14) Prime numbers are greater than 1, and it is easily shown that the product of two or more such numbers is again greater than 1.

(18) Use Complete Induction on the minimum of j and k or, if you prefer, on the number $j + k$. If you wish, there's no harm in assuming that $j \leq k$. Also, you might need to break the induction step up into the two cases $j = k$ and $j \neq k$.

Exercises 8.3

(3) (a) These matrices don't have to be complicated. With a bit of trial and error, you can probably find appropriate matrices with only one nonzero entry each.

(4) Show that if $M_2(2\mathbb{Z})$ has a multiplicative identity, then $M_2(\mathbb{Z})$ has *two* multiplicative identities; this violates Proposition 8.1.

(7) What is the degree of the polynomial x? Of the polynomial 1?

(13) Use Theorem 8.17(b).

(15) (a) This will take some trial and error. Call the elements 0, 1, a, and b, set up tables, and start filling in whatever you can. Often you may have to use a process of elimination. Theorem 8.25(e) tells you that this is a field of characteristic 2. This means that $1 + 1 = 0$, which implies that $x + x = 0$ for every x in the field.

(b) What is the characteristic of this field? Also, it is helpful to know that in a field with nine elements, there is an element k such that all nine elements are of the form $ak + b$, where a and b can be 0, 1, or $1 + 1$.

Exercises 8.4

(3) (b) Start with the equation $x + -x = 0_A$.

(9) You may use the result of Theorem A-5 in Appendix 2, which holds for all rings. Also use that $x + (-x) = y + (-y) = 0$.

(10) Try a product ring.

(15) Use the definition of isomorphism and the function f^{-1} to define these operations on any pair of members of B.

(18) (b) Theorem 7.27(a) provides most of the content of this result.

Exercises 9.1

(1) (e) $\sqrt[3]{7}$ and undefined. (f) 1 and 0.

(5) For instance, under this alternate definition, what would the set of upper bounds of the set $\{4\}$ be?

(6) (a) You are asked to prove that if $x > 0$ and $x^2 < 2$, then $x \leq 2$. Consider the contrapositive.

(11) Does this result hold in the special case that $A = B$?

(12) Let $u \in A + B$. That means $u = x + y$, for some x in A and y in B. By definition of supremum, we must have $x \leq a$ and $y \leq b$. Therefore, $u \leq a + b$. This proves that $a + b$ is an upper bound of $A + B$. Of course, the problem asks you to prove more than this.

(13) (a) It suffices to prove this for open intervals (a, b), with $a < b$. By Theorem 9.4, there must be an $n \in \mathbb{N}$ such that $n > 1/(b - a)$. Then, using the well-ordering property of $\mathbb{N}$, it can be shown that (a, b) contains a rational number of the form m/n.

(14) Given an interval, the easiest way to prove this is to apply Exercise 13(b) to a *different* interval.

Exercises 9.2

(2) (b) What does each term of the right side converge to? To find n in terms of ε, you could use the idea of the proof of Theorem 9.8(a).

(3) (a) Modify the sequence $a_n = n$.
(b) Apply a theorem in this section.

(7) (a) Use indirect proof, and note that $a_n = (a_n + b_n) - b_n$.

(9) The core of the proof is a high school algebra problem, namely: in order to assure that $1/b_n$ is within ε of $1/L_2$, how close does b_n need to be to L_2?

(12) (c) Not geometric. You may recall from calculus that this is a **harmonic** series, and that harmonic series are always divergent.

(16) (b) Consider the first decimal place where the two decimals differ. Do not try to make this proof very rigorous.

(17) It is probably easiest to define one-to-one functions in each direction and then apply the CSB theorem. To go from $\wp(\mathbb{N})$ to $\mathbb{R}$, a function like $f(A) = \sum_{n \in A} (0.1)^n$ works nicely. For the other direction it suffices (by Exercise 20 of Section 7.5) to define a one-to-one function from the (0, 1) to $\wp(\mathbb{N})$. For this, you may use Theorem 9.11 with the word "decimal" (base 10) replaced by "binary" (base 2). The point of this is that the binary expansion of a number between 0 and 1 is an infinite sequence of 0's and 1's.

Exercises 9.3

(1) Parts (a) and (b) are both possible.

(6) The first step in this problem is to carefully write what it means to say that $\lim_{x \to a} f(x)$ does not exist (and similarly for one-sided limits).

(11) The point of this problem is that every member of a finite set or of $\mathbb{Z}$ is an isolated point of the set. One way to do this problem is to define this notion, and then show that a function must be continuous at every isolated point of its domain.

(12) (b) Use the triangle inequality (Theorem A-15(b)).

(16) (c) If a function is not continuous, there is no reason to expect its graph to be "nice," even if it is one-to-one.

(d) Suppose that f is continuous and one-to-one. If we also assume that f is not strictly monotone, then f must "reverse direction. That is, there must be numbers a, b, and c with $a < b < c$, such that $f(a) < f(b)$ and $f(b) > f(c)$ or vice versa. Now apply the intermediate value theorem to this situation.

(22) It is probably easiest to prove the inverse rather than the converse. If you carefully examine what it means to say f is discontinuous at a, it shows you how to define an appropriate sequence (b_n). Technically, the axiom of choice is needed to define (b_n).

Exercises 9.4

(1) (e) This set can be described more simply.

(f) What are the accumulation points of $\mathbb{Z}$? The answer to this should make it clear whether $\mathbb{Z}$ is closed.

(5) (c) Consider the union of an arbitrary collection of open sets under this definition. If one of the sets in the collection is $\mathbb{R}$, then the union is also $\mathbb{R}$. Otherwise, consider the set of b's for which (b, ∞) is in the collection. Proceed by cases according to whether this set of b's is bounded below. You will need to use completeness.

(7) The reverse direction is trivial. For the forward direction, the assumption that x is an accumulation point of A makes it possible to define an infinite sequence of elements of A whose limit is x. However, the axiom of choice is needed to define this sequence!

(9) You may use the results of Exercises 13 and 14 of Section 9.1.

(11) It might seem natural to define this concept in terms of an arbitrary infinite subset of $\mathbb{N}$, and this can be done. But since a sequence is technically a function, it is easier to define subsequences in terms of a certain type of function from $\mathbb{N}$ to $\mathbb{N}$.

(17) Is it true in general that $C \cup D$ is infinite iff C is infinite or D is?

(18) Is it true in general that $C \cap D$ is infinite iff C and D are both infinite?

Exercises 9.5

(1) (b) For instance, addition in $\mathbb{Z}$ should be defined by $[m - n] + [a - b] = [(m + a) - (n + b)]$. To define multiplication, use the usual distributive laws.

(2) Given r, let n be the smallest positive denominator of all the fractions in r. Then show that r has only one fraction with denominator n, and that the denominator of every fraction in r is a multiple of n.

(4) (a) Use the fact that trichotomy is in fact an exclusive or. (Since axiom V-11 is not used to prove Theorem A-9, this theorem is guaranteed to be true in $\mathbb{Z}$.)

(b) Remember which number system we are working in at this point.

(8) To show that A is a cut, you may use the result of Exercise 30 of Section 4.4. The proof that $A^2 = 2$ is almost identical to the proof of Theorem 9.3.

(15) The condition involving addition is implicit in the proof of Lemma 9.37. For the multiplication part, you need to use several cases.

Exercises 10.1

(1) (a) -10 (c) $-i$

(2) Use Theorem A-11 of Appendix 2.

(8) (c) $16i$

(11) (c) By part (b), z is a primitive nth root of unity iff z is an nth root of unity and z is *not* an mth root of unity for any $m < n$. Knowing this, it is easy to use Theorem 10.7 to define a primitive nth root of unity.

(14) It is possible to define this function, but it will necessarily be discontinuous and therefore not very practical to work with.

Exercises 10.2

(2) Given a nonempty subset A of $\mathbb{N} \cup \{0\}$, use cases based on whether $0 \in A$.

(7) (a) With factors in $\mathbb{R}$, this polynomial becomes a difference of squares. With factors in $\mathbb{C}$, all polynomials, including this one, split into first-degree factors.

(9) (b) (i) This number is algebraic because one of its powers is an integer.
(ii) First show that the square of this number is algebraic. The desired conclusion follows easily.

(11) (a) Use Theorem 10.14.

(12) A slight adaptation of the proof of Corollary 10.17(b) establishes this result.

(13) For the reverse direction, the first assumption tells us that $f(x)$ can be written in the form $(x - r)\,g(x)$. Then the division algorithm allows us to put $g(x)$ in the form $(x - r)\,h(x) + s$, where s is a number. Now evaluate $f'(r)$.

References

Baker, Alan. 1984. *A Concise Introduction to the Theory of Numbers*. Cambridge: Cambridge University Press.

Boas, Ralph P., Jr. 1987. *Invitation to Complex Analysis*. New York: McGraw-Hill.

Burn, R. P. 1982. *A Pathway into Number Theory*. Cambridge: Cambridge University Press.

Burton, David M. 1997. *Elementary Number Theory*, 4th ed. New York: McGraw-Hill.

Copi, Irving M., and Cohen, Carl.1997. *Introduction to Logic*. Engelwood Cliffs, NJ: Prentice-Hall

Davis, Philip J., and Hersh, Reuben. 1987. *The Mathematical Experience*. Cambridge, MA: Birkhauser Boston.

Davis, Philip J., and Hersh, Reuben. 1986. *Descartes' Dream: The World According to Mathematics.* San Diego: Harcourt Brace Jovanovich.

Devlin, Keith. 1993. *The Joy of Sets: Fundamentals of Contemporary Set Theory*. New York: Springer-Verlag.

Enderton, Herbert B. 1972. *A Mathematical Introduction to Logic*. New York: Academic Press.

Eves, Howard. 1997. *Foundations and Fundamental Concepts of Mathematics*, 3rd ed. New York: Dover.

Feferman, Solomon. 1989. *The Number Systems: Foundations of Algebra and Analysis*. New York: Chelsea.

Fraleigh, John B. 1994. *A First Course in Abstract Algebra,* 5th ed. Reading, MA: Addison-Wesley.

Grimaldi, Ralph. 1994. *Discrete and Combinatorial Mathematics, an Applied Introduction,* 3rd ed. Reading, MA: Addison-Wesley.

Hamilton, A. G. 1988. *Logic for Mathematicians.* Cambridge: Cambridge University Press.

Hardy, G. H., and Wright, E. M. 1980. *An Introduction to the Theory of Numbers*, 5th ed. New York: Oxford University Press.

Herstein, I. N. 1996. *Abstract Algebra*, 3rd ed. Englewood Cliffs, NJ: Prentice-Hall.

Hofstadter, Douglas R. 1989. *Gödel, Escher, Bach: An Eternal Golden Braid.* New York: Random House.

Kline, Morris. 1981. *Mathematics and the Physical World.* New York: Dover.

Kline, Morris. 1982. *Mathematics, the Loss of Certainty.* New York: Oxford University Press.

Lakatos, Imre. 1976. *Proofs and Refutations, the Logic of Mathematical Discovery.* Cambridge: Cambridge University Press.

Mendelson, Elliott. 1987. *Introduction to Mathematical Logic*, 3rd ed. Monterey, CA: Wadsworth & Brooks/Cole Advanced Books & Software.

Nagel, Ernest, and Newman, James R. 1958. *Gödel's Proof.* New York: New York University Press.

Niven, Ivan, and Zuckerman, H. S. 1991. *An Introduction to the Theory of Numbers,* 5th ed. New York: Wiley.

Pfleeger, Shari Lawrence and Straight, David W. 1985. *Introduction to Discrete Structures.* New York: Wiley.

Polya, George. 1945. *How to Solve it; a New Aspect of Mathematical Method.* Princeton, NJ: Princeton University Press.

Polya, George. 1965. *Mathematical Discovery: On Understanding, Learning and Teaching Problem Solving*. 2 vols. New York: Wiley.

Polya, George. 1954. *Mathematics and Plausible Reasoning*. Princeton, NJ: Princeton University Press.

Ross, Kenneth A., and Wright, Charles R. B. 1992. *Discrete Mathematics*, 3rd. ed. Englewood Cliffs, NJ: Prentice-Hall.

Royden, H. L. 1988. *Real Analysis,* 3rd ed. New York: Macmillan; London: Collier Macmillan.

Rudin, Walter. 1976. *Principles of Mathematical Analysis*, 3rd ed. New York: McGraw-Hill.

Rueff, Marcel, and Jeger, Max. 1970. *Sets and Boolean Algebra*. New York: American Elsevier.

Shoenfield, Joseph R. 1967. *Mathematical Logic.* Reading, MA: Addison-Wesley.

Smullyan, Raymond. 1992. *Gödel's Incompleteness Theorems*. New York: Oxford University Press.

Sominskii, I. S. 1961. *The Method of Mathematical Induction.* Oxford: Pergamon Press.

Stabler, E. R. 1953. *An Introduction to Mathematical Thought.* Cambridge, MA: Addison-Wesley.

Stoll, Robert R. 1979. *Set Theory and Logic.* New York: Dover.

Suppes, Patrick. 1960. *Axiomatic Set Theory*. New York: Dover.

Tucker, Alan. 1994. *Applied Combinatorics,* 3rd ed. New York: Wiley.

Vaught, Robert L. 1995. *Set Theory: An Introduction*, 2nd ed. Boston: Birkhauser.

Wilder, Raymond Louis. 1965. *Introduction to the Foundations of Mathematics,* 2nd ed. New York: Wiley.

List of Symbols and Notation

Where more than one page number is shown, the *last* one contains the full definition or explanation of that symbol or notation. Earlier references are to incidental uses, usually in examples or exercises. For the most part, notation which is an abbreviation or acronym for words (such as GCD for greatest common divisor or *Dom* for domain) is not included in this list.

P, Q, R, ...	Propositional variables	17
$P \wedge Q$	Conjunction	18
$P \vee Q$	Disjunction	18
$\sim P$	Negation	18
$P \rightarrow Q$	Conditional	18
$P \leftrightarrow Q$	Biconditional	18
$P \veebar Q$	Exclusive or	26
■	End of proof	28
$A \subseteq B$	Subset	41, 147
$\lvert x \rvert$	Absolute value	41, 384
$x \in A$	Set membership	41, 133
$P \mid Q$	Sheffer stroke	44
$f(x)$	Function notation	46, 195
$\forall x\, P$	Universal quantifier	48
$\exists x\, P$	Existential quantifier	48
$P(x)$, $P(x, y)$, and so on	Propositional functions	50
$\mathbb{N}$	Natural numbers	59, 115
$\mathbb{Z}$	Integers	59, 136
$\mathbb{R}$	Real numbers	59, 297
$\mathbb{C}$	Complex numbers	59, 352
t	Term	64
$\forall x < t$, $\forall x \in t$, $\exists x > t$, $\exists x \in t$, and so on	Restricted quantifiers	64
$\exists! x\, P$	Unique existence	68
$A, B, C, \ldots$	Variables for sets	133
$\{ \ldots \}$	Roster method for defining a set	133
$\{ .. \mid \ldots \}$	Set-builder method for defining a set	134
$A = B$	Equality of sets	135
$\mathbb{Q}$	Rational numbers	136

Symbol	Meaning	Page
(a, b), $[a, b]$, $(a, b]$, $[a, b)$	Standard (bounded) intervals	136
(a, ∞), $[a, \infty)$, $(-\infty, b)$, $(-\infty, b]$, $(-\infty, \infty)$	Unbounded intervals	136
$A \cup B$	Union	142
$A \cap B$	Intersection	142
$B - A$	Relative complement	142
$\varnothing$	Empty set	142
U	Universal set	143
A' or $\overline{A}$	Complement	143
$A \subset B$	Proper subset	147
$A \supseteq B$, $A \supset B$	Superset	147
$A \Delta B$	Symmetric difference	152
$\wp(A)$	Power set	153
$\mathscr{A}, \mathscr{B}, \mathscr{C}, \ldots$	Variables for collections of sets	156
$\{A_i \mid i \in I\}$	Indexed family of sets	156
$\bigcup \mathscr{A}$, $\bigcup_{i \in I} A_i$	Generalized union	157-158
$\bigcap \mathscr{A}$, $\bigcap_{i \in I} A_i$	Generalized intersection	157-158
(a, b)	Ordered pair	164
$A \times B$	Cartesian product	165
(a, b, c)	Ordered triple	167
$(a_1, a_2, \ldots, a_n)$	Ordered n-tuple	167
$A^2, A^3, \ldots$	Abbreviations for $A \times A$, $A \times A \times A$, etc.	168
$R, S, T, \ldots$	Variables for relations	169
xRy	Membership in a relation	169
R^{-1}	Inverse relation	170
$x \equiv y$, $x \approx y$ or $x \sim y$	Membership in an equivalence relation	175
id_A	Identity relation	175
$x \equiv y \pmod{c}$	Congruence modulo c	176
$\lfloor x \rfloor$	Greatest integer function	177
$[x]_R$, $[x]$	Equivalence class	178
A/R	A modulo R	178
$R \times S$	Product ordering	184
$m \mid n$	m divides n	188, 270
f, g, F, G	Function variables	193
$f\colon A \to B$	f is a function from A to B	193
χ_B	Characteristic function	199
$f + g$, fg, $f - g$, $-f$	Operations on functions	203
$f\colon A \xrightarrow{1-1} B$ or $f\colon A \xrightarrow{\text{inj}} B$	One-to-one function	205
$f\colon A \xrightarrow{\text{onto}} B$ or $f\colon A \xrightarrow{\text{sur}} B$	"Onto" function	205

Symbol	Meaning	Page
$f\colon A \xrightarrow{\text{bij}} B$ or $f\colon A \xrightarrow[\text{onto}]{1-1} B$	One-to-one correspondence	205
$f \circ g$, $R \circ S$	Composition	207
f^{-1}	Inverse function	208
$f\vert_A$	Restriction of a function	210
$f(C)$	Image of C under f	219
$f^{-1}(C)$	Inverse image of C under f	220
(a_n), $(a_1, a_2, a_3, \ldots.)$	Sequence notation	223-224
$n!$	n factorial	225
$A \sim B$	Equivalence of sets (in cardinality)	230
$A \preceq B$, $B \succeq A$, $A \prec B$, $B \succ A$	Cardinality inequalities	231
$\mathbb{N}_k$	Initial segment of $\mathbb{N}$	232
Card(A), $\vert A\vert$, $\overline{\overline{A}}$, or $\#A$	Cardinality	232
B^A	Set of all functions from A to B	241
$P(n, r)$ or ${}_nP_r$	Number of permutations	247
$C(n, r)$, ${}_nC_r$, or $\binom{n}{r}$	Number of combinations, binomial coefficients	249
$A \sqcup B$	Formal disjoint union	252
a, b, c, i, j, k, m, n	Integer variables	261
$n\mathbb{Z}$	Subring of $\mathbb{Z}$	266
M_n, $M_n(A)$	Rings of matrices	283
$A[x]$	Ring of polynomials	284
$A(x)$	Field of rational functions	285
$\mathbb{Z}_n$	Integers modulo n	286
$\lim_{n \to \infty}(a_n) = L$ or $a_n \to L$	Limit of a sequence	304
$\sum_{n=1}^{\infty} a_n$ or $\sum a_n$	Infinite series	310
$\lim_{x \to a} f(x) = L$	Limit of a function	314
$\lim_{x \to a^-} f(x) = L$, $\lim_{x \to a^+} f(x) = L$	One-sided limits	317
$\lim_{x \to +\infty} f(x) = L$, etc.	Limits at infinity	326
$\lim_{x \to a} f(x) = +\infty$, etc.	Infinite limits	326
$\mathbb{Q}^*$	Fractions	340
r^*	Dedekind cut corresponding to r	345
i	Imaginary unit	352
z	Complex variable	352
$\overline{z}$	Complex conjugate	354
cis θ	Cos $\theta + i \sin \theta$	357

Index

A boldface page number by a person's name indicates a biography. For other entries, a boldface page number indicates a primary reference, possibly a definition.

Absolute value, 41, **384**
  of a complex number, 355
Accumulation point, 331
Algebra
  abstract, 261
  linear, 204
  of sets, 144
Algebraic closure, 368
Algebraic element (over a field), 368
Algebraic structure, 261
And, 18
Antecedent, 28
Antisymmetry, 114, **183**
  strong, 189, 382
Archimedean property, 301, 303
Archimedes, 8, **302**
Argument, 38
Aristotle, **17**
Associative property, 374
Assumption, 73, 81-82
Atomic statement, 18
Axiom, 8, 72-73
  logical, 72
  of choice, 253-257, 374
  proper, 72
Axiom schema, 376
Axiom system, 73
Axiomatic method, 8, 73
Banach-Tarski paradox, 255
Barber's Paradox, 137
Becker, Joseph, 282
Berry's Paradox, 141
Bertrand's Postulate, 54, 278
Biconditional statement, 19, **31**
Biconditional rule, 85
Bijection, 205
Binary relation, 168
Binomial coefficient, 249
Binomial theorem, 249
Bolzano-Weierstrass theorem, 332-334
Boolean algebra, 10, 147, 153
Boolean ring, 269, 296
Bounded sequence, 305
Bounded subset
  of $\mathbb{N}$, 237
  of $\mathbb{R}$, 298
Cantor, Georg, **235**, 257
Cantor set, **158-159**, 163, 331
Cantor's theorem, 240
Cantor-Schröder-Bernstein theorem, 233
Cardinal, 231, 253
Cardinality, 230, 232
Cartesian product, 165, 167
Catalan's conjecture, 280
Cauchy sequence, 343
Characteristic (of a ring), 289
Characteristic function, 199
Choice function, 254
Closed set, 331
Closed statement, 50
Closure property, 374
Codomain, 193, 197
Cohen, Paul, 257

Collection of sets, 153
Combination, 248-249
Combinatorics, 245
Commutative property, 374
Comparable elements, 184
Comparable cardinals, 255
Complement, 142-143
Complete induction, 122, 227, **274**
Completeness axiom, 297
Complex conjugate, 354
Complex number, 352
  trigonometric representation of, 355
Complex plane, 355
Complex *n*th root of unity, 359
  primitive, 361
Composition, 206
Compound statement, 19
Comprehension axiom, 133, 135
Conclusion, 28, 38
Conditional statement, 19, **27**
Conditional proof, 81
Congruence (geometric), 176
Congruence modulo c, 176
Conjecture, 7-8
Conjunction, 19
Conjunction rule, 72, 88
Connected set, 335
Connective, 18
Consequence
  logical, 60
  propositional, 38
Consequent, 28
Consistent theory, 137
Continuous function, 319
  nowhere, 321
  piecewise, 323
Continuum hypothesis, 257
Contradiction, 21
  proof by, 82
Contrapositive, 28
Contrapositive conditional proof, 88
Convergence, 304
Converse, 28
Convex set, 339
Corollary, 73
Cost function, 200
Coterminal angles, 358
Countable set, 237
Counterexample, 91, **101**
Course-of-values induction, 174
CSB theorem, 233
Cut, 344
Decimal representation, 302
Dedekind cut, 344
Definition, 69, 73, 74
  by cases, 199, 218
  inductive, 225
De Moivre's formula, 358
De Morgan's laws, 36, 37
  for quantifiers, 62, 92
  for sets, 145
De Polignac's conjecture, 12, 101, 280
Denumerable set, 237
Descartes, René, **165**
Determinant, 215
Diagonalization argument, 241
Direct product of rings, 285
Direct proof, 81
Direct sum of rings, 285
Dirichlet's theorem, 278
Discontinuity, 320, 321
Discrete set, 298
Disjoint sets, 142
Disjunction, 19
Disjunctive Normal Form, 44
Distributive property, 374
Divergence, 304
Divides, 188, 270
Division, 380
Division algorithm, 120, **270**
  for polynomials, 364
Divisor, 270
Domain, 45, 169, 193
Double induction, 118, 122, 282
Duality, 147, 153
Edison, Thomas, 8
EG, 99
Einstein, Albert, 10
Element, 133
Ellipsis, 119, 134, 225

Empirical method, 5
Empty set, 142
Epistemology, 4
Epsilon, 133, 304
Equality, 66, 107, 175
Equivalence, 19, 27
  logical, 60
  of sets, 230
  propositional, 21
Equivalence class, 178
Equivalence relation, 175
ES, 97
Euclid, 8, **9**
Euclidean domain, 285
Euclidean geometry, 48
Euler, Leonhard, 278, 337, **353**, 362
Existence theorem, 98
Existential generalization, 99
Existential quantifier, 48
Existential specification, 69, **97**
Extension, 266
  algebraic, 368
  of a function, 210
  transcendental , 368
Extensionality axiom, 133, 135, 373
Extraneous solution, 111
Factor, 270
Factor theorem, 365
Factorial, 225
Family of sets, 153
Fermat's last theorem, 278
Fermat's "little" theorem, 278
Fermat, Pierre de, **279**
Fibonacci sequence, 228
Field, 261, 377
  algebraically closed, 363
  finite, 288
  of quotients, 285, 343
  of rational functions, 285
  ordered, 377
Field axioms, 113, **374**, 377
Finitary operation, 159
Finite set, 134, **232**
First-order language, 48
Fixed costs, 200
Formal disjoint union, 252
Formalism, 10
Foundation axiom, 374
Fraction, 340
Fraenkel, Abraham, 139
Function, 193
  continuous, 319
  decreasing, 327
  increasing, 327
  inverse, 208
  monotone, 327
  one-to-one, 205
  onto, 205
  partial inverse, 210
  propositional, 50
  rational, 285
Function addition, 203
Function multiplication, 203
Function notation, 195
Function symbol, 47
Fundamental counting principle, 168, 245
Fundamental theorem of algebra, 17
Fundamental theorem of arithmetic, 275
Gauss, 122, **123**
GCD, 272
Generalized continuum hypothesis, 257
Given, 38, 73
GLB, 299
Gödel's incompleteness theorem, 139
Gödel, Kurt, 257
Goldbach's conjecture, 7, 101, 280
Gravitational constant, 200
Greatest common divisor, 272
Greatest element, 191
Greatest lower bound, 299
Group, 217-218, 292
Hilbert Hotel, 245
Homomorphism, 292
Horizontal line test, 204
Hypothesis, 4, 28, 38, 73
Idempotent, 269
Identity axioms, 374-375, 378

Identity relation, 175
If and only if, 18
If ... then, 18
Image of an element, 197
Image of a set, 219
Imaginary part, 354
Implication, 19
Inclusion map, 198
Incomparable elements, 184
Inconsistent theory, 137
Independence, 257
Indexed family, 156, 224
Indirect proof, 37, **82**
Induced set operation, 219
Induction, mathematical, 115
    variations of, 121-122
Induction hypothesis, 118
Induction step, 118
Inductive definition, 225
Inductive reasoning, 5, 117
Inequality, 381-382
Infimum (also Inf), 299
Infinitary operation, 159
Infinite set, 134, **232**
Infinite sequence, 156, **223**
Infinity Hotel, 245
Injection, 205
Instantiation, 98
Integer, 136, 263, 340
Integral domain, **284**, 343
Interior point, **319**, 328
Intermediate value theorem, 324
Intersection, 142
    generalized, 157
Interval, 136-137
Inverse (of a relation), 170
Inverse axioms, 375, 378
Inverse function, 208
Inverse image, 220
Inverse square law, 200
Inverse trigonometric function, 211
Irreflexive ordering, 188, 382
Irreflexive property, 375, 382
Isolated point, 163, 331
Isomorphic embedding, 292
Isomorphism, 288, **291**
Königsberg bridge problem, 337-338
Largest set, 264
Lattice diagram, 185
Law of logic, 60
Law of propositional logic, 21
Least common multiple (or LCM), 272
Least element, 191
Least upper bound, 299
Lemma, 73
Lexicographic ordering, 187
Limit of a function, 314
    one-sided, 317
Limit of a sequence, 304
Linear function, 204
Linear algebra, 204
Linear ordering, 183, 382
Logic, 16
    first-order, 48
    predicate, 48
    propositional, 16
    quantifier, 48
    sentential, 16
Lower bound, 298
LUB, 299
Mapping (or Map), 193
Mapping to the equivalence classes, 203
Mathematical Induction, 115. *See also* Induction, mathematical
Mathematical variable, 17, 23, **45**
Matrix, 129, **283-284**
Maximal element, 199
Mean Value Theorem, 98
Member, 133
Minimal element, 199
Model, 10
Modular arithmetic, 176, 286
Modulo, 176, 178
Modulus, 355
Modus Ponens, 37, **79**
Modus Tollens, 37, **88**
Monotone convergence theorem, 307
Multiple, 55, **270**
Naive set theory, 133, 137

Negation, 19
Neighborhood, 328
Newton, Isaac, 200
Nim, 14, 281
Non-Euclidean geometry, 10
Not, 18
Number
  algebraic, 368
  imaginary, 352
  natural, 114-115, 375
  pure imaginary, 354
  rational, 136, 266, 340, 341
  real, 343, 374
  transcendental , 368
Number theory, 7, **269**
One-to-one correspondence, 205
Open sentence, 50
Open set, 319
Open statement, 45, 50
Operator symbol, 47
Or, 18
  exclusive, 20
  inclusive, 20
Ordered field axioms, 374, 377
Ordered *n*-tuple, 167
Ordered pair, 164
Ordered triple, 167
Ordering, 183
  irreflexive, 188, 382
  lexicographic, 187
  linear, 183, 382
  partial, 183
  product, 184
  reflexive, 183
  simple, 183, 382
  total, 183, 382
  well, 149, 191
Order-preserving function, 292
Overhead, 200
Pairing axiom, 373
Paradox, 137-139
Parallel postulate, 10, 70
Partial ordering, 183
Partition, 178
  coarsest, 179
  finer, 182
  finest, 179
Pascal, Blaise, **251**
Pascal's triangle, 250, 252
PC, 78
Permutation, 217, 247
Permutation group, 218
Pigeonhole principle, 232
PMI, 114. *See also* Induction, mathematical
Pointwise, 267, 285
Postulate, 8, 72
Power set, 153
  axiom, 374
Predicate, 50
Predicate calculus, 48
Predicate symbol, 47
Preimage, 197
Premise, 38, 73
Preordering, 188
Prime number(s), 7, 11-12, **270**
  relatively, 270
Prime number theorem, 278
Principal square root function, 210
Product ordering, 184
Product rule for counting, 168, 245
Proof
  good, 77
  informal, 75
  formal, 40, 41, **73**, 75
  in law, 5
  in mathematics, 7
  in science, 4
Proof by cases, 37, 42, **84**
Proof by contradiction, 82
Proof preview, 32, 41, 58, 64, 68
Proper subset, 147
Proposition, 16, 73
Propositional calculus, 16
Propositional function, 50
Propositional variable, 17
Pythagoras' theorem, 14
Pythagoras, **277**, 340
Pythagorean triple, 14
Quantifier, 48

Quotient structure, 287, 341,343
Random-number generator, 201
Range, 169, 193
Rational number, 136, 266, 340, 341
Ray, 136
Real number, 343, 374
Real analysis, 297
Real part, 354
Reductio ad absurdum, 82
Refinement, 182
Reflexive ordering, 183
Reflexivity, 67, **174-175,** 373
Relation, 168
  binary, 168
  equivalence, 174-175
  identity, 175
  inverse, 170
  ordering, 183
Relative complement, 142
Relatively prime numbers, 270
Remainder theorem, 365
Replacement axiom, 374
Restricted quantifier, 64
Restriction of a function, 210
Reversibility, 110
Ring, 261
  Boolean, 269, 296
  of functions, 284, 290
  noncommutative, 283
  polynomial , 284
  product, 285
  quotient, 287
Root of a polynomial, 365
  multiple, 365-366
Root of unity, 359
  primitive, 361
Roster method, 133
Rule of inference, 72
  derived, 79
Russell's paradox, 137
Russell, Bertrand, 137, **138**
Scientific method, 5
Sentential calculus, 16
Separation axiom, 136, 374
Sequence, infinite, 156, **223**
  bounded, 305
  Cauchy, 343
  convergent, 304
  decreasing, 306
  divergent, 304
  increasing, 306
  monotone, 306
  of partial sums, 310
  unbounded, 305
Series, 120, **310**
  alternating geometric, 311
  arithmetic, 120
  geometric, 120, 310
Set-builder notation, 134-135
Set, 133
Sheffer Stroke, 44
Similarity, 176
Simple ordering, 183, 382
Simple statement, 18
Smallest set, 264
Sort, 45
Statement, 17
Step, 73
Step function, 324
Subgroup, 182
Subring, 263
Subsequence, 332
Subset, 147
Substitution, 86, 372
Substitution of equals, 67, 108, 373
Subtraction, 380
Sum rule for counting, 149, 245
Supremum (or Sup), 299
Surjection, 205
Symmetric difference, 152
Symmetry, 67, 174, 373
Tautology, 21, 387-388
Telescoping series, 313
Term, 47, 64, 93
Theorem, 73
Topological space, 331
Topology, 328
  discrete, 337
  trivial, 337
Total ordering, 183, 382

Transcendental element (over a field), 368
Transitive property, 67, 174, 373
Translate, 181
Triangle inequality, 384
Trichotomy, 188, 375, 382, 383
Truth function, 19
Truth table, 20
Twin prime conjecture, 280
UG, 95
Uncountable set, 237
Union, 142
  formal disjoint, 252
  generalized, 157
Uniqueness, **67**, 205
Universal generalization, 95
Universal law of gravitation, 200
Universal quantifier, 48
Universal set, 143
Universal specification, 93
Universe, 45
Upper bound, 298
US, 93
Valid argument, 38
Variable,
  bound, 50
  dependent, 194
  free, 50
  independent, 194
  mathematical, 45
  propositional, 17
Vector, **204**, 356, 357
Venn diagram, 142
Vertical line test, 195, 204
Weierstrass, Karl, **333**
Well-defined operation, **287**, 341
Well-ordering, 149, 191
Well-ordering principle, 255
Well-ordering property of $\mathbb{N}$, 148-149
Wiles, Andrew, 279
Wilson's theorem, 278
Zermelo-Fraenkel set theory, 139
Zermelo, Ernst, 139
Zero-divisor, 284